CLIMATE CHANGE AND BUSINESS

Climate change is the greatest market failure in world history and thus the most urgent challenge in the world of business for the foreseeable future. This book brings together experts to shed light on the historical impact of business on climate change and of climate change on business.

The book shows how corporate responses to climate change emerged out of earlier environmental concerns, regulatory frameworks, and in many cases already established business strategies. Contributors to the book analyse the evolution of business strategies to navigate environmental challenges even before climate crisis was widely recognised as an urgent concern.

The historical insights presented by this book will be essential reading for business historians, as well as students, scholars, and reflective practitioners with an interest in the environment, political economy, business strategy, and risk management.

Teresa da Silva Lopes is Professor of International Business and Business History at the School for Business and Society, and Director of the Centre for the Evolution of Global Business and Institutions (CEGBI) at the University of York, UK.

Paul Duguid is Adjunct Professor Emeritus in the School of Information at the University of California, Berkeley, USA.

Robert Fredona is Research Associate at Harvard Business School, USA.

"This thoughtful, stimulating, collection of original papers, written by accomplished business historians, excels in exploring both the impact and responses of business over time to the evolving evidence and proliferating policies on climate change. Here there is a fresh perspective that includes but extends beyond the debates on environmental and fossil fuel programs to encompass agriculture (palm oil), industry (automobiles, alcoholic beverages), insurance, and tourism. The coverage is global, including special chapters on China and India. The result is a rich and ambitious set of contributions that reveals the complex, nuanced reciprocal historical influences of business and climate change."

Mira Wilkins, *Professor of Economics Emeritus; Florida International University, USA*

"As a scientist and business owner, I have found that for something real and game-changing to be implemented in the climate change sphere, we need businesses to incorporate climate thinking into every area, from production, to human resources, to sales and marketing. This book documents the trajectory of these initiatives in various industries and regions. It is an essential piece of reading for all of us today."

Laura Catena, *Managing Director of Bodega Catena Zapata and founder of the Catena Institute of Wine in Mendoza, Argentina*

"This edited volume, by Teresa da Silva Lopes, Paul Duguid, and Robert Fredona investigates the history of one of the biggest problems in the world today, climate change. The collected essays provide a critical perspective on the complex evolution of the problem itself, from nineteenth-century awareness of weather changes and pollution hazards to modern efforts to collect data and enact policy changes to curb dependence on fossil fuels. These essays, many of which draw on archival sources, recast traditional histories of economic development by focusing on the consequences of industrial, agricultural, and technological change. They present a complex history of ways that firms have adapted to climate changes in the past and the ways that individual entrepreneurs have sought to abate environmental devastation in their industries. The essays also reveal how the effects of climate change have altered industries such as tourism and insurance and made other industries, such as winemaking, more vulnerable. Together the essays help to tell the multifaceted, non-linear, history that has led to this enormous global challenge."

Walter Friedman, *Director of the Business History Initiative at Harvard Business School and co-editor of the journal* Business History Review

CLIMATE CHANGE AND BUSINESS

Historical Perspectives

Edited by Teresa da Silva Lopes, Paul Duguid, and Robert Fredona

Routledge
Taylor & Francis Group
LONDON AND NEW YORK

Designed cover image: m.malinika / iStock / Getty Images Plus

First published 2025
by Routledge
4 Park Square, Milton Park, Abingdon, Oxon OX14 4RN

and by Routledge
605 Third Avenue, New York, NY 10158

Routledge is an imprint of the Taylor & Francis Group, an informa business

British Library Cataloguing-in-Publication Data
A catalogue record for this book is available from the British Library

ISBN: 978-1-032-76339-2 (hbk)
ISBN: 978-1-032-76336-1 (pbk)
ISBN: 978-1-003-47808-9 (ebk)

DOI: 10.4324/9781003478089

Typeset in Times New Roman
by Apex CoVantage, LLC

CONTENTS

ABOUT THE EDITORS

Teresa da Silva Lopes is Professor of International Business and Business History at the School for Business and Society and Director of the Centre for the Evolution of Global Business and Institutions (CEGBI) at the University of York, UK. Her research interests focus on the long-term impact of global business on society. Her publications include *Global Brands: The Evolution of Multinationals in Alcoholic Beverages*.

Paul Duguid is Adjunct Professor Emeritus in the School of Information at the University of California, Berkeley. His work focuses on information in history. Recent publications include the co-edited *Information: A Historical Companion* and *Information: A Short History*.

Robert Fredona is a research associate at Harvard Business School. His interests include the legal and business history of premodern Europe as well as contemporary historiography and theory. He has published in *Business History Review*, *History of Political Economy*, and other journals and edited books.

ABOUT THE CONTRIBUTORS

Grace Ballor is Assistant Professor of International Economic History at Bocconi University, where she is also a faculty fellow at the Institute for European Policymaking. Her research historicises European political economy and the intersections of global capitalism and global governance.

Ann-Kristin Bergquist is Professor of Economic History at Uppsala University and Director of Uppsala Center for Business History (UCBH). Her research interest is focused on the history of business and the environment. Her current research project addresses the role of international business in shaping the architecture of global climate governance, financed by the Swedish Research Council.

Marten Boon is a business historian, specialising in the history of the oil industry, and a lecturer in the history of international relations at Utrecht University. His latest book, *National Champion: Statoil and Equinor 2001–2022*, charts the recent history of Norwegian state-owned oil company Equinor.

Bart Elmore is Professor of environmental history and core faculty member of the Sustainability Institute at Ohio State University. He is the author of three books: *Citizen Coke: The Making of Coca-Cola Capitalism*; *Seed Money: Monsanto's Past and Our Food Future*; and *Country Capitalism: How Corporations from the American South Remade Our Economy and the Planet.*

Niels Viggo Haueter is an affiliated member of the Centre for Evolution of Global Business and Institutions at the University of York and serves as Deputy Chairman of the Academic Council at the European Association for Banking and Financial History in Frankfurt.

Geoffrey Jones is Isidor Straus Professor of Business History at Harvard Business School. He researches the history and ecological and social impact of global business. His recent books include *Profits and Sustainability: A History of Green Entrepreneurship* (2017) and *Deeply Responsible Business: A Global History of Values-Driven Leadership* (2023).

Andrea Lluch is a researcher at the National Council of Scientific and Technical Research of Argentina (CONICET, Argentina). Andrea is also an associate professor and a member of the Business History Research Group (GHE) at the School of Management, University of Los Andes, Colombia. She is the co-editor of *A Business History of Latin America* (2025).

Ghassan Moazzin is Assistant Professor at the Hong Kong Institute for the Humanities and Social Sciences and the School of Humanities (Department of History), the University of Hong Kong. He is the author of *Foreign Banks and Global Finance in Modern China: Banking on the Chinese Frontier, 1870–1919.*

Simon Mowatt is Principal Lecturer at the University of Bedfordshire, and has published on business history of organic farming with Geoffrey Jones in *Varieties of Green Business: Industries, Nations and Time* (2018) and sustainable strategies in journals such as *Business History* (2016), and the *Journal of Business Strategy and the Environment* (2015, 2020).

Mattias Näsman is Assistant Professor of Economic History at Umeå University. His research uses history as a lens to explore questions about ongoing sustainability transitions, with a specific interest in the historical political economy of the car industry.

Sabine Pitteloud is Assistant Professor of contemporary history at UniDistance Switzerland. She is specialised in business history and the political history of capitalism. Her current research project focuses on the role of business associations in shaping environmental governance.

Natalie Rauscher is a post-doctoral researcher at Heidelberg University and project leader at the Heidelberg Academy of Sciences and Humanities. Her research includes discourse studies, the future of work, American philanthropy, natural disaster mitigation, and the study of the polycrisis.

Tirthankar Roy is Professor in Economic History at the London School of Economics and Political Science and is the author of *Monsoon Economies: India's History in a Changing Climate*, among other books and articles on economic and business history.

Judith Stroehle is Assistant Professor of Sustainability Governance at the Accounting and the Sustainability Institute, University of St. Gallen (HSG) in Switzerland. Her research and teaching focuses on how tools of measurement, accounting, and finance can help to implement sustainable business practice.

Laurent Tissot is Emeritus Professor of Contemporary History at the University of Neuchâtel. He is the co-founder of the International Commission of the History of Travel and Tourism and former treasurer of the International Committee of Historical Sciences. His main themes of research are history of transport, mobility and tourism, and history of industrialisation (business history, regional history).

Chinmay Tumbe is Associate Professor in the Economics Area at the Indian Institute of Management Ahmedabad with research interests in business and economic history and migration studies. He is the author of *India Moving: A History of Migration* (2018) and *The Age of*

Pandemics, 1817–1920 (2020). He has published widely in leading journals and newspapers and helped set up the IIMA Archives.

Welf Werner is Professor of American Studies at the Faculty of Economics and Social Sciences at Heidelberg University and director of the Heidelberg Center for American Studies (HCA). His research and teaching focus on U.S. domestic and foreign economic policies and how these are intertwined with history and political science.

Shakila Yacob currently serves as a professor at the Jeffrey Cheah Institute on Southeast Asia, Sunway University. As a business historian and political scientist, her research interests include business history, international business, comparative politics, and public policy.

0

CLIMATE CHANGE AND BUSINESS

A New Perspective and Research Agenda

Teresa da Silva Lopes, Robert Fredona, and Paul Duguid

0.1 Wrestling with Climate Change

In 2023, as the authors worked on their contributions to this book, the year became the hottest on record.[1] The following year, as the book went to press, held two of the hottest days ever recorded and was predicted to be yet hotter than 2023. Both years will probably contribute to the hottest decade on record, and that will continue to increase the amount of CO_2 in the atmosphere, which in 2019 was measured as higher than at any period in the past two million years.[2] As a result of this growing heat, the climate is changing: glaciers are melting; sea levels rising; oceans acidifying; and hurricanes, fires, floods, and droughts destabilising societies around the world.

In these and many other ways, the presence of climate change has become ever more inescapable. The conclusion that its cause is anthropogenic—driven by human activity—has become similarly unavoidable. Thus, to understand and respond to this climate transformation, it is essential to explore the human contribution. One critical contribution comes from business activity. On the one hand, businesses are prey to climate change, which, in 2009, Nicholas Stern labelled the "greatest market failure the world has ever seen."[3] But, on the other hand, business output has to be understood as a major causal factor. Hence, as society tries to understand how climate change developed in the past and how it might transform (or be transformed in) the future, the field of business history has potentially a great deal to offer to society's understanding. Yet the field has not, until recently, provided much towards that understanding. Its conventional, finance-shaped perspective has made climate change appear as an "externality" to business behaviour and deterred research leading to climate-related criticism of business activity. This book, however, building on the pioneering work of business historians in the twenty-first century, reflects an increasing awareness of the importance of the issue and the potential significance of the field's insights.

Business history is distinct from other subfields of history and the social sciences due to its focus on firms, other institutional forms, and entrepreneurs, even when examining industries and nations. It develops causal interpretations to address core issues in business and management, including their connections to major societal challenges. To achieve this, it combines original historical business sources with comparative and international analysis involving multiple and varied institutional forms.[4]

DOI: 10.4324/9781003478089-1

0.2 Climate Change as a Business History Priority

Historical research on the relationship between business and climate change has so far predominantly focused on the fossil fuel industry's responsibility for causing and influencing the climate crisis.[5] Business historians have paid much less attention to the variety of business responses of different sectors to climate change and to understanding why some corporations and industries started to respond to scientific evidence and the public debate around climate change earlier than others.

Furthermore, existing historical scholarship on business and climate change has almost entirely overlooked the impact of a changing global climate and related weather conditions on business strategies and operations. It has also failed to examine how businesses have adapted to dramatic weather changes in the past. Certain industries such as agribusiness, forestry, and insurance are particularly vulnerable to a changing climate and extreme weather events, and investigating how they have responded to these changes can reveal a great deal about the subject of climate change more generally.

Since the mid-1980s, however, the issue of business and the environment has received increased attention in academic research, particularly within the field of business administration. The rapid advances in this area in the 1990s and the 2000s are highlighted in *The Oxford Handbook of Business and the Natural Environment*.[6] This included 38 overview chapters on various subfields such as business strategy, organisational theory, marketing, accounting, international business, finance, and more. Social science research on climate change and business is extensive and spans many different subfields. However, this research has rarely examined and synthesised changes over time and has often focused on North America alone. For instance, Hoffman's (2001) seminal work on the history of corporate social responsibility covers only the United States.[7]

Meanwhile, environmental history has developed into a diverse and vibrant subject with global coverage.[8] A prominent subfield within the history of science and environmental history has emerged, focusing specifically on the historical role of businesses in causing climate change. This research has documented the harmful activities of companies such as Exxon, Shell, Total, and other fossil fuel companies in fostering climate change scepticism and concealing their own research from the public.[9] This research agenda has unfortunately not been widely taken up in mainstream business history. What is most lacking there is a detailed examination of the varieties of business responses, extending beyond the fossil fuel industry and oil companies.

For a long time, business history was largely silent on the relationship between business and the natural environment, with only a few scholars engaging with the topic. In 2011, Friedman and Jones called this the "time for debate."[10] This perspective echoes a long-standing concern of Alfred Chandler, who, as early as 1980, urged business historians to study the successes and failures of earlier institutional arrangements dependent on fossil fuels.[11]

The first attempt to integrate the natural environment into business history came in a special issue of *Business History Review* in 1999.[12] While the collection did not focus on climate change, contributors addressed early industry responses to pollution from the late nineteenth century to the mid-twentieth century. A special issue of *Enterprise and Society* in 2007 on "Business and Nature" also failed to empirically address business and climate change.[13] A decade later, in 2017, a special issue of *Entreprise et Histoire,* "Entreprises et Changement Climatique," covered natural resources, energy industries, agriculture, and the impact of search engines on global warming.[14] Unfortunately, this significant issue has not been widely recognised in English-language research.

In 2019, a special issue on "Business and the Environment Revisited" in *Business History Review,* edited by Bergquist, called for the integration of environmental issues into business history scholarship, urging business historians to focus more on the time period after the 1960s, which, when compared to previous time periods, introduced new complexities to business, in particular, climate change.[15] Although several articles in that issue addressed climate change, they lacked empirical studies on cases outside the West and research on how businesses have adapted to the risks and realities of climate change and extreme weather conditions.[16]

Few monographs in business history have integrated climate change into their approach. Jones' books on the history of green entrepreneurship provide the first comprehensive work in this field, offering new historical evidence on the complex role market institutions have played in supporting fossil energy growth while hindering the development of renewable energy companies until the 1990s.[17] Also, in 2017, the edited volume by Berghoff and Rome united business history and environmental history research under the umbrella of "green capitalism," but it did not explicitly focus on historical research on climate change.[18] A significant amount transpired between the "environmental awakening" of the 1960s and the mainstreaming of business sustainability in the twenty-first century, all against the backdrop of an escalating climate crisis.[19] Consequently, there remains great potential for developing a coherent approach and a well-formulated research agenda within the field of business history.

0.3 Overview of This Volume

This volume adheres to established business history standards but allows for broader industry-wide research. It aims to deliver a comprehensive overview of relevant topics and help explore a research agenda that has not yet been fully developed and that addresses big questions with the potential to have a strong impact on the field.[20]

To present a business history perspective on climate change, the editors have brought together researchers who have innovatively engaged with the issue. Some are among the pioneers who have engaged with the topic in the decade before this book was published. Others are key business historians, experts on the industries, regions, or phenomena being addressed, and who have used their expertise to look at those topics using a climate change lens. In doing so, they allow this book not only to advance knowledge on the history of climate change but also to contribute to developing a new stream of research, making it mainstream in the field of business history.

The chapters of this volume draw on the developing work of these scholars to provide insight into business strategies and responses as climate change has unfolded. They reveal the complex adaptations of businesses to what can be seen as threats to the planet in the long term but also as threats to their business activities in the short term. And from a long-term, historical perspective, these accounts put business responses to the current crisis in relation to previous, enduring business practices such as resistance to government regulation and oversight, underlying faith in market and technological solutions, and the ability to protect their own environments by polluting in other countries—usually developing countries with weak institutional environments.

To try to draw a broad picture of business activity in the context of the climate, chapters address these issues from several perspectives. They look across the globe, offering studies based in countries and regions such as China, Europe, India, Indonesia, Malaysia, North and South America, and Oceania. They examine widespread concerns about heat, drought, fire, floods, hurricanes, deforestation, melting glaciers, and rising sea levels, investigating how these both arise from and, in turn, affect business activity. They explore particular sectors that have made significant contributions to climate change, including agribusiness, car production, energy supply, and tourism. And they examine reactions to government intervention and political activists. From all these perspectives, the book seeks to map both business responses and business responsibilities.

In Chapter 1, Ann-Kristin Bergquist and Geoffrey Jones explore tensions between government regulation and corporate innovation, showing how in recent years the two have merged, allowing corporations to influence government policy involving climate change while presenting themselves in a positive environmental light. Chapter 2 focuses on the car industry. Here, Grace Ballor and Mattias Näsman follow that industry's response to the emergence of climate change and to growing resistance to fossil fuel burning. They use the development of the industry and the emergence of electric vehicles to explore the significant differences between corporate and national strategies in the United States, Europe, and Asia. Chapter 3 examines China's responses to environmental concerns. Ghassan Moazzin sets these in the context of the country's longer engagement with pollution, particularly that produced by bus companies and power providers. In Chapter 4, Shakila Yacob explores the implications of commercial deforestation, looking in particular at palm oil agribusiness in Southeast Asia. She points to claims by some businesses to be highly sustainable but shows how questionable such claims are when the deforestation behind their products is brought to light.

Chapter 5 then moves to the topic of hurricanes and floods and the costs of related disasters. Natalie Rauscher and Welf Werner look at how disaster policies developed in response to increased flooding across the second half of the twentieth century and how inadequate "solutions" have inadvertently supported damaging behaviour by individuals and corporations. In Chapter 6, Tirthankar Roy and Chinmay Tumbe use the concept of "seasonality" to explore the combination of floods and droughts in monsoon-affected regions of India. They note in particular the emergence of an industry producing water-mining "borewells" in response to water shortages and the unintended consequences, as these wells have allowed water-intensive crops to replace drought-resistant agriculture while depleting the groundwater reserves such crops rely on. Chapter 7 stays with the topic of agriculture. Looking in particular at New Zealand, Simon Mowatt tracks the industrialisation of agriculture, the resulting increase in its carbon footprint, and the challenges facing those who have attempted to "green" the food chain.

In Chapter 8, Marten Boon examines the response of the energy sector to climate change and the sector's influence on national and international policy by looking in particular at the Norwegian state oil company Equinor and comparing its distinctive response to climate change with those of other multinational corporations in the oil sector. Chapter 9 investigates the reaction, individually and collectively, of firms to climate change. Exploring the wine trade in South America, Teresa da Silva Lopes and Andrea Lluch note the challenges to agriculture of climate change and the ways in which wine areas have formed collaborations to adapt to such challenges.

In Chapter 10, Laurent Tissot investigates the effect of climate change on the tourism industry. He looks in particular at Alpine tourism, its development since the nineteenth century, and the response of the skiing sector when faced with the loss of snow in the mountains: some firms developed snow-manufacturing technologies, while others explored alternative forms of "ecotourism." Chapter 11 shows how the history of the insurance and reinsurance industry casts a distinctive light on climate change in the context of business strategy. Here, Niels Viggo Haueter traces how Swiss-based reinsurance has responded to changing climate disasters, leading the sector to acknowledge awareness of climate change and its threat to the industry and their clients.

In Chapter 12, Sabine Pitteloud examines businesses' response to growing climate activism. Looking at Germany and Switzerland in particular, the chapter explores how national business associations developed strategies in response to anti-corporate activism and related regulatory proposals. To fight back, these associations formed astute political alliances and green public relations (PR). Chapter 13 examines "greenwashing," the practice of shaping a climate-friendly public identity without necessarily changing climate-unfriendly practices. Paul Duguid and Teresa da Silva Lopes set greenwashing in the larger history of PR and its prevalent practice of "whitewashing," using the work of PR pioneer Edward Bernays to explore the development of climate-related PR in the twenty-first century.

In Chapter 14, Judith Stroehle follows corporations that have come to recognise climate change as a business risk and sought to affirm their behaviour as responsible. Such recognition produced numerical measures of "sustainable" behaviour, putting firms in pursuit of high, positive rankings. But, as Stroehle's research shows, these rankings are not as objective in their measurement of environmental behaviour as they are often assumed to be. Finally, in Chapter 15, the author uses climate-change litigation to investigate corporate claims to "sustainability." Bart Elmore looks at court cases that challenge corporate behaviour and shows how these reveal activities that corporations are eager to hide.

The historical perspective of these chapters reveals recurring business strategies and innovative new ones. Among the most enduring, the research shows businesses ignoring, denying, and suppressing criticism of their activities while spreading doubt and uncertainty about unwelcome evidence. But they also show businesses engaging with such criticism by developing complex alliances and intricate standards by which, they claim, their behaviour should be judged, and loudly endorsing their "green" commitments, "corporate social responsibility," and "sustainability," while washing their hands and profiles in the public eye. The research repeatedly shows resistance to regulation, which corporations have fought back against with plans of self-regulation, partnerships in non-governmental organisations, and varieties of short-term solutions which are shown often to have unintended consequences and induce moral hazard. Some of the case studies, however, show businesses, both small and large, making important, positive responses to the challenges society faces, putting the public interest before private interest.

Overall, the book seeks to use business history to illuminate climate change, examining the evolution over time of corporate strategies and, in turn, the effects of such strategies on the world today and the unfolding future of tomorrow. It draws on conventional resources of business history, in particular corporate archives. But as these are often inaccessible for many years after their creation, and corporations have been reticent about revealing their responses to the issues under discussion, the researchers have innovatively drawn on other resources such as interviews, court records, legislative proceedings, government hearings, market data, and the like. Inevitably, such a collection is only a selection. There are other sectors, companies, and topics that could be included. But its authors hope that such a selection will be more broadly informative, indicating appropriate approaches to other topics involving business history and climate change for future research. And in giving a picture of the past and the recent emergence of business-driven climate concerns, they also hope to provide a glimpse of the future and appropriate interpretations, anticipation, and responses to contemporary activity and its implications. A historical perspective can, we hope, inform and inspire more responsible business practices and policies in the face of our ongoing climate crisis.

The editors would like to thank the contributors to this volume for their work and for their cooperation in bringing the book together. We would also like to thank Swiss Re, and in particular Niels Viggo Haueter, for contributing invaluable insights during the early stages of this project and for hosting a conference in Zurich in May 2023 at the Centre for Global Dialogue. This gave the authors the opportunity for productive discussions around the themes developed in this book. We would also like to thank Walter Friedman, Mira Wilkins, Mary Yeager, and Laura Catena Zapata, each of whom provided valuable insights and very generous endorsements for the volume. And we would also like to express our gratitude to Routledge and its editors for overseeing the publication of the book.

Notes

1 Copernicus Programme, "Copernicus: 2023 Is the Hottest Year on Record."
2 IPCC, "Climate Change 2023: Synthesis Report," 4.
3 Stern, *The Economics of Climate Change*, xviii.
4 Lopes, "'The Nature of the Firm,'" 753; Lopes, "Business History in International Business."
5 Oreskes and Conway, *Merchants of Doubt*; Supran and Oreskes, "Assessing ExxonMobil's Climate Change Communications (1977–2014)," 084019; Frumhoff, Heede, and Oreskes, "The Climate Responsibilities." See also the often-overlooked book by Marchand, *Creating the Corporate Soul*.
6 Bansal and Hoffman, *The Oxford Handbook of Business and the Natural Environment*.
7 Hoffman, *From Heresy to Dogma*.
8 See the literature surveys of the field in McNeill, "The State of the Field of Environmental History"; Isenberg, *The Oxford Handbook of Environmental History*.
9 Oreskes and Conway, *Merchants of Doubt*; Supran and Oreskes, "Assessing ExxonMobil's Climate Change"; Franta, "Early Oil Industry Knowledge of CO_2 and Global Warming"; Franta, "Early Oil Industry Disinformation on Global Warming"; Franta, "Weaponizing Economics"; Bonneuil, Choquet, and Franta, "Early Warnings and Emerging Accountability."
10 Friedman and Jones, "Business History."
11 Chandler, "Industrial Revolutions and Institutional Arrangements."
12 Rosen and Sellers, "The Nature of the Firm."
13 Rosen, "Doing Business History in the Age of Global Climate Change."
14 Aggeri and Cartel, "Le Changement Climatique et les Entreprises."
15 Bergquist, "Renewing Business History in the Era of the Anthropocene."
16 See, for example, Boon, "A Climate of Change"; Bergquist et al., "Understanding and Overcoming Roadblocks to Environmental Sustainability."
17 Jones, *Profits, and Sustainability*.
18 Berghoff and Rome, *Green Capitalism?*
19 Bergquist, "Renewing Business History."
20 Friedman and Jones, "Business History."

Bibliography

Aggeri, Franck, and Mélodie Cartel, eds. "Le Changement Climatique et les Entreprises: Enjeux, Espaces D'action, Régulations Internationales." *Entreprise et Histoire* 86, no. 1 (2017): Special Issue—*Entreprise et Changement Climatique*, 6–20.

Bansal, Pratima, and Andrew J. Hoffman, eds. *The Oxford Handbook of Business and the Natural Environment*. Oxford University Press, 2012.

Berghoff, Hartmut, and Adam Rome, eds. *Green Capitalism? Business and the Environment in the Twentieth Century*. Philadelphia: University of Pennsylvania Press, 2017.

Bergquist, Ann-Kristin. "Renewing Business History in the Era of the Anthropocene." *Business History Review* 93, no. 1 (2019): 3–24.

Bergquist, Ann-Kristin, Shawn A. Cole, John Ehrenfeld, Andrew A. King, and Auden Schendler. "Understanding and Overcoming Roadblocks to Environmental Sustainability: Past Roads and Future Prospects." *Business History Review* 93, no. 1 (2019): 127–48.

Bonneuil, Christophe, Pierre-Louis Choquet, and Benjamin Franta. "Early Warnings and Emerging Accountability: Total's Responses to Global Warming, 1971–2021." *Global Environmental Change* 71 (2021): 102386. https://doi.org/10.1016/j.gloenvcha.2021.102386.
Boon, Marten. "A Climate of Change? The Oil Industry and Decarbonization in Historical Perspective." *Business History Review* 93, no. 1 (2019): 101–25.
Chandler, Alfred. "Industrial Revolutions and Institutional Arrangements." *Bulletin of the American Academy of Arts and Sciences* 33, no. 8 (1980): 33–50.
Copernicus Programme. "Copernicus: 2023 Is the Hottest Year on Record." https://climate.copernicus.eu/copernicus-2023-hottest-year-record.
Franta, Benjamin. "Early Oil Industry Knowledge of CO2 and Global Warming." *Nature Climate Change* 8, no. 12 (2018): 1024–25.
Franta, Benjamin. "Early Oil Industry Disinformation on Global Warming." *Environmental Politics* 30, no. 4 (2021): 663–68.
Franta, Benjamin. "Weaponizing Economics: Big Oil, Economic Consultants, and Climate Policy Delay." *Environmental Politics* 31, no. 4 (2022): 555–75.
Friedman, Walter A., and Geoffrey Jones. "Business History: Time for Debate." *Business History Review* 85, no. 1 (2011): 1–8.
Frumhoff, Peter C., Richard Heede, and Naomi Oreskes. "The Climate Responsibilities of Industrial Carbon Producers." *Climatic Change* 132, no. 2 (2015): 157–71.
Hoffman, Andrew J. *From Heresy to Dogma: An Institutional History of Corporate Environmentalism*. Stanford University Press, 2001.
IPCC. "Climate Change 2023: Synthesis Report." https://www.ipcc.ch/report/ar6/syr/downloads/report/IPCC_AR6_SYR_SPM.pdf.
Isenberg, Andrew C., ed. *The Oxford Handbook of Environmental History*. Oxford University Press, 2017.
Jones, Geoffrey. *Profits and Sustainability: A History of Green Entrepreneurship*. Oxford University Press, 2017.
Jones, Geoffrey. *Varieties of Green Business: Industries, Nations and Time*. Edward Elgar Publishing, 2018.
Jones, Geoffrey. *Deeply Responsible Business: A Global History of Values-Driven Leadership*. Harvard University Press, 2023.
Lopes, Teresa da Silva. "'The Nature of the Firm'—and the Eternal Life of the Brand." *Enterprise and Society* 20, no. 4 (2019): 752–76.
Lopes, Teresa da Silva. "Business History in International Business." In *Oxford Research Encyclopaedia of Business and Management,* edited by Michael A. Hitt. Oxford University Press, 2023.
Marchand, Roland. *Creating the Corporate Soul: The Rise of Public Relations and Corporate Imagery in American Business.* University of California Press, 1998.
McNeill, John R. "The State of the Field of Environmental History." *Annual Review of Environment and Resources* 35, no. 1 (2010): 345–74.
Oreskes, Naomi, and Erik M. Conway. *Merchants of Doubt: How a Handful of Scientists Obscured the Truth on Issues from Tobacco Smoke to Global Warming*. Bloomsbury Publishing, 2011.
Rome, Adam, and Hartmut Berghoff. "The Ecology of Commerce: Environmental History and the Challenge of Building a Sustainable Economy." In *Green Capitalism—Business and the Environment in the Twentieth Century*, edited by Hartmut Berghoff and Adam Rome. University of Pennsylvania Press, 2017.
Rosen, Christine Meisner, and Christopher C. Sellers. "The Nature of the Firm: Towards an Ecocultural History of Business." *Business History Review* 73, no. 4 (1999): 577–600.
Stern, Nicholas. *The Economics of Climate Change: The Stern Review*. Cambridge University Press, 2007.
Supran, Geoffrey, and Naomi Oreskes. "Assessing ExxonMobil's Climate Change Communications (1977–2014)." *Environmental Research Letters* 12, no. 8 (2017): 084019.

1

INSTITUTIONAL ENTREPRENEURSHIP AND CLIMATE CHANGE[1]

Ann-Kristin Bergquist and Geoffrey Jones

1.1 Introduction

Private regulatory governance systems are the norm in the contemporary global economy.[2] Such systems are primarily voluntary efforts by firms to restrain their own behaviour and include those that target business operations, ranging from environmental, social, and governance (ESG) reporting and carbon accounting to certification schemes for sustainable agriculture, fisheries, and forestry practices. These are now the primary form of global environmental governance. Since the Paris Agreement in 2015, international treaties, including the United Nations Framework Convention on Climate Change (UNFCCC)—first negotiated at a conference in Rio de Janeiro in 1992—have largely relied on the private sector and its willingness to voluntarily combat climate change. The role of the financial sector was crucial in this regard.[3]

Yet it is significant that more than half of all industrial emissions of CO_2 emitted since the Industrial Revolution occurred between 1988, when climate change emerged as a political issue, and the Paris Agreement in 2015.[4] This reliance on private regulation is therefore contested in the academic literature. On the one hand, it has been seen as exercising a role in pressuring businesses to green their operations, including the mitigation of greenhouse gases. On the other hand, critics have argued that it has enabled firms to use modest private regulations as a political strategy to pre-empt or delay more stringent public regulations. The aim of this chapter is to provide a historical foundation and evolution of private sector initiatives in global climate governance and to explore why this happened. It will examine how far this private regulation represents a significant, if incremental, attempt by corporations to reduce their negative impact on climate change or whether it should be seen as forms of corporate greenwashing, green-hushing, and transition-washing.

This chapter, after providing a brief literature review in Section 1.2, explores two different historical paths in private regulation and why they came about. As Section 1.3 shows, the first path involved certification and standards programmes designed to facilitate the growth of green industries and the development of the early stages of ESG investing. Section 1.4 discusses the second path, which developed from the 1970s, growing out of the interest of big business, particularly multinationals, who sought an alternative route to governmental regulations which

DOI: 10.4324/9781003478089-2

they regarded as costly and as a threat to international trade. A key agent was the International Chamber of Commerce (ICC) during the 1990s. The alternative routes launched by business organisations in the 1980s and 1990s came to be based on self-regulation, which global institutions such as the United Nations Environment Programme (UNEP) increasingly came to support and even co-create with business. Section 1.5 demonstrates why and how these two paths came together, partly as a result of a conflict between the push for increased government intervention to reduce greenhouse gas emissions from the late 1980s, mainly by limiting fossil fuel consumption, and the associated economic restrictions and costs. Section 1.6 concludes and suggests that, while private regulation has significant limitations, it may have played a crucial role in climate action since government measures have remained blocked or weak. Throughout, we employ the concept of institutional entrepreneurship—individuals who drive change within established institutions and/or create entirely new ones—to explain the momentum behind the foundation of institutions and the framing of debates.

1.2 Literature Overview

The role of business in shaping global environmental and climate governance rules has received attention in several disciplines, most notably in international political economy literature. Parts of this literature have been concerned with growing corporate influence over environmental governance rules since the 1990s and the role played by transnational corporations in the negotiations of global treaties such as the Montreal Protocol on ozone, negotiated in 1987, and the UNFCCC in 1992.[5] A stream of research has also focused on multilateral environmental governance mechanisms designed to influence the environmental behaviour of multinationals. This research has looked at formal endeavours to promote better environmental business practices, developed by both industry-driven efforts and different kinds of partnerships between businesses and environmental organisations and/or international organisations such as the United Nations.[6] Overall, the political economy literature has been broadly concerned about the increasing power of corporate interests in the governance system, arguing that international business can shape the rules to avoid being constrained by states and international regulations.[7]

There is a vibrant business administration literature concerned with the development of new market institutions created by the private sector itself to deal with climate change issues.[8] The scholarship on corporate environmentalism contains different subfields which include studies of self-regulation programs, certification schemes, and new metrics supporting voluntary action.[9] This research has been both positive and sceptical about the actual impact of the array of certifications, standards, and frameworks to guide companies to integrate "sustainability," including climate change, into core business strategies.[10]

In contrast, the subject of business and climate change has only attracted the attention of a modest number if impactful group of business historians. The negative reason for this is that the literature has been excessively focused on firms as creators of wealth rather than on the costs to the environment. A more defensible reason is that the central role of climate change in environmental challenges is recent. It only emerged as a subject requiring regulatory action in the late 1980s. So the limited amount of business historical literature has focused on other environmental issues, including air and water pollution and organic food.[11] Although there are some exceptions,[12] most of these studies focus on national case studies rather than the global situation. Only recently has attention turned to the history of certification and the role of the transnational business associations at the United Nations.[13] Boon's research on the oil industry and climate change, by contrast, provides a rare study in the field of the impact of private regulation.[14] He concludes that the voluntary nature of climate governance has resulted in a situation where the oil industry took very little meaningful action to achieve decarbonisation before 2010. Focusing on the case of the United States, Rome provided an important case study of how market institutions, including Wall Street expectations, limited the chemical firm DuPont's ability to seriously address climate change.[15] The larger story of how businesses came to become active actors in shaping the architecture of global climate governance has remained largely unexplored in the business history literature.

There have been a growing number of historical studies beyond formal business history that have explored the role of the fossil fuel industry in shaping both national and global climate policies.[16] In an important study, the historian of science Oreskes and her co-author Conway demonstrated how the emergence of climate science scepticism in the 1990s was orchestrated by some oil companies and conservative think tanks in the United States.[17] Yet, beyond the fact that major fossil fuel companies succeeded in causing delays in climate action, historical research on how businesses have shaped rules and institutions for global climate governance is scarce.[18]

1.3 The Development of Certification and Metrics

The creation of certification schemes and environmental reporting that emerged in the last decades of the twentieth century was the work of institutional entrepreneurs. The settings were emerging new categories whose definition was unclear and which lacked legitimacy, certainly to incumbents and many potential customers. Policymakers, at least initially, showed little interest in these new categories, leaving it to the entrepreneurs within the categories to establish the institutions necessary for building credibility and legitimacy. This section will look at the cases of organic food, which was a pioneer in this new wave of certification, buildings, and sustainable investing.

It is important to note that the foundational work on certification predated the era when climate change became the most pressing environmental issue, which can be dated from the late 1980s, even though strong ecological and social concerns were already present. It was only in the early 1970s that scientists began to define anthropogenic climate change—climate change caused by human activities—as an environmental issue.[19] At the United Nations Conference on the Human Environment (UNCHE) in Stockholm (hereafter the Stockholm Conference) in 1972, the potential danger of anthropogenic "climatic change" was highlighted in the conference's Action Plan (recommendation 79, a), but the issue was still characterised by great scientific uncertainty and was not seen as a threat in the near future. Instead, other immediate environmental challenges, including pollution from toxic chemicals, acidifying substances (including acid rain), hazardous wastes, and the depletion of natural resources in general, were on the agenda in Stockholm.[20] The 1970s saw, in general, a strengthening of governmental environmental regulation in the Western world, based on command-and-control regulatory approaches (permission, prohibition, and enforcement by law) that took different forms in different countries.[21] However, these regulations at the time did not address climate change but rather a range of other problems such as sulphur dioxide (SO_2) emissions, heavy metals, chemical substances, waste management, and many other pollution issues.

Neither climate change nor the negative environmental impact of conventional agriculture was known during the foundational period of organic agriculture. The first certification scheme which took environmental aspects into consideration was in organic food and was an outgrowth of the Anthroposophical movement created by Rudolf Steiner in the early twentieth century. In 1924, Steiner first outlined the principles of biodynamic agriculture, a form of organic farming, and already in 1927, his followers created Demeter. This organisation advised farmers to try out Steiner's methods and engaged in the marketing of biodynamic products. A Demeter trademark was introduced in 1928 to facilitate the marketing of biodynamic produce. Demeter was the only international organic food trademark for the next half century. Anthroposophy was highly esoteric, and biodynamic farming—involving the application of nine secret "preparations" and co-ordination with the movement of the moon and stars—was highly time-consuming and complex. As a result, the number of working biodynamic farms remained small.[22]

There were only sporadic attempts to create new certification schemes until environmental concerns rose during the 1960s, sometimes referred to as the "second wave" of environmentalism following a first wave in the nineteenth century which petered out during the interwar years.[23] Over the 1970s, there were ad hoc attempts in different countries to develop certification schemes, primarily led by organic farmers who disliked and distrusted capitalism but who wanted standards to give their category legitimacy, not least with consumers. The health of the soil and the health of people were the primary concerns, rather than climate change.[24]

The turning point for certification in organic agriculture came in Europe in 1980. An international standard emerged from the French biodynamic organisation Nature et Progrés (N & P), whose leadership had attended fringe meetings at the Stockholm conference in 1972. Led by the visionary Roland Chevriot, one of many institutional entrepreneurs in organic food, N & P released its first set of guidelines for organic agriculture in France and also launched the International Federation for Organic Agriculture Movements (IFOAM), persuading organic organisations from Sweden, Britain, South Africa, and the United States to join. The N & P leadership believed that if organic food was to be seen as legitimate, there had to be a standard definition between countries. By 1980, the first draft of organic standards was in place. The IFOAM standards signalled a transition away from a system which had rested on personal relationships between farmers and consumers, towards transferring the powers of verification to a separate certification and inspection system. In a movement initially based on strong values, it began to make formal rules the primary form of regulation.[25]

IFOAM's mission continued to evolve. As Asian and then Latin American societies joined in the 1980s, IFOAM's agenda expanded to reflect their concerns, including social justice and development. In 1992, IFOAM participated in the UN Food and Agriculture Organization's Rio Conference in Brazil. Finally, by the 2010s, it began to make the case for organic agriculture as a solution to climate change.[26]

Although certification in organic food began as self-regulation, during the 1980s several European governments came to be involved in organic standards or developed logos for organic products.[27] Public policy motivations ranged widely, from enthusiasm for environmental and social benefits to wanting to support what was seen as a potentially lucrative infant industry. A further layer of complexity was added as regional integration in Europe resulted in a new set of public agencies becoming involved. In 1991, William Vidal—another institutional entrepreneur—created a limited liability company called ECOCERT in France for the provision of certification services. Subsequently, it diversified beyond France, opening its first international subsidiary in Portugal and acquiring large U.S. and Swiss certifiers in 2011 and 2015 respectively. It also diversified from organic food into cosmetics, perfumes, and textiles. The company remains owned by Vidal's family holding company.[28] In 2024, ECOCERT operated in more than 130 countries.[29]

The certification schemes put in place for organic food helped to legitimise and define the new category. However, many uncertainties remained, including competing certification schemes and poor enforcement. While the United States Department of Agriculture began certifying organic food products in 1990, the execution of the process using contracted certifying agents was problematic. Between 2005 and 2014, 38 of the 81 USDA certifying agents failed at least once to uphold basic USDA standards.[30] Nor was possessing certification a guarantee of environmental sustainability. The growth of large organic retailers—such as Whole Foods in the United States (and their equivalents elsewhere)—meant that certified organic products were shipped from around the world to satisfy consumers, completely negating any positive impact on climate change.

The creation of green building standards and certification systems also had its origins during the 1980s and was the result of interactions between architects, conventional property developers, public agencies, and NGOs. In the wake of the 1973 oil crisis, the primary focus was on low-energy design. In Britain, during the late 1980s, a number of property developers involved in London's Canary Wharf redevelopment project became interested in the idea of building commercial buildings that incorporated effective environmental improvements, especially low-energy performance. The outcome was a pioneering building assessment and certification system by a government agency called the Building Research Establishment Environmental Assessment Method (BREEAM), launched in 1990. The standard focused not only on energy efficiency and resource scarcity but also on broader environmental challenges, including the effects of chlorofluorocarbons used in aerosols and air conditioning. It was an entirely voluntary system for developers, who used it either to understand best practices or to highlight their achievements to the market.[31]

Although BREEAM was pioneering, it did not remain unique. The most globally important certification scheme became the Leadership in Energy and Environmental Design (LEED) system, based in the United States. This was primarily the work of another institutional entrepreneur, David Gottfried, a real estate entrepreneur and environmentalist who sought to create a non-profit organisation to promote green buildings. He persuaded environmentally inclined building materials companies to join the venture as dues-paying members by laying out a programme for developing environmental standards. Sixty firms and non-profit organisations became founding members of the non-profit U.S. Green Building Council (USGBC) in 1993. The central strategy was to include members from across industrial sectors, as well as environmental organisations, in the hope of securing consensus among key stakeholders.[32] The first rating system, called LEED, was piloted in 1998. A more sophisticated version, known as LEED 2.0, was launched two years later.[33]

The endeavour initially met much scepticism, but it had considerable momentum. Hundreds of volunteers drafted LEED rating systems for commercial interiors, existing buildings, healthcare, schools, neighbourhood developments, and other markets. The U.S. government's General Services Administration, which owned a huge portfolio of federal buildings, became an early adopter. They were joined by municipalities and state governments. In 2008, California became the first state to adopt a statewide green building code. The Green Building Council (GBC) model spread to other countries. The first World GBC conference was held in 2002. By 2015, the World GBC included national green building councils in 75 countries.[34]

The building certification schemes were important examples of private regulatory governance. The institutionalisation of building performance was a major influence, encouraging architects and builders to think harder about sustainability, including energy conservation, than they had previously. Yet critics argued that both LEED and BREEAM failed to measure post-construction performance, for instance of carbon emissions, to see if tenant behaviour changes expectations expressed in the certification rating. LEED's point system, touted for its flexibility, also attracted critics, especially among ecological architects. As in other certification systems, the use of metrics in the standards provided an opportunity for gaming the system. There were also incentives for developers and builders to game it because certification could be, and was, used in marketing. Environmental activists talked explicitly of "LEEDwashing." Furthermore, the Green Building Councils, although non-profits, needed to earn revenues in order to fund the organisation and therefore had no incentive to make standards so rigorous that certification was too difficult to obtain. At best, the certificate schemes captured only part of the problem. Critics noted that a new home could receive a top LEED certification even if it was an over-sized house built in an environmentally stressed area, like a desert.[35]

The development of metrics in sustainable investment had some similarities and also some differences from organic foods and building materials. There was no government involvement in the early stages of ESG investing. Instead, its origins can be traced back to the policies of certain Christian groups such as Quakers and Methodists not to invest in industries considered sinful such as slavery and alcohol, but it was again in the 1980s that there was a new momentum by ecologically and socially motivated individuals.[36]

During that decade, a number of female activists and institutional entrepreneurs in Boston, and later elsewhere, were instrumental in scaling this kind of investment beyond the niche and the religious. Joan Bavaria was a value-driven pioneer in what was then called Socially Responsible Investment. She formed Franklin Research and Development Corporation (FRDC, later Trillium Asset Management) in 1982 as a cooperative owned by its (mainly female) employees. FRDC employed traditional financial analysis, research by non-profits and NGOs, and interviews with managers to rank companies that both showed promising financial upside and passed FRDC's screens for evaluating investments. These included specific criteria such as no investment in South Africa and broader considerations such as "sensitivity to the environmental impact" and "product/service quality and social usefulness."[37]

FRDC published its ratings and invested in the equity of companies that it believed would deliver attractive financial and social returns in the long term. Bavaria saw no tension between the two different types of returns. She believed that strong environmental and social policies would increase the profitability of companies. Two-thirds of FRDC's investors were high-net-worth investors, and the remainder were churches and small foundations.[38]

Bavaria sought to build a community that would develop and share metrics. She moved early to build relationships with other like-minded investment houses, including Pax World and Calvert. By 1985, this group evolved into a non-profit trade association named the Social Investment Forum (later U.S. SIF). It began to develop a set of principles regarding corporate conduct on environmental matters to echo the Sullivan Principles, a code of conduct introduced in 1977 by Bishop Leon Sullivan, who sat on the Board of General Motors, regarding investment in South Africa. In 1989, a huge oil spill in Alaskan waters from the tanker Exxon Valdez, owned by the largest U.S. oil company, served as a catalyst for elevating these efforts. The spill energised the ongoing discussions about how to improve the environmental practices of big businesses. SIF wrote the "Valdez Principles" (subsequently renamed the "Ceres Principles") to catalyse the release of standardised information on environmental performance, designed to operate similarly to the standards of the Financial Accounting Standards Board in the United States. The principles included wide-ranging environmental goals such as the protection of the biosphere and biodiversity, and energy efficiency, but nothing on climate change.[39]

Bavaria formed the non-profit organisation Ceres (Coalition for Environmentally Responsible Economies) in 1989 to manage the Ceres Principles. While SIF focused solely on investors, Ceres sought to gather investors, environmental organisations, and public interest groups to persuade companies to adopt and practice the Ceres Principles. Signatories of the Principles paid an upfront fee and an annual administrative fee for the cost of Ceres monitoring and verification of compliance. In the first few years, it attracted a handful of value-driven companies, including Ben & Jerry's. In 1993, Sunoco, an oil refiner, became the first Fortune 500 company to endorse the Ceres Principles. Sunoco remained the only oil company signatory as of 2023.[40]

In 1996, Bavaria recruited Bob Massie, an Episcopal priest, unsuccessful Democratic politician, and social justice campaigner, as president of Ceres. He developed a set of environmental and social reporting standards. In 1997, Ceres launched the Global Reporting Initiative (GRI), whose guidelines aimed to prompt disclosure and transparency. GRI was intended to be international, particularly since European businesses had been more receptive than U.S. companies to the concepts of SRI. In 2002, Ceres spun GRI off as an independent entity organised as a non-profit institution based in the Netherlands. By 2008, over 1,500 companies around the world had adopted its reporting standard, although less than 20% of the companies were based in the United States. It was mainly used by larger, publicly traded companies, and although professionally designed, it remained a voluntary system which allowed selected reporting of some metrics and not others.[41]

Amy Domini was a second, unconventional institutional entrepreneur hailing from Massachusetts who also exercised a formative impact on the socially responsible investing movement. Inspired by Bavaria as well as strong religious beliefs, Domini co-founded KLD Research & Analytics, which launched the Domini 400 Social Index (DSI) in 1990. This was a portfolio of 400 companies identified as socially responsible using proprietary screening criteria. The DSI included approximately 250 companies from the Standard & Poor's 500 Index, 100 other large- and mid-capitalisation companies selected to diversify the sectors represented in the DSI, as well as 50 smaller companies that were chosen for their "exemplary" social and environmental practices. Companies had to demonstrate the potential for strong financial performance to be included.[42]. Domini co-founded KLD Research & Analytics, which launched the Domini 400 Social Index (DSI) in 1990. This was a portfolio of 400 companies identified as socially responsible using proprietary screening criteria. The DSI included approximately 250 companies from the Standard & Poor's 500 Index, 100 other large- and mid-capitalisation companies selected to diversify the sectors represented in the DSI, as well as 50 smaller companies that were chosen for their "exemplary" social and environmental practices. Companies had to demonstrate the potential for strong financial performance to be included.[43]

KLD experienced limited interest when it attempted to license the index to large mutual fund companies as a way to earn revenue. To earn revenue on the index, Domini established her own mutual fund in 1991, which was called the Domini Social Equity Fund. The new fund launched with a $600,000 investment from the retirement fund of an affluent colleague. The fund's initial growth was slow until it significantly outperformed the S&P 500 Index between 1995 and 1999. This was less proof that social responsibility pays, however, than a reflection that the screening criteria led to a portfolio of technology stocks, which boomed at that time.[44]

The impact on the ecological and social performance of firms included in the index was impossible to measure. The screening criteria certainly resulted in unlikely firms being included in DSI. In 2003, the environmentalist and entrepreneur Paul Hawken waged a crusade against Domini for calling McDonald's fast-food business—"a company whose mission harms children and its workers," in Hawken's words—socially responsible. He pondered whether SRI was best seen as "a way for upper-class people to launder their money." Domini responded to Hawken's argument by stating that McDonald's had made progress in reducing its output of solid waste, promoting racial and gender diversity among middle and upper management positions, and being responsive to other social and environmental concerns. It therefore fell into the category of one of her "better" companies.[45]

By the 2000s, then, as climate change rose to prominence, institutional entrepreneurs had put in place private regulatory schemes in the form of certification schemes and bodies, and indexes measuring social and ecological performance. Governments had a supporting role in some of these cases. These certification schemes had defined categories and set standards, and they were globalised. The primary downside was their potential to become corporate vehicles for greenwashing. The issue of greenwashing through these types of metrics was exacerbated when large business corporations and their industry associations began to develop a business case for voluntary commitments to environmental protection, including self-regulatory schemes, from the late 1980s, a trend that exploded in the 1990s. This trend of private regulation in global environmental governance was, as will be explained in the next part of the chapter, made possible by another set of institutional entrepreneurs that focused on big business and was supported by various international organisations, and, most importantly, the United Nations.

1.4 The United Nations and Business-Led Initiatives

The United Nations, which emerged as the key institution in global environmental governance from the 1970s, came to work more closely with business. It preferred to collaborate through umbrella organisations of international industry, initially the ICC, rather than with individual corporations.[46] This section explains why and how the UN arena opened up for business-led initiatives that stressed voluntarism in global climate governance, and in what direction business wanted to shape the rules. When the harsh conflict between the call for increased government interventions to control greenhouse gas emissions (meaning limiting the consumption of fossil fuels) and free-market globalisation became a reality from the late 1980s, a feasible way forward was to pass the solution of the problem to the market actors. It was within this "institutional void" that private regulation, initiated by business or in partnership with governments or non-government actors, became widely accepted as a practicable means to address a wide range of complex sustainability problems, including climate change. This happened between the late 1980s and the early 1990s, which, as we will argue, also caused the two paths of private regulation to integrate.

When big businesses such as Shell and others initially began engaging in the development of private regulation within environmental governance, it was at first in response to the United Nations' initiative to convene their first conference dedicated to environmental issues in Stockholm in 1972. The Stockholm Conference marked the first comprehensive assessment of humanity's global environmental impact, calling upon governments to collaborate in establishing institutions and regulatory frameworks to safeguard the environment. It was recognised that global environmental degradations, including "climatic change," could not be addressed by single nations alone. This conceptualisation of environmental problems and their solutions meant that nation-states were viewed as the appropriate agents of environmental action, but that international regimes were the suitable governance mechanism.[47]

In some countries, the pressure to curb pollution from manufacturing firms and transport predated the Stockholm Conference due to domestic environmental regulations. This pressure led to burgeoning tensions between businesses and governmental bodies over the associated compliance costs. Notably, in the United States, the enactment of the Clean Air Act Amendments of 1970 ignited contentious debates between the Environmental Protection Agency (EPA) and the automobile industry, including General Motors, Ford, and Chrysler.[48] Representing global industry interests at the Stockholm Conference, the ICC highlighted the discordant nature of environmental legislation across nations as a hindrance to international trade. The ICC contended therefore that a market economy offered the most conducive environment for addressing environmental concerns and advocated for harmonising environmental standards among nations. In 1973, Ian MacGregor, the CEO of the U.S. mining company AMAX, proposed that the ICC therefore should elaborate an international code of environmental practice to govern fair and equitable industry conduct concerning environmental issues. The Environmental Guidelines for the World Industry, adopted in 1973, reflected the ICC's inclination towards business self-regulation in environmental governance. In essence, the ICC argued against the increasingly prevalent command-and-control approaches favoured by national governments, advocating instead for restrictions on such methods and the supplementation of regulatory frameworks with business-driven self-regulation measures.[49]

Yet the Stockholm Conference had the ambition to enforce new institutions and an international legal governance structure. It produced a Declaration on the Human Environment (the so-called Stockholm Declaration) which outlined 26 principles of international environmental law, along with an Action Plan that contained 109 recommendations to be implemented. Recommendation 79 (a, b, and c) called attention to "climatic" change and recommended states to set up approximately 10 baseline stations (with consent from states involved) in order to monitor long-term global trends in atmospheric constituents and properties "which may cause changes in the meteorological properties, including climatic change" (UN, 1972).[50] UNEP, the first-ever international environmental management body, was created in 1973 and was given the overall responsibility for the implementation of the Action Plan's recommendations.

Maurice Strong, a self-made Canadian oil and gas billionaire and, among other things, President of Power Corporation of Canada until 1966, served as General Secretary of the Stockholm Conference as well as UNEP and would later become the member of so-called Brundtland Commission and the secretary-general of the UN Conference on Environment and Development in Rio in 1992 (hereafter the Rio Conference). As recently demonstrated by Huf, Sluga, and Selchow (2022), Strong mobilised his business networks to incorporate industry and multinationals into the work of international institutions which were developing approaches to environmental governance in the early 1970s. He created UNEP's Industry Office with the purpose of identifying, classifying, and evaluating international environmental concerns on an industry-by-industry basis. In building the Industry Office, Strong selected the ICC as UNEP's exclusive partner.[51] With a platform at the United Nations, Strong emerged from the early 1970s as an institutional entrepreneur in global environmental governance, with distinct abilities to network with both government representatives and researchers, as well as non-governmental organisations, including businesses. Over time, he managed to support a development where business leaders became increasingly closely intertwined with UNEP's agenda for environmental governance.[52]

While the Western-based business community represented by the ICC remained critical about environmental regulation that would complicate trade and other economic operations, UNEP struggled. The implementation of the Stockholm Declaration and the Action Plan was characterised by complex conflicts of interest, not the least between the developed and developing countries. According to Grey, the philosophical conflict between these countries, as well as innumerable other political disputes, forced UNEP into compromises and avoidance of tough issues throughout the 1970s.[53] The tenth anniversary of the Stockholm Conference in Nairobi in 1982, for example, was characterised by major political disputes such as Berlin, the Falkland Islands, Israel and Palestine, Kampuchea, and Cyprus.[54] As argued by Borowy, the Nairobi conference demonstrated overall that the world had experienced a massive increase in environmental degradation and a huge decline in key environmental indicators, not the least in developing countries.[55] There was therefore a considerable rationale for UNEP to turn to big businesses through the ICC in order to bring about some change in a context that otherwise seemed locked.[56] The failed Nairobi Conference was also the reason why the UN General Assembly agreed to create the World Commission on the Environment and Development, chaired by the former Norwegian Prime Minister, Gro Harlem Brundtland (hereafter the Brundtland Commission).[57]

In this context, UNEP, under the leadership of the Egyptian scientist Mustafa Tolba, continued with Strong's agenda to seek more cooperation with the world business community in the 1980s. Thus, in 1983, UNEP took the initiative to co-organise the first World Industry Conference on Environmental Management (hereafter WICEM) in collaboration with the ICC. The event was symbolically held at Versailles, Paris, and brought more than five hundred government and business representatives together from developed and developing countries to discuss how companies could meet the growing challenges and concerns with respect to the long-term effects of pollution on the planet. Industry had by then emerged as an important partner of government and international organisations such as the World Wildlife Fund in the arena of international environmental governance policy discussions. A key issue addressed time and again at WICEM was the efficacy of self-regulation by industry. The reason, as Hans König, Secretary-General of the ICC, argued, was the virtual impossibility of drafting treaties that are equally appropriate to the many nations asked to ratify them.[58] The Brundtland Commission, in which Maurice Strong was a member, took an impression from the event, and stated in their final report:

> The world industry has taken some significant steps through voluntary guidelines concerning industry practices on environment, natural resources, science and technology. Although few of these guidelines have been extended or applied regionally in Africa, Asia, or Latin America, industry continues to address these issues through various industry associations. [. . .] These efforts were advanced significantly by the 1984 World Industry Conference on Environmental Management (WICEM).[59]

The WICEM meeting in 1984 led to a follow-up event in Rotterdam in 1991, WICEM II, where the ICC launched its Business Charter for Sustainable Development. The Charter was a self-regulatory scheme that set 16 principles for environmental management. It became influential in the development of voluntary business codes across different sectors, which, according to Jessica Green, demonstrated ICC's entrepreneurial authority in private environmental governance, as it came to dictate behaviour and practice that were adopted by others.[60] Climate change was discussed at WICEM II,[61] but the principles in ICC's Business Charter were formulated in very generic terms and did not mention climate change or any other specific environmental issue. Instead, wordings such as "environmental impacts," "environmental performance," and "sustainable use" were applied in the principles.[62] It is important to note that WICEM II was explicitly focused on preparations for the Rio Conference in 1992, which would represent the largest political gathering in history. WICEM II was co-organised between the ICC, UNEP, and in fact also the Rio Conference secretariat and was attended by more than 200 corporations and industry associations "operating at all continents."[63]

The Rio Conference in 1992 was held on the occasion of the 20th anniversary of the Stockholm Conference, with its primary objective to produce a broad agenda and a blueprint for international action on environmental and development issues. Among other things, the UN Climate Framework Convention on Climate Change was up for signature, along with the Rio Declaration, which ended up containing 27 universal principles for governance. Two influential business organisations had prepared for and were present at the Rio Conference. Strong, who was made secretary-general of the Rio Conference, appointed Stephan Schmidheiny, a Swiss asbestos industrialist close to the ICC, as his special adviser. Strong gave Schmidheiny the mission to act as the representative of business at the Rio Conference and to encourage and promote a broad expansion of interests in "sustainable development" in business and industry. With the help of Edgard Woolard, Chairman of DuPont, Schmidheiny created the Business Council for Sustainable Development (BCSD) with 48 business leaders from around the world, representing companies such as 3M, ALCOA, Dow Chemical, ENI, Shell, Chevron, ABB, Volkswagen, Nippon Steel, and others.[64] In parallel, the ICC founded a new organisation—the World Industry Council for the Environment (WICE)—to lobby environmental issues for business interests at the Rio Conference.[65]

BCSD, under the leadership of Schmidheiny, wrote the manifesto titled *Changing Course: A Global Business Perspective on Development and the Environment* in preparation for the Rio Conference. The document underscored the efficacy of self-regulation in effecting significant enhancements in the environmental practices of businesses and industries, both presently and in the foreseeable future. It contended that self-regulatory measures could potentially offer greater cost-effectiveness to society at large compared to traditional regulatory approaches such as command-and-control policies, including carbon taxes or tradable permits. The BCSD maintained that self-regulation has been, and would continue to be, instrumental in fostering notable advancements in mitigating the environmental impacts of business operations. However, it stressed the necessity of establishing a transparent framework that clarified expectations and requirements through negotiation between participating industries and governmental bodies. This was seen as important to ensure the efficacy and credibility of such self-regulatory initiatives.[66] Schmidheiny's most successful ploy, which would prove to be highly influential, was the introduction of the notion of *eco-efficiency*, which in simple terms pointed at firm-level strategies to produce more with less. BCSD maintained that climate change was one of many environmental challenges that could be addressed through energy efficiency measures. *Changing Course* argued that business action to improve energy efficiency should be initiated without a delay since it made both economic and environmental sense "whether or not global warming becomes a dangerous reality."[67]

The BCSD strategies were to demonstrate action by weaving together concerns about climate change into the overall rhetoric about sustainable development and how to make a business case for sustainability through eco-efficiency measures. The main policy recommendation that the business sector made in the run-up to the Rio Conference was to focus on so-called "no regrets" measures. These included attempts to increase energy efficiency, reduce energy use in manufacturing, and promote technological change and diffusion.[68] The essential point with the notion of "no regrets" was to only undertake measures that yielded firm-level benefits, even in absence of climate change. But as Robert Falkner has argued, no major global firm or industry spoke out in support of a global limit on greenhouse gas emissions under the UNFCCC at the

Rio Conference.[69] In fact, there was considerable business opposition by major oil and coal companies, especially in the United States, to suggestions that a reduction in carbon emissions was necessary.[70] In 1989, ExxonMobil, Chevron, BP, Shell, and others formed the Global Climate Coalition to represent industry interests and counteract growing concerns and regulations related to climate change. Its mission was to influence policy and public opinion by questioning the scientific consensus on global warming and highlighting the economic costs of climate action (see Boon's chapter in this volume). This position was thus not taken because of lack of awareness of the issue, but rather because of considerable costs associated with cutting down on greenhouse gas emissions.

According to Chatterjee and Finger, the Rio Conference process was set up in a way that meant that big business and industry would turn out to be winners. They argue that the BCSD and the ICC took over the way the relationship between the environment and development should be looked at. This reflected the two organisations' own considerable lobbying skills and resources, but the spread of pro-market neoliberal policies in many Western countries.[71] Agenda 21, one of the main documents arising from the Rio Conference, included a chapter on the benefits and potentials of voluntary programs. As Jennifer Clapp and Jason Thistlethwaite noticed, voluntary programmes were viewed as consistent with the broader goals of sustainable development, although a number of state-based environmental agreements were also negotiated at the time.[72] The Rio Conference at large, institutionalised the view that liberalisation in trade and finance was consistent with, and even necessary for, international environmental protection, and that both were compatible with the overarching goal of sustained economic growth.[73] When the UN Framework Convention on Climate Change—which eventually had received signatures from 154 countries at the Rio Conference—was negotiated in Kyoto in 1997, Maurice Strong noted that "though governments must establish the policy and regulatory frameworks for emissions trading, the prime actors in developing and applying the agreements will be private-sector companies, financial institutions and individuals, and most of the capital will be private."[74]

After Rio in 1992, Schmidheiny's BCSD merged with ICC's WICE to form the World Business Council for Sustainable Development (WBCSD). WBCSD developed according to some studies into the most central corporate-funded climate and environmental policy group because of its organisational structure, as a forum of CEOs or chairs of the world's largest corporations.[75] The official mission of the WBCSD was to provide business leadership as a catalyst for change towards sustainable development and promote the attainment of eco-efficiency through high standards of environmental and resource management. In fact, energy efficiency became the WBCSD's only strategy to address climate change. According to the WBCSD's first policy declaration, the objective would be to address climate issues through eco-efficiency and define policies and suggest projects that promoted these issues. WBCSD declared that their objective was to "help companies to understand and implement eco-efficiency through seminars, workshops to promote the eco-efficiency concept. WBCSD would also produce guidelines "based on expertise and input from banks, insurers and industry."[76] Yet no policy recommendations mentioned any need to substitute fossil fuels with alternative energy sources.

At the same time, efforts to develop environmental management systems (EMS) were emerging. Besides ICC's Business Charter on Sustainable Development, the International Organization for Standardization (ISO) who, since 1946 had worked on international standardisation frameworks, set up a Strategic Advisory Group on the Environment in 1991 in order to examine the need for internationally harmonised standards dealing with sustainability issues. These discussions gained momentum after the Rio Conference and eventually led to the adoption of the ISO 14000 series in 1996. This provided a global standard for environmental management that became widely adopted by multinational corporations.[77] Other examples of standardisation organisations included the creation of the Forest Stewardship Council (FSC), founded in 1993. The initial motivation for the FSC was a frustration amongst environmental NGOs, some timber producers, and high-end furniture retailers over the failed efforts to sign a global forest convention at the 1992 Rio Conference and the perceived incapability of the International Tropical Timber Organization (ITTO) to advance a robust approach to sustainable forest management.[78] The examples of the ICC's Business Charter, ISO-14000, and the FSC all demonstrate the growing trend of private regulation as economies globalised in the 1990s. It was a major shift in policy preferences for environmental governance compared with the Stockholm Conference 20 years earlier.

Some industries had, at the same time, more vested interest in delaying climate action than others.[79] It has been demonstrated that some oil industry leaders in both the United States and Europe had been aware that their products were causing CO_2 pollution to accumulate in the planet's atmosphere in a "potentially dangerous fashion" long before the Rio Conference in 1992. Exxon established internal research programmes on climate change in the late 1970s and privately acknowledged that prompt action would be required to avoid severe damage. And Shell had commissioned private research on climate change by 1981. But both remained silent about their findings.[80] During the run-up to the Rio Conference and INC negotiations of the UNFCCC, U.S.-based oil companies intensified a campaign against climate science. With considerable support from the fossil fuel industry and conservative think tanks, contrarian scientists were employed to create uncertainty and doubt to promote scepticism regarding climate change.[81] Coal companies such as the Western Fuels Association from the United States also deliberately sought to prevent other energy firms from accepting the growing consensus on climate change in the late 1980s. The overall aim was to block the door to regulation, both at national level and internationally at Rio.[82]

The industry's anti-regulatory stance resonated strongly in the United States during the preparations for the Rio Conference. This alignment was facilitated by the ideological commitment of George Bush's Republican administration to an anti-regulatory agenda.[83] The European Commission initially proposed a combination of "no-regret" measures alongside a CO_2 energy tax. However, the CO_2 tax faced substantial opposition from European industry lobbies. This resistance resulted in a compromised solution that included a conditionality clause, stipulating that the tax would only be implemented if the EU's major competitors enacted similar measures. Despite the resistance, industry lobbies within the EU could not prevent the adoption of emission targets. Nonetheless, they successfully delayed the implementation of the CO_2 tax and shifted the focus towards voluntary agreements. Over time, the European industrial sector aligned with the broader consensus, recognising the necessity of international climate action. However, they remained reluctant to shoulder the financial burden associated with European leadership through unilateral binding regulations.[84] In the end, the Rio agreement on climate change took the form of a framework convention that excluded binding targets and timetables for greenhouse gas reductions, which has been defined as a victory for the business lobby.[85]

After the Rio Conference, it was not certain whether the industry lobby could once again block national and regional delegations from agreeing on binding regulations on global greenhouse gas emissions. Several large corporations, including DuPont, the U.S. chemical company that had supported the Montreal Protocol on ozone layer depletion in 1987, turned their attention to climate change and created a new business lobbying organisation: the International Climate

Change Partnership (ICCP) in 1992.[86] The most obvious distancing from the anti-regulatory business lobby was the global insurance industry and the world's largest reinsurers, Munich Re and Swiss Re. As Lechner, Haueter, and Kenny note, reinsurance "was one of the few industries whose models reached out into the post-2050s, when the forecast effects of rising sea levels and shifts in precipitation and temperature would potentially lead to major shifts in population and loss of insured values."[87]

With no binding agreement under the UN Framework Convention on Climate Change, the UN continued its efforts to energise the private sector to address sustainability issues, including climate change, on a voluntary basis. A key factor that eventually caused the two paths of private regulation to start integrating was when the United Nations, along with other private sector initiatives, started to pay more attention to the role of the finance sector.

1.5 The Two Paths of Business Self-Regulation Merge

While individuals such as Maurice Strong and Stephan Schmidheiny had facilitated the increasing role of business and business self-regulation in global climate governance until the early 1990s, another set of institutional entrepreneurs emerged after the Rio Conference in 1992. They developed ideas, arguments, and new metrics, which they believed would, or at least could, reshape capitalism into harmony with the natural world.

The British management consultant John Elkington was one of the most important of these figures. In 1987, he co-authored a book titled *The Green Capitalists,*[88] and a decade later, he launched the concept of a "triple bottom line" of profitability, environmental quality, and social justice (or people, planet, and profit).[89] In 1987, Elkington (together with Julia Hailes) co-founded a for-profit business consultancy and think tank called SustainAbility to advise firms on how to become more sustainable. Elkington's vision was that "if we wanted to try to change the world, working with the private sector" rather than with governments was key. Sustainability became an early mover in trying to develop data on environmental performance. Two years after its founding, it began work on a report called "The Environmental Audit" with the World Wildlife Fund, which raised awareness of the potential of corporate reporting and auditing. In 1994, Sustainability and UNEP launched regular surveys of corporate environmental reporting called the Engaging Stakeholders Programme.[90] Environmental reporting was intended to guide investors. This was the first stage in which the self-regulatory space between the two paths of business self-regulation started to merge, and this was particularly evident when the financial sector joined global climate governance on the initiative of UNEP. Elkington denounced the triple bottom line concept in 2018 on the grounds that it had simply become an accounting system rather than a means of promoting system-wide change.[91]

The financial sector was brought into global environmental governance by UNEP, which started to work closely with the finance sector in the early 1990s. The reason was that UNEP saw that the finance sector as powerful enough to make a considerable, even decisive, impact. The concept of the so-called UNEP Finance Initiative (UNEP-FI) was launched in 1991 when a small group of commercial banks joined forces with UNEP to catalyse the banking industry's awareness of the environmental agenda before the Rio Conference. In 1992, the UNEP Statement by Banks on the Environment and Sustainable Development was launched, and the Banking Initiative was formed. In 1995, UNEP also joined forces with insurance and reinsurance companies, which merged with the UNEP Financial Initiative in 1997 under the name UNEP Financial Institutions Initiative, UNEP-FII (see also the chapter by Niels Viggo Haueter in this volume). At the same time, Ceres, the non-profit organisation founded by Bavaria (see Section 1.3), created the Global Reporting Index (GRI) in 1996, in which UNEP became a partner in 1997. The GRI standards were designed to guide the voluntary preparation of sustainability reports, based on the Triple Bottom Line framework.[92] The GRI guidelines provided a set of reporting principles and a recommended reporting format and content.

To further create linkages between UNEP-FI and other business sectors on a global scale, the United Nations Global Compact (UNGC) initiative was launched in 2000. UNGC sought to engage business in the voluntary delivery of global social and environmental policy measures. In 2004, the UN Secretary-General Kofi Annan gathered financial institutions to consider how to better integrate environmental, social, and corporate governance factors into asset management and other financial products and services. The report resulting from this gathering mainstreamed the language of ESG and forced out the term "SRI." Corporations were now also invited to voluntarily align their strategies with the ESG goals of the Global Compact.[93] In partnership with the UN Global Compact, the UNEP-FI launched in 2006 the UN Principles for Responsible Investments (PRI). By doing so, UNEP-FI established a notable global sustainability framework that aimed at helping the finance industry address ESG challenges. The PRI developed into the largest global network of institutional investors committed to considering ESG issues in their investment processes. In 2019, PRI had reached more than 2,500 signatories.[94]

The ESG reporting landscape quickly fragmented, forming what has been called an alphabet soup of acronyms, with no common framework to guide corporate disclosures.[95] Several frameworks that only focus on climate-related indicators such as the Climate Disclosure Standards Board (CDSB), the Carbon Disclosure Protocol (CDP), and the recommendations of the Task Force on Climate-related Financial Disclosures (TCFD) have emerged. In parallel with investor initiatives to broaden the performance metrics of companies, the United Nations also pursued a process of expanding the quantitative measures of social and environmental performance. This culminated in the creation of the Sustainable Development Goals (SDGs) in 2015.[96] But the problem was that the ESG trend was right from the start challenged by greenwashing, given the voluntary approach and confusion of definitions.[97]

The question is why a voluntary global regime to address climate change took hold after Rio and why finance emerged as the key. One part of the answer lies in the fact that it was uncertain from the start if the UNFCCC would be based on a top-down "command-and-control" regulation or a market-based system with tradable emissions permits, or some combination.[98] The Kyoto Protocol included a commitment by industrialised countries to reduce greenhouse gas emissions by an average of 5.2% below 1990 levels and within the commitment period of 2008–12. But many details of how this was to be achieved, including the rules for reporting and verifying emission reductions and measures to ensure compliance, were left to be negotiated at future conferences of the parties (COP). As the United States did not ratify the Kyoto protocol, a fragmented governance regime was created. This fragmentation was supposed to be resolved in 2009 in the Copenhagen negotiations, the most business-friendly COP of the UNFCCC. As argued by Ciplet and Roberts, the Copenhagen negotiations pushed through a bottom-up regime, manifested in the so-called Copenhagen Accord. Besides pledges of climate finance, the regime established in Copenhagen embodied an almost entirely a voluntary approach to addressing climate change.[99] In this context, private regulation, based on businesses' voluntary involvement and partnerships with the United Nations, had already been formed as a key feature in global environmental governance.

From the Copenhagen Accord in 2009 to the Paris Agreement in 2015, the process of voluntarism took, according to Ciplet and Roberts, a more institutionalised and structured form. The discourse in the year leading up to and during the Paris negotiations was that the transformation of the economy away from fossil energy could not be funded by governments alone. Rather, it would require a shift in large-scale private investments by creating incentives for markets and investors to quickly move to renewable energy and climate resilience.[100]

Although the reason why the financial sector became important in the landscape of private regulation and voluntary commitments in global climate governance is due to its potential to redirect investments to more climate-friendly sectors, critical questions exist about the degree of greenwashing in the ESG estimates, or whether the world is truly witnessing a structural change toward decarbonisation in the way investors allocate resources. As concluded by Pucker and King, ESG ratings are built on comparative rankings of industry peers, not on universal standards.[101] This is why fossil fuel companies can have better ESG ratings than makers of electric vehicles.

1.6 Conclusions

The growing interest in private regulation coincided both with the issue of climate change rising on the international policy agenda during the late 1980s and early 1990s but also with the diffusion of neo-liberal policies of deregulation and privatisation. After the Rio Conference in 1992, the belief that markets would solve climate change was widely diffused. Initially the followers of the first path of private regulation were not aligned with the intentions of the institutions and people behind the second path. But as this chapter has demonstrated, the boundaries between these two development paths subsequently dissolved. However, although private regulation helped firms to expand markets for greener products and services and inspired some incumbent firms to commit to energy-efficiency programmes and reducing carbon in value chains, structures within the political economy militated against action to limit carbon emissions. The creators of the second path had ended up in a global climate governance system based on voluntarism, under the motto that it must "pay to be green." Following the negotiations in Paris in 2015 showed an even more growing focus on how to effectively leverage the private sector to engage on climate change.

The absence of systemic governmental interventions, including global carbon pricing in some form, made it challenging for companies to live up to their voluntary commitments and visions, even as the scientific alarms about global warming accelerated from the 1990s onwards. Yet the problems of self-regulation are apparent, including confusing metrics, lack of transparency, and sometimes blatant greenwashing. Perhaps except for reinsurance, the "business case" for seriously addressing environmental challenges was not compelling. In the era when quarterly reporting was the key metric and maximising shareholder value the primary motivation, a predicted 3% rise in global temperatures by the end of the century was hardly a priority for most corporate boards. Moreover, the businesses of the five largest carbon emitters—China, the United States, India, the Russian Federation, and Japan—which account for 55% of global carbon emissions—are either entirely unconcerned with climate change or, at the very least, hostile to the idea of taking a serious hit to their bottom lines by doing anything about it.

Would binding global regulation have done a better job? In theory, the answer must be yes, but it is difficult to see how it could have happened. The sharp differences between developed and developing countries were evident from the beginning. Even if it had happened, it may not have been effective, given that most governments missed even the targets they had signed up to reach. The governments of democracies as a whole prioritise generating wealth over the environment because it translates into votes, while autocracies such as China and Iran (the seventh largest carbon emitter) prioritise geopolitical power. Private regulation may be flawed, but given the existing barriers to regulatory reform, it has been more feasible than government alternatives.

Notes

1 A first draft of this chapter was presented at a Swiss Re conference on Climate Change on May 25–26, 2023, organised by Bob Fredona, Niels Viggo Haueter, and Teresa da Silva Lopes at the Centre for Global Dialogue in Rüschlikon, Switzerland. We are very grateful for all the comments we received from the organisers and participants. An earlier version of the chapter was published as Bergquist, Ann-Kristin, and Geoffrey Jones. "Private Regulation, Institutional Entrepreneurship and Climate Change: A Business History Perspective." Harvard Business School Working Paper, No. 24-041, January 2024.
2 Bartly, "Power and the Practice of Transnational Private Regulation."
3 Ciplet and Roberts, "Climate Change"; Peake and Ekins, "Exploring the Financial and Investment Implications of the Paris Agreement."
4 Frumhoff, Heede, and Oreskes, "Climate Responsibilities."
5 Clapp, Privatization; Levy and Newell, *The Business of Global Environmental Governance*; Vogel, *Trading Up*; Falkner, The Business of Ozone; Falkner, *The Business Power.*
6 Clapp, "Privatization," 23–34.
7 See Newell, "The Political Economy"; Levy and Newell, *The Business of Global Environmental Governance.*
8 Hsueh, "Calling All Volunteers."
9 For a comprehensive overview, see Bansal and Hoffman, *The Oxford Handbook.*
10 Chrun, Dolsak, and Prakash, Corporate Environmentalism; Wright and Nyberg, An Inconvenient Truth.
11 For example, Rosen, "Businessmen Against Pollution"; Uekötter, "Age of Smoke"; Bergquist and Söderholm, "Green Innovation Systems"; Jones and Lubinski, "Making 'Green Giants'." On organic food, see, for example, Jones and Mowatt, "National Image"; Jones, *Profits and Sustainability.*
12 Jones, "Capitalism and the Environment."
13 On certification, see Jones, *Profits and Sustainability*, 233–69; on the ICC, see Bergquist and David, "Beyond Planetary."
14 Boon, "A Climate of Change."
15 Rome, "DuPont and the Limits."
16 Supran and Oreskes, "Rhetoric and Frame"; Franta, "Early Oil Industry Disinformation"; Bonneuil, Choquet, and Franta, "Early Warnings."
17 Oreskes and Conway, *Merchants of Doubt.*
18 Huf, Sluga, and Selchow, "Business and the Planetary"; Bergquist and David, "Beyond Planetary Limits."
19 Hart and Victor, "Scientific Elites," 661–62. As they state, the greenhouse effect hypothesis—that changes in the abundance of atmospheric CO_2 would affect climate—was formulated already in 1824, while the Swedish scientist Svante Arrhenius predicted that temperature would rise if enough CO_2 was added to the atmosphere at the turn of the twentieth century. From World War II until the mid-1970s, research proceeded along two scientific lines: one on the carbon cycle and the other on atmospheric modelling.
20 United Nations, *Report.*
21 Gunningham, Kagan, and Thornton, *Shades of Green.*
22 Jones, *Profits and Sustainability*, 35–38; Jones, *Deeply Responsible Business*, 272–73.
23 Jones, *Profits and Sustainability*, 86–88.
24 Guthman, *Agrarian Dreams*; Jones, *Profits and Sustainability*, 98–100.
25 Jones, *Profits and Sustainability*, 236–37.

26 IFOAM, "Climate Change"; Biodynamic Federation, "Climate Action."
27 Jones, *Profits and Sustainability*, 237–38.
28 Jones, *Profits and Sustainability*, 239.
29 ECOCERT.
30 Jones, *Profits and Sustainability*, 243.
31 Jones, *Profits and Sustainability*, 250–51.
32 Gottfried, *Greed to Green*.
33 Jones, *Profits and Sustainability*, 253.
34 Jones, *Profits and Sustainability*, 254.
35 Bowers, Boyd, and McGoun, "Greenbacks, Green Banks"; Jones, *Profits and Sustainability*, 253.
36 Jones, *Deeply Responsible Business*, 309–11.
37 Jones and Amble, "Joan Bavaria."
38 Jones, *Deeply Responsible Business*, 312–13.
39 Jones, *Deeply Responsible Business*, 313.
40 Jones, *Deeply Responsible Business*, 313.
41 Jones, *Deeply Responsible Business*, 314.
42 Domini and Kinder, *Ethical Investing*.
43 Jones, *Deeply Responsible Business*, 316–17.
44 Jones, *Deeply Responsible Business*, 317.
45 Jones, *Deeply Responsible Business*, 317–18.
46 Bergquist and David, "Beyond Planetary Limits."
47 Lemos and Agrawal, "Environmental Governance."
48 McCarthy, *Auto Mania*, Bergquist and Näsman, "Safe Before Green," 69.
49 Bergquist and David, "Beyond Planetary Limits."
50 United Nations, *Report*.
51 Bergquist and David, "Beyond Planetary Limits."
52 Bergquist and David, "Business (In)- Action."
53 Gray, "The United Nations Environment Programme."
54 Gray, "The United Nations Environment Programme," 314.
55 Borowy, *Defining Sustainable Development*, 49.
56 Bergquist and David, "Beyond Planetary Limits," 493–94.
57 Borowy, "*Defining Sustainable Development*."
58 Bergquist and David, "Beyond Planetary Limits," 496.
59 WCED, Chapter 12, section 4.2.
60 Green, *Rethinking Private Authority*, 79.
61 Bergquist and David, "Business In Action," 14.
62 Maurice Strong Papers, Box 443, Folder 4212, "ICC Business Charter for Sustainable Development—Principles for Environmental Management." Adopted by the ICC Executive Board, November 27, 1990.
63 Maurice Strong Papers, Box 443, Folder 4212, "Final Declaration of the Second World Industry Conference on Environmental Management," Rotterdam, April 10–12, 1991.
64 Schmidheiny, *Changing the Course*, Appendix II.
65 Bergquist and David, "Beyond Planetary Limits," 505–7; see also Chatterjee and Finger, *The Earth Brokers*.
66 Schmidheiny, *Changing the Course*, 20.
67 Schmidheiny, *Changing the Course*, 35.
68 Schmidheiny, *Changing the Course*.
69 Falkner, *Business Power*, 107.
70 Falkner, *Business Power*, 104.
71 Chatterjee and Finger, *Earth Brokers*, 111–12.
72 Clapp and Thistlethwaite, "Private Voluntary Programs," 64.
73 Bernstein, "Liberal Environmentalism."
74 Strong, *Where on Earth*, 375.
75 Sapinsiki, "Climate Capitalism."
76 Maurice Strong Papers, Box 588, Folder XI: World Business Council for Sustainable Development, 5–6.
77 Falkner, *Business Power*, 6–7.

78 van der Ven and Cashore, "Forest Certification."
79 Pulver, "Organising Business."
80 Supran and Oreskes, "Rhetoric and Frame Analysis"; Franta, "Early Oil industry Disinformation"; Bonneuil, Choquet, and Franta, "Early Warnings."
81 Oreskes and Conway, *Merchants of Doubt.*
82 Falkner, *Business Power*, Chapter 2.
83 Falkner, *Business Power*, 104.
84 Michaelowa, "Impact of Interest Groups."
85 Falkner, *Business Power*, 109.
86 Rome, "DuPont and the Limits"; Falkner, *Business Power*, 111.
87 Lechner, Haueter, and Kenny, "Continuity and Change," 293.
88 Elkington and Burke, *The Green Capitalists.*
89 Elkington, *Cannibals With Forks.*
90 Jones, *Profits and Sustainability*, 179.
91 Elkington, "25 Years Ago."
92 Jones, *Profits and Sustainability*, 290–94; Jones, *Deeply Responsible Business*, Chapter 10.
93 Jones, *Deeply Responsible Business*, Chapter 10.
94 Matos, "ESG and Responsible Institutional Investing."
95 Davies, Dudek, and Wyatt, "Recent Developments in ESG Reporting."
96 See Bose, "Evolution of ESG," For a comprehensive overview of the characteristics of all these frameworks.
97 Jones, *Deeply Responsible Business*, Chapter 10.
98 Levy and Newell, "Business Strategy."
99 Ciplet and Roberts, "Climate Change," 152.
100 Ciplet and Roberts, "Climate Change," 152–53.
101 Pucker and King, "ESG Investing."

Bibliography

Agrawal, Arun, and Maria Carmen Lemos. "Environmental Governance." *Annual Review of Environment and Resources* 31, no. 1 (2006): 297–325.

Bansal, Pratima, and Andrew J. Hoffman, eds. *The Oxford Handbook of Business and the Natural Environment*. Oxford: Oxford University Press, 2012.

Bartley, Tim. "Power and the Practice of Transnational Private Regulation." *New Political Economy* 27, no. 2 (2022): 188–202.

Bergquist, Ann-Kristin, and Thomas David. "Beyond Planetary Limits! The International Chamber of Commerce, the United Nations, and the Invention of Sustainable Development." *Business History Review* 97, no. 3 (2023): 1–31.

Bergquist, Ann-Kristin, and Thomas David. "Business (In-) Action: The International Chamber of Commerce and Climate Change from Stockholm to Rio." *Uppsala Papers in Economic History* 3 (2024): 1–26.

Bergquist, Ann-Kristin, and Mattias Näsman. "Safe Before Green! The Greening of Volvo Cars in the 1970s–1990s." *Enterprise & Society* 24, no. 1 (2023): 59–89.

Bergquist, Ann-Kristin, and Kristina Söderholm. "Green Innovation Systems in Swedish Industry, 1960–1989." *Business History Review* 85, no. 4 (2011): 677–98.

Bernstein, Steven. "Liberal Environmentalism and Global Environmental Governance." *Global Environmental Politics* 2, no. 3 (2002): 1–16.

Biodynamic Federation. "Climate Action Through Agriculture." 2021. https://demeter.net/climate-action.

Bodansky, Daniel. "The History of the Global Climate Change Regime." In *International Relations and Global Climate Change*, edited by Urs Luterbacher and Detlef F. Sprinz, 23–40. Cambridge, MA: The MIT Press.

Bonneuil, Christophe, Pierre-Louis Choquet, and Benjamin Franta. "Early Warnings and Emerging Accountability: Total's Responses to Global Warming, 1971–2021." *Global Environmental Change* 71 (2021): 2–10.

Boon, Marten. "A Climate of Change? The Oil Industry and Decarbonization in Historical Perspective." *Business History Review* 93, no. 1 (2019): 101–25.

Borowy, Iris. *Defining Sustainable Development for Our Common Future: A History of the World Commission on Environment and Development (Brundtland Commission)*. London: Routledge, 2013.

Bose, Satyajit. "Evolution of ESG Reporting Frameworks." In *Values at Work: Sustainable Investing and ESG Reporting*, edited by Daniel Esty and Todd Cort, 13–33. Cham: Palgrave Macmillan, 2020.

Bowers, Brittany, Neil Boyd, and Elton McGoun. "Greenbacks, Green Banks, and Greenwashing via LEED: Assessing Banks' Performance in Sustainable Construction." *Sustainability: The Journal of Record* 13, no. 5 (2020): 208–17.

Chatterjee, Pratap, and Matthias Finger. *The Earth Brokers: Power, Politics and World Development*. London: Routledge, 1994.

Chrun, Elizabeth, Nives Dolšak, and Aseem Prakash. "Corporate Environmentalism: Motivations and Mechanisms." *Annual Review of Environment and Resources* 41 (2016): 341–62.

Ciplet, David, and J. Timmons Roberts. "Climate Change and the Transition to Neoliberal Environmental Governance." *Global Environmental Change* 46 (2017): 148–56.

Clapp, Jennifer. "The Privatization of Global Environmental Governance: ISO 14000 and the Developing World." *Global Governance* 4, no. 3 (1998): 295–316.

Clapp, Jennifer, and J. Thistlethwaite. "Private Voluntary Programs in Environmental Governance: Climate Grange and the Financial Sector." In *Business and Climate Policy: The Potentials and Pitfalls of Private Voluntary Programs*, 43–47. New York: United Nations University Press, 2012.

Davies, Paul A., Paul M. Dudek, and Kristina S. Wyatt. "Recent Developments in ESG Reporting." In *Values at Work: Sustainable Investing and ESG Reporting*, edited by Daniel Esty and Todd Cort, 161–79. Cham: Palgrave Macmillan, 2020.

Dodds, Felix. "Maurice Strong–From Poverty to UN Under-Secretary General and Father of Sustainable Development–The Rio Earth Summit." In *Heroes of Environmental Diplomacy*, edited by Felix Dodd and Chris Spence, 111–32. London: Routledge, 2022.

Domini, Amy L., and Peter D. Kinder. *Ethical Investing*. Addison-Wesley, 1984.

ECOCERT. 2024. https://www.ecocert.com/en-US/home.

Elkington, John. *Cannibals with Forks: The Triple Bottom Line of 21st Century Business*. Gabriola Island: New Society Publishers, 1997.

Elkington, John. "25 Years Ago, I Coined the Phrase 'Triple Bottom Line': Here's Why It's Time to Rethink It." *Harvard Business Review*, June 25, 2018.

Elkington, John, and Tom Burke. *The Green Capitalists*. London: Victor Gollancz, 1987.

Falkner, Robert. "The Business of Ozone Layer Protection: Corporate Power in Regime Evolution." In *The Business of Global Environmental Governance*, edited by David L. Levy and Peter J. Newell, 105–34. Cambridge, MA: The MIT Press, 2005.

Falkner, Robert. *Business Power and Conflict in International Environmental Politics*. London: Palgrave Macmillan, 2008.

Franta, Benjamin. "Early Oil Industry Disinformation on Global Warming." *Environmental Politics* 30, no. 4 (2021): 663–68.

Frumhoff, Peter C., Richard Heede, and Naomi Oreskes. "The Climate Responsibilities of Industrial Carbon Producers." *Climatic Change* 132, no. 2 (2015): 157–71.

Gottfried, David. *Greed to Green: The Transformation of an Industry and a Life*. Berkeley: WorldBuild Publishing, 2004.

Gray, Mark Allan. "The United Nations Environment Programme: An Assessment." *Environmental Law (Portland Ore.)* 20, no. 2 (1990): 291–319.

Green, Jessica F. *Rethinking Private Authority: Agents and Entrepreneurs in Global Environmental Governance*. Princeton: Princeton University Press, 2013.

Gunningham, Neil, Robert A. Kagan, and Dorothy Thornton. *Shades of Green: Business, Regulation, and Environment*. Stanford: Stanford University Press, 2000.

Guthman, Julie. *Agrarian Dreams*. Berkeley and Los Angeles: University of California Press, 2004.

Hart, David M., and David G. Victor. "Scientific Elites and the Making of US Policy for Climate Change Research, 1957–74." *Social Studies of Science* 23, no. 4 (1993): 643–80.

Hsueh, Lily. "Calling All Volunteers: Industry Self-Regulation on the Environment." In *Handbook of US Environmental Policy*, edited by David M. Konisky, 243–56. Cheltenham: Edward Elgar Publishing, 2020.

Huf, Ben, Glenda Sluga, and Sabine Selchow. "Business and the Planetary History of International Environmental Governance in the 1970s." *Contemporary European History* 31, no. 4 (2022): 553–69.

IFOAM. "Climate Change." 2017. https://www.ifoam.bio/climate-change.

Jones, Geoffrey. *Profits and Sustainability: A History of Green Entrepreneurship*. Oxford: Oxford University Press, 2017.

Jones, Geoffrey. "Capitalism and the Environment." In *Evolutions of Capitalism: Historical Perspectives 1200–2000*, edited by Catherine Casson and Philipp Robinson Rössner, 187–211. Bristol: Bristol University Press, 2022.

Jones, Geoffrey. *Deeply Responsible Business: A Global History of Values-Driven Leadership*. Cambridge, MA: Harvard University Press, 2023.

Jones, Geoffrey, and Seema Amble. "Joan Bavaria and Multi-Dimensional Capitalism." Harvard Business School Case 9-390-027. Revised December 4, 2018.

Jones, Geoffrey, and Christina Lubinski. "Making 'Green Giants': Environment Sustainability in the German Chemical Industry, 1950s–1980s." *Business History* 56, no. 4 (2014): 623–49.

Jones, Geoffrey, and Simon Mowatt. "National Image as a Competitive Disadvantage: The Case of the New Zealand Organic Food Industry." *Business History* 58, no. 8 (2016): 1262–88.

King, Andrew A., and Michael J. Lenox. "Industry Self-regulation Without Sanctions: The Chemical Industry's Responsible Care Program." *Academy of Management Journal* 43, no. 4 (2000): 698–716.

Lechner, Roman, Niels Viggo Haueter, and Lawrence Kenny, "Continuity and Change in Reinsurance 1990–2016." In *Managing Risk in Reinsurance: From City Fires to Global Warming*, edited by Niels Viggo Haueter and Geoffrey Jones, 277–300. Oxford: Oxford University Press, 2017.

Lemos, Maria Carmen, and Arun Agrawal. "Environmental Governance." *Annual Review of Environment and Resources* 31, no. 1 (2006): 297–325.

Levy, David L., and Peter J. Newell. "Business Strategy and International Environmental Governance: Toward a Neo-Gramscian Synthesis." *Global Environmental Politics* 2, no. 4 (2002): 84–101.

Levy, David L., and Peter J. Newell, eds. *The Business of Global Environmental Governance*. Cambridge, MA: The MIT Press, 2004.

Matos, Pedro. *ESG and Responsible Institutional Investing Around the World: A Critical Review*. CFA Institute Research Foundation, 2020.

McCarthy, Tom. *Auto Mania: Cars, Consumers, and the Environment*. New Haven: Yale University Press, 2007.

Michaelowa, Axel. "Impact of Interest Groups on EU Climate Policy." *European Environment* 8, no. 5 (1998): 152–60.

Newell, Peter. "The Political Economy of Global Environmental Governance." *Review of International Studies* 34, no. 3 (2008): 507–29.

Newell, Peter, and Matthew Paterson. "The Politics of the Carbon Economy." In *The Politics of Climate Change: A Survey*, edited by Maxwell Boycoff, 80–99. London: Routledge, 2009.

Oreskes, Naomi, and Erik M. Conway. *Merchants of Doubt: How a Handful of Scientists Obscured the Truth on Issues from Tobacco Smoke to Global Warming*. London: Bloomsbury Publishing, 2011.

Peake, Stephen, and Paul Ekins. "Exploring the Financial and Investment Implications of the Paris Agreement." *Climate Policy* 17, no. 7 (2017): 832–52.

Pucker, Ken, and Andrew King. *ESG Investing Isn't Designed to Save the Planet.* Boston University Open Access Library, 2022. https://open.bu.edu/handle/2144/45424.

Pulver, Simone. "Organising Business: Industry NGOs in Climate Debates." In *The Business of Climate Change: Corporate Responses to Kyoto*, edited by Kathryn Begg, Frans Van der Woerd, and David Levy, 47–60. London: Routledge, 2018.

Rome, Adam. "Dupont and the Limits of Corporate Environmentalism." *Business History Review* 93, no. 1 (2019): 75–99.

Rosen, Christine Meisner. "Businessmen Against Pollution in Late Nineteenth Century Chicago." *Business History Review* 69, no. 3 (1995): 351–97.

Rugman, Alan M., and Alain Verbeke. "Corporate Strategies and Environmental Regulations: An Organizing Framework." *Strategic Management Journal* 19, no. 4 (1998): 363–75.

Sapinski, Jean Philippe. "Climate Capitalism and the Global Corporate Elite Network." *Environmental Sociology* 1, no. 4 (2015): 268–79.

Schmidheiny, Stephan. *Changing Course: A Global Business Perspective on Development and the Environment*. Cambridge, MA: The MIT Press, 1992.

Strong, Maurice. *Where on Earth Are We Going?* Toronto: Alfred A. Knopf Canada, 2000.

Supran, Geoffrey, and Naomi Oreskes. "Rhetoric and Frame Analysis of ExxonMobil's Climate Change Communications." *One Earth* 4, no. 5 (2021): 696–719.

Uekötter, Frank. *The Age of Smoke*. Pittsburgh: University of Pittsburgh Press, 2009.

United Nations. *United Nations Conference on the Human Environment*, Stockholm, June 5–16, 1972. https://documents.un.org/doc/undoc/gen/nl7/300/05/pdf/nl730005.pdf?token=qDK1Nh4KtnaaV3fUS7&fe=true.

van der Ven, Hamish, and Benjamin Cashore. "Forest Certification: The Challenge of Measuring Impacts." *Current Opinion in Environmental Sustainability* 32 (2018): 104–11.
Vogel, David. *Trading Up: Consumer and Environmental Regulation in a Global Economy*. Cambridge, MA: Harvard University Press, 2009.
Watson, Michael, and Anthony R.T. Emery. "Environmental Management and Auditing Systems: The Reality of Environmental Self-Regulation." *Managerial Auditing Journal* 19, no. 7 (2004): 916–28.
World Commission on Environment and Development (WCED), Special Working Session. "World Commission on Environment and Development." *Our Common Future* 17, no. 1 (1987): 1–91.
Wright, Christopher, and Daniel Nyberg. "An Inconvenient Truth: How Organizations Translate Climate Change into Business as Usual." *Academy of Management Journal* 60, no. 5 (2017): 1633–61.

2

CARS AND CLIMATE CHANGE

The Historical Political Economy of a Green Transition[1]

Grace Ballor and Mattias Näsman

2.1 Introduction

The car industry has been one of the most environmentally consequential industries since its emergence in the early twentieth century, and its historical dependence on fossil fuels underscores its cumulative and negative impact on greenhouse gas emissions and global warming.[2] Climate change emerged widely on political agendas only in the late 1980s,[3] and policies aimed at climate change mitigation were introduced in the 1990s, especially following the signing of the Kyoto Protocol in late 1997.[4] Despite efforts to reduce the climate impact of the road transportation sector, including through electrification, greenhouse gas emissions from the car sector have been growing faster than those of most other sectors since the 1970s, trailing only the growth in emissions from industrial processes, electricity, and heating.[5] In 2019, road transportation was responsible for 12.6% of global greenhouse gas emissions and roughly 17% of all carbon dioxide (CO_2) emissions, on par with emissions from the manufacturing and construction sectors.[6]

Since the 2010s, governments around the world have applied mounting pressure to decarbonise the car industry by eliminating the CO_2 emissions it produces.[7] The car industry has become the focus of green transition policy, largely because of its forward and backward linkages to other industries and the enormous size of the consumer car market. With government support and pressure, the car industry has seemingly begun to replace the fossil fuel-powered internal combustion engine (ICE) with engines powered by electricity, which promise no pollutants or greenhouse gas emissions.[8] Consequently, the car industry has become a high-profile case for a wider green transition. It is important to recognise, however—and this is potentially instructive for other areas of the economy—that the industry's shift towards EVs is the joint product of both policy choices and consumer demand. As the following sections show, since the 1970s, governments have viewed electrification in the car industry as a means of addressing the issue of climate change on the one hand, and as a means of achieving a competitive edge in intensifying geoeconomic competition on the other. Such developments have highlighted business strategies for responding to the global governance of the climate and environmental policy regulation, at both the national and international levels.[9] They also demand careful consideration of the entangled history of cars and climate change.

DOI: 10.4324/9781003478089-3

This chapter takes a business-historical approach to cars and climate change by examining how the car industry and carmakers have contributed to greenhouse gas emissions and responded to increasing climate change mitigation.[10] It is motivated by the relative lack of attention historians have paid to the developments of recent decades, particularly the effects of industries on climate change, and it highlights the importance of history for understanding the ways business and governments/governance institutions might approach the future. We focus on passenger cars specifically, rather than automobiles broadly, since they represent the highest volume (roughly 70% globally) of road vehicles producing emissions, although we also briefly discuss light trucks, since their role in private transportation has increased dramatically since the 1990s.

Case studies on carmakers have long been the backbone of classical business history scholarship.[11] Historians of science and technology have also traced the development of scientific research on anthropogenic climate change and the early effects of climate change on developing countries and have also uncovered the ways some actors sowed public doubt about scientific findings and traced.[12] There is furthermore a growing body of scholarship on the history of business and environmental regulation.[13] But, as Ann-Kristin Bergquist notes, little historical scholarship has directly studied the car industry and the environment together: "[a] painful gap exists in the historical literature, the content of which could give a sense of the historically shaped inertias and difficulties in transforming this industry."[14] This chapter aims to fill that gap by tracing the evolution of the car industry, its negative effects on the environment and climate, national and international efforts to regulate vehicle emissions, and the industry's shift towards new technologies as part of a wider energy transition. Such an approach underscores the importance of analysing industry perspectives and strategies in the process of the green transition.

This chapter briefly reviews the history and historiography of the car industry and climate change, from the formative decades of automobility in the late nineteenth century to the twenty-first century (Section 2.1) and historicises business strategies to climate change (Section 2.2). We then take stock of contemporary developments, including the car industry's shift towards electrification in the context of industrial policy and geopolitical competition (Section 2.3). This history is especially important for debates in the 2020s since Europe, the United States, and China are all currently pushing for and competing over the future of the car industry. In what has been called a "second automotive revolution," the world's long-standing dependence on fossil fuels and the ICE is progressively giving way to electrification and the possibility of using renewable energy to power personal transportation.[15] Before concluding, we briefly assess private and public strategies to limit cars' CO_2 emissions since the 1970s to draw conclusions on progress on climate mitigation up to the present.

2.2 Brief Global History and Historiography of the Car Industry

Since the beginning of the twentieth century, a growing global reliance on cars for personal transportation transformed the global energy landscape and created an ever-increasing demand for petroleum products. Thus, the "automotive age," beginning with the emergence of mass markets for cars in the United States in the 1920s and in Western Europe in the 1960s, was thus also the "age of gasoline."[16] Historical research on energy transitions has emphasised that the addition of new sources of energy has not replaced existing ones.[17] Hence, the automotive age did not mean that oil replaced coal but rather that oil became an additional integral part of global energy systems.

As a consequence of increasing car drivership in the United States in the early twentieth century, oil became a strategic resource of national interest, raising concerns and scientific debates in the 1920s about the potential consequences and probability of depleting known crude oil reserves.[18] Adding to such fears were the technical limitations of refining petrol from crude oil. Together with the discovery of new crude sources, including in the Middle East, advancements in refining technologies significantly lowered the price of gasoline in the mid-twentieth century while also igniting new geopolitical tensions.[19] Consumer culture around cars, whose creation has been promoted by the oil and car industry, deepened global dependence on oil to such an extent that cars and cheap access to oil and petrol became means of social and economic inclusion in developed countries.[20]

The U.S. industry's global dominance in car production endured until the 1960s, when competition intensified—first from European and later from Japanese manufacturers. Consolidation was an integral part of the rise of the U.S. car industry. While the United States was home to several hundred carmakers in the early twentieth century, it quickly became dominated by the "Big Three"—General Motors, Ford, and Chrysler—which together claimed 85% of the U.S. market by 1946.[21] In the United States, as elsewhere, government policies after World War II facilitated the growth of the car industry and the expansion of car use because domestic car production was seen by many governments as a proxy for modernisation as well as for economic and social development, reflected in national employment rates and living standards. Across the West, government investment was instrumental in creating the dense road networks that increased the use value of passenger car ownership, and government policies for wealth and income distribution facilitated the expansion of car use and contributed to a self-sustaining process of continued car industry growth in industrialised countries until 1974 and the wake of the first Oil Crisis.[22]

2.3 Business Strategies and the Political Economy of Technological Choice

2.3.1 Carmaker Strategies on Climate Change

In the mid-to-late 1980s, climate change became a global political issue, with culmination in the UN Framework Convention on Climate Change (UNFCCC) adopted in Rio de Janeiro in 1992, according to law scholar Daniel Bodansky.[23] Carmakers were quick to notice the potential business implications of climate policies. Notably, the European Economic Community (EEC), the United States, and Japan all considered the implementation of carbon taxes at the time.[24] In 1989, the Global Climate Change Coalition (GCC), a notorious actor organising U.S. business opposition to climate change, was founded under the aegis of the U.S. National Association of Manufacturers, which counted Ford, Chrysler, and General Motors, the Automobile Importers of America, and the Motor Vehicle Manufacturers Association among its members.[25] Scholarship on how the oil industry and oil companies created doubt about the science of climate change and how they organised climate opposition is quickly expanding.[26] But historians have yet to explore how the car industry and carmakers organised and positioned themselves in relation to the climate issue, including their role in the GCC vis-à-vis the oil, gas, and coal companies whose products were more directly affected by carbon taxes than cars.

Drawing on public statements, regulatory emissions performance data, and interviews, scholars of international political economy have explored how carmakers' strategic responses to climate change have varied according to their domestic institutional contexts. David Levy and Sandra Rothenberg, for example, argue that perceptions of climate science have shaped car companies' strategies for climate mitigation, with U.S. companies like Ford publicly and aggressively challenging the need for binding international CO_2 emission controls.[27] According to Levy and Rothenberg, European companies have been less publicly vocal, instead accommodating regulatory demands for significant emissions reductions, as exemplified by the German Automobile Industry Association's (VDA) 1994 agreement to reduce CO_2 emissions from new cars by 25%. Levy and Rothenberg ascribed these differences in corporate responses to their different regional institutional frameworks: the U.S. political process has been described as adversarial and legalistic, whereas European political processes have been characterised by tripartite bargaining and rely on intergovernmental consensus.[28] As the next section discusses in detail, U.S. carmakers have historically produced large, fuel-demanding cars for the U.S. consumer market, which could provide another explanation of U.S. producers' more hostile position on climate change relative to their European and Japanese counterparts.

In his seminal study on "greening the car industry," political economist John Mikler found that firms in co-ordinated market economies (CMEs), characterised by cooperation between firms and regulators; that is, Japan (Toyota) and Germany (Volkswagen, BMW), have better internalised environmental externalities than firms in liberal market economies (LMEs), that is, the United States (Ford), which typically develop their strategies to match market demands and generally have a more contentious relationship with regulators.[29]

Consumers and scholars alike have grown sceptical of published emissions data and started to examine the relationships between cars and climate change. Historian Mattias Näsman traced the development of international regulatory emission standards in Europe and found that the technical, scientific, regulatory, and competitive underpinnings of these standards have diverged from those of U.S. standards since the 1960s due to different compromises made within and between nationally specific knowledge creation systems. Näsman's findings underscore the problems of relying on published emissions performance data, which lack transparency, are not easily comparable between the United States and Europe, and often do not represent actual road performance.[30] Motivated by economies of scale, multinational carmakers typically work to optimise their product ranges across markets with diverse regulatory requirements, subjecting export-oriented manufacturers to conflicting pressures. In joint work, Bergquist and Näsman used the case of Swedish Volvo to show that, because of the company's export orientation, regulations in export markets had an equal or greater role in shaping Volvo's environmental and innovation strategies than regulations in its home market.[31] While many European producers, the subsidiaries of American companies like Ford Europe and GM Europe, and Japanese carmakers have all historically been export-oriented, U.S. Ford, GM, and Chrysler have sold the absolute majority of their cars in the U.S. domestic market. As a result, carmakers' positions on environmental and climate regulations are shaped in relation to the relative importance of different markets.[32] Carmakers have also used national and international peak-level associations to exercise political power when trying to shape regulation and markets because, as Neil Rollings has shown, the collective action of business interest associations augments business power.[33]

2.3.2 *Indirect Climate Action Before Climate Politics: Fuel Efficiency and Alternative Fuels*

Even before the entrance of climate change into political agendas in the late 1980s,[34] carmakers, responding to various pressures, indirectly addressed one important aspect of cars' climate impacts, namely fuel consumption. In the interwar period, European producers responded to domestic policies adopted to protect them from U.S. competition by designing smaller, lower-powered, and more fuel-efficient cars than those of their U.S. competitors. Behind high tariff protections, U.S. producers, meanwhile, constructed large and powerful cars to serve their larger and more affluent domestic market.[35] Consumers did not always prefer added horsepower and weight, however. As the historian Tom McCarthy has shown, U.S. consumers began trading large, U.S.-made "gas-guzzlers" for smaller European cars of higher quality in the latter half of the 1950s. The "Big Three" U.S. carmakers responded by introducing "compacts," cars smaller than the usual U.S. models but slightly larger than European imports, which managed to stave off imports until the mid-1960s.[36] In the 1970s, however, U.S.-made cars increased in size again, while European and, specifically, Japanese imports picked up speed in the U.S. market.[37] This trend accelerated when the Oil Crises of 1973–74 and 1978–79 limited petrol availability and raised petrol prices, motivating American consumers to look for smaller, more fuel-efficient cars. In the wake of the crises and in response to market and regulatory pressures, U.S. manufacturers increasingly prioritised fuel economy, and small, fuel-efficient Japanese models became increasingly popular. The U.S. car industry invested $80 billion in the development of new fuel-efficient cars, $45 billion of which was contributed by General Motors alone.[38] Importantly, environmental activism also surged globally in the 1970s, augmenting market pressures on the car industry to produce smaller, more fuel-efficient models.

Governmental regulation in the United States pushed carmakers to develop fuel-efficient cars. In 1975, the U.S. Congress adopted the Energy Policy and Conservation Act, which mandated carmakers to meet corporate average fuel economy (CAFE) standards by 1978. Japan adopted mandatory fuel economy standards in 1979 as well, to take effect in 1985.[39] In Europe, fuel conservation also became an important issue in relation to the oil shock, and some governments such as those in Sweden and the UK adopted voluntary fuel economy standards in collaboration with the car industry, but none adopted mandatory standards.[40] Such fuel conservation efforts were complicated by the fact that in the late 1960s and early 1970s, governments in the United States, Europe, and Japan had begun implementing standards for controlling pollutant emissions. Because of the technical trade-offs of controlling exhaust emissions and increasing fuel economy, the oil shock of 1973–74 put the goals of energy conservation and clean air in conflict, since most technologies carmakers used to control emissions also increased fuel use.[41]

Carmakers in Europe also responded to rising fuel prices and governmental concern for energy conservation and air pollution mitigation by introducing new technologies, notably diesel engines, as these were more fuel-efficient than petrol ones. Having been used almost exclusively in heavy-duty vehicles, diesel-fuelled passenger cars made significant inroads in European markets in the late 1970s and early 1980s.[42] Since the 1990s, European carmakers such as Volkswagen, Peugeot-CSA, and Renault[43] pursued diesel technology, made possible by the EEC's relatively less stringent emission standards for oxides of nitrogen (NOx) in general and even less stringent standards for diesels compared to petrol-powered cars. EEC policymakers and the car industry mutually reinforced the diesel trend in relation to the binding emission reduction targets set by the UNFCCC's Kyoto Protocol, when the European Commission, the EEC's executive body, and the newly reconstituted European sectoral business interest association, the European Automobile Manufacturers Association (ACEA), struck a voluntary deal to reduce average carbon emissions—and consequently average fuel consumption—from new cars in 1998. As a result of the deal, diesels increased from 25% of sales in 1998 to roughly half of EU sales between 2006 and 2015. Since the 2015 Dieselgate scandal, which uncovered carmakers' use of electronics to cheat regulatory emission tests, diesels have fallen out of favour even in the European diesel stronghold, dropping to just 23% of market share in 2021.[44]

As early as the 1970s, carmakers also began searching for alternatives to potentially replace oil products as transportation fuel. High oil prices and industrial policies, especially in countries like Brazil, explored the potential of methanol and ethanol. As oil prices began falling during the second half of the 1980s, however, carmakers and policymakers abandoned most investments in fuel alternatives and returned to petrol cars.[45]

2.3.3 *Towards Heavier, Larger, and More Expensive Cars*

According to historian Elisabeth Bini, new forms of consumerism emerged in the United States in the second half of the 1970s, marked by virulent individualism and a firm belief in both the right to access consumer goods—like petrol—in a free market and scepticism of federal regulations trying to limit personal transport.[46] This U.S. shift materialised in rising sales of big light trucks such as minivans, pickup trucks, and sport utility vehicles (SUVs), in which Ford has long been a leader: its F-150 pickup truck has claimed the title of best-selling vehicle in the United States every year since 1981.[47] Light trucks accounted for a third of all passenger vehicle sales and leases in the United States in 1990, close to half in the 2000s, and a staggering 79% in 2022.[48]

Low petrol taxes and the design of fuel economy standards helped facilitate this shift to larger vehicles: the 1978 US CAFE standards for light trucks were laxer than those set for passenger cars, where the gap between the categories widened until 1985 and remained substantial during the 1990s.[49] Although oil prices remained relatively high in the early twenty-first century, which had the potential to orient consumers towards smaller and more fuel-efficient cars, interest rates remained relatively low after the 2008 Global Financial Crisis until the aftermath of the Covid-19 pandemic, instead lowering the costs of leasing and purchasing expensive light trucks. Carmakers were keen on selling light trucks, too, because their profit margins were much larger than those of sedans or wagons. Production costs were marginally higher for larger vehicles, but dealership prices were also substantially higher.[50] By retaining the leniency of CAFE standards towards light trucks and keeping petrol taxes low, U.S. regulators have contributed to legitimising the trend towards larger cars. Modest gains in lowering the car fleet's climate emissions have been achieved despite the increasing size. Between 2005 and 2021, U.S. manufacturers reduced average CO_2 emissions per mile by 25%. In comparison, between the adoption of the CAFE standards in 1975 and 1987, CO_2 emissions decreased by 41%, and even increased by 14% between 1988 and 2004.[51]

In addition to the macro- and microeconomic factors behind the increasing size of cars, regulations in Europe also legitimised the making of heavier and larger cars, as technology scholar Tommaso Pardi has shown.[52] After carmakers failed to meet the CO_2 emission reduction targets established in the 1998 voluntary agreement, the EU set mandatory standards in 2009. These standards, however, were differentiated by weight, meaning that carmakers with a legacy of marketing large and heavy vehicles, notably Daimler-Benz and BMW, were subject to less strict norms, thus disincentivising the meeting of emissions goals by reducing vehicle size. While maintaining or increasing vehicle weight, carmakers' only option for reducing CO_2 emissions in the early 2000s was to use diesel and direct-injection gasoline engines, both costly premium technologies. The sport-utility vehicle (SUV) in Europe, similar to light trucks in the United States, has been the only expanding vehicle category in Europe, growing eightfold since 2000 and comprising 51% of new registrations in 2023.[53]

Not only did carmakers not have to trade increased efficiency for lower climate impact during the late twentieth and early twenty-first centuries, but consumers also demanded larger and more powerful cars. Although large cars and light trucks created new revenue streams for carmakers since the 1990s, they were also major drivers of increasing carbon emissions.[54] The U.S. Environmental Protection Agency acknowledged that the trend towards larger vehicles "has offset some of the fleetwide benefits that otherwise would have been achieved" if only support for smaller and more fuel-efficient cars had been stronger.[55]

2.3.4 Towards a Green Transformation?: The Return of the Electric Vehicle

Under regulatory pressures, the global car industry progressively shifted away from a dependence on petroleum towards electric vehicles in the early twenty-first century. Although the industry explored several technological options as replacements for petrol and diesel engines such as engines using biofuels or hydrogen (fuel cells), government intervention pushed carmakers to settle on electrification. Global EV sales, including both battery cars and plug-in hybrids, grew exponentially over the 2010s, thanks in large part to government subsidies and incentives. By the early 2020s, EVs dominated the petrol-alternative market; global EV sales had increased from 120 thousand in 2012 to 10.2 million in 2022, accounting for 0.2% of global sales in 2012 and 14% in 2022.[56] EVs promised "zero" greenhouse gas emissions, since they do not use ICEs powered by fossil fuels. EV critics, however, point to the environmental costs of their production and the problem of disposing of used lithium-ion batteries in an environmentally safe way.[57]

The breakthrough of EVs in international markets in the 2010s obscures the technology's long history. Individual inventors, carmakers, and governments had been experimenting with and innovating EV technologies for over a century. According to Matthew Eisler's book on EVs, the technology's history has been marked by a series of "false starts."[58] A substantial body of literature has studied the long historical development of EVs, focusing especially on the period of EV development between the 1880s, when the German Flocken Elektrowagen was introduced, and the 1920s, after which petrol-powered cars achieved total market domination.[59] Scholars have offered various explanations for why electric vehicles initially "lost" the competition against fossil fuel-powered cars and ICEs. In their review of this literature, economic historians Josef Taalbi and Hana Nielsen noted that explanations have focused on three technological, cultural, and practical factors: the technical inferiority of EVs (low operating range, bulky batteries accounting for much of the vehicle weight, and higher production costs); socio-cultural dynamics (the prioritisation of speed and individualism over reliability and low pollution, the gendering of EVs as female and petrol vehicles as male, that EVs were marketed to the upper class and petrol cars to mass markets); and the lack of electricity and charging infrastructure outside of large cities.[60]

After 1920, carmakers focused on the production of petrol-powered cars. Then, in response to growing environmental and energy concerns in the 1960s and 1970s, large manufacturers, firms in auxiliary businesses such as electricity generation, and independent entrepreneurs dabbled with EV experiments again, either by developing new electric cars from scratch or by installing electric motors in adapted petrol models. Investments were small overall, however, and production of EV models remained very low, rarely exceeding a few thousand units.[61] As innovation scholars Robin Cowan and Staffan Hultén have argued, most of these EV development projects rested on the assumption that battery capacity and performance could be improved rapidly—and that the fate of EVs was contingent on technological advancements. The requisite improvements to make EVs competitive against ICEs did not materialise quickly, however.[62] Because of batteries' relatively low capacity for energy storage, EVs were plagued by low performance (low range, low top speeds) and high costs compared to conventional ICE cars. Investment in research and development (R&D) for EVs remained low, dwarfed by investments in fossil fuel technologies, although the U.S. Congress, for example, did provide $160 million for the development of batteries through the 1976 Electric and Hybrid Vehicle Research and Development Act, and the U.S. "Big Three" carmakers collectively invested $80 billion to downsize and increase the fuel economy of petrol-powered cars.[63] Throughout the twentieth century, skewed investment patterns, reflecting public concern about energy supplies/costs and the environment, favoured petroleum-based technologies and slowly made conventional ICEs cleaner and more efficient, thus maintaining the performance gap between petrol and diesel cars relative to EVs.[64]

The "performance argument" that EVs lost the automotive race to fossil-fuelled engines because of lower speeds and range has been challenged by Kirsch and Mom, and more recently by Eisler.[65] Instead, Eisler highlighted a "temporal mismatch" in EV technology: that batteries have a shorter life compared to electric motors, which required finding ways of servicing and replacing batteries during a car's operational lifespan. Traditional car manufacturers, moreover, were reluctant to switch from ICEs to EVs in the late twentieth century because their business models relied on revenue from servicing and providing spare parts for ICEs. Since battery cell production was a separate line of business and dominated by Asian firms, traditional carmakers were at risk of losing a substantial and profitable stream of revenue by embracing EVs.[66]

In 2020, the Chinese government embraced EVs in its New Energy Vehicle Industrial Development Plan, through which it supported domestic EV producers such as BYD—which has transformed from the "laughingstock" of the industry into a global market-share leader—and NIO, which similarly claims considerable domestic and international market share. By 2022, Chinese carmakers like these produced about 60% of global electric vehicle sales.[67] The growth of Chinese EV production and the global market share of Chinese producers accelerated when China responded quickly to regulatory actions in international markets, including the European Union's (EU) 2022 decision to prohibit the sale of ICEs by 2035, with the important caveat that it will still allow ICEs running strictly on electrofuels—that is, petrol made from hydrogen and captured CO_2.[68] Such developments motivate the following discussion of industrial policy and EVs.

2.3.5 The Role of Government Policy and the Issue of Climate Change

The rise of EVs has a much shorter chronology than the longer history of ICE and petrol competition with the alternatives. Moreover, the history of EVs is connected to the technological development of batteries since the 1980s, as well as the development of clean air regulations and the emergence of the issue of climate change in the 1990s.[69] In short, the rise of EVs is the result of both technological advancements and government regulatory and incentive interventions.

Ultimately, governmental policies, specifically in California, the world's largest car market, fuelled the rise of EVs starting in the 1990s. Notably, the emergence of batteries gained prominence in 1997 when Toyota introduced the Prius hybrid model, equipped with nickel-metal-hydride batteries, in the Japanese market. In 2000, Toyota launched the Prius globally with great success in the U.S. market. Honda followed Toyota's lead, launching the Insight model in Japan in 1999 and internationally in 2001.[70]

In 1990, before the launch of the Prius and Insight models, the California Air Resources Board (CARB) adopted its Low Emissions Vehicle programme, requiring major carmakers to sell vehicles with zero pollutant emissions. Aiming to combat the air pollution problem that had plagued California for several decades, the state initially required 2% of total car sales to be zero-emission vehicles in California from 1998, with a planned increase to 10% in 2003.[71] Under continued lobbying and litigation pressure, especially by firms such as GM and Ford, CARB modified the regulation several times and created additional vehicle categories.[72] Between 1999 and 2003, CARB delayed the deadline by two years, allowing carmakers in the California market—such as the U.S. "Big Three" and Japanese Toyota, Nissan, Mazda, and Honda—to comply with the mandate by offering hybrid-electric and hybrid-fuel-cell vehicles.[73] Although numerous manufacturers showcased electric and hybrid vehicles to meet the programme's criteria, many of these models remained experimental concepts and were not intended for commercialisation. Examples include GM's Impact and Ford's Ecostar, which were primarily developed to contest CARB's policies.[74] Companies adopted what Eisler has called the "better battery discourse," arguing that contemporary batteries could not match the energy and cost of petrol engines.[75]

The 1990 California zero-emissions policy motivated Japanese companies reliant on the U.S. market such as Toyota, Nissan, Honda, and Mazda to make substantial R&D investments in electric and hybrid-electric vehicles and hydrogen fuel cells, as research on patent data has shown.[76] Because of their limited domestic market, Japanese producers remained export-oriented and anticipated international regulations and consumer preferences for EVs well.

Toyota's development of its hybrid Prius model reflects the evolution of corporate environmentalism and how public and corporate perceptions of business responsibility for addressing environmental challenges changed in the late 1980s and 1990s.[77] Economist Masaru Yarime and colleagues noted that Toyota was aware of its perception as an environmental laggard in the 1970s, compared to the efforts of Honda and Mazda to address air pollution.[78] Wanting to change this negative image, Toyota set up an environmental committee chaired by the company's president in 1992 and published a global environmental charter in 1993, following the general trend of multinational corporations in the 1990s to boost green communication, now often criticised as "greenwashing."[79] In 1993, Toyota set up a concept study called the "Global 21st Century" (G21) committee, which outlined visions for the company's planned advanced propulsion vehicle. After the G21 committee delivered initial fuel efficiency gains by 50% using a direct injection petrol engine, Toyota management responded by requiring engineers to double fuel economy, or it would end the G21 programme. Because realising such a fuel economy goal was not possible using conventional engines, the committee began testing new technologies and settled on electric-petrol hybrid technology, which the company had developed for a sports vehicle already in 1977.

Toyota's development of the Prius was largely kept secret until the model's production launch, timed to coincide with the third climate change conference (COP 3) in Kyoto in 1997, hosted by Japan, to prove Toyota's environmental responsibility.[80] While it is clear that regulatory pressure was a key driver of the innovation and the commercialisation of petrol-electric hybrids, especially in relation to California's zero-emission vehicle mandate, other policies reinforcing technological shifts towards radically more fuel-efficient cars were generally lacking. Consequently, it was only in Toyota and Honda's home market, Japan, that petrol-electric hybrids, like the Prius, achieved significant market shares in the 2000s.[81]

Europe, the United States, and China made headway on electrification, especially during the Covid-19 pandemic. As Table 2.1 illustrates, global car markets shifted towards electrification in the late 2010s and early 2020s, when industrial policy fuelled the rise of EVs. These regions are responsible for the majority of EV adoption. Out of a total of 13.7 million EVs sold globally in 2023, only 1 million or 7% were sold outside Europe, the United States, and China.[82] Despite the rapid adoption of EVs, the impact on climate change is less clear due to the fossil fuel content in electricity as well as the composition of the EV fleet. In 2021, electricity generated from

TABLE 2.1 Annual passenger car registrations/sales (in million units) and share (%) of EV sales (including plug-in hybrid and battery electric cars) in the United States, Europe, Japan, and China, 2010–2022

	2010	*2012*	*2014*	*2016*	*2018*	*2020*	*2022*
USA[a]	11.6 (<0.1%)	14.5 (0.4%)	16.5 (0.9%)	17.6 (1.0%)	17.3 (2.0%)	14.6 (2.2%)	13.8 (7.7%)
Europe[b]	13.3 (<0.1%)	12.0 (0.2%)	10.0 (0.6%)	12.0 (1.0%)	12.7 (1.9%)	10.0 (10%)	9.7 (21%)
China	12.5 (<0.1%)	15.2 (<0.1%)	19.4 (0.4%)	24.6 (1.5%)	23.1 (4.9%)	19.7 (5.8%)	22.0 (29%)

Source: Constructed by the authors using the following data: For registrations, see ICCT, European Vehicle Market Statistics, various years, for Europe; US BTS, National Transportation Statistics, for the United States; China Statistics Press, China Statistical Yearbook, for China. See IEA, Global EV Data, for EV sales shares.

Note: The last entry on registrations for Europe and China is from 2021.

[a] *New vehicle sales include passenger cars, light trucks, and leases.*

[b] *EU 27 includes Croatia but not the UK.*

fossil fuels accounted for 60% of electricity generation in the United States, 67% in Japan, 77% in China, and 39% in the EU.[83]

Moreover, the data in Table 2.1 include both plug-in hybrids and battery electric cars. Whereas petrol-electric hybrids are integrated systems and charged through the ICE and through braking, studies by the International Council on Clean Transportation (ICCT) suggest that drivers do not sufficiently charge the batteries of plug-in hybrids to keep emissions low, and that real-world emissions from plug-in hybrids are close to ICE cars.[84] According to the International Energy Agency (IEA), plug-in hybrids made up 44% of EV registrations in the EU and 40% in Japan in 2022, whereas they comprised only 25% in China and 19% in the United States.[85] The seemingly widespread transition towards EVs could be a powerful tool in mitigating climate change, but only with a parallel decarbonisation of the utilities sector providing energy to recharge EV batteries. While EVs are more efficient than ICEs in transforming fuel into motion power, the high fossil content of electricity and the share of plug-ins offset more of the potential climate gains from an EV transition than otherwise could have been possible by using renewable energy.

2.4 Industrial Policy and the Struggle Over EV Markets

Since the 2008 Global Financial Crisis, globalisation has been in slow retreat.[86] The Covid-19 pandemic and the Russian invasion of Ukraine in 2022 further contracted overseas investment and trade. At the same time, policies for addressing climate change have gained prominence in many policy agendas, particularly after the 2015 Paris Climate Agreement. Governments "rediscovered" industrial policy to bolster their industries relative to global competition, often with a green or environmental focus. In 2009, the Obama administration combined policies for supporting the U.S. car industry in the form of bailouts with setting higher fuel economy standards and earmarked stimulus funds for advanced batteries and vehicles, extending these policies with EV purchase incentives in 2015.[87] Tesla Motors, the exclusively EV carmaker whose name has dominated discussions of EVs in the West since the 2010s, scaled its business through federal and state support for EVs, including a California programme through which it has sold credits to other carmakers who failed to sell sufficient zero-emission cars—and from which it made nearly 9 billion USD.[88] Meanwhile, U.S. policymakers treated Tesla as a national champion, even as other states courted Tesla's FDI.[89] Tesla also lobbied Chinese officials to adopt a similar credits scheme as California, a condition it used in negotiating the opening of its Shanghai factory in 2019.[90] In exchange for the 1 billion USD of revenue China has paid Tesla for its EV credits through this new scheme, China has benefitted from Tesla's inadvertent training of a generation of Chinese EV engineers and manufacturers. Furthermore, emissions trading systems around the world allow carbon-producing companies to pay to pollute and are often slow to lower emissions caps sufficiently to meet mitigation targets.

While the EU only turned its focus to electric vehicles after the Dieselgate scandal in 2015, it has also faced institutional challenges in responding to U.S. and Chinese industrial policy.[91] Since the 1980s and until the mid-2010s, EU economic policy encouraged international and internal competition through market forces, with governmental intervention restricted by new state aid policies implemented to complete the internal Single Market, and competition policy taking a precedence over industrial policy interventions by EU institutions.[92] After the 2008 financial crisis, as the EU lagged behind the United States in technological development and Chinese companies became globally dominant and increasingly invested in Europe—such as Geely's acquisition of Volvo Cars from Ford in 2010—EU policymakers began arguing for more active state involvement in restoring Europe's industrial competitiveness.[93]

In 2020, the EU passed the European Green Deal to support industrial development, while aiming to be climate-neutral by 2050.[94] In addition to cutting pollution, cultivating an international climate coalition, and securing the region's access to industrial raw materials, the Green Deal aimed to support the R&D of environmentally friendly transport, energy, and industrial production, and to position the European car industry to benefit from a green transition. Of course, carmakers took interest in this new policy agenda, especially since the European Commission, the EU's executive body, made explicit its intention to move the EU away from fossil fuels and ICEs by 2035, and, as Anu Bradford has shown, EU policymaking often has a global "Brussels effect."[95] According to Stellantis CEO Carlos Taveres, the institutional pressure on EVs has had a clear impact on manufacturers' electrification strategies in Europe: "The regulatory framework is clear: You sell EVs or you die."[96] In 2022, the United States passed the Inflation Reduction Act, which, in addition to curbing inflation, reducing the federal debt, and lowering drug prices, also aimed to aggressively promote building North American production chains for batteries. And in early 2024, the US Environmental Protection Agency (EPA) announced new exhaust emissions requirements aimed at shifting the majority of U.S. cars and light trucks to all-electric or hybrid by 2032, a slight delay from President Biden's initial target of 2030.[97]

This "rediscovery" of industrial policy in Europe and the United States should be viewed against the backdrop of, and in competition with, the long and successful industrialisation strategy of the Chinese Communist Party (CCP).[98] Once devoid of any car production, China became the world's largest car producer since the 1990s. As in other centrally planned economies during the Cold War, commercial vehicles initially dominated Chinese production targets, whereas passenger car production served only the CCP elite. Since the 1980s and in various stages afterward, the Chinese government has aimed to make the domestic industry more efficient and internationally competitive by encouraging foreign car companies to invest in the Chinese car industry through joint ventures with state-owned enterprises such as the Shanghai Automobile Industry Corporation's joint venture with Volkswagen in 1984.[99]

The Chinese state retained control of its industry, however, by limiting the approval of joint ventures, by team-partnering state-owned enterprises with two foreign companies, and through domestic sourcing requirements. It also implemented high tariffs to shield its state-owned enterprises, while private domestic companies were kept out of China by high entry requirements.[100]

According to Niewnhius and Xiao, the political structure of the Chinese car market created a mutual dependence, in which foreign carmakers were dependent on high sales in China's large market, and Chinese partner companies were dependent on foreign technology.[101] Since its accession to the World Trade Organisation (WTO) in 2001, Chinese industrial policy has aimed to resolve the problem of foreign-technology dependence through the creation of national brands, as well as the development of so-called new energy vehicles, cars powered entirely or predominantly by electricity. Forced to open its car market and relax the restrictions on new entrants per the terms of its WTO membership, the Chinese government gradually granted vehicle assembly permission to five domestic companies until 2003—Chery, Geely, Hafei, and Brilliance, and the battery maker BYD—out of which only Geely and BYD were fully private.[102]

Since no established carmakers had a specific or significant competitive advantage in producing vehicles using alternative fuels or powertrains, the Chinese government recognised the business opportunity in developing new energy vehicles and non-ICE cars like EVs. This Chinese plan began in 2001 with R&D support for pure EVs, hybrid electric, and fuel cell vehicles, as well as key components like multi-energy powertrain controllers, drive motors, and power batteries.[103] Despite state support for R&D, however, the Chinese industry's total investments in R&D remained small compared to that of Western and Japanese firms. Chinese carmakers sought to remain competitive mainly on price rather than through technological competition, and domestic brands focused on producing cheap cars.[104] Commercialisation of EVs began in the 2010s, however, largely due to the Chinese government's targets for EV and battery production and its stimulus policies in the wake of the 2008 Financial Crisis, including tax breaks, laxer registration policies for EVs, the building of charging infrastructure, and public procurement of new energy vehicles.[105]

As of 2024, China's competitive edge in EV production lay in controlling the entire value chain: Chinese mining companies monopolised access to the strategic minerals used in batteries, as well as mineral processing and refinement.[106] China also worked to command these strategic mineral resources through international standards.[107] The CCP's policies and EV production targets further propelled the global rise of Chinese battery companies like BYD and the Contemporary Amperex Technology Co (CATL). In 2021, China housed three-fourths of global battery cell production, an industrial policy success similar to China's early dominance in the production of solar panels.[108] Moreover, China, together with Japan, had a competitive advantage in global charging infrastructure, since most car battery chargers used Chinese or Japanese standards.[109]

Although China was leading the world in the transformation of the car industry towards electrification as of 2023, it is important to keep in mind that the Chinese industry was still mainly producing ICEs, with the capacity to produce 15 million gasoline-powered vehicles per year. Amid a slumping domestic economy, these Chinese ICE cars were "dumped" on international markets with lax emissions regulations.[110] Moreover, China's built-up ICE production capacity is not going away as long as demand for fossil-fuelled cars remains: "What are you going to do, close a factory?" the former executive of Chrysler China, Bill Russo, remarked.[111]

Together, Chinese competition, the challenges of the 2008 Financial Crisis and the Covid-19 pandemic, and climate change motivated many governments in the United States and Europe to use industrial policy to promote EV production. Many industrialised countries have car industries with strong legacies, in which carmakers became national champions by facilitating industrialisation and modernisation. To reach the goal of net-zero emissions from global road transportation by 2050, the IEA has estimated that 30 million EVs need to be sold by 2025 and 70 million by 2030, or a projected 30% and 60% respectively of total vehicle sales.[112] Regardless of the accuracy of those estimates, the dual aims of replacing the ICE but maintaining private motoring are creating demand for EV production volumes that exceed current capacity, making it challenging to determine which firms and countries will benefit the most from this transition. Currently, Western and legacy Asian car firms have a competitive edge in producing larger and high-end EVs, whereas Chinese firms are focusing on smaller and cheaper EVs. Nevertheless, it is safe to say that whoever succeeds in capturing large parts of the growing global market for EVs has a lot to gain. Meanwhile, established firms, national industries, and states with legacies of producing ICE cars that are failing to innovate and transition into EV technologies will face economic losses, which helps explain the widespread resurgence of industrial policy.

That China redoubled its support for its national champions in the 2020s prompted the European Commission to launch an investigation into China's subsidisation of EVs, especially through the EV carmaker Nio.[113] These developments in the geopolitical contest of industrial policy and EVs reflect the history of Japanese competition in Europe and the EEC/EU's solution to restricting Japanese exports into its liberalised Single Market through a secretive agreement.[114]

Developing countries without legacy car industries have also recognised business opportunities in the changing landscape of the green transition. For infrastructural, economic, and geopolitical reasons, two-wheeled EVs promise industrial development and solutions for personal transportation in countries across Africa and Southeast Asia. Twenty-seven million two-wheelers travelled the roads of sub-Saharan Africa in 2022, reflecting the large market for micro-mobility. With the introduction of EVs in Africa in recent years, countries like Kenya have experienced a "Two-wheeled Revolution," shifting from diesel-powered motors to electric ones. This revolution developed in parallel to Nairobi's embrace of electric public buses, prompted by Kenya's desire to move towards energy independence and away from its reliance on petrol imports. Kenya's government also celebrated the launch of local manufacturer Roam's new two-wheeled EV factory as an economic opportunity for the firm to supply East Africa with clean personal transportation.[115] And Uber announced plans to shift 20% of its fleet in Kenya to electric bikes.[116] Most two-wheeled EV models are green replacements for small motorcycles, mopeds, and scooters. Some, however, like the SHANE, are two-wheeled concept cars in a category of their own. These smaller, cheaper EVs have become another battleground for international competition, especially within and between developing countries.[117]

In India, for example, the most polluted and densely populated country in the world, two-wheeled EVs can be cheaply produced and promise a path to net-zero emissions. Once an iconic petrol motorcycle, the Bajaj Chetak has been redesigned as an EV. Other popular models in India include the Ampere, OLA, and Hero EVs. Even legacy petrol two-wheel producers, like Harley-Davidson, BMW, and Vespa, have introduced electric versions of their classic bikes and scooters, both cleaner and quieter than the traditional models.

2.5 Assessing the Outcomes of Climate Change Mitigation

Since the 1970s, efforts by regulators and carmakers to improve fuel economy and embrace new technologies have only somewhat reduced emissions from road transportation. Nor did they curb the global trend of increasing emissions, largely because widespread automobility became a strong consumer demand and an economic development goal for many governments.[118] Figure 2.1 shows fossil CO_2 emissions from road transportation in the United States, European OECD countries, Japan, and China and uses U.S. road traffic volumes as a benchmark.

Figure 2.1 reveals that temporary emissions reductions have coincided with external economic shocks such as those related to the 1970s energy crises and the recessions in the early 1990s and in 2007–2008. Regarding emission trends, it seems that advances in fuel efficiency, beginning with efforts in the 1970s, have somewhat offset the marked increase in road traffic since the early 1980s. As is well known in the economics literature, however, better fuel economy lowers driving costs, which in turn encourages drivers to drive more and longer: the so-called rebound effect.[119]

Yet, since 2007, real decoupling seems to have taken place in the United States and European OECD, indicating that fuel economy improvements have managed to plateau emissions despite increasing traffic. The trend towards larger cars discussed earlier adds additional roadblocks to reducing the industry's carbon footprint. Japan and China stand out at opposite ends of this spectrum. Japanese emissions actually decreased by 25% between 2001 and 2019, partly owing

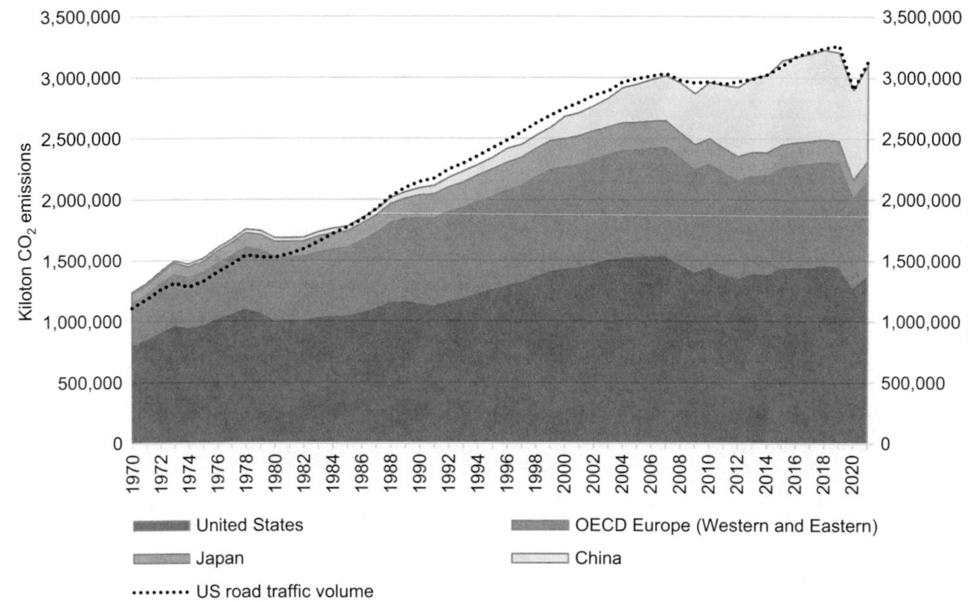

FIGURE 2.1 Fossil CO_2 from road transportation and road traffic in selected countries, 1970–2021

Source: Constructed by the authors using the following data: For Fossil CO_2 emissions, EDGAR database, "CO_2 Emissions." For vehicle miles travelled in the United States, U.S. DOT, Highway statistics series.

to the penetration of fuel-efficient hybrids and partly to a decrease in road traffic due to stricter regulations on personal transportation and improvements in public transportation infrastructure. Chinese emissions, on the other hand, grew by 380% between 2001 and 2019. While China, a global latecomer to automobility, is focusing most intensively on transitioning to EVs, most of its car production capacity and active fleet remain reliant on fossil fuels.

2.6 Conclusions

This business history of cars and climate change shows both how cars contributed to climate change and how the development of alternatives to ICEs and fossil fuels—and the current primacy of EVs—does not reflect the limits of technology but is instead the outcome of an entangled history of political choices and business strategies. That policies and business strategies evolved differently across time and space underscores our point that the current relationship between cars and climate change is not inevitable but is instead the result of regulatory, business, and consumer choices. Likewise, choices will continue to shape the future of cars and climate change; as governments make industrial policy decisions, consumers shape markets, and champion firms compete for global market share, the resulting geopolitical contest will continue to shape the industry and its impact on the climate.

Carmakers will continue to adapt to regulatory and competitive pressures to align their business strategies with the markets in which they operate—markets that are increasingly oriented around policies promoting zero emissions as a key objective for the car industry's "green transition" to mitigate the causes and consequences of climate change. Since EVs emerged as a favourite alternative to ICEs in the 2020s in response to both policy incentives and market pressures, competition between EV producers has become the primary battleground of geopolitics, and EV producers have predictably demonstrated their commitment to profits over climate. Consequently, while the historical consequences of cars and climate change are clear, there is not yet an ideal solution for the future, demanding further research on the future of cars and climate change.

Notes

1 We thank the editors of this volume, Teresa da Silva Lopes, Robert Fredona, and Paul Duguid, for the opportunity to contribute and for their helpful feedback and comments. We are also grateful to Routledge Press for supporting this project, to Ann-Kristin Bergquist for encouraging our collaboration, and to an anonymous reviewer for their suggestions. Mattias Näsman's work has been generously funded by Riksbankens Jubileumsfond, project no. M22-0029. Grace Ballor thanks Rasmus Salén for his meticulous research assistance.

2 Creutzig et al., "Transport."

3 Bodansky, "The History."

4 UNFCC, "Kyoto Protocol."

5 Creutzig et al., "Transport"; Lamb et al., "Countries."

6 Climate Watch, *Historical GHG Emissions.*

7 In 2023, for example, the European Union (EU) passed a regulation to limit new car registrations to vehicles emitting zero CO_2 emissions. See EU, Regulation 2023/851. This regulation is set to take effect in 2035. Importantly, a backdoor for vehicles running on e-fuels was left open. See Section 2.3.5.

8 IEA, *Global EV Outlook 2023.*

9 On the historical relationship between business and global governance, see Ballor and Pitteloud, "Capitalism and Global Governance."

10 On business and environmental activism, see the chapter in this volume, as well as other ongoing work by Sabine Pitteloud.

11 Chandler, *Strategy and Structure*; Chandler, *Giant Enterprise*; Wilkins and Hill, *American Business Abroad*; Bardou et al., *The Automobile Revolution.*
12 Oreskes and Conway, *Merchants of Doubt.*
13 Bergquist and David, "Beyond Planetary Limits!"; Huf et al., "Planetary History"; Klebaner, *Normes environnementales européennes*; Bergquist and Näsman, "Safe Before Green!"; Näsman, "The Political Economy"; Klebaner and Ramírez Pérez, "Managing Technical Changes"; Ballor, "Liberal Environmentalism." In addition to this volume, see also the forthcoming edited volume by Bott et al., *Environmental Regulation.*
14 Bergquist, "Renewing Business History," 11.
15 Freyssenet, "Second Automobile Revolution." Freyssenet also includes autonomous driving capability in this revolution.
16 Yergin, *The Prize*, Chapter 11; Wilkins, "Multinational Automobile Enterprises."
17 Bergquist and Lindmark, "Energy Transitions."
18 Yergin, *The Prize*, Chapters 11, 13.
19 Vast oil reserves were discovered in the Arabian Peninsula in 1932, prompting Western powers, who were increasingly dependent on oil, to form alliances with the Kingdom of Saudi Arabia (Yergin, *The Prize*, 284–92). See also Garavini, *The Rise and Fall.* On advances in refining, see Volti, *Cars and Culture*, 37–38.
20 Bini, "Back to the Future"; Mattioli et al., "Car Dependence."
21 Bardou et al., *The Automobile Revolution*, 176.
22 Freyssenet, "The Second Automobile Revolution," 443–54.
23 Bodansky, "The History," 26–28. See also Bulkeley and Newell, *Governing Climate Change.*
24 Franta "Weaponizing Economics."
25 Brulle, "Advocating Inaction." For the membership of GCC in 1989, see Climate Files, "1989 GCC Membership."
26 See, for example, Franta, "Weaponizing Economics"; Bonneuil et al., "Early Warnings." Also, see Chapter 9 of this volume.
27 Levy and Rothenberg, "Heterogeneity and Change."
28 Levy and Rothenberg, "Heterogeneity and Change."
29 Mikler, *Greening the Car Industry.*
30 Näsman, "The Political Economy."
31 Bergquist and Näsman, "Safe Before Green!"
32 Näsman, "The Political Economy," 312; Bergquist and Näsman, "Safe Before Green!," 82.
33 Peak associations include the German VDA, the Swedish Association of Car Manufacturers and Importers, and the European Committee for Common Market Automobile Constructors (CCMC). See Rollings, "The Vast"; Rollings, idem, "Transnational Business Associations." For recent studies focusing on national and international peak-level associations and car emissions policy in Europe, see Ballor, "Liberal Environmentalism"; Näsman and Pitteloud, "The Power and Limits"; Näsman, "The Political Economy"; Klebaner and Ramírez-Pérez, "Managing Technical Changes."
34 Bodansky, "The History."
35 Wilkins, "Multinational Automobile Enterprises," 238.
36 McCarthy, *Auto Mania.*
37 Wilkins, "Multinational Automobile Enterprises," 239.
38 Jones, *Maturity and Crisis.*
39 Iguchi, *Divergence and Convergence.*
40 Sorell, "Fuel Efficiency"; Näsman, "The Political Economy," 241.
41 For a detailed account of this, see Näsman, "The Political Economy."
42 Neumaier, "Eco-Friendly."
43 Berggren et al., "Hybrids, Diesel, or Both?"
44 Hooftman et al., "A Review"; ICCT, *European Vehicle Market Statistics.*
45 On Volvo and Ford's methanol projects, see Mårald, "Methanol as Future Fuel"; Nichols, "The Methanol Story." On Brazil's ethanol programme and carmaker responses, see Amatucci and Spers, "Institutional."
46 Bini, "Back to the Future," 285–86.
47 Luke, "SUVs"; Talay, "The Ford F-150."
48 US BTS, *National Transportation Statistics.*
49 Godek, "Regulation of Fuel Economy."

50 Snyder, "Crossovers." For estimates of profit margins, see Talay, "How the Ford F-150."
51 Hula et al., "EPA Automotive Trends Report."
52 Pardi, "Prospects and Contradictions."
53 Padeanu, "More Than Half"; ICCT, *European Vehicle Market Statistics.*
54 Bergquist and Näsman, "Safe Before Green!," 79–81.
55 Hula et al., "EPA Automotive Trends Report," 17. Cf. Pardi, "Prospects and Contradictions."
56 IEA, *EV Data Explorer.*
57 Riofrancos et al., "Achieving Zero Emissions"; Greimel, "Toyota."
58 Eisler, *Auto Electric.*
59 Flink, *The Automobile Age*; Kirsch, *The Burden of History*; Mom, *The Electric Vehicle.*
60 Taalbi and Nielsen, "Energy Infrastructure."
61 Westbrook, *The Electric Car*, 20–27.
62 Cowan and Hultén, "Escaping Lock-In," 70.
63 Cowan and Hultén, "Escaping Lock-In," 69, 75.
64 Cowan and Hultén, "Escaping Lock-In," 75.
65 Kirsch, *Burden of History*; Mom, *The Electric Vehicle*; Eisler, *Auto Electric.*
66 Eisler, *Auto Electric*, 9.
67 Bradsher, "How China Built BYD."
68 EU, Regulation 2023/851.
69 Eisler, *Auto Electric.*
70 Yarime et al., "The Strategies."
71 Collantes and Sperling, "The Origin."
72 Eisler, "Public Policy"; Fredrickson, "Ecostar."
73 Wesseling et al., "Exploring."
74 Eisler, "Public Policy," 786; Fredrickson, "Ecostar," 139–44.
75 Eisler, *Auto Electric*, 5–6.
76 Yarime et al., "The Strategies."
77 See, for example, Yarime et al., "The Strategies"; Magnusson and Bergren, "Environmental Innovation"; Eisler, "Public Policy."
78 Yarime et al., "The Strategies."
79 Bergquist, "Renewing Business History."
80 Eisler, "Public Policy," 799; Kameyama, *Climate Change Policy*, 49.
81 In 2010, hybrids made up 11.4% of new passenger car registrations, calculated from JAMA, *The Motor Industry of Japan 2023.*
82 IEA, *Global EV Data.*
83 IEA, *Energy Statistics Data Browser.*
84 Plötz et al., "Real-World Usage."
85 IEA, *Global EV Data.*
86 On deglobalisation trends, see Baldwin and Evenett, *COVID-19 and Trade Policy.*
87 Turner, *Charged*, 137–38.
88 Trudell, "Tesla."
89 Eiseler, *Auto Electric*, chapters 12–13.
90 Hvistendahl et al., "A Pivot to China."
91 Klebaner and Ramírez Pérez, "The European Automotive Industry."
92 On European competition policy, see the work of Laurent Warlouzet, including "Competition Versus Planning" and "The Implementation of the Single Market Programme."
93 For an overview, see Defraigne et al., "Past Lessons, Current Challenges."
94 Leonard et al., "The Geopolitics of the European Green Deal"; Matthijs and Meunier, "Europe's Geoeconomic Revolution."
95 Bradford, *The Brussels Effect.*
96 Motavalli, "Stellantis CEO."
97 Biden, "Strengthening American Leadership."
98 Wenten, "The Automotive Industry"; Nieuwenhuis and Lin, "China's Car Industry." See also the work of Yuan Jia Zheng on the Chinese car industry.
99 Wenten, "The Automotive Industry," 282.
100 Wenten, "The Automotive Industry," 281.

101 Niewnhius and Xiao, "China's Car Industry," 113.
102 Wenten, "The Automotive Industry," 285; Niewnhius and Xiao, "China's Car Industry," 117.
103 Niewnhius and Xiao, "China's Car Industry," 119.
104 Niewnhius and Xiao, "China's Car Industry," 117.
105 Niewnhius and Xiao, "China's Car Industry," 119–24.
106 See Business Sweden, *The Nordic Battery.*
107 On critical mineral standards, see Ballor, "Ruling the Natural World."
108 Business Sweden, *The Nordic Battery*, 61–62.
109 Wenten, "The Automotive Industry," 288.
110 Bradsher, "China."
111 Bradsher, "China."
112 IEA, *Global EV Outlook 2023*, 110.
113 Bradsher, "China's EV Threat."
114 Ballor, "Liberalisation"; Milor, "Non-Documents."
115 Gunderson, "Kenya's Two-Wheeled Revolution."
116 Miriri, "Uber Rolls Out Electric Bike Fleet in Kenya."
117 Gupta et al., "The Real Global EV Buzz Comes on Two Wheels."
118 On this point, see Bini, "Back to the Future"; Mattioli et al., "Car Dependence"; Freyssenet, "The Second Automobile Revolution."
119 In this vast literature, see, for example, Linn, "The Rebound Effect."

Bibliography

Amatucci, Marcos, and Eduardo Eugênio Spers. "Institutional, Technological and Commercial Innovations in the Brazilian Ethanol and Automotive Industries." In *The Greening of the Automotive Industry*, edited by Guiseppe Calabrese, 164–84. Basingstoke: Palgrave Macmillan, 2012.

Baldwin, Richard, and Simon Evenett, eds. *COVID-19 and Trade Policy: Why Turning Inward Won't Work*. London: CEPR Press, 2020.

Ballor, Grace. "Liberal Environmentalism: The Public-Private Production of European Emissions Standards." *Business History Review* 97, no. 3 (2023): 575–602. https://doi.org/10.1017/S0007680523000272.

Ballor, Grace. "Liberalisation or Protectionism for the Single Market? European Automakers and Japanese Competition, 1985–1999." *Business History* 65, no. 2 (2023): 302–28. https://doi.org/10.1080/00076791.2021.2025218.

Ballor, Grace. "Ruling the Natural World: The Historical Political Economy of Standards and Natural Resources." In *The Economic History of Natural Resources*, edited by Espen Storli, Madeleine Dungy, and Audrey Gerrard. Routledge, forthcoming.

Ballor, Grace, and Sabine Pitteloud. "Introduction: Capitalism and Global Governance in Business History." *Business History Review* 97, no. 3 (2023): 459–80. https://doi.org/10.1017/S0007680523000855.

Bardou, Jean-Pierre, Jean-Jacques Chanaron, Patrick Fridenson, and James M. Laux. *The Automobile Revolution: The Impact of an Industry*. Translated by James M. Laux. Chapel Hill: The University of North Carolina Press, 1982.

Berggren, Christian, Thomas Magnusson, and Dedy Sushandoyo. "Hybrids, Diesel or Both? The Forgotten Technological Competition for Sustainable Solutions in the Global Automotive Industry." *International Journal of Technology and Management* 9, no. 2 (2009): 148–173. https://doi.org/10.1504/IJATM.2009.026395.

Bergquist, Ann-Kristin. "Renewing Business History in the Era of the Anthropocene." *Business History Review* 93, no. 1 (2019): 3–24. https://doi.org/10.1017/S0007680519000369.

Bergquist, Ann-Kristin, and Thomas David. "Beyond Planetary Limits! The International Chamber of Commerce, the United Nations, and the Invention of Sustainable Development." *Business History Review* (April 28, 2023): 1–31. https://doi.org/10.1017/S0007680522001076.

Bergquist, Ann-Kristin, and Magnus Lindmark. "Economic History and the Political Economy of Energy Transitions: A Research Overview." Uppsala papers in Papers in Economic History, November 2023, 1–16. https://doi.org/10.33063/upeh.v2i.576.

Bergquist, Ann-Kristin, and Mattias Näsman. "Safe Before Green! The Greening of Volvo Cars in the 1970s–1990s." *Enterprise & Society* 24, no. 1 (2023): 59–89. https://doi.org/10.1017/eso.2021.23.

Biden, Joseph R. "Strengthening American Leadership in Clean Cars and Trucks." Executive Order 14037, August 5, 2021. https://www.federalregister.gov/documents/2021/08/10/2021-17121/strengthening-american-leadership-in-clean-cars-and-trucks.

Bini, Elisabetta. "Back to the Future: Changes in Energy Cultures and Patterns of Consumption in the United States, 1973–86." In *Counter-Shock: The Oil Counter-Revolution of the 1980s*, edited by Duccio Basosi, Giuliano Garavini, and Massimiliano Trentin. London: I. B. Tauris & Company, Limited, 2018, ProQuest Ebook Central.

Bodansky, Daniel. "The History of the Climate Change Regime." In *International Relations and Global Climate Change*, edited by Urs Lutterbacher and Detlef F. Sprinz, 23–40. Cambridge, MA: MIT Press, 2001.

Bonneuil, Christophe, Jean-Baptiste Choquet, and Benjamin Franta. "Early Warnings and Emerging Accountability: Total's Responses to Global Warming, 1971–2021." *Global Environmental Change* 71 (2021): 102386. https://doi.org/10.1016/j.gloenvcha.2021.102386.

Bott, Sandra, Sabine Pitteloud, and Janick Marina Schaufelbuehl, eds. *Environmental Regulation and the History of Capitalism: The Role of Business from Stockholm 1972 to the Climate Crisis.* Routledge, forthcoming.

Bradford, Anu. *The Brussels Effect: How the European Union Rules the World.* Oxford University Press, 2020.

Bradsher, Keith. "China Is Flooding the World with Cars." *The New York Times*, September 6, 2023.

Bradsher, Keith. "China's EV Threat: A Carmaker That Loses $35,000 a Car." *New York Times*, October 5, 2023.

Bradsher, Keith. "How China Built BYD, Its Tesla Killer." *New York Times*, February 12, 2024.

Brulle, Robert J. "Advocating Inaction: A Historical Analysis of the Global Climate Coalition." *Environmental Politics* 32, no. 2 (2023): 185–206. https://doi.org/10.1080/09644016.2022.2058815.

Bulkeley, Harriet, and Peter Newell. *Governing Climate Change*. 3rd ed. Abingdon: Routledge, 2023.

Business Sweden. *The Nordic Battery Value Chain—Market Drivers, the Nordic Value Proposition, and Decisive Market Necessities.* Innovation Norway, Business Finland, Business Sweden, and the Swedish Energy Agency, 2023. https://www.eba250.com/wp-content/uploads/2023/02/NordicBatteryReport.pdf.

Chandler, Alfred D. *Strategy and Structure: Chapters in the History of the American Enterprise*. Cambridge, MA: MIT Press, 1962.

Chandler, Alfred D. *Giant Enterprise Ford, General Motors and the Automobile Industry: Sources and Readings*. New York: Harcourt, Brace and World, 1964.

China Statistics Press. "China Statistical Yearbook." https://www.stats.gov.cn/sj/ndsj/2022/indexeh.htm.

Climate Files. "1989 GCC Membership." Last viewed April 10, 2024. http://www.climatefiles.com/denial-groups/global-climate-coalition-collection/1989-membership/.

Climate Watch. *Historical GHG Emissions (1990–2020)*. Washington, DC: World Resources Institute, 2023. Last viewed September 11, 2023. https://www.climatewatchdata.org/ghg-emissions.

Collantes, Gustavo, and Daniel Sperling. "The Origin of California's Zero Emission Vehicle Mandate." *Transportation Research Part A* 42, no. 10 (2008): 1302–13. https://doi.org/10.1016/j.tra.2008.05.007.

Cowan, Robin, and Stuffan Hultén. "Escaping Lock-In: The Case of the Electric Vehicle." *Technological Forecasting and Social Change* 53 (1996): 61–79. https://doi.org/10.1016/0040-1625(96)00059-5.

Creutzig, Felix, Patrick Jochem, Oreane Y. Edelenbosch, Linus Mattauch, Detlef P. van Vuuren, David McCollum, and Jan Minx. "Transport: A Roadblock to Climate Change Mitigation?" *Science* 350, no. 6263 (2015): 911–12. https://doi.org/10.1126/science.aac8033.

Defraigne, Jean-Christophe, Edoardo Traversa, Jan Wouters, and Dimitri Zurstrassen. "Past Lessons, Current Challenges and Future Scenarios." In *EU Industrial Policy in the Multipolar Economy*, edited by Jean-Christophe Defraigne, Jan Wouters, and Edoardo Traversa, 1–44. Cheltenham: Edward Elgar, 2022. ProQuest Ebook Central.

EDGAR Database. "CO2 Emissions of All World Countries." JRC/IEA/PBL 2022 Report. https://edgar.jrc.ec.europa.eu/report_2022#data_download.

Eisler, Matthew N. "Public Policy, Industrial Innovation, and the Zero-Emission Vehicle." *Business History Review* 94, no. 4 (2020): 779–802. https://doi.org/10.1017/S0007680520000719.

Eisler, Matthew N. *Age of Auto Electric: Environment, Energy and the Quest for the Sustainable Car.* Cambridge, MA: MIT Press, 2022.

EU. Regulation (EU) 2023/851 of the European Parliament and of the Council of April 19, 2023 Amending Regulation (EU) 2019/631 as Regards Strengthening the CO2 Emission Performance Standards for

New Passenger Cars and New Light Commercial Vehicles in Line with the Union's Increased Climate Ambition.

Flink, James J. *The Automobile Age.* Cambridge, MA: MIT Press, 1988.

Franta, Benjamin. "Weaponizing Economics: Big Oil, Economic Consultants, and Climate Policy Delay." *Environmental Politics* 31, no. 4 (2022): 555–75. https://doi.org/10.1080/09644016.2021.1947636.

Fredrickson, Leif. "The Rise and Fall of an Ecostar: Green Technology Innovation and Marketing as Regulatory Obstruction." In *Green Capitalism? Business and the Environment in the Twentieth Century*, edited by Hartmut Berghoff and Adam Rome, 132–48. Philadelphia: University of Pennsylvania Press, 2017.

Freyssenet, Michel. "The Second Automobile Revolution—Promises and Uncertainties." In *The Second Automotive Revolution: Trajectories of the World Carmakers in the XXI Century*, edited by Michel Freyssenet, 443–54. Basingstoke: Palgrave Macmillan, 2009.

Garavini, Giuliano. *The Rise and Fall of OPEC in the Twentieth Century.* Oxford: Oxford University Press, 2019.

Godek, Paul E. "The Regulation of Fuel Economy and the Demand for 'Light Trucks'." *The Journal of Law & Economics* 40, no. 2 (1997): 495–510.

Goodwin, Grace Eliza. "Tesla Found a Way to Get Out of Environmental Regulations at Its Texas Gigafactory." *Business Insider*, April 24, 2024.

Greimel, Hans. "Toyota, Go Ahead and Say It: You Told Them so on EVs." *Automotive News*, February 8, 2024.

Gunderson, Otto. "Kenya's Two-Wheeled Revolution." *Clean Technica*, January 2024.

Gupta, Rahul, Patrick Hertzke, Vivek Lath, and Gandharv Vig. "The Real Global EV Buzz Comes on Two Wheels." *McKinsey & Co.*, August 17, 2023. https://www.mckinsey.com/industries/automotive-and-assembly/our-insights/the-real-global-ev-buzz-comes-on-two-wheels.

Hooftman, Nils, Maarten Messagie, Joeri Van Mierlo, and Thierry Coosemans. "A Review of the European Passenger Car Regulations—Real Driving Emissions vs Local Air Quality." *Renewable and Sustainable Energy Reviews* 86 (2018): 1–21. https://doi.org/10.1016/j.rser.2018.01.012.

Huf, Ben, Glenda Sluga, and Sabine Selchow. "Business and the Planetary History of International Environmental Governance in the 1970s." *Contemporary European History* 31, no. 4 (November 2022): 553–69. https://doi.org/10.1017/S0960777322000546.

Hula, Aaron, Andrea Maguire, Amy Bunker, Tristan Rojeck, and Sarah Harrison. "The 2022 EPA Automotive Trends Report: Greenhouse Gas Emissions, Fuel Economy, and Technology Since 1975." EPA-420-R-22-029, 2022. https://www.epa.gov/system/files/documents/2022-12/420r22029.pdf.

Hvistendahl, Mara, Jack Ewing, and John Liu. "A Pivot to China Saved Elon Musk: It Also Binds Him to Beijing." *The New York Times*, March 28, 2024.

ICCT. *European Vehicle Market Statistics: Pocketbook 2022/23.* Berlin: International Council on Clean Transportation Europe, 2023. https://theicct.org/publication/european-vehicle-market-statistics-2022-23/.

ICCT. *Trends of New Passenger Cars in China: Air Pollutant and CO2 Emissions and Technologies, 2012–2021.* Washington: International Council on Clean Transportation, 2023.

IEA. *Global EV Data Explorer.* Paris: International Energy Agency, 2023. https://www.iea.org/data-and-statistics/data-tools/global-ev-data-explorer.

IEA. *Global EV Outlook 2023.* Paris: International Energy Agency, 2023. https://iea.blob.core.windows.net/assets/dacf14d2-eabc-498a-8263-9f97fd5dc327/GEVO2023.pdf.

IEA. *Energy Statistics Data Browser—Electricity Generation by Source.* Paris: International Energy Agency. https://www.iea.org/data-and-statistics/data-tools/energy-statistics-data-browser?country=WEOEUR&fuel=Energy%20supply&indicator=ElecGenByFuel.

Iguchi, Masahiko. *Divergence and Convergence of Automobile Fuel Economy Regulations: A Comparative Analysis of EU, Japan and the US.* Cham: Springer, 2015.

JAMA. *The Motor Industry of Japan 2023.* Tokyo: Japanese Automobile Manufacturers Association. https://www.jama.or.jp/english/reports/.

Jones, Daniel T. *Maturity and Crisis in the European Car Industry: Structural Change and Public Policy.* Sussex: Sussex European Research Centre, 1981.

Jones, Geoffrey. *Profits and Sustainability: A History of Green Entrepreneurship.* Oxford University Press, 2017.

Kameyama, Yasuko. *Climate Change Policy in Japan: From the 1980s to 2015.* London: Routledge, 2017.

Kirsch, David A. *The Electric Vehicle and the Burden of History.* New Brunswick: Rutgers University Press, 2000.

Klebaner, Samuel. *Normes environnementales européennes et stratégies des constructeurs automobiles: Un jeu coopératif aux résultats ambigus*. Paris: Presses des Mines, 2019.

Klebaner, Samuel, and Sigfrido M. Ramírez Pérez. "Managing Technical Changes from the Scales of Legal Regulation: German Clean Cars Against the European Pollutant Emissions Regulations in the 1980s." *Management & Organizational History* 14, no. 4 (2019): 442–68. https://doi.org/10.1080/17449359.2019.1702885.

Klebaner, Samuel, and Sigfrido M. Ramírez Pérez. "The European Automotive Industry: A Strategic Sector in Search of a New Industrial Policy." In *EU Industrial Policy in the Multipolar Economy*, edited by Jean-Christophe Defraigne, Jan Wouters, and Dimitri Zurstrassen, 304–31. Cheltenham: Edward Elgar, 2022, ProQuest Ebook Central.

Lamb, William F., Michael Grubb, Francesca Diluiso, and Jan C. Minx. "Countries with Sustained Greenhouse Gas Emissions Reductions: An Analysis of Trends and Progress by Sector." *Climate Policy* 22, no. 1 (2022): 1–17. https://doi.org/10.1080/14693062.2021.1990831.

Leonard, Mark, Jean Pisani-Ferry, Jeremy Shapiro, Simone Tagliapietra, and Guntram Wolff. "The Geopolitics of the European Green Deal." European Council on Foreign Relations, 2021. http://www.jstor.org/stable/resrep29130.

Levy, David L., and Sandra Rothenberg. "Heterogeneity and Change in Environmental Strategy: Technological and Political Responses to Climate Change in the Global Automobile Industry." In *Organizations, Policy, and the Natural Environment: Institutional and Strategic Perspectives*, edited by Andrew J. Hoffman and Marc Ventresca, 173–93. Stanford: Stanford University Press, 2002.

Linn, Joshua. "The Rebound Effect for Passenger Vehicle." *The Energy Journal* 37, no. 2 (2016): 257–88. http://www.jstor.org/stable/24696756.

Luke, Timothy W. "SUVs and the Greening of Ford." *Organization & Environment* 14, no. 3 (2001): 311–35. https://www.jstor.org/stable/26161667.

Magnusson, Thomas, and Christian Bergren. "Environmental Innovation in Auto Development—Managing Technological Uncertainty Within Strict Time." *International Journal of Vehicle Design* 26, no. 2–3 (2001): 101–15. https://doi.org/10.1504/IJVD.2001.001932.

Mårald, Erland. "Methanol as Future Fuel: Efforts to Develop Alternative Fuels in Sweden After the Oil Crisis." *History and Technology: An International Journal* 26, no. 4 (2010): 335–57. https://doi.org/10.1080/07341512.2010.523175.

Matthijs, Matthias and Sophie Meunier. "Europe's Geoeconomic Revolution: How Europe Learned to Wield Its Real Power." *Foreign Affairs* 102 no. 5 (2023): 168–179.

Mattioli, Giulio, Cameron Roberts, Julia K. Steinberger, and Andrew Brown. "The Political Economy of Car Dependence: A Systems of Provision Approach." *Energy Research & Social Science* 66 (2020): 101486. https://doi.org/10.1016/j.erss.2020.101486.

McCarthy, Tom. *Auto Mania: Cars, Consumers, and the Environment.* New Haven: Yale University Press, 2007.

Mikler, John. *Greening the Car Industry: Varieties of Capitalism and Climate Change*. Cheltenham: Edward Elgar, 2009.

Milor, Alice. "Des normes automobiles pour l'environnement, l'économie ou l'Europe? (années 1980)." *Encyclopédie d'histoire numérique de l'Europe*, online May 12, 2022. https://ehne.fr/fr/node/21838.

Milor, Alice. "Non-Documents for Big Decisions: The Commission and the EEC–Japan Automotive Agreement (1991)." *Journal of Common Market Studies* (2024). https://doi.org/10.1111/jcms.13578.

Mom, Gijs. *The Electric Vehicle: Technology and Expectations in the Automobile Age.* Baltimore: John Hopkins Press, 2004.

Motavalli, Jim. "Stellantis CEO Gets Surprisingly Candid About Electrification." *Autoweek*, March 1, 2023.

Näsman, Mattias. "The Political Economy of Emission Standards: Politics, Business and the Making of Vehicle Emission Regulations in Sweden and Europe, 1960–1980s." PhD diss., Umeå University, 2021. https://urn.kb.se/resolve?urn=urn:nbn:se:umu:diva-186828.

Näsman, Mattias, and Sabine Pitteloud. "The Power and Limits of Expertise: Swiss-Swedish Linking of Vehicle Emission Standards in the 1970s and 1980s." *Business and Politics* 24, no. 3 (2022): 241–60. https://doi.org/10.1017/bap.2022.3.

Neumaier, Christopher. "Eco-Friendly Versus Cancer-Causing: Perceptions of Diesel Cars in West Germany and the United States, 1970–1990." *Technology and Culture* 55, no. 2 (2014): 429–60. https://www.jstor.org/stable/24468912.

Nichols, Roberta J. "The Methanol Story: A Sustainable Fuel for the Future." *Journal of Scientific & Industrial Research* 62 (2003): 97–105.

Oreskes, Naomi, and Erik M. Conway. *Merchants of Doubt: How a Handful of Scientists Obscured the Truth on Issues from Tobacco Smoke to Global Warming*. Bloomsbury Publishing, 2010.

Padeanu, Adrian. "SUVs Accounted for More Than Half of Sales in Europe for the First Time Ever." *Motor1.com*, September 11, 2023.

Pardi, Tommaso. "Prospects and Contradictions of the Electrification of the European Automotive Industry: The Role of European Union Policy." *International Journal of Automotive Technology and Management* 21, no. 3 (2021): 162–79. https://doi.org/10.1504/IJATM.2021.116620.

Plötz, Patrick, Steffen Link, Hermann Ringelschwendner, Marc Keller, Cornelius Moll, Georg Bieker, Jan Dornoff, and Peter Mock. "Real-World Usage of Plug-In Hybrid Vehicles in Europe: A 2022 Update on Fuel Consumption, Electric Driving, and CO2 Emissions." ICCT White Paper, June 2022. https://theicct.org/wp-content/uploads/2022/06/real-world-phev-use-jun22-1.pdf.

Riofrancos, Thea, Alissa Kendall, Kristi K. Dayemo, Matthew Haugen, Kira McDonald, Batul Hassan, and Margaret Slattery. "Achieving Zero Emissions with More Mobility and Less Mining." *Climate + Community Project*, UC Davis, January 2023. https://www.climateandcommunity.org/_files/ugd/d6378b_b03de6e6b0e14eb0a2f6b608abe9f93d.pdf.

Rollings, Neil. "The Vast and Unsolved Enigma of Power: Business History and Business Power." *Enterprise & Society* 22, no. 4 (2021): 893–920. https://doi.org/10.1017/eso.2021.53.

Rollings, Neil. "The Development of Transnational Business Associations During the Twentieth Century." *Business History* 65, no. 2 (2023): 235–59. https://doi.org/10.1080/00076791.2021.1958783.

Snyder, Jesse. "Crossovers, SUVs Are 'Margin Builders'." *Automotive News*, July 24, 2017, 4.

Sorell, Steve. "Fuel Efficiency in the UK Vehicle Stock." *Energy Policy* 20, no. 8 (1992): 766–80.

Taalbi, Josef, and Hana Nielsen. "The Role of Energy Infrastructure in Shaping Early Adoption of Electric and Gasoline Cars." *Nature Energy* 6 (2021): 970–76. https://doi.org/10.1038/s41560-021-00898-3.

Talay, M. Berk. "How the Ford F-150 Became King of Cars." *The Conversation*, June 8, 2018.

Trudell, Craig. "Tesla Rakes in $9 Billion from Carmakers Failing to Sell Enough EVs." *Bloomberg*, February 9, 2024.

Turner, James Morton. *Charged: A History of Batteries and Lessons for a Clean Energy Future*. Seattle: University of Washington Press, 2022.

UNFCC (United Nations Framework Convention on Climate Change). *Kyoto Protocol*. UNFCC, December 11, 1997.

US BTS (Bureau of Transportation Statistics). "National Transportation Statistics." https://www.bts.gov/product/national-transportation-statistics.

US DOT (Department of Transport). "Highway Statistics Series." Washington, 2023. https://www.fhwa.dot.gov/policyinformation/statistics.cfm.

US EIA (Energy Information Agency). "Monthly Energy Review." Washington, September 2023. https://powerlibrary.theelectricityhub.com/2023/10/18/monthly-energy-review-september-2023/.

Volti, Rudi. *Cars and Culture: The Life Story of a Technology*. Baltimore: The John Hopkins Press, 2004.

Warlouzet, Laurent. "The Implementation of the Single Market Programme (1985–92): The Examples of the Car Emission and of Competition Policy." In *Reshaping Europe: Towards a Political, Economic and Monetary Union, 1984–1989*, edited by Michael Gehler and Wilfried Loth, 247–62. Baden-Baden: Nomos, 2020.

Warlouzet, Laurent. "Competition Versus Planning: A Battle That Shaped European Integration (1945–2022)." In *The Cambridge History of the European Union. Volume 2, European Integration Inside-Out*, edited by Mathieu Segers and Steven Van Hecke, 234–60. Cambridge, MA: Cambridge University Press, 2023.

Wenten, Frido. "The Automotive Industry in China: Past and Present." In *New Frontiers of the Automobile Industry: Exploring Geographies, Technology, and Institutional Challenges*, edited by Alex Covarrubias and Sigfrido M. Ramírez Perez, 279–300. Cham: Palgrave McMillan, 2020.

Wesseling, Joeri, Jacco Farla, and Marko Hekkert. "Exploring Car Manufacturers' Responses to Technology-Forcing Regulation: The Case of California's ZEV Mandate." *Environmental Innovation and Societal Transitions* 16 (2015): 87–105. https://doi.org/10.1016/j.eist.2015.03.001.

Westbrook, Mike H. *The Electric Car: Development and Future of Battery, Hybrid and Fuel-Cell Cars*. London: Institution of Engineering & Technology, 2001.

Wilkins, Mira. "Multinational Automobile Enterprises and Regulation: An Historical Overview." In *Government, Technology and the Future of the Automobile*, edited by Douglas H. Ginsburg and William J. Abernathy, 221–58. New York: McGraw-Hill, 1980.

Wilkins, Mira, and Frank Ernest Hill. *American Business Abroad: Ford on Six Continents*. Detroit: Wayne State University Press, 1964.

Yarime, Masaru, Hideaki Shiroyama, and Yusuke Kuroki. "The Strategies of the Japanese Auto Industry in Developing Hybrid and Fuel-Cell Vehicles." In *Making Choices About Hydrogen: Transport Issues for Developing Countries*, edited by Lynn K. Mytelka and Grant Boyle, 187–212. Tokyo: United Nations University Press, 2008.

Yergin, Daniel. *The Prize: The Epic Quest for Oil, Money and Power*. New York: Simon and Schuster, 1991.

3

POLLUTION AND BUSINESS

Case Studies From Modern Chinese Business History[1]

Ghassan Moazzin

3.1 Introduction

China today is a major contributor to climate change. In 2022, the Chinese economy was responsible for more than a quarter of global CO_2 emissions specifically and greenhouse gas emissions generally. At the same time, similar to other industrialising countries in the past, pollution has become a major problem for China. While China's reliance on coal is the major culprit behind its problem with pollution, emissions by cars also play a major role.[2]

Like other chapters in this volume, this chapter tries to discuss the interaction between business history and climate change. The specific focus of this chapter is twofold: On the one hand, it concentrates on the connection between business and pollution. On the other hand, it focuses on the specific case of China. Given the important role China plays within global climate change and the significant pollution challenge it faces, it only seems adequate to explore the historical relationship between business and pollution in China. Moreover, this chapter's discussion of the interaction between business and the environment in China also aims to respond to recent calls for the writing of an "alternative business history" that takes into account the specific conditions of emerging markets and addresses the problem that the field of business history hitherto has largely concerned itself with the regions of North America, Western Europe, and Japan.[3] More specifically, in the introduction to the recent *Business History Review* special issue, "Business and the Environment Revisited," Bergquist also notes the need for more business history scholarship on the relationship between business and the environment in regions outside the United States and Western Europe.[4]

This chapter unfolds as follows: First, I discuss historical scholarship on the connection between business and the environment and business and pollution, and I contrast it with the insufficient coverage of the relationship between the environment and business in the literature on Chinese business history. In the same section, I also discuss several examples of studies employing the perspectives of energy and environmental history that nevertheless touch upon the interaction between business and the environment in general and business and pollution in particular and can be instructive in terms of developing a future research agenda for historians of Chinese business that incorporates the environment. To complement these examples, the chapter then

DOI: 10.4324/9781003478089-4

zooms in on the example of early twentieth-century Shanghai. First, it explores the response of the Shanghai Municipal Council (SMC) to air pollution in the 1920s and 1930s. Then, it closes with a discussion of two concrete businesses and polluters in early twentieth-century Shanghai: The Shanghai Power Company and the China General Omnibus Company. Overall, this chapter hopes to encourage historians of Chinese business history to engage with the environmental side of the history of Chinese business, including the connection between business and pollution.

3.2 Historiographical Discussion

Business history as a larger field has started to interrogate the historical interactions and connections between business and business history and the environment, including the connection between business and pollution.[5] Indeed, for the nineteenth and twentieth centuries, business historians have extensively studied the relationship between entrepreneurs, business, and pollution. Rosen uses the example of the meatpacking industry in Chicago, New York, and New Jersey in the late nineteenth century to show how protests, lawsuits, and new laws driven by concerns about pollution not only helped curtail the negative influence of businesses on the environment but could also foster the modernisation of enterprise.[6] Elsewhere, focusing on Chicago in the 1890s, she demonstrates that historically, the attitudes of businesspeople towards the abatement of pollution could vary considerably. Besides opposition, there were also those that agitated for measures for protection against pollution.[7] For the other side of the Atlantic, Desrochers examines debates around the reduction of industrial waste and the profitable use of by-products in Victorian Britain.[8]

Bergquist and Lindmark study the case of the Swedish Boliden mining company, which, during the first half of the twentieth century, took active measures to prevent pollution, pursuing both profitability and environmental sustainability. The case of Boliden also shows the importance of global connections, whereby the United States served both as a negative example for potential environmental damage and as a source of technology for environmental protection.[9] Focusing on the beginning of the twentieth century, Stradling and Tarr show that pollution also became an issue in the American transportation sector. Railway operators were confronted with public criticism against the pollution railways created and regulatory pressure and took various measures to abate smoke. At the same time, Stradling and Tarr demonstrate that such operators could at times also successfully stall demands for further potentially costly measures for smoke abatement.[10] In the petroleum industry in the United States, the wish of firms to be more economically efficient drove business-led efforts of self-regulation and pollution control during the first half of the twentieth century. However, government regulation came to replace such efforts from the mid-twentieth century after self-regulation had reached its limits in controlling pollution.[11]

Smith's study of DuPont in the United States and its dealing with pollution between the 1930s and 1960s highlights the efforts DuPont made in this regard but also the significant challenges it faced, including the high cost of acting against pollution.[12] Focusing on the case of the chemical industry and pollution in Sarnia, Canada, in the 1960s, Temby shows how local chemical firms monopolised pollution-related information and worked with other local stakeholders to stifle criticism and possible action against air pollution.[13] Finally, in their study of the environmental policies of the German chemical firms Bayer and Henkel, Jones and Lubinski demonstrate how these two firms took incremental steps against pollution during the 1950s and 1960s. During this period, environmental strategies of the two companies did not differ from their U.S. counterparts. In the 1970s, in response to pressures at both the regional and national levels, Henkel and Bayer started to adopt more far-reaching and impactful environmental strategies, diverging from the behaviour of American chemical firms during this period.[14]

While this is a mere sample of the literature produced by business historians on the connection between business and pollution, it nevertheless shows the breadth of the scholarly enquiry on this topic in recent decades. At the same time, it also shows a clear geographical focus on the West. This confirms Bergquist's aforementioned criticism of the geographical limitations of existing scholarship on the relationship between business and the environment.[15] Indeed, the scholarly literature on Chinese business history has neglected this relationship.[16] Given that China saw significant industrial growth both before and after 1949, this is particularly surprising.[17] However, this does not mean that the connection between Chinese business and China's environment has not been studied at all. Rather, similar to what Rosen showed in an article in 1997 in which she encouraged business historians to take ecological perspectives into account, we discover that studies by historians that do not primarily work in the field of Chinese business history can provide examples of how we can explore the connection between business and the environment in modern China. Following Rosen's example, I will discuss four recent studies on the history of energy and the environment in modern China that touch upon the nexus of business and the environment, including pollution.[18]

Arguably the most instructive example for the field of Chinese business history is a recent article by Micah Muscolino. In this article, Muscolino uses the perspective of energy history to explore the connections between business and the environment in 1920s and 1930s China and shed new light on the activities of the well-known and well-studied Chinese industrialist Liu Hongsheng. Specifically, he focuses on the Shanghai Portland Cement Company and the China Coal Briquette Company, two enterprises Liu was involved in, to show how Liu channelled waste produced by some of his businesses into others for reuse. Muscolino also emphasises that Liu's own enterprises and his customers significantly contributed to the release of greenhouse gas and pollution. Unlike the three other studies that follow, Muscolino directly engages with existing scholarship on Chinese business history and can be read as encouraging scholars of Chinese business history to engage with environmental aspects of the history of Chinese business.[19]

Another recent example is Ying Jia Tan's *Recharging China in War and Revolution, 1882–1955*. In this book, Tan traces the history of Chinese electrification between the late nineteenth and mid-twentieth centuries. While Tan mainly writes from the perspective of the history of energy, the electrical power industry takes centre stage in Tan's account. Most importantly for our understanding of the nexus between business and the environment, Tan shows how the turmoil of war and revolution that China went through in the decades after the introduction of electricity shaped the country's power sector and electricity grid, leading to a disregard for sustainability and the excessive reliance on fossil fuels.[20]

Also writing from the angle of energy history, Victor Seow's *Carbon Technocracy: Energy Regimes in Modern East Asia* focuses on the history of coal mining in Fushun in Northeast Asia between the early twentieth century and the early People's Republic of China after its establishment in 1949. Under successive Japanese and Chinese regimes, Fushun became a major site for the excavation of coal. In his exploration of Fushun's coal industry, Seow writes with a focus on the changes to and the impact on the environment that the construction of mines and the mechanisation of mining brought about. He sheds light on the environmental dangers such as gas explosions or floods created by the mining operations and their impact on Fushun workers and those living in the vicinity of the mines. Seow's account also stresses how the enterprise of excavating coal on a massive scale in Fushun was ultimately unsustainable and has left its mark on the area until today, including in the form of water pollution.[21]

Importantly, Tan's and Seow's accounts speak to several of the themes of the alternative business history recently advocated by Austin, Jones, and Davila. Explicitly, among the specific factors they list as having shaped the development of business in emerging markets are instability, colonialism, and state intervention.[22] Tan's account shows not only how such instability affected China's power sector but also how this in turn had important environmental consequences. Moreover, he also highlights the primacy of the state in the production of electrical energy in China. Finally, in Tan's case, we also see the involvement of imperialism in the history of the Chinese electrical power industry in the form of foreign power production in the treaty ports and under Japanese occupation during World War II. As Seow's book shows, Japanese imperialism was the central force that shaped the excessive extraction of coal in Fushun between the early twentieth century and the end of World War II, when Japan controlled the mines at Fushun. Thereafter, as Seow describes, first the Nationalist and then the Communist state took control of the mines. Finally, the outbreak of war in 1937 led to a further acceleration of the extraction of coal in Fushun by Japan. Thus, we see that both Tan's and Seow's works not only explore the nexus of business and the environment in modern China but also demonstrate characteristics of business development that have been noted as specific to emerging markets. This shows that writing the environment into the business history of developing countries like China also involves taking into account the special circumstances that impacted business in these regions.

A final example of recent scholarship that shows the potential for exploring the connection between business and the environment in China is an article by Pei Guangqiang that deals with the problem of air pollution in modern Shanghai caused by the growing industrialisation and urbanisation in the city. Pei writes from the perspective of environmental history. His focus is the assessment of air pollution as opposed to the specific interaction between businesses and the environment. Indeed, industrial facilities are only one of several sources of pollution he identifies and are not discussed in detail. Moreover, while Pei mentions other businesses such as the China General Omnibus Company, which I will discuss further later, as causes of pollution as well, they are also not discussed in any depth. Likewise, we find very little on how government institutions or businesses tried to address the problem of pollution. Nevertheless, Pei's article is important insofar as it clearly shows the severe level of air pollution modern Shanghai suffered from, particularly during the first half of the twentieth century. Moreover, it clearly identifies a variety of businesses, ranging from bus and shipping companies to factories and power companies, as being a major source of pollution.[23]

Taken all together, these studies clearly show the potential for further investigations into the interaction between business and the environment in modern China. Moreover, we have also seen that China's specific circumstances as an emerging market will need to be taken into account by business historians in future explorations of the role of the environment in Chinese business history.

3.3 Business and Air Pollution in Early Twentieth- Century Shanghai

This section of the chapter zooms in on the case of early twentieth-century Shanghai to further illustrate the potential for research into the connection between Chinese business history and the environment. More specifically, this section will focus on Shanghai's International Settlement during the 1920s and 1930s before 1937. The International Settlement spanned part of

Shanghai and was the result of a merger of the British and American settlements that had been established after Shanghai's opening as a treaty port in 1842. It was run largely independently by the SMC, which was dominated by foreign and—from the time of the first admission of Chinese councilors in 1928—Chinese businesspeople and elected not by the entire population of the Settlement but by a small, wealthy elite. During the 1870s and thereafter, the majority of the population in the International Settlement was Chinese.[24] In the early twentieth century, Shanghai saw rapid industrial growth and was China's industrial center, with a significant number of factories situated in the International Settlement.[25] Significantly, Shanghai became an important hub for China's all-important textile industry.[26]

Pei's work mentioned earlier ably discusses the problem of air pollution in modern Shanghai, and the main reasons he identifies for this problem—apart from large-scale emissions—are a lack of green space in the city, the mixing of industrial and residential spaces, and dense and problematic living conditions. However, he does not focus specifically on the SMC and the International Settlement or on how the SMC or businesses dealt with the problem of air pollution.[27] Scholarship on the SMC has also largely neglected if and how the SMC treated air pollution connected to industrialisation and businesses.[28] This section will try to shed some light on this issue by probing how the SMC in general dealt with the problem of air pollution from the 1920s to 1937, with a particular eye on the connection between air pollution and industry, and by focusing in on the SMC's dealing with the pollution caused by the Shanghai Power Company and the China General Omnibus Company. My specific focus when exploring air pollution in this section is on the problem of smoke or what in the SMC's annual reports is often called "smoke nuisance."[29]

If we look at the SMC's annual reports, the first general mention of the problem of smoke as a general phenomenon from 1920 onwards can be found in 1927. The report by the Commissioner of Public Health in the annual report states that "complaints of smoke nuisance are steadily increasing." While this problem fell under the purview of the SMC's Public Health Department, the report stresses that the SMC lacked "any really effective means of preventing smoke nuisances in Shanghai, or of dealing with them when they are proved to exist." This was because the SMC did not have any legal basis to move against air pollution. So no real solution was suggested apart from encouraging better separation between residential and industrial areas in the future and relying on persuasion and accommodation between those inhabiting the Settlement.[30] In spite of its insufficient legal powers, the SMC managed to deal with the problem of smoke to a certain extent, though. As Figure 3.1 shows, the statistics given in the SMC's annual report clearly show that some action was taken against smoke. At the same time, we see that there is no significant growth in the cases addressed after a jump from 1926 to 1927.[31]

Between 1928 and 1932, the SMC's annual reports continued to mention the problem of pollution with smoke. As the 1929 report put it, smoke "arises principally from steamships and steam launches on the river and from the greater industrial works but is aggravated by a large number of small factories which in many instances are adapted premises totally unsuitable for their purposes." In 1928, the report by the Commissioner explained that even if theoretically certain existing laws could be used to address the problem of smoke, "it is improbable that in an International community success would attend legal action until the state of public opinion demanded the suppression of all such nuisances," which suggests that the public at large might not have had much of an issue with the problem of smoke. Likewise, in 1931, the Commissioner reported that "legal powers [were] not only slender but difficult in application." The reports between 1928 and 1932 show no evidence that any significant new measures to address the

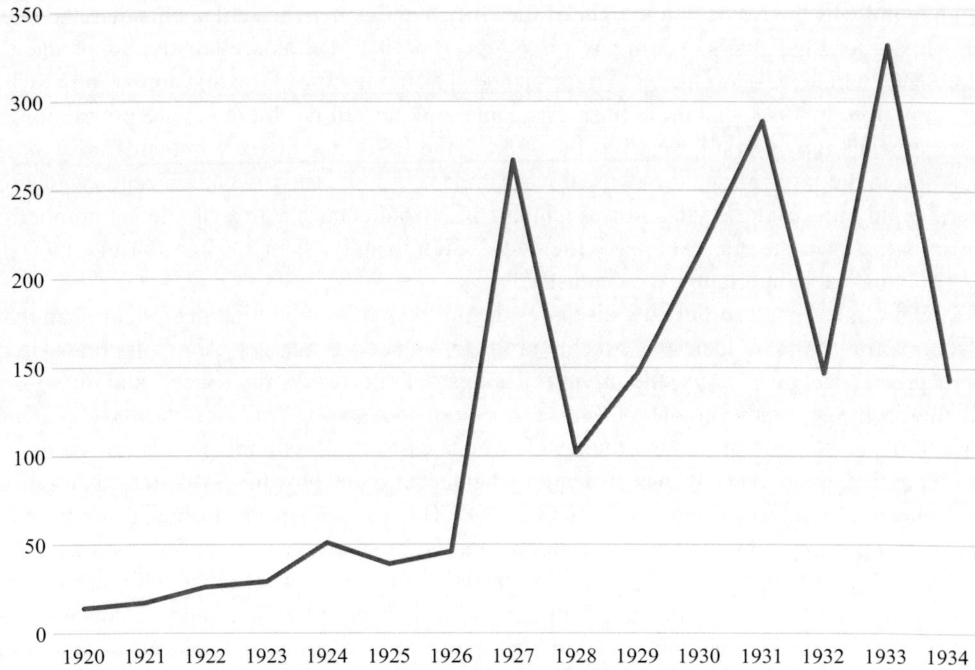

FIGURE 3.1 Number of "Smoke Nuisances Suppressed," 1920–1934

Source: Shanghai Municipal Council Annual Reports, 1920–1934

problem of smoke were enacted. Instead, the Commissioner of Public Health continued to mention the deficiency of legal powers to address the problem and suggested once again that a better functional partition of the Settlement could help, but also that certain requirements for buildings could hypothetically be useful.[32]

Besides the cases of the Riverside Power Station and the China General Omnibus Company, which will be dealt with in some more detail later, the SMC's annual reports for the years 1933 to 1937 show a continuation of the problem of air pollution. For instance, as the 1934 report explained, "The Smoke Nuisance is increasing with the growth of factories and other buildings." In its 1935 report, the SMC mentions that "several smoke nuisance complaints have been received from the public during the year. These concerned either large factories and buildings where improper fuel is combined with faulty and unsupervised stoking, or small shops and house factories where the chimney is often of insufficient length or completely absent." As the SMC's report for 1936 records, at the meeting of the SMC's ratepayers in 1936, it was noted that rapid growth of industry "contribut[ed] to the smoke cloud hanging over the city." This was part of a larger discussion on a resolution regarding the rising population, quality of life, and growth of factories in the Settlement. In this context, the problem of houses being converted into factories, which was in contravention of the regulations of the Settlement but had been neglected by the SMC due to the "difficulty of fixing a dividing line between home-crafts and manufacturing on a larger scale," also came up. However, the resolution that was eventually passed only called for an investigation into the "Housing Situation" in the Settlement. In the final annual report from the period investigated here, namely the annual report of 1937, the Commissioner of Public Health explained that complaints about smoke continued, but that these were mainly "due to imperfect stoking of furnaces." However, he remained silent on the larger connection between industry and pollution. That is despite the fact that just a year before, different kinds of industrial facilities had featured prominently as producers of smoke and other emissions in a list of examples of "nuisances and offensive trades" the Public Health Department had to deal with mentioned in the report of the Commissioner of Public Health in the SMC's annual report. In general, the reports from 1933 to 1937 once again show little in terms of larger solutions to the problem of air pollution apart from minor measures or suggestions such as the use of longer chimneys.[33]

In sum, if we look at the discussion and treatment of the issue of smoke and air pollution in the SMC's annual reports in the 1920s and 1930s up to 1937, we see that while the SMC clearly understood the connection between industrialisation and pollution, little was done in the way of addressing this issue.

If we wish to trace the reasons for this inaction, in addition to the annual reports and the issue of the smoke problem, we need to look at the SMC more broadly. First, the aforementioned reports mention the lack of a legal basis for dealing with the problem of air pollution. This speaks more generally to the problem that the SMC had great difficulties in adding new legislation due to the complex array of parties that needed to be involved.[34] As we have seen earlier, the Public Health Department was responsible for dealing with the problem of "smoke nuisance."[35] However, the lack of sufficient legal powers was a constant problem for the Department.[36] In 1933, the SCM established an Industrial Section, which was specifically concerned with the factories in the Settlement and thus could have plausibly concerned itself with the problem of air pollution caused by industry. However, it also suffered from its lack of sufficient legal authority. In addition, its dependence on the cooperation of Chinese courts made things even more difficult.[37] Clearly, insufficient legal means made it more difficult for the SMC to confront the problem of air pollution.

A second factor that needs to be taken into account is the limited scope of activities of both the Public Health Department and the Industrial Section. In the case of the Public Health Department, the SMC did not make major investments in public health. While the Department recognised that "a smoke-laden atmosphere is detrimental to health," its activities seem to have focused on other aspects of public health such as vaccination and not on air pollution.[38] In the case of the Industrial Section, the SMC was generally not particularly concerned with the state of factories in the Settlement. The Industrial Section was not only established relatively late but was also, certainly before 1937, relatively small in size. Similar to the Public Health Department, with which it did cooperate, the Industrial Section does not seem to have specifically dealt with the problem of smoke.[39] In addition to the legal limitations both the Public Health Department and the Industrial Section had to deal with, this limited support from the SMC at large for the two institutions likely also contributed to the neglect of the problem of air pollution.[40] This limited support from the SMC for the Public Health Department and the Industrial Section in turn can mainly be traced to its general tendency toward fiscal restraint, relative disinterest in public welfare, and its generally giving priority to the economy and opposing intervention in business.[41]

In general, it is, of course, not surprising that we find that a governing body dominated by business interests would not have much of an interest in concerning itself with the problem of smoke and air pollution caused by industry. However, as Jackson has shown, despite the SMC's initial limited and non-existent engagement in the fields of public health and industrial matters respectively, outside pressure from parties within and outside (including the Chinese government and international parties) of the International Settlement caused the SMC to expand its role in these fields.[42] Nevertheless, it seems that similar pressures did not exist in the case of air pollution. On the contrary, as we saw in the annual reports of the SMC discussed earlier, public opinion seems to have played a role in determining whether the problem of smoke could be addressed. Similarly, in the case of the Industrial Section, popular opposition could stymie attempts at regulating factories even if a legal basis existed.[43]

Moreover, there does not seem to have been pressure from the side of the Chinese authorities. China's own Factory Law (*Gongchang fa*) of 1929 did not specifically touch upon air pollution. The supplementary Ordinances for the Administration of the Factory Law (*Gongchang fa shixing tiaoli*), published in 1930, deal briefly with pollution in two articles. Article 26 explains that official approval was necessary for factories and workshops "when the gases or liquids generated in the process of manufacturing are detrimental to public health." Furthermore, Article 27 states that

> all poisonous gases, liquids, or other substances produced in the process of manufacturing shall be filtered, purified or dissolved, and disposed of by the factory according to their nature and quantity. They shall not be thrown into rivers, creeks, ponds, or wells, or spread about freely.[44]

However, the actual implementation of the Factory Law turned out to be difficult. As a contemporary later remembered: "The difficulty lay in the application of the code. Provisions regarding reform in factory conditions, for example, were so far reaching that early and complete conformity to the standards set could not be hoped for."[45] In an official statement on August 1, 1931, the SMC commented that "the Factory Laws promulgated by the National Government are of an extremely far reaching character, and . . . a long time must necessarily elapse before the standards at which they aim can be attained." It set out certain areas that first should be focused on, none of which pertained to air pollution.[46] Indeed, there is no evidence that the parts of the supplementary ordinances quoted earlier or the Chinese government more generally created pressure for the SMC to deal with air pollution.[47]

Furthermore, while the SMC was aware of efforts in Britain and North America to implement legal measures to control air pollution,[48] there is no evidence that it was under much international pressure to deal with air pollution in the Settlement. Finally, while complaints from people in Shanghai about business-related smoke existed and can be traced back to the late nineteenth century, they do not seem to have led to much, given how even in 1936 the ratepayers meeting at the SMC still discussed the smoke problem, as mentioned earlier.[49] Furthermore, as we have seen, the SMC's annual reports show an awareness of complaints about smoke, but this did not lead to major action being taken against air pollution. Thus, it seems that the lack of outside pressure related to the problem of air pollution was also a decisive factor in why more was not done about the problem of smoke.

3.4 The Shanghai Power Company and the China General Omnibus Company

We can further illustrate the relationship between the SMC and businesses that caused pollution if we look at two concrete examples: The Shanghai Power Company (SPC) and the China General Omnibus Company (CGOC).[50] The SPC provided electricity in the International Settlement and had originally been the SMC's Electricity Department but was sold into American private ownership in 1929. One of its main power plants was the Riverside Power Station located in the Eastern part of the International Settlement.[51] The power plants of the SPC were a major source of air pollution in Shanghai.[52] However, it is only in 1936 that the annual reports of the SMC mention the smoke emissions of the SPC. The 1936 report explains that "several complaints were received concerning the nuisance created by the smoke from the Riverside Station of the Shanghai Power Company, and in the interest of public health this difficult problem has received close attention."[53] Indeed, an article in the *North-China Herald* from July 1936 mentions the many complaints about the "nuisance created by smoke from the Riverside Station." The response of the SPC to the criticism was to explain that it was aware of the problem and had already taken measures to fix it. However, it also mentioned that "nowhere else in the world, as far as they are aware, does the problem of combating the nuisance present the same difficulties" and that it first wished to conduct experiments before committing to "large expenditure." It proposed to use washers and possibly extended chimneys to deal with the smoke problem and promised that "the experiments should be completed by November, and if successful, there will thereafter be gradual improvement in the situation over a period of 18 months." The SMC requested regular updates from the SPC.[54]

However, the experiments of the SPC with water washers proved unsuccessful. While the "most costly step" would have been the installation of chemical washers, following the example of some English power plants, the SPC opted to erect a "giant chimney" to deal with the problem and informed the SMC accordingly on June 8, 1937. Remaining defensive, the SPC in a statement explained that the smoke issue went back ten years and blamed the most recent problems "on the imposition by the Chinese government of an import duty on coal," which meant that more Chinese coal, which was richer in sulphur, had to be used.[55]

The response of the SMC to the actions of the SPC was not necessarily positive. During a Council meeting in June 1937, a Council member broke into a tirade against the SPC. He criticised that

> the trouble started in or before 1929, and . . . the Power Company stated then that the improvements contemplated would probably abate the nuisance; they have been continuing to say so ever since, but conditions now are worse than before, owing to the use of inferior coal with higher sulphur content.

In general, he felt that the SPC was merely using "technicalities" as a delaying strategy towards the SMP, but that "the result so far is precisely nil" and brought up the possibility of suing the power provider. However, the chairman was more understanding of the position of the SPC and rejected the idea of moving against the SPC legally. Not only was it unlikely to succeed, but it might mean that the SPC would change "its present co-operative attitude and . . . make no attempt to do anything further in the matter."[56]

While a long chimney of 110 metres was indeed erected in 1938, this did not solve the problem at hand.[57] Importantly, this episode shows both the legal limitations that the SMC had to work with, which we already discussed earlier, and the fact that it was dependent on cooperation with businesses like the SPC in tackling air pollution.[58] At least once a related report also mentions that the use of washers might lead to "an increase in the municipal rates."[59] Given that the SMC was interested in keeping the cost of electricity low,[60] this might also explain why the issue of smoke pollution was not pushed further.

Another instructive case study is that of the China General Omnibus Company (CGOC). The CGOC was established in 1923. It received the right to operate public buses in the International Settlement, and its passenger numbers grew steadily in the following years (Figure 3.2).[61] In its annual report for 1933, the SMC praised the CGOC for making public bus transport available to a large portion of Shanghai residents due to the low fares it charged. However, the CGOC's diesel engines, which made the cheap fares possible, had the disadvantage of generating "a certain amount of smoke from the exhausts," though the report suggests that this problem had been solved.[62] However, next year's report noted the "recurrence of [the] smoke nuisance," while expressing optimism that a solution would be found soon. It also once again stressed the positive connection between diesel engines and low fares.[63]

However, the hopes of the SMC were not fulfilled. In 1935, the smoke problem connected to the CGOC continued, and the SMC in July decided to give the CGOC three months to resolve the problem. The SMC's Commissioner for Public Works noted though that the use of better fuel, which was seen as likely being able to stop the nuisance, would mean that the CGOC would need to spend significantly more money.[64] One possible implication of such higher costs could have naturally been an increase in fares. By autumn, the CGOC's Engineer-in-Chief made promises to the SMC's Acting Commissioner of Police that "he was confident that the nuisance generally would cease." In the discussion of the Council, the issue that the use of a better fuel might help was brought up, but it was also explained that this might lead to an increase in costs for passengers. The Council ordered no concrete measures apart from a continuing investigation to monitor the progress made by the CGOC in solving the smoke problem.[65] In January 1936, the *Municipal Gazette* reported that the investigation had come to the conclusion that the CGOC was "making a genuine attempt to overcome the nuisance caused by the smoke emitted from many of their vehicles." The trouble was partly blamed on the cold temperatures. The CGOC committed to the use of a different kind of fuel with the hope that this would improve things.

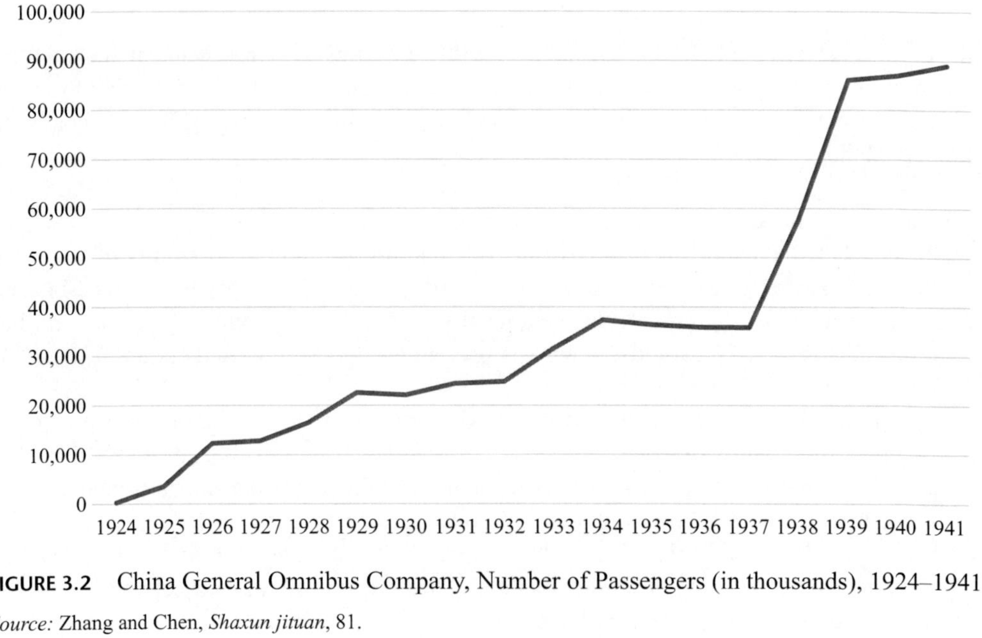

FIGURE 3.2 China General Omnibus Company, Number of Passengers (in thousands), 1924–1941

Source: Zhang and Chen, *Shaxun jituan*, 81.

However, it is unclear whether this was the better-quality fuel mentioned earlier. The Council decided not to further pursue the matter.[66]

The problem of the smoke emanating from the CGOC's vehicles did not end here, however. In July 1937, it was reported that "war has been declared by the First Special District Citizens' Association on the exhaust fumes from Settlement buses." The smoke generated by the CGOC was marked as "a grave menace to public health," and the CGOC was accused of not having taken any action to remedy the situation. The SMC was asked to pressure the CGOC to solve the problem.[67] However, by the end of 1938 and into the 1940s, the smoke problem of the CGOC's buses still remained.[68]

In the case of the CGOC, we once again see the SMC confronted with air pollution caused by a business. In its dealings with the CGOC, the SMC overall took a conciliatory stance. Mentioning of the issue of the impact the use of better-quality fuel might have on the fares the company would demand from its customers combined with the praise from the SMC for the cheap provision of public bus transport by the CGOC also suggests that the SMC might have been more concerned with maintaining a cheap public bus service than solving the problem of smoke emissions by the buses.[69] Incidentally, the possibility of bringing legal action against the CGOC seems to have been considered only once, in July 1935, and there is no evidence that this was ever pursued.[70] This might also, once again, hint at the legal limitations the SMC faced in dealing with businesses.

3.5 Conclusion

In sum, this chapter has attempted to probe the potential for research into the connection between Chinese business history and the environment, with a specific focus on the issue of pollution. As we have seen, the field of Chinese business history has so far neglected environmental factors. However, not only the examples of scholars writing in the tradition of energy and environmental history, but also the case studies from early twentieth-century Shanghai, in particular the cases of the Shanghai Power Company and the China General Omnibus Company, show the potential for incorporating the environment as a factor in studies of Chinese business history. It is thus hoped that historians working on the history of Chinese business and entrepreneurship will follow the example of their colleagues working on the connection between business and the environment in general and business and pollution in particular in Western contexts and explore the environmental side of Chinese business history. China's current problems with pollution and its role as a major contributor to climate change make the study of the history of the connection between business and the environment in China particularly pressing. At the same time, a better understanding of developments in the Chinese context with its unique conditions as an emerging market would also greatly enrich the broader historical scholarship on the relationship between business and the environment.

Notes

1 The author would like to thank Teresa da Silva Lopes, Paul Duguid, Robert Fredona, and Niels Viggo Haueter for their comments on previous drafts of this chapter.

2 Kroeber, *China's Economy*, chapter 10. For China's share of emissions, I have referred to the data and measurement methodologies for both measures (CO_2 and greenhouse gases) in Ritchie, Rosado, and Roser, "CO_2 and Greenhouse Gas Emissions." As explained there, the measurement of "greenhouse gas emissions include carbon dioxide, methane, and nitrous oxide from all sources, including land-use

change. They are measured in tonnes of carbon dioxide-equivalents over a 100-year timescale." As for CO_2 emissions, the measurement includes "carbon dioxide (CO_2) emissions from fossil fuels and industry. Land-use change is not included." More specifically, "Fossil emissions measure the quantity of carbon dioxide (CO_2) emitted from the burning of fossil fuels, and directly from industrial processes such as cement and steel production. Fossil CO_2 includes emissions from coal, oil, gas, flaring, cement, steel, and other industrial processes. Fossil emissions do not include land use change, deforestation, soils, or vegetation."

3 Austin, Dávila, and Jones, "Alternative Business History."

4 Bergquist, "Renewing Business History," 18–19. Bergquist's focus is on the second half of the twentieth century, though.

5 Bergquist, "Renewing Business History." For earlier calls for a greater inclusion of the environment in studies by business historians, see, for example, Rosen, "Industrial Ecology"; Rosen, "Doing Business History in the Age of Global Climate Change."

6 Rosen, "Pollution Regulation and Litigation."

7 Rosen, "Businessmen Against Pollution."

8 Desrochers, "Victorian Pioneers of Corporate Sustainability."

9 Bergquist and Lindmark, "Sustainability and Shared Value."

10 Stradling and Tarr, "Environmental Activism."

11 Gorman, "Oil Field Brines."

12 Smith, "Turning Silk Purses into Sows' Ears."

13 Temby, "Control and Suppression."

14 Jones and Lubinski, "Making 'Green Giants'."

15 The only exception would be that Bergquist does not explicitly mention Canada as a focus of previous work on the relationship between business and the environment, though her criticism is clearly aimed at the dearth of scholarship on the topic in non-Western contexts. See Bergquist, "Renewing Business History," 18–19.

16 See, for instance, the recent review of the relevant literature by Adam Frost: Frost, "Reframing Chinese Business History." Referencing other literature reviews and the literature on Liu Hongsheng's businesses, Muscolino, whose study of Liu Hongsheng's businesses is an important exception to this rule that will be discussed further later (including endnote 19), also notes the problem that "the historiography of modern Chinese business has heretofore paid scant attention to the ecological implications of business activities and their connections to the shifting environmental settings in which they occur." See Muscolino, "Energy and Enterprise," 160.

17 For a good overview of Chinese industrialisation since the nineteenth century, see Brandt, Ma, and Rawski, "Industrialization in China."

18 Rosen used three examples of works by environmental historians. See Rosen, "Industrial Ecology." The four examples given here are, of course, not exhaustive. As I point out later, the examples I choose largely do not engage with the literature in Chinese business history, the only exception being Muscolino's work discussed later. However, while Muscolino's research interests span several fields, including environmental and energy history, by his own description on his faculty website, business history is not among them: See "Micah Muscolino." Importantly, none of the examples listed here were published in business history journals or book series.

19 Muscolino, "Energy and Enterprise." As Muscolino notes, Liu's business activities have been previously studied by several business historians, who, however, did not use an environmental perspective. See Muscolino, "Energy and Enterprise," 160. Also, see endnote 16 for his critique of the Chinese business history scholarship more broadly.

20 Tan, *Recharging China*.

21 Seow, *Carbon Technocracy*.

22 Austin, Dávila, and Jones, "Alternative Business History."

23 Pei, "Jindai Shanghai de kongqi wuran."

24 The other parts of Shanghai were the French Concession, run by its own council, and the Chinese municipality. On the foreign presence in Shanghai and the International Settlement, see Jackson, *Shaping Modern Shanghai*; Nield, *China's Foreign Places*, 8, 196–209. On the population of the International Settlement, see Henriot et al., *The Population of Shanghai*. As the following endnotes show, in my understanding of the workings of the SMC, I am particularly indebted to the excellent work done by Isabella Jackson on the topic.

25 On industrialisation in Shanghai, see Ma, "Rise of Modern Shanghai"; Lieu, *Growth and Industrialization*.
26 Coble, *Shanghai Capitalists*, 149–50; Paulès, *Republic of China*, 166.
27 Pei, "Jindai Shanghai de kongqi wuran." Pei does mention the Shanghai Power Company and the China General Omnibus Company, but only does so briefly in passing (in the case of the Shanghai Power Company, very briefly the use of chimneys to attempt to deal with pollution is mentioned). A short book chapter by Hu Yong on air pollution in modern China only mentions the SMC briefly in passing: Hu, "Zhongguo jindai chengshi daqi wuran jiqi zhili." A government publication on the protection of the environment in Shanghai that mainly focuses on the post-1949 period also mentions the SMC occasionally briefly in its discussion of air pollution. See Lü, *Shanghai huanjing baohu zhi*, especially 192–215.
28 Jackson, *Shaping Modern Shanghai*. Jackson mentions the role of the SMC's Public Health Department in the field of water pollution, though. But it is unclear to what extent this was connected to business. See Jackson, *Shaping Modern Shanghai*, 167–68.
29 See, for example, Shanghai Municipal Council, *Report for the Year 1929*, 160. While Pei's (Pei, "Jindai Shanghai de kongqi wuran") focus is on soot (*meiyan*), he covers a wide range of air pollution, including the problem of smoke (which he seems to translate into Chinese as *yanwu*. See his translated quotation from "The Smoke Nuisance in Shanghai," *North-China Herald*, August 14, 1896, in Pei, "Jindai Shanghai de kongqi wuran," 54).
30 Shanghai Municipal Council, *Report for the Year 1927*, 191–92. Jackson also mentions that the Industrial Section and Public Health Department had to rely on cooperation and persuasion due to their limited legal authority. See Jackson, *Shaping Modern Shanghai*, introduction and Chapters 4–5.
31 Between 1935 and 1937, the relevant statistics given in the annual reports of the SMC only refer to "Smoke Nuisances Investigated," so I have not included these years in Figure 3.1. See Shanghai Municipal Council, *Report for the Year 1935*, 191; Shanghai Municipal Council, *Report for the Year 1936*, 170; Shanghai Municipal Council, *Report for the Year 1937*, 180.
32 For the annual reports covered in this paragraph, see Shanghai Municipal Council, *Report for the Year 1928*; Shanghai Municipal Council, *Report for the Year 1929*; Shanghai Municipal Council, *Report for the Year 1930*; Shanghai Municipal Council, *Report for the Year 1931*; Shanghai Municipal Council, *Report for the Year 1932*. The issue of water vehicles emanating smoke is also mentioned in other annual reports covered in this section. However, due to space reasons and because this does not necessarily pertain to businesses, I do not further cover these here. Also see endnote 50.
33 For the reports discussed in this paragraph, see Shanghai Municipal Council, *Report for the Year 1933*; Shanghai Municipal Council, *Report for the Year 1934*; Shanghai Municipal Council, *Report for the Year 1935*; Shanghai Municipal Council, *Report for the Year 1936*; Shanghai Municipal Council, *Report for the Year 1937*. The issue of houses being converted into factories had already been touched upon in Shanghai Municipal Council, *Report for the Year 1929*.
34 Jackson, *Shaping Modern Shanghai*, 175. The legal basis for the International Settlement was the Land Regulations and the attached bye-laws. See Jackson, *Shaping Modern Shanghai*, 5; Hinder, *Life and Labour*, 3–4; Kotenev, *Shanghai: Its Mixed Court and Council*, 557–75.
35 On the Commissioner of Public Health and the Department of Public Health being responsible for the "abatement of nuisances" in general, including "smoke nuisances," also see Shanghai Municipal Council, *Report for the Year 1933*, 134–202. On the Public Health Department, also see Jackson, *Shaping Modern Shanghai*, Chapter 3. Jackson also notes that the Public Health Department originated in part due to the need for a "nuisance department." See Jackson, *Shaping Modern Shanghai*, 167.
36 Jackson, *Shaping Modern Shanghai*, Chapter 4.
37 Jackson, *Shaping Modern Shanghai*, Chapter 5.
38 The quote is from the Shanghai Municipal Council, *Report for the Year 1929*, 160–61. On the low priority of public health for the SMC and the activities of the Department, see Jackson, *Shaping Modern Shanghai*, Chapter 4. Jackson does not discuss the problem of air pollution in her examination of the Public Health Department, so it does not seem to have been a major task of the Department.
39 On the activities of the Industrial Section, see *Shaping Modern Shanghai*, Chapter 5. Jackson does not mention smoke as a task of the Industrial Section. The SMC annual report for 1936 suggests that cooperation between the Public Health Department and the Industrial Section existed but also that by then the Industrial Section took the lead in issues relating to industry. See Shanghai Municipal Council, *Report for the Year 1936*, 174.

40 As discussed earlier, in her treatment of the SMC, Jackson does not discuss air pollution in Shanghai. However, in her discussion of the Public Health Department and the Industrial Section, she mentions not only the limitations of the law towards their activities but also the fact that they were limited by the limited support from the SMC. See Jackson, *Shaping Modern Shanghai*, Chapter 4–Chapter 5.

41 On the SMC's attitudes towards finance and public welfare, see Jackson, *Shaping Modern Shanghai*, 28–31, 173–74. In her chapter on the Public Health Department, Jackson also stresses the SMC's "ever-present concern for economy," that "public health provision was never a priority for the laissez-faire Council" and how relatively little funding it allocated to the Department (as also mentioned earlier). In her chapter on the Industrial Section, Jackson does not comment directly on funding for the Industrial Section but (as already discussed earlier) shows the general disinterest of the SMC in the conditions of industry and explains that only in 1940 was the Industrial Section transformed into a proper division "with its own committee to formulate policy and oversee its operations." Moreover, she mentions the SMC's dislike of interference in business due to its main championing of commercial interests. More broadly, she sees this dislike among ratepayers (and thus of the SMC) as one of the "major blocks to socially progressive policies throughout the existence of the Settlement" (the other factors being the SMC's legal framework and "apathy, [and] conservatism" of ratepayers). See Jackson, *Shaping Modern Shanghai*, Chapter 4–Chapter 5 (quotations on 201, 214, 228).

42 Jackson, *Shaping Modern Shanghai*, introduction and chapters 4–5. As already pointed to in previous endnotes, Jackson's book in general highlights the tension between the SMC representing business interests and other interests, and its reluctance on burdening business.

43 Jackson, *Shaping Modern Shanghai*, 219–20.

44 For the Chinese text and an English translation of the Factory Law and the Ordinances for the Administration of the Factory Law, see Koo, *China's Labor Laws* (translations of the names of the two laws, "Factory Law" and "Ordinances for the Administration of the Factory Law," and of the quoted passages are also from Koo). On the impact the Factory Law had on the SMP and its industrial work in other areas, see Jackson, *Shaping Modern Shanghai*, Chapter 5. Earlier, "Chinese Provisional Factory Regulations, promulgated in 1923, had been a dead letter and had never been actually enforced in any part of China." See Shanghai Municipal Council, *Report for the Year 1931*, 324.

45 Hinder, *Life and Labour*, 6. Also see Jackson, *Shaping Modern Shanghai*, 216. Jackson writes that "the city government's Bureau of Social Affairs declared 17 clauses of the law impracticable . . . those measures that they considered the most desirable and achievable were published for the information of factory owners." She cites the *Shenbao* issue from November 12, 1931, for this information. However, I have not been able to find a relevant article in this issue of *Shenbao*. However, an article on the Factory Law appeared in *Shenbao* on the following day. See "Xingzheng Yuan jieshi benshi dui Gongchang fa yijian," *Shenbao*, November 13, 1931. In the article, the Bureau of Social Affairs raises certain issues with the Factory Law, but pollution or environmental factors are not raised.

46 "Factory Laws," *The Municipal Gazette: Being the Official Organ of the Executive Council for the Foreign Settlement of Shanghai* 24, 337, August 7, 1931.

47 In 1935, the Chinese government issued a set of detailed regulations on factory health and safety regulations. These are mentioned in the SMC annual reports twice. In its 1935 report, the SMC mentions the promulgation of the regulations but also notes that "achievement of any better standard of safety or of hygiene will, in any case, be a long process." In its 1936 report, the SMC mentions that these regulations had been disseminated and read out to factories. These regulations touch upon pollutants, but only insofar as they pertain to working conditions within the factory and not to public health. The only potential exception one might see as being aimed at pollution is paragraph 50, which explains that factories should not allow the accumulation of waste that might lead to a foul odor, but I have, in any case, not found evidence of this clause impacting the SMC. See Gu, *Zhongguo laodong faling*, 132–41; Shanghai Municipal Council, *Report for the Year 1935*, 37–38; Shanghai Municipal Council, *Report for the Year 1936*, 32.

48 Shanghai Municipal Council, *Report for the Year 1927*, 191–92. Apart from discussing legal measures, the report only draws on the British example of urban zoning, which was already mentioned as a suggestion earlier. Jackson explains in the context of public health that the West was something the SMC looked at as a standard but did not necessarily plan to reach. See Jackson, *Shaping Modern Shanghai*, 168.

49 For complaints about business-related smoke, see, for instance, "The Smoke Nuisance in Shanghai," *North-China Herald*, August 14, 1896; "The Smoke Nuisance," *North-China Herald*, August 27, 1921; Pei, "Jindai Shanghai de kongqi wuran," 51–53 also notes instances of complaints against smoke (one

of which also goes back to the nineteenth century) that seemed not to have led to any remedy (including that of the CGOC discussed later). Hu also mentions two incidents of business-related pollution for the late nineteenth century (in addition, it also mentions the CGOC case discussed later). See Hu, "Zhongguo jindai chengshi daqi wuran jiqi zhili," 609–10, 612. Also see some of the cases mentioned in Lü, *Shanghai huanjing baohu zhi*, 12–19, 192–215. The last two sources mention certain cases where the sources of pollution could be persuaded to take measures against pollution (which, however, were not necessarily successful). More broadly, Pei sees a generally ascending trend in air pollution in Shanghai from the first Opium War (1839–1842) to 1937, with the early 1930s representing the apex of air pollution in pre-war Shanghai. See Pei, "Jindai Shanghai de kongqi wuran," 76.

50 As mentioned earlier, steamships also seem to have been a major generator of the smoke problem. However, for space reasons and because steamships are not definitely business-related, I do not touch upon them here.

51 Tan, *Recharging China*, Chapter 1; North-China Daily News and Herald, Limited, *Plan of Shanghai.*

52 Pei, "Jindai Shanghai de kongqi wuran," 49–50. Tan also suggests that the Riverside Power Station would have caused air pollution. See Tan, *Recharging China*, 18.

53 SMC, *Report for the Year 1936*, 174.

54 "Smoke Nuisance at Power Station," *North-China Herald*, July 29, 1936.

55 "Giant Chimney for Power Plant," *North-China Herald*, June 23, 1937.

56 Shanghai shi danganguan, *Gongbuju dongshihui huiyilu*, 160–63.

57 Pei, "Jindai Shanghai de kongqi wuran," 50–51.

58 This chimes with Jackson's emphasis that the Industrial Section and Public Health Department also had to rely much on persuasion and cooperation. See note 29.

59 "Giant Chimney for Power Plant," *North-China Herald*, June 23, 1937.

60 Both Tan and Jackson suggest that the SMC was interested in supplying cheap electricity: See Tan, *Recharging*, 32; Jackson, *Shaping Modern Shanghai*, 47. Jackson states that investment in the SPC allowed the SMC to push for lower rates. However, the controlling interest of the SPC lay with American investors. See Remer, *Foreign Investments in China*, 286–88. Still, Remer also explains that according to the franchise given by the SMC to the SPC, "Dividends above 10 per cent are to be divided into two parts, one of which is to bring about reduction in rates." This, again, emphasises the SMC's interest in keeping electricity costs low.

61 On the CGOC, see Zhang and Chen, *Shaxun jituan*, 73, 79–80.

62 Shanghai Municipal Council, *Report for the Year 1933*, 56. Pei suggests that the problem of the CGOC's smoke emissions went back to the late 1920s. See Pei, "Jindai Shanghai de kongqi wuran," 53.

63 Shanghai Municipal Council, *Report for the Year 1934*, 36.

64 "Bus Company Is Ordered to Abolish Fumes," *The China Press*, July 11, 1935.

65 "Omnibus Smoke Nuisance," *Municipal Gazette of the Council for the Foreign Settlement of Shanghai* 18, 1572, November 8, 1935.

66 "Emission of Smoke From Omnibuses," *Municipal Gazette of the Council for the Foreign Settlement of Shanghai* 19, 1581, January 3, 1936.

67 "Smoke Nuisance in Shanghai Said Going from Bad to Worse," *The China Press*, July 24, 1937.

68 "Leading Article Criticized," *North-China Herald*, December 14, 1938. On the manifold complaints against the CGOC that, however, did not lead to a solution of the problem, see also Pei, "Jindai Shanghai de kongqi wuran," 53. Hu shows that the problem persisted into the 1940s, and the SMC asked the CGOC to invest more money in technological improvements. See Hu, "Zhongguo jindai chengshi daqi wuran jiqi zhili," 610.

69 When briefly discussing the CGOC case, Pei suggests that the dragging on of the CGOC's smoke problem was "due to it involving the interests of several parties" and because "one needed to take into account the economic costs of the operation of the buses and the technical difficulties of realising the substitution of energy." See Pei, "Jindai Shanghai de kongqi wuran," 53. However, he does not further elaborate and does not seem to make this statement specifically with regard to the SMC. From the evidence I have been able to find, the maintenance of cheap fares seems to have been the most likely reason causing the conciliatory stance of the SMC.

70 It is mentioned in "Bus Company Is Ordered to Abolish Fumes," *The China Press*, July 11, 1935. Incidentally, the traffic regulations of the International Settlement included a clause against "excessive discharge of smoke from the exhaust of a motor vehicle on a highway." See Shanghai Municipal Council, *Municipal Report for 1920*, 81A–84A.

Bibliography

Austin, Gareth, Carlos Dávila, and Geoffrey Jones. "The Alternative Business History: Business in Emerging Markets." *Business History Review* 91, no. 3 (2017): 537–69.

Bergquist, Ann-Kristin. "Renewing Business History in the Era of the Anthropocene." *Business History Review* 93, no. 1 (2019): 3–24.

Bergquist, Ann-Kristin, and Magnus Lindmark. "Sustainability and Shared Value in the Interwar Swedish Copper Industry." *Business History Review* 90, no. 2 (2016): 197–225.

Brandt, Loren, Debin Ma, and Thomas G. Rawski. "Industrialization in China." In *The Spread of Modern Industry to the Periphery Since 1871*, edited by Kevin O'Rourke and Jeffrey Williamson, 197–228. Oxford: Oxford University Press, 2017.

The China Press.

Coble, Parks M. *The Shanghai Capitalists and the Nationalist Government, 1927–1937*. Cambridge, MA: Harvard University Press, 1980.

Desrochers, Pierre. "Victorian Pioneers of Corporate Sustainability." *Business History Review* 83 (2009): 703–29.

Frost, Adam K. "Reframing Chinese Business History." *Business History Review* 96, no. 2 (2022): 245–87.

Gorman, Hugh S. "Efficiency, Environmental Quality, and Oil Field Brines: The Success and Failure of Pollution Control by Self-Regulation." *Business History Review* 73, no. 4 (1999): 601–40.

Gu Bingyuan [Koo Ping Yuen]. *Zhongguo laodong faling huibian*. 3rd ed. Shanghai: Shanghai faxue bianyishe, 1937.

Henriot, Christian, Shi Lu, and Charlotte Aubrun. *The Population of Shanghai (1865–1953): A Sourcebook*. Leiden: Brill, 2019.

Hinder, Eleanor M. *Life and Labour in Shanghai: A Decade of Labour and Social Administration in the International Settlement*. New York: International Secretariat, Institute of Pacific Relations, 1944.

Hu Yong. "Zhongguo jindai chengshi daqi wuran jiqi zhili." In *Zhongguo shilunji*, edited by Chen Feng, 603–15. Beijing: Shehui kexue wenxian chubanshe, 2019.

Jackson, Isabella. *Shaping Modern Shanghai: Colonialism in China's Global City*. Cambridge: Cambridge University Press, 2017.

Jones, Geoffrey, and Christina Lubinski. "Making 'Green Giants': Environment Sustainability in the German Chemical Industry, 1950s–1980s." *Business History* 56, no. 4 (May 19, 2014): 623–49.

Koo, Ping Yuen. *China's Labor Laws, 1929–1931. Chinese Text with English Translation*. Shanghai: The Commercial Press, 1931.

Kotenev, A.M. *Shanghai: Its Mixed Court and Council*. Shanghai: North-China Daily News & Herald, 1925.

Kroeber, Arthur R. *China's Economy: What Everyone Needs to Know*. 2nd ed. Oxford: Oxford University Press, 2020.

Lieu, D.K. *The Growth and Industrialization of Shanghai*. Shanghai, 1936, reprint. Beijing: Shangwu yinshuguan, 2015.

Lü, Shuping, ed. *Shanghai huanjing baohu zhi*. Shanghai: Shanghai shehui kexue yuan chubanshe, 1998.

Ma, Debin. "The Rise of Modern Shanghai, 1900–1936: An Institutional Perspective." In *The Treaty Port Economy in Modern China: Empirical Studies of Institutional Change and Economic Performance*, edited by Billy K.L. So and Ramon H. Myers, 33–46. Berkeley: Institute of East Asian Studies, 2011.

"Micah Muscolino." https://history.ucsd.edu/people/faculty/muscolino.html.

The Municipal Gazette: Being the Official Organ of the Executive Council for the Foreign Settlement of Shanghai. Municipal Gazette of the Council for the Foreign Settlement of Shanghai.

Muscolino, Micah. "Energy and Enterprise in Liu Hongsheng's Cement and Coal-Briquette Businesses, 1920–1937." *Twentieth-Century China* 41, no. 2 (2016): 159–79.

Nield, Robert. *China's Foreign Places: Foreign Presence in China in the Treaty Port Era, 1840–1943*. Hong Kong: Hong Kong University Press, 2015.

North-China Daily News and Herald, Limited. *Plan of Shanghai Published by the North-China Daily News and Herald, Limited*. Shanghai: North-China Daily News and Herald, 1928.

Paulès, Xavier. *The Republic of China: 1912 to 1949*. Translated by Lindsay Lightfoot. Cambridge: Polity Press, 2023.

Pei Guangqiang. "Jindai Shanghai de kongqi wuran jiqi yuanyin tanxi—yi meiyan wei zhongxin de kaocha." *Zhongyang yanjiuyuan jindaishi yanjiusuo jikan* 97 (2017): 45–86.

Remer, C.F. *Foreign Investments in China*. New York: Macmillan, 1933.

Ritchie, Hanna, Pablo Rosado, and Max Roser. "CO_2 and Greenhouse Gas Emissions." *OurWorldInData*, 2023. https://ourworldindata.org/co2-and-greenhouse-gas-emissions.

Rosen, Christine Meisner. "Businessmen Against Pollution in Late Nineteenth Century Chicago." *Business History Review* 69, no. 3 (1995): 351–97.

Rosen, Christine Meisner. "Industrial Ecology and the Greening of Business History." *Business and Economic History* 26, no. 1 (1997): 123–37.

Rosen, Christine Meisner. "Doing Business History in the Age of Global Climate Change." *Enterprise & Society* 8, no. 2 (2007): 221–26.

Rosen, Christine Meisner. "The Role of Pollution Regulation and Litigation in the Development of the U.S. Meatpacking Industry, 1865–1880." *Enterprise & Society* 8, no. 2 (2007): 297–347.

Seow, Victor. *Carbon Technocracy: Energy Regimes in Modern East Asia*. Chicago: University of Chicago Press, 2023.

Shanghai Municipal Council. *Municipal Report for 1920 and Budget for 1921*. Shanghai: Kelly & Walsh, 1921.

Shanghai Municipal Council. *Municipal Report for 1922 and Budget for 1923*. Shanghai: Kelly & Walsh, 1923.

Shanghai Municipal Council. *Report for the Year 1921 and Budget for the Year 1922*. Shanghai: Kelly & Walsh, 1922.

Shanghai Municipal Council. *Report for the Year 1923 and Budget for the Year 1924*. Shanghai: Kelly & Walsh, 1924.

Shanghai Municipal Council. *Report for the Year 1924 and Budget for the Year 1925*. Shanghai: Kelly & Walsh, 1925.

Shanghai Municipal Council. *Report for the Year 1925 and Budget for the Year 1926*. Shanghai: Kelly & Walsh, 1926.

Shanghai Municipal Council. *Report for the Year 1926 and Budget for the Year 1927*. Shanghai: Kelly & Walsh, 1927.

Shanghai Municipal Council. *Report for the Year 1927 and Budget for the Year 1928*. Shanghai: Kelly & Walsh, 1928.

Shanghai Municipal Council. *Report for the Year 1928 and Budget for the Year 1929*. Shanghai: Kelly & Walsh, 1929.

Shanghai Municipal Council. *Report for the Year 1929 and Budget for the Year 1930*. Shanghai: Kelly & Walsh, 1930.

Shanghai Municipal Council. *Report for the Year 1930 and Budget for the Year 1931*. Shanghai: Kelly & Walsh, 1931.

Shanghai Municipal Council. *Report for the Year 1931 and Budget for the Year 1932*. Shanghai: Kelly & Walsh, 1932.

Shanghai Municipal Council. *Report for the Year 1932 and Budget for the Year 1933*. Shanghai: Kelly & Walsh, 1933.

Shanghai Municipal Council. *Report for the Year 1933 and Budget for the Year 1934*. Shanghai: Kelly & Walsh, 1934.

Shanghai Municipal Council. *Report for the Year 1934 and Budget for the Year 1935*. Shanghai: Kelly & Walsh, 1935.

Shanghai Municipal Council. *Report for the Year 1935 and Budget for the Year 1936*. Shanghai: North-China Daily News & Herald, Ltd., 1936.

Shanghai Municipal Council. *Report for the Year 1936 and Budget for the Year 1937*. Shanghai: North-China Daily News & Herald, Ltd., 1937.

Shanghai Municipal Council. *Report for the Year 1937 and Budget for the Year 1938*. Shanghai: North-China Daily News & Herald, Ltd., 1938.

Shanghai shi danganguan, comp. *Gongbuju dongshihui huiyilu di ershi qi ce. The Minutes of the Shanghai Municipal Council. Vol. 27*. Shanghai: Shanghai guji chubanshe, 2001.

Smith, John K. "Turning Silk Purses into Sows' Ears: Environmental History and the Chemical Industry." *Enterprise & Society* 1, no. 4 (2000): 785–812.

Stradling, David, and Joel A. Tarr. "Environmental Activism, Locomotive Smoke, and the Corporate Response: The Case of the Pennsylvania Railroad and Chicago Smoke Control." *Business History Review* 73, no. 4 (1999): 677–704.
Tan, Ying Jia. *Recharging China in War and Revolution, 1882–1955*. Ithaca: Cornell University Press, 2021.
Temby, Owen. "Control and Suppression in Sarnia's Chemical Valley During the 1960s." *Enterprise & Society* 21, no. 2 (2020): 380–412.
Zhang Zhongli, and Chen Zengnian. *Shaxun jituan zai jiu Zhongguo*. Beijing: Renmin Chubanshe, 1985.

4

FORESTS FOR SALE

Multinationals and Civil Society's Oversight in the Palm Oil Agribusiness—Wilmar International

Shakila Yacob

4.1 Introduction

The following passage is an extract taken from an open letter addressed to the European Commission in 2019. It was signed by 200 stakeholders based in Indonesia, including indigenous peoples' organisations, farmers' unions, civil society organisations (CSOs), traditional communities, farmers, labourers, human rights defenders, and environmental activists.

> We read and receive many contradictory messages and statements that seem to ignore the impact of the palm oil industry in the form of deforestation, land grabs, human rights violations, labor exploitation, corruption, socio-economic and political problems and ecological problems . . . the palm oil plantation and industry have indeed destroyed and eliminated forests on a large scale, . . . We have also lost our village lands, our food sources, livelihood, sources of herbal medicine, rare vegetation and . . . our collective ways of life based on our local wisdom, . . . identity and cultural heritage . . . we experience conflicts, harassment, criminalization and human rights violations, the denial of labor rights and socio-economic and environmental problems . . . suffer long-term ecological disasters, such as droughts, forest fires, water pollution and the decline of fish stocks, loss of food security all of which diminish the quality of life of the local people, especially women and children.[1]

Indonesia, in 2023, maintained its position as the largest producer and exporter of palm oil globally, controlling a significant 54% share of the world market.[2] The extract also highlights the critical condition of the palm oil agribusiness which, if not duly addressed, has detriments on climate change, the society, and the biosphere.[3] The palm oil agribusiness has been identified as a significant driver of deforestation which releases stored carbon and increases environmental challenges.[4] In the late 1990s, the situation was particularly grave in Indonesia, affecting the once well-preserved tropical rainforests of Kalimantan, Papua, and some parts of Sumatra. For instance, it caused the marginalisation of smallholders in the palm oil supply chain, who had no control over pricing and land ownership, and no access to capital and knowledge. Furthermore, funds intended for smallholders were used to subsidise large corporations.[5] The palm oil

DOI: 10.4324/9781003478089-5

industry has caused various impacts on the environment, farmers, labourers, and the rights of villagers and indigenous people. With big investors exerting significant control over palm oil resources, the government appears to prioritise the interests of financiers over those of smallholders and communities.

The open letter referenced reports from CSOs: TuK Indonesia (2015, updated in 2018)[6] and the Auriga Foundation (2020),[7] both of which exposed the dominance of a few wealthy multinational enterprises (MNEs) and tycoons noted in the palm oil supply chain such as Sinar Mas, Musim Mas, and Wilmar. Due to the influence gained from their global operations in Singapore, Malaysia, and Indonesia, these MNEs and tycoons (individuals with great influence, power, and wealth) were able to exert control on both the upstream and downstream activities of the palm oil value chain.[8] Consequently, they also received preferential treatment from the Indonesian government; they were viewed as important contributors to the foreign exchange revenue. Even subsidies provided by the Plantation Fund Management Agency were extended to these corporations, one of which was the Wilmar Group.[9] Evidently, this practice came at the expense of smallholders, indigenous communities, and the environment.[10] The letter noted that Wilmar and the Sinar Mas Group were controlling large landholdings in Indonesia because of state support and illegal practices. They were among the 25 MNEs and 29 tycoons controlling more than five million hectares of land in Indonesia.[11] The figure quoted surpasses the land ownership restrictions imposed by the Agriculture Minister's Regulation of No. 26 of 2007, which limits corporate organisations from acquiring no more than 100,000 ha per province.[12]

The discourse highlights tensions in the palm oil supply chain, including government-business collusion, conflicts between big businesses and communities, and citizen-government tensions. It also underscores the challenge of balancing business interests with social and environmental responsibilities, emphasising the crucial role of CSOs in monitoring the industry amidst these disputes.[13] CSOs oversee corporate and government activities, campaign for marginalised groups such as smallholders and indigenous peoples,[14] and collaborate with or challenge corporate leaders and government officials.[15] However, their pressure alone has been insufficient to ensure accountability. Between 1998 and 2000, Indonesia's democratisation triggered a surge in civil societies, totalling 70,000[16] in number, including environmental and women's rights groups,[17] but internal conflicts,[18] competition for funding,[19] ineffective coalitions, and reliance on foreign donors[20] had weakened their collective efforts for accountability.

The rise of palm oil in Indonesia and Malaysia, another important producer of palm oil, which together account for 83% of the world's palm oil production,[21] is set against the backdrop of broader societal changes that are linked to the notion of success and economic identity. The palm oil industry is a symbol of triumph for both countries, serving as a strategic commodity for economic growth. In the context of Malaysia, its integral role in the nation's economic fabric is even emphasised on the face of the Malaysian 50 *ringgit* banknote. Its success story, ironically, has roots in colonialism in the early twentieth century and activism against trans-fat oils,[22] underscoring the interconnectedness of palm oil's prominence and environmental activism.

To navigate this complex interplay between palm oil agribusiness and environmental activism, it is necessary to delve into its historical evolution to understand the symbiotic relationship existing between local environmental movements and their global counterparts. Criticism of Western CSOs imposing environmental policies on Southeast Asian nations, particularly in the palm oil sector, intersects with the struggle of Indonesian environmentalists against the nation's powerful oligarchy. This reflects the complex dynamics within the industry, where concerns about colonial legacies and economic consequences intertwine with domestic power struggles, shaping the environmental discourse and activism in the region.[23] These dynamics have led to numerous academic works addressing these regional concerns.[24]

In the post-independence era of Malaysia and Indonesia, palm oil was utilised to address poverty and social inequality, amidst overarching concerns of environmental preservation and climate change.[25] Undoubtedly, MNEs, government-linked companies (GLCs), smallholders, and other stakeholders have contributed to the multifaceted nature of this issue, but this chapter specifically centres on Wilmar International and its palm oil business practices since the early 1990s, examining its impact on biodiversity, climate change, and deforestation allegations. Thus, the aim of this chapter is to answer the following question: How significant has civil society oversight been in ensuring the accountability of MNEs in terms of sustaining environmental responsibility and preventing deforestation?

Founded in 1991 by William Kuok Khoon Hong and Martua Sitorus, Wilmar International Limited (hereafter Wilmar) quickly became a powerhouse in the palm oil agribusiness. Originally focusing on marketing edible oils and oilseeds, Wilmar rapidly integrated backward into palm oil plantations across Indonesia and Malaysia.[26] By 2006, Wilmar debuted on the Singapore Exchange (SGX),[27] witnessing its market capitalisation surge from SGD 100,000 in 1991 to SGD 20.3 billion by 2007, driven by the Kuok Group's strategic merger, particularly its stake in PBB Oil Palms Berhad, making them Wilmar's top shareholder with an 18.3% share.[28] Wilmar has been a member of the Roundtable on Sustainable Palm Oil (RSPO) since 2004 and was the first to adopt the "No Deforestation, No Peat, No Exploitation Policy" (NDPE) in 2013.[29]

Leveraging the Kuok Group's business networks, Wilmar expanded globally, with ventures like Adani Wilmar in India and Yihai Kerry Arawana Holdings Co. Ltd. in China. By 2022, over half of Wilmar's sales and 40% of its profits came from China.[30] As the world's largest palm oil supplier, responsible for 40% of global trade since 2019,[31] Wilmar closely collaborates with Musim Mas and Sinar Mas, palm oil producers in Indonesia. Collectively, these three business groups owned over 50% of refinery capacity by 2020, effectively controlling Indonesia's palm oil exports.[32] A recent study suggests that a cartel among these groups controlled at least 60% of the Indonesian cooking oil market, producing over 150 million litres per month in 2021.[33] By 2023, Wilmar held a palm oil landbank of 354,330 ha and cultivated 232,053 ha globally, an area roughly the size of Luxembourg, with two-thirds of this land situated in Indonesia.[34] Wilmar secured the 174th spot on the Fortune Global 500 list in 2023,[35] second only to Cargill in the agribusiness sector. Its market capitalisation surged further to SGD 19.48 billion by May 2024.[36] It boasts an extensive distribution network covering Indonesia, China, India, and 50 other countries.[37] Given its dominant position, Wilmar naturally faced intense scrutiny from the CSOs, notably Greenpeace.

This chapter draws on the reports of civil societies, ranging from local to international green advocacy groups like Greenpeace, Mongabay, the Zoological Society of London, and China Dialogue. Evidence from academic research is also provided as an impartial assessment of Wilmar's sustainability commitments and initiatives. The documents being examined focus on the period from the mid-1990s to the present, highlighting Wilmar's active cultivation of palm oil and its control over the supply chain during this time frame.

Section 4.1 looks at the historical perspective of the colonial legacy and its impact on deforestation of the rubber industry in Malaya since the 1920s. Section 4.2 briefly discusses post-colonial initiatives aimed at reducing poverty such as the promotion of oil palm monoculture. Homegrown businesses also played an instrumental role during this period. Section 4.3 presents a literature review on the recent history of business, deforestation, and climate change within the palm oil agribusiness sector. Section 4.4 reviews the role of CSOs, primarily green advocacy groups, including both local and international chapters of CSOs like Greenpeace, in monitoring palm oil agribusiness activities. Since the late 1990s, these advocates have been actively holding palm oil giants accountable.[38] Section 4.5 provides the history of Wilmar's deforestation efforts and climate change concerns. It examines criticisms directed at Wilmar from 1991 to the present, including allegations of negative impacts on biodiversity and local communities' livelihoods, monopolistic behaviour, and greenwashing practices. The chapter concludes by providing insights into the role of CSOs and their interactions with major palm oil corporations such as Wilmar in the context of climate change mitigation.

4.2 Colonial Legacy and Deforestation

The East India Company founded the first botanical garden in Penang in 1794, followed by the establishment of the Singapore Botanic Gardens in 1859.[39] Initially reminiscent of private British parks, these gardens transitioned under government oversight in the late 1800s due to financial constraints, eventually becoming hubs for imperial botany and plant acclimatisation.[40] Plants from other British colonies such as Ceylon and Nigeria were introduced, and climate suitability experiments were conducted.[41] In 1883, the Forest Department of the Straits Settlements was established, and under the guidance of its second director, H. N. Ridley, a centralised forest department was formed in 1901 to manage forests in the Federated Malay States (FMS) and Straits Settlements.[42] The establishment of the Department of Agriculture in 1903 promoted agricultural techniques favouring large estates, leading to extensive deforestation and sparking conflicts over environmental management between bureaucrats and botanists amidst a shift from scientific to industrial priorities in the FMS.[43]

British officials prioritised commercialisation by supporting rubber corporations, with profit-driven motives favouring capitalist investments over scientific knowledge, leading to ecological consequences.[44] Between 1900 and 1914, rubber acreage in the FMS surged from 2,428 ha to 472,672 ha under the Department of Forestry, with over 914,590 ha of forests cleared for rubber plantations by 1922, ultimately resulting in biodiversity loss and the decline or extinction of many species.[45] The drive for industrial dominance integrated the FMS into the global economy from the 1920s onwards and profoundly shaped environmental exploitation throughout the twentieth century.[46]

Colonial power, rather than capital, technology, or plant selection, facilitated Southeast Asia's rise as the leading palm oil exporter by the late 1930s by providing land and labour.[47] By 1930, 96% of British investments in Malaya focused on tin and rubber, supported by a liberal investment regime[48] offering leases ranging from 75 years (Sumatra) or even 999 years to perpetuity (Perak, FMS).[49] The FMS land policy favoured foreign investors, particularly the British, offering attractive land allocation and loan terms while discriminating against native Malays.[50] This fuelled Malaya's rapid economic growth, with GDP doubling from 1870 to 1929, although industrialisation remained minimal.[51] British administrators generated "super-profits" through extensive exploitation, notably in the rubber industry, transforming British Malaya into a "cash cow" for British investors.[52] A study of 17 companies operating plantations in the Netherlands Indies (currently Indonesia) from 1919 to 1958 showed that there was a significantly higher rate of return on foreign investments than on domestic investments during the interwar period.[53] Colonial institutions brought premium returns for the European investors, but those high returns were based on risky investment environments, shaped by fluctuating rubber and tin prices.[54]

By 1921, the rubber price hit its lowest point since 1900 due to U.S. import restrictions.[55] Officials in the FMS and Sumatra urged diversification into oil palm as rubber prices fell in the early 1920s.[56] Incentives such as pioneer status, discounted rents, and subsidies were offered. However, caution was applied, given past exploitation in the rubber industry.[57] No such incentives were extended to native rubber smallholders. Unlike the Dutch government, British administrators enforced the Stevenson's Restriction Scheme (1922–8) and the International Rubber Regulation Agreement (1934–42) to control production and raise prices, which resulted in discrimination against native smallholders.[58] British estate interests persistently advocated for relocating peasants to less competitive and less profitable crops. A ban on new planting hindered peasant entry, and even after its lifting in 1947, the government declined to allocate additional land for rubber smallholdings.[59] Another group disregarded by the British bureaucrats and business owners was the rights of indigenous peoples to forest lands from the early 1900s, viewing such considerations as disruptive.[60] MNEs followed colonial precedents, with weak enforcement of regulations protecting native smallholders and indigenous populations.[61]

4.3 Post-colonialism: Smallholder, Poverty Alleviation, and Palm Oil Monoculture

During the colonial era, British authorities viewed Malay peasants as low-cost labour, perceiving them as conservative and lacking in traits for economic progress. However, decolonisation measures promoting the rubber smallholding sector showed the Malay peasants' capability for innovation and driving rural Malaya's development.[62] The Malaysian government's emphasis on land reforms and development to alleviate poverty among the poor led to the establishment of institutions such as the Federal Land Development Authority (FELDA) in 1956.[63] With an initial capital of RM10 million, FELDA provided state-sponsored land settlement schemes to landless farmers and actively promoted oil palm cultivation. It governed the longest-running and largest land resettlement programme in Malaysia for over 32 years (1956–90), with 422 schemes, 470,998 ha of settler smallholdings, and 340,142 ha of commercial plantations.[64] FELDA has remained Malaysia's largest investor in the public sector, maintaining a significant presence. Currently, it oversees a total area of 344,714 hectares planted with oil palms in Malaysia and Indonesia,[65] and by 2020, controls 39% of Malaysia's oil palm plantations owned by smallholders.[66] It played a crucial role in poverty reduction efforts, decreasing poverty from 50% in the 1960s to less than 5% by 2020.[67] With half a million of direct workers on palm oil plantations, and an additional 290,000 workers supporting downstream businesses, the palm oil industry has propelled the nation's economic growth for a century and, among other spillovers, provided a vast network of employment opportunities.[68]

Similar developments were also unfolding in the private sector plantations, reflecting competing interests. A key player in the early 1900s tea trade, Harrisons & Crosfield (H&C) expanded into rubber plantations in British Malaya in 1903 and the Dutch East Indies in 1906. By 1914, it had acquired equity shares in around 40 plantation companies.[69] The company faced stiff competition from its Malayan rival, Guthrie Corporation (Guthries), which entered the palm oil industry in 1924. By 1931, Guthries dominated much of Malaya's oil palm industry, controlling one-third of the planted area. They also established research facilities on their oil palm estate, introducing a new species of oil palm after 1945.[70] H&C became Malaysia's major player in the rubber industry in the 1970s, after Guthries. In 1921, it established a rubber research station in Prang Besar, followed by an oil palm research station in Banting, Selangor, in 1955. By the early

1970s, H&C covered extensive plantations of over 95,101 ha across Malaysia, Indonesia, India, Sri Lanka, and Papua New Guinea.[71]

By 1970, foreign-owned firms held a 43% equity stake in the agribusiness sector, with palm oil as a major product. The "Big Three"—Guthries, H&C, and Sime Darby—were the largest plantation companies, collectively overseeing 29.1% of the total plantation areas. Of their businesses, 32% was dedicated to palm oil cultivation and rubber.[72] Other British-owned corporations included Kuala Lumpur Rubber Co. Ltd. (1906; renamed Kuala Lumpur Kepong Amalgamated in 1960) and Kulim Rubber Plantations Ltd. (1933; renamed Kulim Group Ltd. in 1970). The Danish-owned Jenderata Rubber Estates, established in 1906, was renamed United Plantations in 1917.[73] By 1981, foreign MNEs collectively owned 337,000 ha of land (see Table 4.1), coinciding with the peak of the takeover of British firms in the plantation sector.

In the 1960s, British rubber businesses feared acquisition by Malayan/Malaysian Chinese speculators who were keen to venture into plantations.[74] Among these speculators was senator Lee Loy Seng, who began a process of acquisitions that would make him one of the country's largest rubber plantation owners by the 1970s. Lee gained control of the British-owned Kuala Lumpur-Kepong between 1969 and 1972 and also acquired over 25% of Highlands and Lowlands, previously under the control of Barlow Holdings.[75] In 1983, the IOI Corporation, established by Lee Shin Sheng, grew through a series of acquisitions, including 27,880 ha from the British-owned Dunlop Estates in 1990. It became one of the largest plantation companies in Malaysia, holding a total planted area of 169,369 ha, primarily for oil palm cultivation.[76] During this period, other Chinese plantation firms like United Malacca, Genting Plantations, and Perlis Plantation Berhad (later known as PPB Group Berhad) diversified their operations to include palm oil cultivation.[77] These firms strategically acquired stakes in or merged with other local plantation companies to establish a significant ownership of palm oil plantations in Malaysia and Indonesia.

Another significant development in the 1970s took place when British plantations were acquired by state corporations. Capital restructuring efforts in the early 1980s led to the localisation of many large British plantation companies in Malaysia. Its aim was to boost Bumiputra[78] equity to at least 30% of corporate shares. The transformation of these companies made them susceptible targets for mergers and acquisitions by GLCs like Pernas and PNB. This trend resulted in PNB's acquisition of the Big Three (Guthrie, H&C, and Sime Darby) by 1990,[79] making it the world's largest palm oil plantation owner, with 200,000 ha of land across Malaysia, Indonesia, and other countries.[80] Sime Darby, another GLC, became one of the world's largest palm oil plantations in 2007 when it merged with Golden Hope Plantations and Kumpulan Guthrie

TABLE 4.1 Land ownership: British and other European plantation firms in Malaysia, c. 1981

Name	*Acreage (hectares)*	*Origins*
Guthrie Corporation Ltd	120,000	1821 (British)
Harrisons & Crosfield	85,000	1844 (British)
Sime Darby Bhd.	77,000	1910 (British)
Socfin Plantations	28,000	1880s (Belgian)
Unilever (Pamol Plantations Sdn. Bhd.)	14,000	1968 (Anglo-Dutch)
United Plantations	13,000	1906 (Danish)

Source: Adapted from Yacob and Md. Khalid, "Adapt or Divest," 465.

Berhad (now Sime Darby Bhd.). By 2022, the combined planted area of palm oil amounted to 578,114 ha, with 296,018 ha in Malaysia and 191,214 ha in Indonesia.[81] Kulim Group, which was listed on the London Stock Exchange in 1970, was acquired by the state agency Johor State Economic Development Corp. (now JCorp) in 1976.[82] United Plantations sold one-third of its shares to the Malaysian partly state-owned company Food Industries of Malaysia (Kumpulan FIMA) in 1982, which regained 50% controlling stake in 1991.[83]

Colonial dominance laid the foundation for the emergence of the palm oil export market. Decolonisation efforts aimed at redistributing wealth have added another layer to the complex interplay, influencing smallholder dynamics, poverty alleviation endeavours, and the rise of large-scale businesses that focus on palm oil monocultures. This dynamic led to the development of the "oil palm complex," an intricate network intertwining socio-economic, environmental, and political factors within Malaysia and Indonesia's oil palm cultivation sectors. In Malaysia, with the rise of national palm oil companies, the industry shifted from supplying raw materials to former colonial powers to value-added downstream manufacturing, primarily refining crude palm oil and diversifying it into oleochemicals and biofuels. However, the enduring consequences of colonialism remained characterised by the exploitation of local populations and resource depletion in Malaysia and Indonesia, and by the 2020s, both countries were still struggling to cope with the ramifications of deforestation.[84]

4.4 Palm Oil Agribusiness Literature

This section explores the multifaceted narratives surrounding climate change, major corporations, and civil society oversight in the palm oil agribusiness sector. It focuses on existing literature that examines the relationship between big businesses and climate change, as well as the role of civil society oversight in the palm oil agribusiness sector. Despite the significant differences between Malaysia and Indonesia, similarities exist, particularly regarding issues such as deforestation and its consequences for climate change.

Malaysia, surpassing Nigeria, was ranked the world's top exporter and producer of palm oil in 1966 and 1971 respectively, but by 2007, Indonesia had emerged as the leading producer.[85] The revenues generated from palm oil helped alleviate poverty and spur economic growth, making it highly attractive to the governments of Malaysia and Indonesia. Palm oil's productivity was exceptional, as it produced high quantities of oil compared to other vegetable oil plants while requiring only a fraction of the land.[86] Indonesia attracted more research interest in the environmental impacts caused by palm oil production because there were higher deforestation rates going on in the country.[87] Additionally, land grabbing (characterised by the acquisition of land through unethical means) has been more prevalent in Indonesia compared to Malaysia. As a result, local communities in both Indonesia and Malaysia were vulnerable to the adverse impacts of climate change, but the effects were particularly pronounced in Indonesia, exacerbating existing challenges within the region's palm oil sector.[88]

The issue of land grabbing began in the 1990s and has since been under intense scrutiny and scholarly debate. Tania Li (2018) likened it to a "Mafia System"[89] characterised by a web of corruption, violence, and exploitation permeating the palm oil sector, and riding on state and business collusions intended to maximise profits, frequently at the expense of the local industry and environment.[90] Berenschot and Dhiaulhaq (2023) investigated conflicts between rural populations and palm oil companies in Indonesia since the early 2000s.[91] They, along with other scholars, highlighted the victimisation of rural communities, often left without access to legal assistance or avenues for redress.[92] The corporations were identified as responsible for the dispossession of locals' land.[93]

Figure 4.1 shows the rapid rise of Indonesia as the primary exporter of both palm oil and palm kernel oil. The period from 1997 to 2007 revealed a significant surge in exports, coinciding with substantial investments from major multinational corporations in Indonesia. The growing demand for palm oil was accompanied by an increase in environmental and social issues that were associated with its production such as deforestation.[94] Between 1975 and 2005, an estimated 40 million hectares of land had been cleared, equivalent to 30% of the total deforestation of Indonesia.[95]

Indonesia clearly had a larger forest cover than Malaysia in the early 1990s. Between 1994 and 1995, its forest coverage was almost equal, but Indonesia's forest cover declined substantially between 1999 and 2000. Malaysia's gradual decline was probably influenced by the major investments of Malaysian and Singaporean palm oil corporations in Indonesia. Between 2000 and 2012, Malaysia faced the world's highest deforestation rate, losing 14.4% of its forest cover, an area larger than Denmark.[96] This was compounded by the expansion of GLCs and national plantation companies in Malaysia, which later expanded into Indonesia. Malaysia's forest cover increased briefly from 2012 onwards, owing to conservation efforts or changes in land-use policies. Between 2014 and 2020, Malaysia achieved a remarkable 70% reduction in primary forest loss.[97] As of 2021, Malaysia's forest cover surpassed Indonesia's by almost 10%.[98] This shift is indicative of Malaysia's deliberate effort to combat deforestation, which is potentially linked to initiatives aimed at curbing oil palm-related deforestation. Global Forest Watch attributed the slower expansion of Malaysia's oil palm plantations to the government's decision to limit the oil palm-planted area to 6.5 million ha from 2019 to 2023.[99] Another contributory factor was the implementation of the Malaysian Sustainable Palm Oil Certification Scheme (MSPO) since 2015 which was aimed at

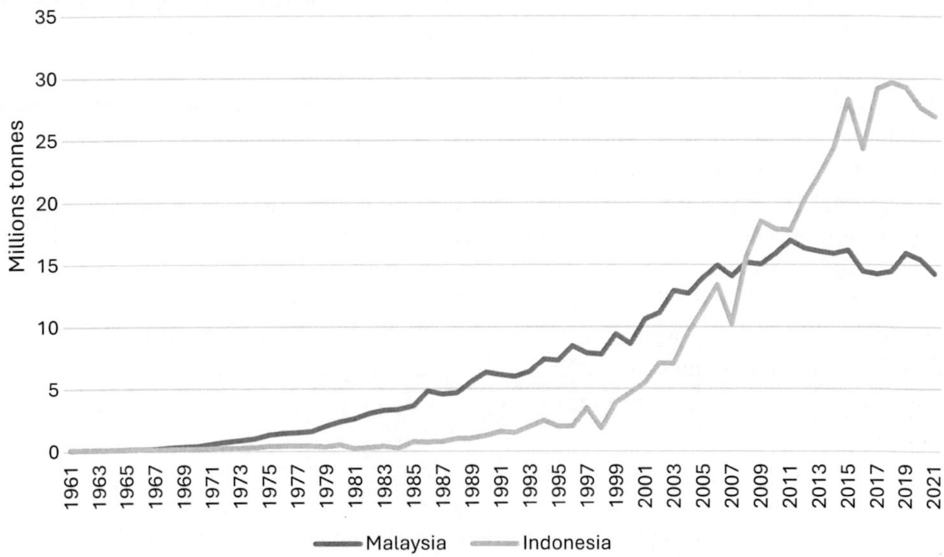

FIGURE 4.1 Export of palm oil and palm kernel oil (total)

Source: FAO. (2023). "Crops and Livestock Products." (https://www.fao.org/faostat/en/#data/TCL) Licence: CC-BY-4.0.

Note: The figures include the exports of palm oil and palm kernel oil.

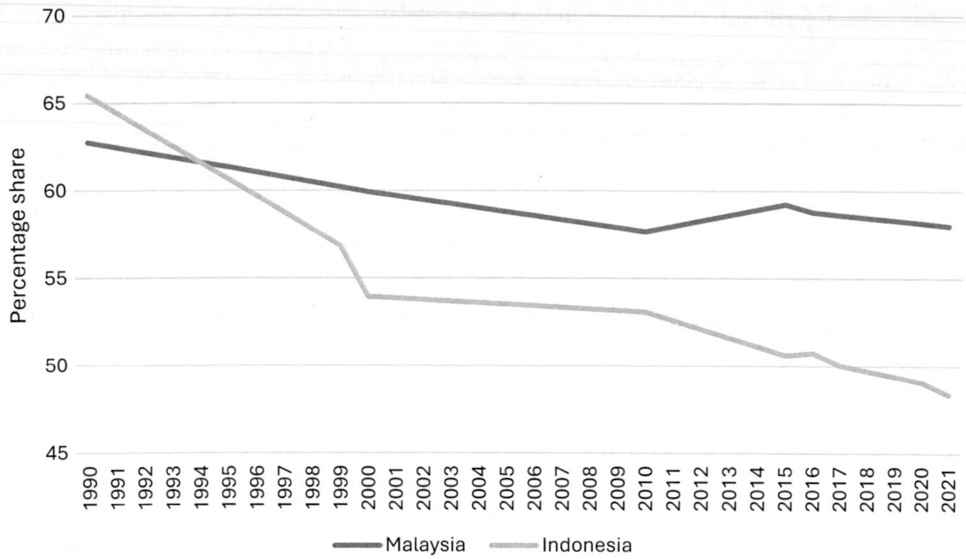

FIGURE 4.2 Area under forest (as a share of the land area)

Source: FAO. (2024). "Land Use." (https://www.fao.org/faostat/en/#data/RL) Licence: CC-BY-4.0.

reducing social and environmental impacts.[100] The overall data indicate differing approaches and outcomes in forest management and conservation between these two nations.[101]

Past literature has primarily focused on the environmental impacts associated with the palm oil agribusiness, particularly deforestation-related issues such as forest fires,[102] haze, human health impacts,[103] flooding risks,[104] and land and environmental degradation on local livelihoods.[105] The expansion of palm oil into peatlands has resulted in the degradation of biodiversity and the emission of carbon while the haze produced by peatland fires posed livelihood and health risks.[106] Flooding not only diminishes land quality and productivity but also poses risks to oil palm plantations, resulting in property damage, loss of life, and substantial economic losses.[107]

In the late 1990s, increased public concern about haze in Singapore, Malaysia, and Indonesia brought CSOs into focus, despite their limited prominence during the extensive expansion of the palm oil sector in the mid-1990s.[108] This chapter contributes to the literature on anti-corporate activism, which aims to influence companies' behaviour on environmental issues.[109] It highlights the growing focus of CSOs on companies, rather than states, employing tactics such as critical reports and media-savvy protests to pressure them for change. These efforts frequently involve transnational advocacy networks, which help draw attention to local injustices. Examining how CSOs like Greenpeace influenced palm oil giants such as Wilmar to adopt sustainable practices sheds light on collective CSO efforts and the complexities of the palm oil supply chain.

4.4.1 Civil Society Organisations and the Palm Oil Agribusiness

Research emphasises the role of CSOs in addressing climate change by establishing connections between local chapters and transnational advocacy networks, which provide information, funding, and expertise to influence government policies.[110] Indonesian CSOs have been actively engaged in monitoring and lobbying for sustainable palm oil production since the late 1990s.[111] Climate change often serves as a dominant environmental frame for local campaigns, with success at the national level often depending on framing concerns as translocal issues and collaborating with similar groups.[112] In this context, local CSOs collaborate with international counterparts to establish equitable and sustainable natural resource management, emphasising the significance of transnational cooperation in addressing regional concerns. For instance, Sawit Watch partnered with Forest Peoples Programmes and other international CSOs to address land acquisition issues and indigenous rights.[113] Greenpeace Indonesia collaborates with local CSOs like WALHI, Sawit Watch, SPKS, and Forest Watch Indonesia to monitor the palm oil agribusiness sector and address climate issues.[114] Their collective efforts aim to foster sustainable practices and align with the UN Sustainable Development Goals (SDGs).

Local CSOs in Indonesia heavily rely on volunteers due to limited resources,[115] which contrasts with larger organisations like Greenpeace that are better equipped and often regarded as "civil society elites,"[116] overseeing diverse activities such as donor-funded projects and awareness campaigns aimed at elevating local concerns to a global audience.[117] However, their activism poses risks, including threats to activists' lives and property, as well as arrests or criminalisation, as documented by CSOs like WALHI and WWF Indonesia.[118] These challenges reflect efforts to suppress dissent and deter conservation groups from challenging government interests in environmental matters.[119]

Since the 1990s, local and international CSOs in Malaysia and Indonesia have been raising awareness about climate change due to deforestation. Their reports expose corrupt practices by major palm oil corporations, government policies favouring business groups, and the absence of environmentally friendly approaches. These annual reports serve as crucial references for other Indonesian CSOs. Detailed assessments of corporate actions, like those of Wilmar, are necessary to determine alignment with environmental claims or identify greenwashing effects. CSOs monitor Wilmar's alleged violations of the No Deforestation, No Peat, and No Exploitation (NDPE) 2013 pledge, which lacks external oversight, unlike the RSPO.[120] While these pledges face challenges, private regulations can support palm oil sustainability, fostering changes in regulatory frameworks. Indonesia and Malaysia serve as examples of such transformations.[121]

Activist groups like Greenpeace, Mongabay, Chain Reaction Research, and China Dialogue have provided valuable insights into the environmental and social impacts of the palm oil industry. However, their vigilant monitoring and assertive viewpoints have faced criticism and pushback from palm oil companies and governmental entities.[122] Concerns about their accountability and advocacy skills have also been raised.[123] For instance, Greenpeace's 2007 report, "How the Palm Oil Industry is Cooking the Climate," attributed forest and peat swamp destruction in Indonesia solely to American and European companies, overlooking the role of local actors, government corruption, and industry-led initiatives.[124]

4.5 Wilmar International and Palm Oil Agribusiness

Wilmar's palm oil agribusiness has received scholarly attention, mainly in the form of brief articles[125] or case studies on corporate social responsibility[126] and sustainable development.[127] While Wilmar's significance in the palm oil industry is well recognised, there is limited academic research on its practices, impact, and role in climate change, except for reports by activist groups that have closely monitored Wilmar since the mid-2000s. They particularly focus on its adherence to sustainable practices and the NDPE policy.[128] Analysing Wilmar's activities in the palm oil agribusiness reveals the intricate relationship between MNE practices and their impact on climate change. MNEs like Wilmar can help governments achieve the UN's SDGs while contributing to economic development, poverty alleviation, and overall growth.[129]

Wilmar's dominant position in the palm oil agribusiness, along with its transnational operations, drew scrutiny from CSOs.[130] In response, Wilmar prioritised sustainability by adopting RSPO certification standards and engaging in stakeholder dialogues[131] to address deforestation, biodiversity loss, and social impacts.[132] This balancing act between growth and sustainability underscores Wilmar's influential role in shaping the future of the palm oil industry through its commitment to responsible practices amidst industry leadership.

CSOs offered several advocacy initiatives, while environmentalists compelled MNEs such as Wilmar to confront and address pressing issues related to deforestation. By 2013, external pressures escalated as Norway's sovereign wealth fund divested shares from 23 palm oil companies, including Wilmar.[133] In October 2013, the Greenpeace report "Licence to Kill" implicated Wilmar as a habitat destroyer of Sumatran tigers caused by illegal deforestation and peatland burning in Indonesia.[134] Wilmar contested these findings, citing an ongoing review of their practices with international supply chain experts.[135] Within two months of the Greenpeace report and mounting external pressure, Wilmar became the first palm oil trader to launch the NDPE policy in December 2013.[136]

Wilmar was closely monitored by several CSOs, including Greenpeace, for its adherence to the NDPE policy, while other CSOs remained vigilant collectively.[137] To ensure the successful implementation of the NDPE, Wilmar worked with Forest Trust for five years (subsequently renamed Earthworm), from 2013 until 2018.[138] This collaboration led to the launch of the Wilmar Sustainability Dashboard in 2015, featuring a map of Wilmar refineries in Indonesia and Malaysia and their supplying mills.[139] The dashboard included information on grievance resolution, community engagement, and sustainability policy implementation, and was hailed as a "gentle revolution"[140] towards sustainability in the palm oil industry.

Wilmar also established a Grievance Procedure which addressed the concerns of individuals, government bodies, and CSOs regarding the implementation of its NDPE Policy[141] towards sustainability in the palm oil industry. In 2018, Wilmar accelerated its implementation of the NDPE policy.[142] Despite engaging in dialogues with various CSOs and stakeholders to address environmental concerns, the authenticity of Wilmar's commitment required verification and assessment. It is important to examine Wilmar's commitment to its sustainable practices.[143] Since adherence to the NDPE was voluntary, there may not be legal repercussions for non-compliance, but reputational damage is a risk.[144] Despite claiming dedication to transparency and supply chain improvements, Wilmar continued trading palm oil with deforestation-linked producers into the 2020s, raising stakeholder concerns.[145] Additionally, Wilmar was criticised for maintaining ties with producer groups that were facing government convictions or sanctions[146] for blatantly refusing to comply with compensation orders aimed at restoring the regions affected by forest fires.[147]

In 2020, Greenpeace accused Wilmar of not living up to its NDPE commitment and its connections with unscrupulous supply chains.[148] Various reports disclosed by Greenpeace Indonesia from 2013 to 2019 highlighted Wilmar's lack of commitment and its transgressions associated with deforestation and oil palm trafficking (see Table 4.2). In 2016, Greenpeace accused Wilmar of destroying over 7,000 hectares of peatland forest in Indonesia. The evidence was substantiated by satellite imagery and field investigations.[149] In 2018, Greenpeace Indonesia labelled Wilmar as "the largest and dirtiest palm oil trader," alleging its involvement in purchasing palm

TABLE 4.2 Greenpeace reports on Wilmar's palm oil practices (2013–2019)

Report	*Publication date*	*Summary*	*Climate change issues*
License to kill	October 2013	The expansion of oil palm plantations between 2009 and 2011 led to the destruction of the Sumatran tiger's habitat, highlighting the affiliation of major brands with Wilmar International.	Habitat
Under fire	November 2015	Wilmar, a palm oil supplier, is accused of deforesting by burning land, causing severe health issues for nearby communities.	Health issues associated with fires and haze
Deadly trade off	September 2016	Identified Wilmar International as an IOI supplier with a systemic failure to implement the NDPE policy.	Deforestation, peatland expansion
Dirty bankers	February 2017	Identified HSBC Holdings Plc as the main provider of financial services to Wilmar and its palm oil suppliers through loans and credit facilities.	Deforestation, peatland expansion
Moment of truth	March 2018	Large brands were not adequately disclosing their palm oil supply chains to support NDPE policies.	Deforestation, peatland expansion
Rogue trader: keeping deforestation in the family	June 2018	Documented Wilmar's relationship with Gama, an Indonesian palm oil business.	Deforestation in Papua
Final countdown: now or never to reform the palm oil industry	September 2018	Wilmar has been identified as responsible for deforestation in Indonesia from 2013 to 2018, coupled with its failure to implement the NDPE policy.	Forest fires/carbon emission and greenhouse gas
Burning down the house	November 2019	Between 2015 and 2018, Wilmar's supply chain was linked to palm oil groups responsible for over 140,000 hectares of burned land and, as of 2019, associated with nearly 8,000 fire hotspots	Forest fires

Source: Greenpeace reports, 2013–2019.

Note: These reports are available at Greenpeace Media website: https://media.greenpeace.org/

oil from 18 local companies that engaged in deforestation.[150] In the same year, another report accused Wilmar of removing more than 30,000 hectares of forest in Indonesia over the preceding five years. The allegations challenged Wilmar's sustainability commitment and accountability for its actions.[151] In 2019, Greenpeace International alleged that Wilmar had sourced its supplies from palm oil businesses that were reportedly responsible for over 140,000 hectares of burned land from 2015 to 2018, and nearly 8,000 fire hotspots up to 2019.[152] It asserted Wilmar's ongoing relationships with palm oil producers connected to forest fires.[153] Greenpeace's investigation put Wilmar as well as other significant trading and consumer goods companies such as Nestlé, which procured palm oil from Wilmar in a bad light.[154]

The issue inevitably gained extensive media coverage. In 2015, Friends of the Earth claimed that Wilmar had violated its own sustainability policy by encroaching on peatlands and destroying high-carbon stock areas. Wilmar was also accused of failing to address forest fires.[155] In 2018, Mongabay, an environmental organisation accused Wilmar of enabling Gama to hide behind "shadow companies," where ownership and destructive activities were concealed from the public.[156] In that same year, Greenpeace International accused Wilmar of collaborating with Gama in environmentally damaging activities while projecting an eco-friendly image.[157] A 2022 study by Chain Reaction Research further implicated Wilmar in the low uptake of certified palm oil in its supply chain.[158] Greenpeace Indonesia's monitoring was crucial in holding Wilmar accountable for sustainability in the palm oil industry. Wilmar pledged to commit to sustainability standards, though its motive was driven more by revenue gains than altruism.[159]

Wilmar has made strides in certifying approximately 81% of its oil palm plantation area as RSPO-certified by January 2024.[160] However, challenges remain in achieving full traceability[161] across its supply chain, despite 27 out of 45 mills and over 98% of its refineries being RSPO-certified.[162] While Wilmar aims to extend its sustainability policy to all third-party suppliers, it does not mandate exclusive adherence to its standards, potentially compromising sustainability efforts.[163] Wilmar reports often attribute negative environmental impacts to suppliers and smallholders, raising concerns about accountability and the responsibility of large palm oil conglomerates.[164] Wilmar has attracted attention for its organisation size, extensive operations, and transparency in reporting. Unlike some palm oil companies that ignored or denied allegations in CSO reports, Wilmar responded to critiques, including Greenpeace's reports on its NDPE policy violations (see Table 4.3).

As a preventive measure, Wilmar collaborated with CSOs, implementing stricter standards such as enhanced NDPE policies and a grievance system. CSOs' efforts in holding Wilmar accountable were instrumental in this improvement. Additionally, Wilmar made efforts to mitigate greenhouse gas emissions and promote forest conservation initiatives, demonstrating its adherence to NDPE standards under CSO scrutiny.[165]

Since joining the RSPO in 2004 and adopting NDPE policies in 2013, Wilmar has implemented various climate mitigation measures in the palm oil sector. These efforts led to its inclusion in the Dow Jones Sustainability Index (DJSI) in 2020.[166] It attained the top spot in the Sustainable Palm Oil Transparency Toolkit (SPOTT), an online assessment platform developed by the Zoological Society of London, consistently from 2021 to 2022.[167] The SPOTT evaluation of Wilmar in 2021 scored a transparency rate of 91% in its disclosure of environmental, social, and governance (ESG) policies and initiatives.[168] In 2023, Wilmar ranked number six with a score of 92.8% and 97.7% for sustainability policy and leadership respectively.[169] Sadly, despite these accolades, the reality on the ground is more complex than what is reflected on paper.

4.6 Concluding Remarks

The history of deforestation in Malaysia and Indonesia dates back to colonial administrations in the early twentieth century, with land commercialisation for rubber and later palm oil, resulting in substantial colonial profits at the expense of deforestation and local development. In the 1970s, Malaysian firms and GLCs transitioned from rubber to palm oil and acquired large British-owned plantations in the 1980s. By the early 1990s, companies such as Wilmar, a leading palm oil MNE producer, ventured into Indonesia, marking a period of intense land-grabbing and accelerated deforestation. This led to adverse environmental

TABLE 4.3 Wilmar NDPE policy and climate change

Year	*Policy*	*Environmental goals*
2013	***No Deforestation, No Peat, No Exploitation (NDPE) Policy***	
	No deforestation	No development of high carbon stock (HCS) forest High conservation value (HCV) forest No burning
	No peat policy	Reduce greenhouse gas emissions on existing plantations. Best management practices for plantations on peat Peat restoration with stakeholders and communities
	No exploitation	Workers' rights Inclusion of smallholders in the supply chain Free, prior, and informed consent (FPIC) from indigenous and local communities
2014 (Quit in 2020, citing governance issues, but still committed to the HCSA methodology)	***High Carbon Stock Approach*** A methodology identifying forests of conservation value and forests open to palm oil cultivation. This group includes organisations such as Greenpeace, the World Wildlife Fund for Nature, the Rainforest Action Network, and the Forest Peoples Programme, as well as palm oil companies like Cargill, Golden Agri Resources (GAR), New Britain Palm Oil, Agropalma, Wilmar, and Asian Pulp and Paper (APP).	Prevent the destruction of high-carbon stock forests.
2017	Supplier reporting tool (SRT), an online self-reporting tool that allows suppliers to report their compliance with environmental and social risk-related issues within Wilmar's supply chain.	Environmentally friendly supply chain
2021	Sectoral roadmap with 13 leading agricultural trading and processing companies to achieve collective emissions reduction goals.	Reduction of emission
	Established a board sustainability committee. Governance over implementation and communications of our sustainability activities with our stakeholders. Inclusion in the Dow Jones Sustainability World Index (DJSI World) Ranked first globally in the 2021 Palm Oil Benchmark Sustainability Policy Transparency Toolkit (SPOTT)/Zoological Society of London (ZSL)	Governance over stakeholders' commitment to sustainable practices
2022	Ranked first globally, Palm Oil Benchmark Sustainability Policy Transparency Toolkit (SPOTT)/ Zoological Society of London (ZSL)	
2022	Maintained inclusion in the Dow Jones Sustainability World Index (DJSI World)	

Source: Dashboards of Wilmar International, Zoological Society of London (2013–22).

consequences, intensifying in the mid-1990s, when CSOs began exerting influence in shaping business responsibility in the palm oil industry towards climate change, advocating for transparency and environmental justice.

History shows that it is important to recognise the interdependence between civil society and businesses which is crucial for promoting collaboration and corporate responsibility. Among other activists, Greenpeace's persistent research and assessment of Wilmar's sustainability commitments played a significant role in pushing the oil palm giant to adopt climate mitigation initiatives. While activism can drive change, it is essential to acknowledge the potential repercussions, some of which may be unintended. The overall influence and effectiveness of these efforts on the palm oil industry are based on ongoing evaluations by various stakeholders, including the government and communities. Balancing the interests of these stakeholders has been crucial for fostering sustainable solutions in complex palm oil ecosystems amid multifaceted challenges. Navigating the complex landscape of the palm oil industry reveals a history and reality that are devoid of simple solutions and characterised by multifaceted challenges.

Notes

1 "Open Letter."
2 "Indonesia," *Antara News*.
3 Schrier-Uijl et al., "Environmental"; Snashall and Poulos, "Oreos Versus Orangutans."
4 Guindon, "Deforestation."
5 "Open Letter."
6 TuK Indonesia, "Tycoons."
7 Auriga Foundation, "Corporate Ownership."
8 TuK Indonesia, "Tycoons," 9.
9 "Open Letter," note 10.
10 "Open Letter," 3; TuK Indonesia, "Tycoons," 3–10.
11 TuK Indonesia, "Tycoons."
12 TuK Indonesia, "Financing."
13 Virgy, Yusa, and Wawan, "Strategi Jaringan."
14 Laebens and Lührmann, "What Halts Democratic Erosion?"
15 Sachs, "From Millennium," 2211.
16 Setiawan and Tomsa, "Defending," 353.
17 Beittinger-Lee, "(Un)Civil Society."
18 McGregor and Setiawan, "Shifting."
19 Aspinall, "A Nation in Fragments."
20 Aspinall, "Conclusion."
21 Siahaan, "Palm Oil Industry."
22 Yacob, "Government."
23 Setiawan and Tomsa, "Defending," 360.
24 Pye, "Civil Society."
25 Fathana, "Palm Oil," 50–51, 59.
26 Wilmar, "History."
27 Wilmar, "History."
28 Teo, "Malaysian Corporations," 28–29.
29 SPOTT, "Wilmar."
30 Rijk, Okita, and Matthew, "Wilmar's Refineries," 1.
31 Phillips, Coca, and Hesling, "From Palm to Plate."
32 Auriga Foundation, "Corporate Ownership," 4; Mahayunan, Tundang, and Tan, "Urgensi Pemberlakuan."
33 Mahayunan, Tundang, and Tan, "Urgensi Pemberlakuan," 145.
34 SPOTT, "Wilmar."
35 Wilmar, "Corporate Profile."
36 Wilmar, "Corporate Profile."
37 Wilmar, "Corporate Profile."

38 Farhan and Hoebink, "Can Campaigns Save Forests?," 27.
39 Abd Malek and Haris, "Isu-isu Alam Sekitar," 21–24.
40 Barnard and Lee, "A Spiteful Campaign," 470.
41 Abd Malek and Haris, "Isu-isu Alam Sekitar," 23–24.
42 Wong, "Gardens Bulletin," 2–3.
43 Barnard and Lee, "Spiteful Campaign," 469.
44 Barnard and Lee, "Spiteful Campaign," 467–69.
45 Barnard and Lee, "Spiteful Campaign," 478, 485.
46 Barnard and Lee, "Spiteful Campaign," 469–70.
47 Robins, "Shallow Roots," 538–39.
48 Yacob, "Model of Welfare Capitalism," 11; Jackson, *Planters and Speculators*, 235, 237; Ghee, *Origins of a Colonial Economy*, 13, 40; Hollen Lees, *Planting Empire*, 43; Stoler, *Capitalism and Confrontation*, 16.
49 Robins, "Shallow Roots," 542.
50 Jackson, *Planters and Speculators*, 235–37.
51 Rönnbäck, Broberg, and Galli, "A Colonial Cash Cow," 152–53.
52 Rönnbäck, Broberg, and Galli, "A Colonial Cash Cow," 166–67.
53 Buelens and Frankema, "Colonial Adventures."
54 Rönnbäck, Broberg, and Galli, "A Colonial Cash Cow"; Hollen Lees, "Planting Empire," 171.
55 Lim Chong Yah, *Economic Development*, Appendix 3.1, p. 323.
56 Robins, "Shallow Roots," 549.
57 Robins, "Shallow Roots," 549–50.
58 Ghee, "Malayan Peasant," 110–11.
59 Rudner, "The Evolving Political Economy."
60 Robins, "Shallow Roots," 549–50.
61 Virgy, Yusa, and Wawan, Strategi Jaringan; Open Letter, 3.
62 Rudner, "The Evolving Political Economy."
63 "No Need to Be Poor," 1–11.
64 Khor, "Political Tussle," 90.
65 FGV Holdings, "Plantations."
66 Malaysian Palm Oil Council, 2020.
67 Malaysian Palm Oil Council, 2020.
68 Annuar, "A Century of Resilient."
69 Jones and Wale, "Diversification," 74.
70 Jones and Wale, "Diversification," 74–77.
71 Jones and Wale, "Diversification," 71.
72 Searle, "Malaysian Capitalism."
73 Teo, "Malaysian Corporations," 24–25.
74 Yacob and Md Khalid, "Adapt or Divest," 462.
75 Yacob and White, The "Unfinished Business," 10, 40.
76 Teo, "Malaysian Corporations," 25, 28.
77 Teo, "Malaysian Corporations," 25, 28–29.
78 "Bumiputera," which translates to "sons of the soil," encompasses the majority population of Malays and indigenous groups in Malaysia.
79 Yacob and Md Khalid, "Adapt or Divest."
80 Segal, "Home to Hard Times," 92–93.
81 Sime Darby Annual Report 2022, 51.
82 "Beyond the Benchmarks."
83 Yacob and Md Khalid, "Adapt or Divest," 462, 468, 471.
84 Barnard and Lee, "Spiteful Campaign," 478, 485.
85 Khera, "The Oil Palm Industry."
86 RSPO, "Why Sustainable Palm Oil?"
87 Austin et al., "What Causes Deforestation"; Austin et al., "Shifting Patterns"; Vijay et al., "The Impacts of Oil Palm."
88 Cramb and McCarthy, *The Oil Palm Complex*.
89 Li, "After the Land Grab."
90 Pye, "Commodifying Sustainability."

91 Berenschot and Dhiaulhaq, "The Production of Rightlessness"; Afrizal and Berenschot, "Land-Use Change Conflicts."
92 Haug, "Resistance, Ritual, Purification"; Colchester and Chao, "Conflict or Consent?"
93 Berenschot and Dhiaulhaq, "The Production of Rightlessness," 3, 8, 15.
94 Nesadurai, "New Constellations," 4.
95 Susanti and Maryudi, "Development Narratives," 130–32.
96 "Causes of Rainforest Deforestation."
97 Global Forest Watch, "Malaysia."
98 Global Forest Watch, "Malaysia."
99 Global Forest Watch, "Malaysia."
100 Law and Wong, "Can Oil Palm."
101 Law, "Navigating the Multiverse."
102 Varkkey and Lee, "The Haze Problem"; Varkkey, "Oil Palm Plantations."
103 Huijnen et al., "Fire Carbon Emissions"; Marlier et al., "Fires, Smoke Exposure."
104 Muhadi, Abdullah, and Muhammad Zahir, "Estimating Flood."
105 Santika et al., "Changing Landscapes"; Santika et al., "Impact of Palm Oil"; Murray Li, "After the Land Grab."
106 Marlier et al., "Fires, Smoke Exposure."
107 Muhadi, Abdullah, and Muhammad Zahir, "Estimating Flood."
108 Khor and Saravanamuttu, "Enflamed."
109 Afrizal and Bernschot, "Anti-Corporate Activism," 40.
110 Pye, "Civil Society," 329–30.
111 McGregor and Setiawan, "Shifting," 353.
112 Rootes, "From Local Conflict."
113 Colchester et al., *Promised Land*, 12.
114 Virgy, Yusa, and Wawan, "Strategi Jaringan," 78.
115 McGregor and Setiawan, "Shifting," 361–62.
116 Johansson and Uhlin, "Civil Society Elites."
117 Setiawan and Tomsa, "Defending," 360.
118 Setiawan and Tomsa, "Defending," 361–62.
119 Setiawan and Tomsa, "Defending," 362.
120 Larsen et al., "Hybrid Governance."
121 Nesadurai, "New Constellations"; Nesadurai, "Transnational Private."
122 Kuok, "Letter"; McGregor and Setiawan, "Shifting"; "After Sarawak."
123 Ramdani and Lounela, "Palm Oil Expansion"; Longhofer et al., "NGOs, INGOs"; Eccleston, "Does North-South Collaboration."
124 "Cooking the Climate."
125 Vijayaraghavan and Desai, "Wilmar International."
126 Kurz, "Corporate Social Responsibility."
127 Rajoana, "Wilmar."
128 Virgy, Yusa, and Wawan, "Strategi Jaringan."
129 Sachs, "From Millennium," 2211.
130 Farhan and Hoebink, "Can Campaigns Save Forests?" 18.
131 Palm Oil Detectives, "Land Grabbing."
132 Purnomo et al., "Fire Economy," 21.
133 Humber and Pakiam, "Palm Oil King."
134 Greenpeace International, "Licence to Kill."
135 Chin, "Greenpeace, Wilmar."
136 Wilmar, "NDPE Policy."
137 Virgy, Yusa, and Wawan, 'Strategi Jaringan."
138 "Change."
139 Cheam and Shah, "Wilmar."
140 Cheam and Shah, "Wilmar."
141 Cheam and Shah, "Wilmar."
142 Wilmar Action Plan.
143 Pye, "Palm Oil Firms."
144 Greenpeace International Press Release, "World's Largest."

145 Kurz, "Corporate Social Responsibility," 40–41.
146 Marks et al., "Institutional Responses."
147 Greenpeace International, "Burning Down the House."
148 Hicks, "Palm Oil Giant."
149 Greenpeace, "A Deadly Tradeoff,"
150 Greenpeace International Press Release, "World's largest,"
151 Greenpeace, "Final Countdown."
152 Greenpeace International, "Burning Down"; Hicks, "Palm Oil Giant."
153 Greenpeace International, "Burning Down."
154 Nestlé, "Responsibly Sourced Palm Oil," https://www.nestle.com/sustainability/sustainable-sourcing/palm-oil.
155 Friends of the Earth Europe, "Up in Smoke," 4.
156 Pye, "Palm Oil Firms."
157 Greenpeace International, "World's Largest."
158 Rijk, Okita, and Matthew, "Wilmar's Refineries," 1.
159 Marks et al., "Institutional Responses," 1.
160 Wilmar, "A Sustainable Approach."
161 Wilmar, "History."
162 Phillips, Coca, and Hesling, "Palm to Plate."
163 Phillips, Coca, and Hesling, "Palm to Plate."
164 Lyons-White and Andrew, "Palm Oil."
165 Jopke and Schoneveld, *Corporate Commitments*, 19.
166 Launched in 1999, the DJSI monitors publicly listed companies committed to sustainability. Felicia Tan, "Wilmar Debuts."
167 SPOTT, "Wilmar."
168 The Edge Singapore, "Wilmar Wins."
169 SPOTT, "Wilmar."

Bibliography

Abd Malek, M.F., and Abu Hanifah Haris. "Isu-isu Alam Sekitar dan Perhutanan di Tanah Melayu Sehingga Tahun 1901." *SEJARAH: Journal of the Department of History* 31, no. 2 (2022): 17–40.

About Forest Watch Indonesia. https://fwi.or.id/en/.

About Global Forest Watch. https://www.globalforestwatch.org/dashboards/global/.

About WALHI. https://www.walhi.or.id/.

Afrizal, A., and Ward Berenschot. "Land-Use Change Conflicts and Anti-Corporate Activism in Indonesia: A Review Essay." *Journal of East Asian Studies* 22, no. 2 (2022): 333–56.

"After Sarawak, Sabah Slams NGO's Projection on Forest Loss." *Malaysia Today*, March 24, 2023.

Annual Reports, Adani-Wilmar, 2017–18.

Annual Reports, Wilmar International, 2017–18.

Annuar, A. "A Century of Resilient Palm Oil Industry." *The Malaysian Reserve*, February 14, 2023. https://themalaysianreserve.com/2023/02/14/a-century-of-resilient-palm-oil-industry/.

Aspinall E. "A Nation in Fragments: Patronage and Neoliberalism in Contemporary Indonesia." *Critical Asian Studies* 45, no. 1 (2013): 27–54.

Aspinall, E. "Conclusion: Social Movements, Patronage Democracy, and Populist Backlash in Indonesia." In *Activists in Transition: Progressive Politics in Democratic Indonesia*, edited by T. Dibley and M. Ford, 187–201. Ithaca: Cornell University Press, 2019.

Auriga Foundation. "Corporate Ownership and Dominance of Indonesia's Palm Oil Supply Chains." 2020. https://auriga.or.id/flipbooks/report/en/66#page/4.

Austin, K.G., A. Mosnier, J. Pirker, I. McCallum, S. Fritz, and P.S. Kasibhatla. "Shifting Patterns of Oil Palm Driven Deforestation in Indonesia and Implications for Zero-Deforestation Commitments." *Land Use Policy* 69 (2017): 41–48.

Austin, K.G., A. Schwantes, Yaofeng Gu, and P.S. Kasibhatla. "What Causes Deforestation in Indonesia?" *Environmental Research Letters* 42, no. 2 (2019): 1–9.

Barahamin, A. "Indonesia, Not the EU, Needs to Make Its Palm Oil Sustainable." *China Dialogue*, May 3, 2022. https://chinadialogue.net/en/food/indonesia-not-the-eu-needs-to-make-its-palm-oil-sustainable/.

Barnard, Timothy P., and Joanna WC Lee. "A Spiteful Campaign: Agriculture, Forests, and Administering the Environment in Imperial Singapore and Malaya." *Environmental History* 27, no. 3 (2022): 467–90.

Beittinger-Lee, Verena. *(Un)Civil Society and Political Change in Indonesia: A Contested Arena*. New York: Routledge, 2013.

Berenschot, W., and Ahmad Dhiaulhaq. "The Production of Rightlessness: Palm Oil Companies and Land Dispossession in Indonesia." *Globalizations* (2023): 1–19.

"Beyond the Benchmarks, Plantation Division Sustainability Report, 2014–15." *Kulim Malaysia Berhad*, 5. https://kulim.com.my/storage/2022/07/KULIM-SR-2014-2015.pdf.

Buelens, F., and E. Frankema. "Colonial Adventures in Tropical Agriculture: New Estimates of Returns to Investment in the Netherlands Indies, 1919–1938." *Cliometrica* 10 (2016): 197–224.

Butler, T. "Deforestation in Borneo." *Mongabay*, April 13, 2005. https://news.mongabay.com/2005/04/deforestation-in-borneo/.

"Causes of Rainforest Deforestation in Malaysia, Internet Geography." https://www.internetgeography.net/topics/causes-of-rainforest-deforestation-in-malaysia/.

Chain Reaction Research. "The Chain: Wilmar and GAMA Group's Zero-Deforestation Policies Threatened by Shadow Companies." https://chainreactionresearch.com/the-chain-wilmar-and-gama-group-zero-deforestation-policy-threatened-by-shadow-companies/.

"Change; The Only Constant." *Thank You, Earthworm Foundation*. https://www.wilmar-international.com/docs/default-source/default-document-library/statement-by-wilmar-discontinuing-membership-with-earthworm-foundation-(formerlytft).pdf?sfvrsn=6dc96de3_0.

Cheam, J., and Vaidehi Shah. "Wilmar: First Palm Oil Giant to Name Suppliers." *Eco-Business*, January 22, 2015. https://www.eco-business.com/news/wilmar-first-palm-oil-giant-name-suppliers/.

Chin, N.C. "Greenpeace, Wilmar Clash Over Palm Oil Supply Chain." *Today*, October 24, 2013. https://www.todayonline.com/singapore/greenpeace-wilmar-clash-over-palm-oil-supply-chain.

Colchester, M., and S. Chao, eds. *Conflict or Consent? The Oil Palm Sector at a Crossroads*. Forest Peoples Program, 2013. http://www.forestpeoples.org/sites/fpp/files/publication/2013/11/conflict-or-consentenglishlowres.pdf.

Colchester, M., J. Norman, Martua S. Andiko, Y.F. Asep, A. Surambo, and H. Pane. *Promised Land: Palm Oil and Land Acquisition in Indonesia—Implications for Local Communities and Indigenous Peoples*. Bogor: Forest Peoples Programme, Perkumpulan Sawit Watch, HuMA and the World Agroforestry Centre, 2006.

"Cooking the Climate or Cooking the Books? Greenpeace's Report on Indonesian Palm Oil." 07JAKARTA3278, November 30, 2007. Public Library of US Diplomacy.

Cramb, R.A., and John F. McCarthy, eds. *The Oil Palm Complex: Smallholders, Agribusiness and the State in Indonesia and Malaysia*. Singapore: NUS Press, 2016.

DataIndonesia.id. "Siapa Produsen Minyak Goreng Terbesar di Indonesia?" https://dataindonesia.id/sektor-riil/detail/siapa-produsen-minyak-goreng-terbesar-di-indonesia.

Eccleston, Bernard. "Does North-South Collaboration Enhance NGO Influence on Deforestation Policies in Malaysia and Indonesia?" In *NGOs and Environmental Policies*, edited by David Potter, 66–89. London: Routledge, 2005.

Farhan, Farwiza, and P. Hoebink. "Can Campaigns Save Forests? Critical Reflections from the Tripa Campaign, Aceh, Indonesia." *Forest Policy and Economics* 105, no. C (2019): 17–27. https://www.foeeurope.org/sites/default/files/corporate_accountability/2015/07_foee_wilmar_report_mr_0_0.pdf.

Fathana, Hangga. "Palm Oil Politics in Malaysia and Indonesia: Competition or Collaboration?" *Jati-Journal of Southeast Asian Studies* 23, no. 2 (2018): 47–64.

FGV Holdings. "Plantations." July 15, 2023. https://www.fgvholdings.com/our-businesses/plantation/.

Friends of the Earth Europe. "Up in Smoke: Failures in Wilmar's Promise to Clean Up the Palm Oil Business."December2015,1–32.https://friendsoftheearth.eu/up-in-smoke-failures-Wilmar%e2%80%99s-promise-clean-up-palm-oil-business.

Ghee, L.T. *Origins of a Colonial Economy*. Penang: Penerbit Universiti Sains Malaysia, 1976.

Ghee, L.T. "Malayan Peasant Smallholders and the Stevenson Restriction Scheme, 1922—28." *Journal of the Malaysian Branch of the Royal Asiatic Society* 47, no. 2 (226) (1974): 105–22.

Global Forest Watch. "Malaysia." https://www.globalforestwatch.org/dashboards/country/MYS/.

Greenpeace International. "Licence to Kill." 2013. https://media.greenpeace.org/archive/Report--Licence-to-Kill-27MZIFVCLODN.html.

Greenpeace International. "Rogue Trader: Keeping Deforestation in the Family." June 25, 2018. https://issuu.com/greenpeaceinternational/docs/report_-_gp_ind_rogue_v6.1_pages.
Greenpeace International. "Burning Down the House: How Unilever and Other Global Brands Continue to Fuel Indonesia's Fires." 2019. https://www.greenpeace.de/publikationen/burning_down_the_house.pdf.
Greenpeace International Press Release. "World's Largest Palm Oil Trader Linked to Rainforest Destruction Twice the Size of Paris." June 25, 2018. https://www.greenpeace.org/international/press-release/17248/worlds-largest-palm-oil-trader-linked-to-rainforest-destruction-twice-the-size-of-paris/.
Guindon, M. "Deforestation Is Slowing, but Palm Oil Still Major Driver." *Dialogue Earth*, February 8, 2022. https://dialogue.earth/en/forests/palm-oil-and-deforestation-today/.
Hapsari, M. "Contesting 'Deforestation': Civil Society Movements and Knowledge Co-Production in Indonesia." *Power Conflict Democracy Journal* 6, no. 1 (2018): 117–46.
Haug, M. "Resistance, Ritual Purification and Mediation: Tracing a Dayak Community's Sixteen-Year Search for Justice in East Kalimantan." *The Asia Pacific Journal of Anthropology* 15, no. 4 (2014): 357–75.
"H&C Plantation Interests in Malaysia." RHM Lindesay, Paper No. 3 Harrisons & Crosfield Plantation Seminar, January 18–20, 1979. London Metropolitan Archives, MS 38275.
Hicks, R. "Palm Oil Giant Wilmar Quits High Carbon Stock Approach Citing Governance Issues." *Eco-Business*, April 3, 2020. https://www.eco-business.com/news.
Hollen Lees, Lynn. *Planting Empire, Cultivating Subjects: British Malaya, 1786–1941*. Cambridge: Cambridge University Press, 2017.
Huijnen, Vincent, Martin Johannes W. Wooster, J. Kaiser, David L.A. Gaveau, Johannes Flemming, Mark Parrington, Antje Inness, Daniel Murdiyarso, B. Main, and Michiel van Weele. "Fire Carbon Emissions Over Maritime Southeast Asia in 2015 Largest Since 1997." *Scientific Reports* 6 (2016): 26886.
Humber, Y., and P. Ranjeetha. "Palm Oil King Goes from Forest Foe to Buddy in Deal with Critics." *Bloomberg*, March 13, 2015.
Jackson, James C. *Planters and Speculators: Chinese and European Agricultural Enterprise in Malaya, 1786–1921*. Oxford: Oxford University Press, 1968.
Johansson, H., and A. Uhlin. "Civil Society Elites: A Research Agenda." *Politics and Governance* 8, no. 3 (2020): 82–85.
Jones, G., and Judith Wale. "Diversification Strategies of British Trading Companies: Harrisons & Crosfield, c.1900–c.1980." *Business History* 41, no. 2 (1999): 69–101.
Jong, H.N. "Indonesia and Malaysia Assail New EU Ban on 'Dirty Commodities' Trade." *Mongabay*, January 31, 2023. https://news.mongabay.com/2023/01/indonesia-and-malaysia-assail-new-eu-ban-on-dirty-commodities-trade/.
Jopke, P., and G.C. Schoneveld. *Corporate Commitments to Zero Deforestation: An Evaluation of Externality Problems and Implementation Gaps*. Occasional Paper 181, 1–37. Bogor, Indonesia: CIFOR, 2018.
Khera, Harcharan Singh. *The Oil Palm Industry of Malaysia: An Economic Study*. Kuala Lumpur: University of Malaya Press, 1976.
Khor, Y.L. "The Political Tussle Over Felda Land Schemes—Umno Strengthens Its Malay Rural Fortress in 13th General Election." *Kajian Malaysia* 32, no. 2 (2014): 89–121.
Khor, Y.L., and J. Saravanamuttu. "Enflamed: Seeing Through Yet Another Haze Crisis." *The Edge Malaysia*, October 16, 2019.
Kuok, K.H. "Letter from Chairman and CEO of Wilmar International to Greenpeace International (Denmark and Netherlands)." http://www.Wilmar-international.com/sustainability/wp-content/uploads/2018/07/WIL-letter-dtd-04-July-2018-to-Greenpeace.pdf.
Kurz, S. "Corporate Social Responsibility in Global Governance—A Driver for Change Towards Environmental Sustainability? An Embedded-Case Study on the Sustainability Discourse in the Palm Oil Industry." MA diss., Malmo University, 2021. https://www.diva-portal.org/smash/get/diva2:1577746/FULLTEXT02.
Laebens, Melis G., and Anna Lührmann. "What Halts Democratic Erosion? The Changing Role of Accountability." *Democratization* 28, no. 5 (2021): 908–28. doi:10.1080/13510347.2021.1897109.
Larsen, R.K., M. Osbeck, E. Dawkins, H. Tuhkanen, Ha Nguyen, A. Nugroho, T.A. Gardner Zulfahm, and P. Wolvekamp. "Hybrid Governance in Agricultural Commodity Chains: Insights from the Implementation of 'No Deforestation, No Peat, No Exploitation' (NDPE) Policies in the Oil Palm Industry." *Journal of Cleaner Production* 183 (2018): 544–54.
Li, T. Murray. "After the Land Grab: Infrastructural Violence and the 'Mafia System' in Indonesia's Oil Palm Plantation Zones." *Geoforum* 96 (2018): 328–37.
Longhofer, Wesley, Evan Schofer, Natasha Miric, and David John Frank. "NGOs, INGOs, and Environmental Policy Reform, 1970–2010." *Social Forces* 94, no. 4 (2016): 1743–68.

Lyons-White, J., and T.K. Andrew. "Palm Oil Supply Chain Complexity Impedes Implementation of Corporate No-Deforestation Commitments." *Global Environmental Change* 50 (2018): 303–13.

Mahayunan, G.R., R.E. Tundang, and J.C. Tan. "Urgensi Pemberlakuan Indirect Evidence pada Penanganan Perkara Kartel di Indonesia." *Jurnal Persaingan Usaha* 3, no. 2 (2023): 142–51, 144.

"Malaysia, a Strong Example of Deforestation Success Story." *New Straits Times*, August 18, 2023. https://www.nst.com.my/business/2023/08/944386/malaysia-strong-example-deforestation-success-story.

"Market Capitalization of Wilmar International." https://companiesmarketcap.com/wilmar/marketcap/.

Marlier, M.E., T. Liu, K. Yu, J.J. Buonocore, S.N. Koplitz, R.S. DeFries, L.J. Mickley, D.J. Jacob, J. Schwartz, B.S. Wardhana, and S.S. Myers. "Fires, Smoke Exposure, and Public Health: An Integrative Framework to Maximize Health Benefits from Peatland Restoration." *Geohealth* 3, no. 7 (2019): 178–89.

Marks, S., Justin L. Shivang M. Mehta, and S. Fernando. "Institutional Responses to Pressures for Sustainable Palm Oil." *Enviro Lab Asia* 1, no. 3 (2017): 1–31.

McGregor, K., and Ken Setiawan. "Shifting from International to 'Indonesian' Justice Measures: Two Decades of Addressing Past Human Rights Violations." *Journal of Contemporary Asia* 49, no. 5 (2019): 837–61.

Muhadi, N.A., A.F. Abdullah, and N.Z. Muhammad Zahir. *Estimating Flood Losses in Oil Palm Plantation Using Flood Modeling*. Edited by L. Mohd Sidek, G. Salih, and M. Boosroh. ICDSME 2019. Water Resources Development and Management. Singapore: Springer, 2020. https://doi.org/10.1007/978-981-15-1971-0_37.

Nesadurai, Helen E.S. "New Constellations of Social Power: States and Transnational Private Governance of Palm Oil Sustainability in Southeast Asia." *Journal of Contemporary Asia* 48, no. 2 (2018): 204–29.

Nesadurai, Helen E.S. "Transnational Private Governance as a Developmental Driver in Southeast Asia: The Case of Sustainable Palm Oil Standards in Indonesia and Malaysia." *Journal of Development Studies* 55, no. 9 (2019): 1892–908.

"No Need to Be Poor, A Policy Statement." Federal Land Development Authority, Kuala Lumpur, 1956.

"Open Letter Highlights Forest Destruction, Land Grabs and Rights Violations in Indonesia." https://www.forestpeoples.org/en/lands-forests-territories-law-policy/news-article/2018/open-letter-highlights-forest-destruction.

Palm Oil Detectives. "Land Grabbing and the Climate Crises Are Strongly Linked." November 20, 2022. https://palmoildetectives.com/2022/11/20/land-grabbing-and-the-climate-crisis-are-strongly-linked-to-palm-oil/.

Phillips, J., Nithin Coca, and Lizi Hesling. "From Palm to Plate: Tracing Sustainable Palm Oil Along the Supply Chain." *China Dialog*, January 26, 2022. https://chinadialogue.net/en/food/from-palm-to-plate-tracing-sustainable-palm-oil-along-the-supply-chain/.

"PPB Group FY22 Net Profit Climbs to RM2.2bil." *The Star*, February 28, 2023.

Purnomo, H., B. Shantiko, S. Sitorus, H. Gunawan, R. Achdiawan, H. Kartodihardjo, and A.A. Dewayani. "Fire Economy and Actor Network of Forest and Land Fires in Indonesia." *Forest Policy and Economics* 78 (2017): 21–31.

Pye, Daniel. "Palm Oil Firms Using 'Shadow Companies' to Hide Their Links to Deforestation: Report. July 9, 2018." https://news.mongabay.com/2018/07/palm-oil-firms-using-shadow-companies-to-hide-their-links-to-deforestation-report/.

Pye, O. "Commodifying Sustainability: Development, Nature, and Politics in the Palm Oil Industry." *World Development* 121 (2019): 218–28.

Pye, O. "Civil Society and Environmentalism: Crossing frontiers of Activism." In *Routledge Handbook of Civil and Uncivil Society in Southeast Asia*, edited by Evan Hansson and Meredith Weiss, 329–45. London: Routledge, 2023.

Rajoana, Jakia. "Wilmar International's Focus on Sustainable Palm Oil Production." In *Singapore Inc.: A Century of Business Success in Global Markets*, edited by Leon Choong, Easwaramoorthy Rangaswamy, Ian Jamieson, and Anne-Marie Kilday, 108–10. London: Routledge, 2023.

Ramdani, R., and A.K. Lounela. "Palm Oil Expansion in Tropical Peatland: Distrust Between Advocacy and Service Environmental NGOs." *Forest Policy and Economics* 118 (2020): 102242.

Robins, J. "Shallow Roots: The Early Oil Palm Industry in Southeast Asia, 1848–1940." *Journal of Southeast Asian Studies* 51, no. 4 (2020): 538–60.

Rönnbäck, K., O. Broberg, and S. Galli. "A Colonial Cash Cow: The Return on Investments in British Malaya, 1889–1969." *Cliometrica* 16 (2022): 149–73.

Rootes, C. "From Local Conflict to National Issue: When and How Environmental Campaigns Succeed in Transcending the Local." *Environmental Politics* 22, no. 1 (2013): 95–114.

Roundtable on Sustainable Palm Oil (RSPO). "Strategy Report on China Sustainable Palm Oil." https://www.rspo.org/articles/RSPOStrategyReport/.

RSPO. "Why Sustainable Palm Oil?" https://rspo.org/why-sustainable-palm-oil/.

Rijk, G., M. Okita, and P. Matthew. "Wilmar's Refineries and Brands Lag in Implementation of ESG Policies." https://chainreactionresearch.com/wp-content/uploads/2022/06/Wilmars-Refineries-and-Brands-Lag-in-Implementation-of-ESG-Policies.pdf.

Rudner, M. "The Evolving Political Economy of Malaya's Rubber Development from Colonial Times to Independence." *Economic History of Malaya Project*. https://www.ehm.my/publications/articles/the-evolving-political-economy-of-malayas-rubber-development-from-colonial-times-to-independence.

Sachs, Jeffrey D. "From Millennium Development Goals to Sustainable Development Goals." *The Lancet* 379 (2012): 2206–11.

Santika, T., A.W. Kerrie, M. Erik, B. Sugeng, E.L. Elizabeth, S. Meindra, S. Matthew, A. Marc, and Tun-Min Poh. "Changing Landscapes, Livelihoods and Village Welfare in the Context of Oil Palm Development." *Land Use Policy* 87 (2019): 1–33.

Santika, T., Kerrie A. Wilson, Elizabeth A. Law, Freya AV St John, Kimberly M. Carlson, Holly Gibbs, Courtney L. Morgans, Marc Ancrenaz, Erik Meijaard, and Matthew J. Struebig. "Impact of Palm Oil Sustainability Certification on Village Well-being and Poverty in Indonesia." *Nature Sustainability* 4, no. 2 (2021): 109–19.

"Sawit Watch: An Indonesian Network Against Oil Palm Plantations." *WRM Bulletin*, August 14–30, 1998. https://www.wrm.org.uy/bulletin-articles/sawit-watch-an-indonesian-network-against-oil-palm-plantations.

Schrier-Uijl, A.P., Marcel Silvius, Faizal Parish, K.H. Lim, S. Rosediana, and Gusti Z. Anshari. "Environmental and Social Impacts of Oil Palm Cultivation on Tropical Peat." In *Reports from the Technical Panels of the 2nd Greenhouse Gas Working Group of the Roundtable on Sustainable Palm Oil*, edited by Timothy J. Killeen and Jeremy Goon, 1–40. RSPO, 2013.

Searle, P. *The Riddle of Malaysian Capitalism: Rent Seekers or Real Capitalists?* Honolulu: New Asian Library, 1999.

Segal, Jeffrey. "Home to Hard Times." *Far Eastern Economic Review*, October 1, 1982.

Setiawan, K., and D. Tomsa. "Defending a Vulnerable yet Resilient Democracy: Civil Society Activism in Jokowi's Indonesia." *Journal of Current Southeast Asian Affairs* 42, no. 3 (2023): 350–71.

Siahaan, Mona. "Palm Oil Industry in Indonesia-Statistics & Facts." *Statista*, March 6, 2024. https://www.statista.com/topics/5921/palm-oil-industry-in-indonesia/.

Snashall, G.B., and M.P. Helen. "Oreos Versus Orangutans: The Need for Sustainability Transformations and Nonhierarchical Polycentric Governance in the Global Palm Oil Industry." *Forests* 12, no. 2 (2021): 252.

SPOTT. "Wilmar." https://www.spott.org/palm-oil/wilmar-international-ltd/.

Stoler, Laura Ann. *Capitalism and Confrontation in Sumatra's Plantation Belt, 1870–1979*. Ann Arbor: University of Michigan Press, 1995.

Susanti, A., and A. Maryudi. "Development Narratives, Notions of Forest Crisis, and Boom of Oil Palm Plantations in Indonesia." *Forest Policy and Economics* 73 (2016): 130–39.

Tan, Felicia. "Wilmar Debuts on Dow Jones Sustainability Index." *The Edge Singapore*, November 17, 2020. https://theedgemalaysia.com/article/wilmar-debuts-dow-jones-sustainability-index.

Teo, C.H. "Malaysian Corporations as Strategic Players in Southeast Asia's Palm Oil Industry." In *The Palm Oil Controversy in Southeast Asia: A Transnational Perspective*, edited by Oliver Pye and Jayati Bhattacharya, 19–47. Singapore: ISEAS, 2013.

TuK Indonesia. "Financing & Land Control of Palm Oil and Sustainable Finance in Indonesia." 2015. https://www.tuk.or.id/wp-content/uploads/2015/07/Policy-Brief-Sustainable-Finance.pdf.

TuK Indonesia. "Tycoons in the Indonesian Palm Oil." 2018. www.Tycoons-in-the-Indonesian-palm-oil-sector_compressed.pdf.

Varkkey, H. "Oil Palm Plantations and Transboundary Haze: Patronage Networks and Land Licensing in Indonesia's Peatlands." *Wetlands* 33 (2013): 679–90.

Varkkey, H., and Lee Onn. "The Haze Problem in Southeast Asia: Palm Oil and Patronage." *Southeast Asian Economies* 34 (2017): 573–74.

Vijay, V., S.L. Pimm, C.N. Jenkins, and S.J. Smith. "The Impacts of Oil Palm on Recent Deforestation and Biodiversity Loss." *PLoS One* 11, no. 7 (2016).

Vijayaraghavan, Nandini, and Umesh Desai. "Wilmar International." *World Scientific Book Chapters* (2017): 31–36.

Virgy, M.R., D. Yusa, and B.D. Wawan. "Strategi Jaringan Advokasi Transnasional Greenpeace Indonesia Terkait Isu Deforestasi Hutan Indonesia oleh Wilmar International." *Journal of Political Issues* 1 (2020): 74–91.

Wilmar. "Corporate Profile." https://www.wilmar-international.com/about-us/corporate-profile.

Wilmar. "NDPE Policy." https://www.wilmar-international.com/sustainability/policies.

Wilmar. "History and Milestones." https://www.wilmar-international.com/about-us/history-milestones.

Wilmar. "Action Plan." https://www.wilmar-international.com/sustainability/wp-content/uploads/2018/10/Wilmar-Action-Plan-Announcement-Final-5-Oct-2018.pdf.

Wilmar. "A Sustainable Approach." https://www.wilmar-international.com/sustainability#:~:text=A%20sustainable%20approach&text=We%20strive%20towards%20a%20palm,integrated%20into%20our%20business%20model.

"Wilmar Wins Top 2021 SPOTT." *The Edge Singapore*, November 22, 2021. https://theedgemalaysia.com/article/wilmar-wins-top-2021-spott-assessment.

Wong, K.M. "A Hundred Years of the Gardens' Bulletin, Singapore." *Gardens' Bulletin Singapore* 64, no. 1 (2012): 1–32.

Yacob, Shakila. "Model of Welfare Capitalism? The United States Rubber Company in Southeast Asia, 1910–1942." *Enterprise & Society* 8, no. 1 (2007): 136–74.

Yacob, Shakila. "Government, Business and Lobbyists: The Politics of Palm Oil in US–Malaysia Relations." *The International History Review* 41, no. 4 (2019): 909–30.

Yacob, Shakila, and Khadijah Md Khalid. "Adapt or Divest? The New Economic Policy and Foreign Businesses in Malaysia (1970–2000)." *The Journal of Imperial and Commonwealth History* 40, no. 3 (2012): 459–82.

Yacob, Shakila, and N.J. White. "The 'Unfinished Business' of Malaysia's Decolonisation: The Origins of the Guthrie 'Dawn Raid'." *Modern Asian Studies* 44, no. 5 (2010): 919–60.

Yah, C.L. *Economic Development of Modern Malaya*. Kuala Lumpur: Oxford University Press, 1967.

Yao-Hua, L., and S.L. Wong. "Can Oil Palm Explain the Lower Forest Loss Here?" *Macaranga*, April 27, 2021. https://pulitzercenter.org/stories/can-oil-palm-explain-lower-forest-loss-here.

Yao-Hua, L. "Navigating the Multiverse of Forest Data." March 3, 2023. https://www.macaranga.org/navigating-the-multiverse-of-forest-data-peninsular-malaysia/.

Yee, S.T. "Agro-Commodity Global Value Chains and Upgrading: The Case of Malaysian Palm Oil." PhD diss., University of Cambridge, 2018.

5

HURRICANES AND FLOODS

The Case of the United States and the National Flood Insurance Program

Natalie Rauscher and Welf Werner

5.1 Introduction

Natural disasters are common in North America—more so than in other developed regions such as Western Europe. North America ranks among the regions where the highest losses due to natural catastrophes occur, especially when it comes to economic losses. Other regions, especially Asia, face a higher number of events and loss of life due to natural disasters than the United States while also facing a high amount of economic losses.[1] Overall, however, over the twentieth century, the world has seen a remarkable increase in resilience[2] against natural disasters. This is especially clear when looking at the fatality rates from disasters, where the world has seen an astonishing decrease over the twentieth century.[3] In the first half of the century, it was common to see the annual death toll due to disasters rise into the millions. Yet improved "food security, resilience to disasters, and better national and international responses mean the world has not experienced death tolls of this scale in many decades."[4] One noteworthy example is Bangladesh, which has been frequently hit by storms and consequent flooding. In 1970, more than 300,000 people died after a strong cyclone in Bangladesh; in 1985, another storm claimed 15,000 lives. In 2020, one of the strongest storms on record hit the Bay of Bengal—yet the death toll was only 26. This example illustrates a long-term trend of improvement in the world's resilience to natural disasters. Many countries around the world are hit frequently by natural disasters, especially storms and flooding, yet even some of the poorest countries have been able to increase resilience to these events.[5] This is crucial as the accelerated pace of man-made climate change will lead to higher frequency and intensity of disasters. To increase resilience worldwide, factors such as the physical environment, as well as economic resources, political systems, technological advances, and cooperation, are vital.[6]

In comparison to the world, the United States has seen a rather low level in fatality rates due to disaster in the early twentieth century (also due to sparse population across a large terrain), yet this fatality rate remained remarkably stagnant over time. Due to factors such as the exceptional economic rise of the United States, technological advancement and a stable democratic political system, which are the basis for a good mitigating capacity in response to natural disasters, one would have expected the death toll to fall more strongly in the United States than

DOI: 10.4324/9781003478089-6

in other parts of the world. Yet this is not the case. Additionally, the number of people affected or injured by disasters in the United States has in fact been on the rise in the late twentieth and twenty-first century.[7] Thus, despite the United States' steep increase in economic growth and wealth compared to the rest of the world,[8] this enormous capacity could not be translated into effective disaster mitigation.[9] The rapidly warming climate will exacerbate this already challenging situation.

Yet, besides human-made climate change, there are other factors to consider when explaining the high vulnerability to natural disasters and extreme weather events in the United States over the twentieth century and into the twenty-first century. In fact, a study of normalised historical data for land-falling hurricanes in the United States shows that over the last century, we cannot observe more intense hurricanes—yet the damages and costs from these storms (and consequent flooding) have been on the rise.[10] This means that factors such as population growth and a massive, more or less systematic domestic migration to disaster-prone areas, especially along the coastlines in the South and West of the United States, as well as government mitigation policies that seem to have largely failed to prevent large and growing damages from natural disasters, are two significant factors in explaining the rise in costs and damages from extreme weather and other disasters throughout the last century. Both these factors have led to an increased exposure of people and their livelihoods to disaster events and show the failure to implement effective mitigation measures.[11]

From a business perspective, one of the sectors that will be affected the most by the warming climate and rising sea levels is the real estate sector, especially in the regions along the coasts where frequent extreme disaster events are common. In North America such assets are especially exposed to risk from rising sea levels and flooding while lacking some of the defences other high-income countries such as Japan, the UK, and the Netherlands have implemented.[12] Climate-related risks will raise costs for all parties involved, specifically for insurance, agriculture, and reinsurance, but also for the construction sector and homeowners. But the question is if these costs are primarily borne by American taxpayers or by the parties directly exposed to climate impacts.[13] Studies show that about 40% of the U.S. population lives in coastal regions and is consequently vulnerable to rising sea levels.[14] It is also apparent that construction in these areas in the early 2020s has continued at an even faster pace than in low-risk areas. Many of these properties are overvalued because flood risk disclosure laws do not exist in many of these areas. The right pricing of assets would be necessary to reflect risks adequately but would necessarily come with immediate negative financial implications for households, communities, and municipalities, which is why local governments quite often shy away from implementing the necessary policy changes.[15]

Florida is the state that is projected to be most affected by rising sea levels and consequent flooding, but most states along the U.S. coasts will be exposed to more flooding in the coming decades.[16] Although commercial development in these regions continues almost unabated, at least some awareness of increasing risks can be observed. Flood risk considerations have started to affect buyer and lender decisions, as mortgage withdrawals in high-risk areas are increasing, yet other climate risks (drought or fire) do not seem to be taken into consideration yet.[17] Nevertheless, by the 2020s some property insurance providers have been pulling out of markets such as Florida or California altogether due to high risks and costs. In affected regions, where insurance is still available, premiums are rising significantly.[18] Yet, despite all this, an ambiguous picture emerges for the early twenty-first-century United States: while some studies argue that the demand for and value of properties in high-risk flood zones are still growing,[19] others show that homes located in Special Flood Hazard Areas[20] decrease in value.[21] There is no doubt that, in the medium and long run, the housing and real estate sector will be affected severely by climate change, especially along the coastlines of the country, which are already affected frequently by severe natural disaster events. Yet the problem remains that the United States does not have an effective disaster mitigation system in the form of government policy and effective (mandatory) insurance which would reflect risks adequately and therefore decrease demand and development in high-risk areas.

To shed light on the quite significant challenges for the United States in disaster mitigation, also compared to the rest of the world, this chapter briefly traces the most important societal and political factors behind the rise of disaster costs and damages before going into more detail on the traditions of disaster policy over the last century. The chapter will then focus on the National Flood Insurance Program (NFIP), considering its predecessors and historical development. As this chapter will discuss, there have been persistent shortcomings of flood insurance in the U.S. Flood insurance can be a crucial mitigation measure, but the United States has not been able to implement it in an effective way since the 1960s. Although the United States was the dominant economic, technological, and political power over the twentieth century on a worldwide scale, the country has unfortunately not developed into a world leader in efforts of disaster mitigation but has stagnated or even worsened in comparison to other regions of the world. Yet, as the impact of climate change advances and affects all regions around the globe, it remains paramount to identify and implement effective mitigation measures.

5.2 Societal and Policy Factors Connected to the Rise of Disaster Costs and Damages from Natural Disasters

Over the twentieth century, the United States has seen enormous economic growth and development, especially in the post-war era.[22] With growing wealth among Americans came a further distribution of the population across the country. The population in the entire United States grew tremendously but most profoundly along the coasts. Economic opportunity as well as a desirable climate attracted more and more people to coastal regions.[23] Especially in states like Florida, the size of the population exploded over the twentieth century, and by 2022, the state constitutes the third largest in the United States.[24] But states like Florida also show the downside of this population development. With more and more housing units close to the coast, as well as in once-natural wildlife areas, the exposure and risk of being affected by extreme weather events have been rising for millions of Americans, and for an increasing amount of private and commercial property values.[25] In effect, even if extreme weather events such as hurricanes are not becoming more severe (yet), damages and costs will continue to rise because of the sheer amount of people and wealth moving into high-risk areas.[26]

In such a situation, the communal, state, and federal governments are challenged to alleviate the damages and costs. And these costs have been on the rise in the twenty-first century. As shown in Tables 5.1 and 5.2, the economic losses from natural disasters as a percent of global GDP have risen in the North American region from 0.14% in 2015 to 0.64% in 2022, a more than four-fold increase in seven years. While the United States faces high economic losses from natural disasters, other world regions such as Oceania and Asia also face enormous damages and economic losses. These regions, together with Africa, also face the highest amount of victims due to natural disasters.[27] Yet, as indicated earlier, the death rate due to disasters around the world has fortunately been falling remarkably over the twentieth century.

TABLE 5.1 Economic losses in USD billion due to disasters by region and percentage of global GDP, 2015[28]

Regions	*USD bn**	*% of GDP*
North America	29	0.14
Latin America and Caribbean	7	0.14
Europe	13	0.06
Africa	1	0.05
Asia	38	0.15
Oceania/Australia	3	0.20
Seas/Space	1	
Total	92	
World total		0.12
10-year average**	192	0.28

* *Rounded numbers*
** *Inflation adjusted*

Source: Swiss Re Economic Research & Consulting and Cat Perils.

TABLE 5.2 Economic losses in USD billions due to disasters by region and percentage of global GDP, 2022[29]

Regions	*USD bn**	*% of GDP*
North America	176	0.64
Latin America and Caribbean	17	0.31
Europe	21	0.09
Africa	8	0.27
Asia	51	0.13
Oceania/Australia	10	0.50
Total	284	0.27
World total		
10-year average**	220	0.27

* *Rounded numbers*
** *Inflation adjusted*

Source: Swiss Re Institute.

So far, mitigation policies in the United States seem to have done little to prevent damages and costs from rising. They most probably have even stimulated growth in disaster-prone regions.[30] In fact, development in high-risk flood zones exceeds development in lower-risk areas in many areas since 2010.[31] Government institutions and policies should have helped prevent people and economic activities from moving systematically into the storm's eye or should have at least introduced policies that reflect the potential costs of such movements for the individual actor. This could happen, for example, through mandatory insurance schemes, rather than burdening the public at large with emergency and relief measures that help only in the short run but establish disincentives to improve resilience over the medium and long run. Federal institutions concerned with disaster risk reduction such as the Federal Emergency Management Agency (FEMA) or the National Flood Insurance Program (NFIP) have been working inefficiently since their inception in the 1970s and 1960s, which will be detailed later. Additionally, federal institutions are not in control of the state and local levels and are thus often financing costly infrastructure projects and ad-hoc relief efforts in places that are hit by natural disaster events time and again and that, after the storm has moved on, do not incentivise necessary changes and preventative measures.[32] Mandatory insurance schemes that have contributed to breakthroughs in risk mitigation in Western Europe in the early twentieth century as for example in the case of flood insurance in the Netherlands or fire insurance in big cities,[33] as well as multi-disaster insurance schemes, have remained unpopular and underdeveloped in the United States. They are often perceived as an excessive and unnecessary form of government intervention into livelihoods and commercial activity by many American citizens and policymakers.[34]

Additionally, once established, inefficient organisational processes such as those in the NFIP are extremely hard to reverse or fundamentally change. The theory of "path-dependence"[35] puts a significant emphasis on the idea that 'history matters,' or, in other words, that the historical origins of institutions or organisations significantly determine their later developments.[36] Such dependencies can be a significant hurdle in any attempts at reforming organisational structures and have to be considered in the case of flood insurance in the United States.[37] As research has shown, government risk management strategies have, in many cases, been identified as path-dependent and often focused on single hazards.[38] Both factors have posed a challenge to effective flood insurance in the United States.

As we will see later, these and other challenges reflect some of the complexity and competing interests when it comes to disaster risk reduction in any given country. Federal, state, and local policymakers, insurers, residents, and the business sector in the United States have rivalling interests in a given region and in order to become more resilient at the community level, these interests need to come together under one overarching goal. Yet, in many parts of the United States, a deep scepticism towards "big government" and specifically towards the federal government, is not making it any easier to address issues of vulnerability and resilience to disaster. In many regions of the country, vulnerability to natural events has also been unevenly spread, as research into the social dimensions of vulnerability of the U.S. population has shown.[39] Additional factors that have made these efforts more complicated in the United States include deeply rooted individualism, political polarisation, ideologically entrenched climate scepticism,[40] and the strength of conservative media—factors that, in many cases, are interwoven with Republican administrations, including the Trump administration, as well as Republicans in Congress and state governments. All these factors have prevented efficient reform. Both the political and technical problems that muddle efforts to improve disaster preparedness, especially in the context of the NFIP, explain the contradictory and ineffective state of American disaster policies to this day.[41]

5.3 Traditions of American Disaster Policy

Besides specific social, cultural, and political factors that have contributed to a low resilience towards natural catastrophes in the United States, there is also a long tradition of specific disaster policy that warrants some scrutiny in the context of today's policy failures.

Before any systematic approach to disaster events was established in the United States after World War II, Congress provided relief funds on a regular but disaster-by-disaster basis since the nineteenth century.[42] Based on this tradition, the New Deal programmes introduced under President Roosevelt in the 1930s saw a constitutional basis rooted in the general welfare clause to define environmental hazards and events such as the Dust Bowl as a disaster that demanded federal government assistance. Additionally, New Deal programmes for land and water conservation directed funds towards the first systematic studies of hazards across the United States.[43] In the aftermath of World War II, new acts such as the Federal Civil Defense Act (1951) established new institutions to prepare not only for an attack from an outside enemy but also for natural disasters which happened frequently across the country. At this time, an executive agency in this field was created, giving the President authority to declare a national disaster. Funding in the area of civil defence significantly added to the emergence of disaster studies and research that, to a certain degree, was also deemed suitable for natural disaster preparation.[44]

Rising costs from disasters such as the deadly earthquake and tsunami hitting Alaska in 1964 also led to an increase in "hazard research" among geographers, land-use planners, economists, and engineers. Gilbert White, a geographer, was a key figure in the emerging academic field even before the 1960s and was later influential in advising the federal government.[45] He worked on crucial studies such as *Human Adjustment to Floods: A Geographical Approach to the Flood Problem in the United States* (1945) and *Changes in Urban Occupancy of Flood Plains in the United States* (1958). Although a lot of research was funded through civil defence dollars, among hazard researchers it became increasingly clear that nuclear attack was not imminent, but that the nation was increasingly threatened by major natural disaster events that needed to be addressed with effective strategies. Natural disaster events seemed a formidable challenge in its own right, and possibly a much more pressing one than any threat of an outside attack. Additionally, the Great Society programmes of the Johnson Administration aimed to tackle poverty (including disaster relief) as well as address environmental protection. The Johnson Administration pursued these efforts much more strongly than previous presidents.[46]

By the end of the 1960s, expertise around flood hazards in the United States had grown enormously. Again, Gilbert White was the most influential figure in this area. In hazard research, White and his colleagues wanted to address the "second environmental crisis"[47] that occurred—the first being the pollution crisis brought into public consciousness by Rachel Carson in the early 1960s. If developments in hazardous terrain were left unmitigated, the argument went, losses from disasters would likely rise significantly over the coming decades[48]—in the light of realities in the early twenty-first century, a remarkable early insight. While such concerns and interventions were already forming among researchers in the mid-twentieth century, the academic community only partly succeeded in instilling them adequately in the establishment of the NFIP.

Another aspect to mention in the area of disaster policy traditions in the United States is that prior to the introduction of specific disaster mitigation institutions, the system was reactionary rather than preventative, and this attitude has outlasted even the creation of more preventative systems such as the NFIP. Politically, elected officials are "encouraged" to spend on relief rather than on preparation and mitigation; sadly, voters would rather reward relief efforts than condone costly preparatory measures when no crisis is at hand.[49] Nevertheless, the United States seems "exceptional" to a certain extent in this regard as other democratically elected governments were able to implement more effective programmes despite high-risk exposure such as in the Netherlands or Japan.[50]

As indicated before, it is likely that certain characteristics of American society such as deeply rooted individualism and distrust of the federal government, as well as path dependence, explain some of the difficulties in reforming the system. Once a path was established—namely a subsidised and non-mandatory system of flood insurance which will be detailed later—it has proven increasingly hard to turn around such a programme, even if it proves inefficient. Moreover, disaster policy is often technocratic in nature and therefore regarded as a "policy without a public,"[51] in which the public is largely disinterested and elected officials disengaged because of its complex nature. Such a policy receives little media coverage compared to well-publicised emergency relief. Thus, the public remains uninformed about the specifics and necessary reforms of a well-functioning flood insurance scheme.[52]

5.4 The National Flood Insurance Program

The NFIP is one of the most important disaster policy measures to be passed in the United States. It was a historic achievement to move from ad-hoc disaster relief to a system that was geared towards prevention, mitigation, and relief. After several historically large disaster events in the United States in relatively short succession in the 1950s and 1960s,[53] the federal government became convinced that it was necessary to support a more elaborate federal programme to address rising costs and damages from natural disaster events.[54]

There were several stages of research, surveying, lobbying, and legislation that preceded the NFIP as a programme to address flood risks throughout the United States. For example, already in 1913, after a flood along the Ohio River Valley killed 415 people and caused millions of dollars in property damages, there was increasing interest in flood control in the public, and several levee associations and lobbying groups were founded.[55] In 1916, the U.S. House of Representatives' Committee on Flood Control was created which became the discussion ground for congressional proponents of flood control.[56] A year later, in 1917, the federal government accepted responsibility for flood control officially for the first time by passing the Flood Control Act (PL 64-367). It established requirements for local financial contributions to flood-control legislation and gave the Corps of Engineers the authority to survey several regions for flood-control improvements. But this new legislation soon proved insufficient. In 1927, the Great Mississippi River Flood killed over 200 people and caused over $200 million in damages, which caused a new Flood Act (1928) to be passed and millions in federal funds to be authorized for the construction of protective measures. Significantly, in 1929, the private insurance industry abandoned coverage of flood losses completely as it was deemed too costly. From that time onwards until the establishment of the NFIP in 1968, Americans could not take out flood insurance even if they wanted to.[57]

Other flood- and natural disaster-related legislation and authorisations followed in the 1930s, including the direct assistance of citizens affected by natural disaster events such as earthquakes or floods. The year 1936 brought a new Flood Control Act which can be seen as a predecessor to the NFIP. It brought together land use management and engineering plans to manage possible floods.[58] After more surveys and studies in the 1940s, the Disaster Relief Act of 1950 was passed to provide assistance by the federal government to state and local governments in disaster situations, including floods. This also established executive authority in disaster management as the president declared a major disaster following a state's request so that federal assistance could be granted.[59]

In the 1950s, major flooding events created never-before-seen amounts of damages (e.g. flooding in Kansas and Missouri with over $800 million in total damages) which prompted President Truman to call for a national system of flood disaster insurance.[60] In his view, such a system could work if it was based on private insurance with reinsurance guarantees by the federal government.[61] Under President Eisenhower, the Federal Flood Insurance Act of 1956 was to establish a programme of federal reinsurance. The programme was supposed to provide up to $10,000 per dwelling and was intended to encourage private companies to provide additional coverage. The cost of coverage for policyholders was the same regardless of location, however.

The programme was thus one of the examples of a severe lack of understanding of the necessity to incentivise disaster mitigation because it did not reflect the actual risk in different locations. At the same time, private insurers remained convinced that flood insurance was still not commercially feasible under the proposed programme.[62] The implementation of the programme was de facto voted down by Congress, as there were too few technical studies supporting the programme's aims,[63] and so the Federal Flood Insurance Act was never enacted.[64]

Interestingly, it soon became clear that problems such as moral hazard that would later plague the NFIP already came to light in the late 1950s. A study on *Changes in Urban Occupancy of Flood Plains in the United States* (1958) by Gilbert White makes clear that despite new measures to address the risk of flooding, the population living in flood zones was in fact increasing because of few incentives to stay away. Rather, there was a perception among communities that if serious flooding happened, the federal government would deal with it (*moral hazard*).[65] White continued to voice his concerns in 1966—before the establishment of the NFIP—to little avail. Falsely administered flood insurance could have effects contrary to what was intended—more rather than less development in flood-plains and a lack of broader flood hazard mitigation[66]—something the United States has been struggling with ever since.

Despite major events such as hurricanes and consequent flooding throughout the United States, there was little progress in flood-plain management in the late 1950s and early 1960s.[67] Although more information about flood risks and hazards was collected and disseminated, for example through the Corps of Engineers, a new policy to address these challenges was not passed. This was because U.S. disaster mitigation policy remained fundamentally reactive rather than preventative.[68] This only changed by the mid-1960s when a political consensus finally developed that there was an acute need for federal intervention in disaster mitigation through a more elaborate programme. This consensus was triggered by major events such as the Alaska earthquake in 1964 and the United States' first billion-dollar disaster, Hurricane Betsy in 1965.[69] The Southeast Hurricane and Disaster Relief Act of 1965 was the first step in this direction.[70] Insurance was a major factor among the ideas for the establishment of a federal programme, and the Department of Housing and Urban Development (HUD) was charged with evaluating possible avenues through an extensive study.[71]

The year 1966 brought to light a crucial report by another task force on the subject, the *Task Force on Federal Flood Control Policy*, chaired by Gilbert White. It recommended that Congress should implement measures to decrease flood losses and control expenditures for flood control. The report concluded that federally subsidised insurance could provide a powerful incentive to local communities to participate in a flood insurance programme. It argued in favour of an integrated national programme of flood insurance that involved federal, state, and local governments, as well as the private sector. However, White's report stressed, as mentioned before, that a perpetually subsidised insurance scheme for flood insurance could have adverse effects and could even incentivise further development in flood-plains because of a lack of reflection of risks.[72] After the HUD report also concluded that a NFIP was feasible and in the public interest, President Johnson submitted this report to Congress in 1966. The report "envisions a program of essentially private character but with continued large-scale participation of the federal government."[73] The report also recommended that all lending entities under federal supervision require flood insurance at unsubsidised rates in high-risk areas on all new mortgages.[74]

In 1968, the National Flood Insurance Act established the NFIP and the Federal Insurance Administration within the Department of Housing and Urban Development to provide flood insurance to communities, which could participate in the programme if they voluntarily adopted and enforced minimum NFIP flood-plain management requirements.[75] Residents were eligible for insurance once the NFIP identified local flood hazards and established rates. The insurance industry formed a group of companies to collaborate with the government on providing insurance, the National Flood Insurers Association.[76] However, crucially, rates in flood-plains would be subsidised. The National Flood Insurance Fund was established at the Department of the Treasury where premiums from the sale of flood insurance were deposited. The NFIP was allowed to borrow up to $1 billion from the Treasury to cover losses,[77] which is only one aspect of the NFIP that would soon prove insufficient.

Another major step besides the inception of the NFIP was the creation of the FEMA in 1979. FEMA was created in order to "streamline the acquisition of disaster relief and to assist states with hazard mitigation efforts."[78] The NFIP was subsequently moved from HUD to FEMA authority. The 1970s was the last time of intense activity in disaster policy and federally funded disaster research. Besides the establishment of FEMA, this decade saw the creation of the Environmental Protection Agency (1970), the Earthquake Hazard Reduction Act (1977), and the creation of hazard and disaster research centres driven by the National Science Foundation, for example the Natural Hazards Center at the University of Colorado.[79]

Although FEMA was refocused somewhat on civilian defence in the 1980s under Ronald Reagan, there was still some improvement in national flood insurance coverage. Private insurance companies were now able to sell and manage flood insurance in their own name, which led to a much wider distribution of insurance policies. In 1988, this meant that the NFIP broke even for the first time since its inception.[80] The 1990s were a mixed period for FEMA and the NFIP, marked by major disaster losses through the Great Midwest Flood of the upper Mississippi and lower Missouri River in 1993 or Hurricane Andrew (1992). FEMA was harshly criticised after these events but also saw new focus and innovation in the agency, as well as growing demand for flood insurance under the Clinton Administration.[81] 9/11 changed the national disaster system profoundly, most of all with the creation of the Department of Homeland Security in 2002. FEMA was removed from cabinet status, and counter-terrorism became a much stronger focus for disaster preparedness.[82] With Hurricanes Katrina, Rita, and Wilma in 2005, FEMA's dysfunction at every level was exposed for the first time since the political shift towards homeland security. The severe storms exposed the weaknesses of the United States' disaster mitigation policy which had not provided structural mitigation nor flood-plain restrictions to keep the population out of risk zones.[83] These events showed that instead of the desired outcomes of the NFIP and FEMA to create structural mitigation coupled with flood-plain restrictions and a safety-minded land-use revolution, they had failed. Since these events, the NFIP has piled up large amounts of debt. In the year 2022, the NFIP's debt stood at $20.5 billion.[84]

5.5 Who Was Influential in Drafting the NFIP as It Was?

As we have seen earlier, there had been a long history of disaster response in the United States before the NFIP was created. But who was actually engaged in its inception?

As mentioned before, there were academic experts involved in the inception of the NFIP. Gilbert White had an exceptional position in this regard and was decisive in issuing several reports before chairing the Task Force on Federal Flood Policy under President Johnson.[85] It can even be argued that the lack of another intellectual and political leader such as Gilbert White in the area of disaster mitigation is another reason why reform of the NFIP has not materialised.[86] Parallel to White's efforts, another task force was founded in the Department of Housing and Urban Development (HUD) which issued a report before the NFIP was established, titled *Insurance and Other Programs for Financial Assistance to Flood Victims (1966)*. Unlike White's report, this report was less concerned about the possible downsides of a federally subsidised flood insurance programme.[87] There were several other federally funded agencies and groups that took part in the foundation of the NFIP, for example, the U.S. Army Corps of Engineers (USACE) that undertook surveys of flood-plains. The USACE came to the conclusion that there were about 5,000 flood-prone communities in the United States (in 1968) and issued Guidelines for Reducing Flood Damages.[88] The Water Resources Council was another ad-hoc agency that was involved in the creation of the NFIP and was to determine the appropriate methodology for flood frequency measurement. The council was also to guide local planning agencies in developing ways to identify flood-plain areas, prepare data on them, and develop flood loss reduction plans through, for example, new regulations.[89] Twenty-six federal agencies adopted a preliminary report by the Water Resources Council in 1967 outlining methods and standards for developing information about flood hazards. This report established the still-used "100-year flood" as the base standard for federal agencies assessing flood risks;[90] there is a 1% likelihood of occurrence of such a flood in any given year. Apart from federal agencies, the states were by then also convinced that some organised form of disaster relief and mitigation was necessary and therefore the programme was approved in both Houses of Congress in 1968.[91]

At this time, the insurance industry in the United States agreed to take part in a federally subsidised programme of flood insurance, which would include an industry pool and federal assistance through congressionally approved funds to cover operational losses every year.[92] Research has shown, however, that the insurance industry was mainly on board because the programme was designed the way it was—in a non-mandatory form, subsidised, and explicitly excluding disaster events outside of floods. The state would ultimately also serve as the reinsurer of last resort.[93] When the NFIP was created, historical records found that studies at the time were heavily influenced by insurance industry representatives who were sceptical of a flood insurance programme.[94] They advocated for voluntary flood risk insurance instead of mandatory insurance. Although there were other studies advocating for mandatory insurance as well as the inclusion of other possible disaster events, the government adopted the NFIP that did not cover other natural disasters besides floods. Thus, the "conservative and—from an insurance technical point of view—less functional program got realized."[95] Deeply held beliefs about independence and self-reliance of American citizens influenced the report of HUD at the time; the underlying rationale was that Americans would protect their property in hazardous zones out of common sense and free will.[96] Additionally, the non-mandatory version of flood insurance was reflective of the notion that comprehensive disaster insurance had never been tried in the United States before and that an insurance programme "mandated by the state would be like administering a tax and limit individual freedom in a way that was not acceptable for the majority of the public."[97] A different version of disaster insurance, discussed by the Office of Emergency Planning at the time, would have covered all sorts of natural disasters and the entire United States—as it would have been mandatory. This would have had the decisive advantage to spread the risk across space and events as well as to keep premiums lower, which would have been more acceptable to homeowners. The problem with flood-only insurance is that premiums are high due to the concentration of flood losses and limited groups of policyholders. The result for the U.S. context shows a corroboration of this insight with high premiums which were ultimately subsidised by the federal government.[98] Until today, all-hazard insurance, as it is practiced in France or Spain, is not available in the United States.[99] Ultimately, the NFIP came into existence because of legal, economic, and cultural factors that did not favour a more comprehensive disaster insurance programme. Instead, the NFIP remains dysfunctional and not self-sustaining to the present day.[100]

Even after the programme was established, several actors tried to slow down and hinder the participation of localities in the programme. These efforts were somewhat successful because the adoption of NFIP standards was completely voluntary. Especially the real estate and construction sectors, but also local municipalities, wanted to sustain tax revenue through unhindered economic expansion and development and were therefore opposed to the adoption of NFIP standards.[101] In the meantime, even though there were some efforts to address the shortcomings of the NFIP, individual policyholders of flood insurance have become accustomed to heavily subsidised rates. This is a benefit that actually stands in the way of further reform, as the recipients of discounted rates are unwilling to give these benefits up lightly.[102] Again, we see here the strong path dependence of a programme such as the NFIP, which makes an organisational overhaul challenging from several perspectives, including those of policymakers, the business community, insurance policyholders, technocrats, and the public at large, who all have different understandings of what the programme should achieve (for them) in the future. Additionally, concrete elements of the programme are "stuck" in the past. For instance, aspects such as the idea of the 100-year flood-plain or the lack of inclusion of new risk factors such as sea-level rise due to climate change[103] show that the historical roots of American disaster mitigation strategies are an obstacle to effective reform of the programme. Path dependence reinforces the existing status quo in policymaking because existing policies result in "1) infrastructural investments ('fixed costs'), 2) knowledge of how to do things a certain way ('learning effects'), 3) governance structures that facilitate the status quo ('coordination effects'), and 4) expectations from the public of how aid will be structured ('adaptive expectations')."[104] Other prominent examples of such detrimental path dependencies can, for example, be found in the U.S. healthcare or welfare system.[105]

5.6 Enduring Issues of the NFIP

Although the NFIP was finally established in the 1960s, it has been plagued by many deficiencies. But what have been the most important elements of dysfunction in the system?

Many experts, including Gilbert White, advocated for the NFIP to be established as an experiment first, given the still-massive lack of reliable information. Yet the officials heading the NFIP at the time did not favour any pilot programme but went ahead with the establishment of the national programme. Soon, one of the most severe problems of the NFIP arose: a lack of participation due to its voluntary nature. Although it was assumed that communities in flood-prone areas would be eager to provide their residents with highly subsidised premiums (for existing dwellings in flood-risk areas; actuarially fair rates for new dwellings) and would pass legislation in this regard, this was not the case. Homeowners and small businesses in these regions were also not eager to purchase flood insurance.[106] Even when the Corps of Engineers identified a special flood hazard area in a community, many localities chose to "take their chances"[107] instead of restricting development. But chances were not good. Many non-complying communities were hit by natural disaster events even in the very early years of the programme. This led to an amendment of the NFIP by Congress in 1970 to let communities enter the programme on an emergency basis as long as they accepted flood-plain control on new construction.[108] With another amendment in the 1970s, the NFIP was strengthened, as it forced local governments to take part or otherwise be ineligible for federal grants and aid. Private households in these areas would also not be able to get insurance and would be ineligible for federal aid. With the amendment, flood-plain property owners (including business owners) who received mortgages from federal institutions were obliged to purchase flood insurance: "Sanctions enacted by the 1973 amendments were sufficient to bring about a marked increase in participation in the NFIP. Within 5 yr, the number of participating local governments increased from <3000 to well over 18,000."[109]

With subsidised rates and a chain of natural disaster events, more communities and residents were interested in flood insurance by the mid-1970s: "However, growth raised serious challenges, and these challenges have remained with the program to the present day: (1) keeping up with new mapping requests and updating old maps, and (2) compelling and enforcing land-use restrictions in flood-prone communities."[110] Through a lack of staff in the NFIP and other organisations, it was not possible to update maps and enrol communities efficiently. Even though state and local governments became more involved in flood-plain land use management overall, many maps remain(ed) in an early and inadequate "emergency phase" even decades after the law had been installed. Almost all states introduced laws that allowed local governments to regulate flood-plains. Yet, even though the NFIP was active, costs from flooding increased over the years, and the law was amended again in 1994 to incentivise even more people to buy flood insurance and foster mitigation of flood hazards.[111] The problem of communities staying in the emergency phase even years after joining the programme and not implementing the flood-plain management standards remained.[112] Mapping has continued to be a major problem of the programme, as it demands continuous updating (and inclusion of new risks) and therefore major funding. This has been one of the most significant weaknesses of the programme since the beginning.[113]

The NFIP was implemented at a time that had witnessed enormous disaster events but also a remarkable shift in population and wealth growth and density in disaster-prone areas, especially along the coasts where hurricanes are likely.[114] In this context, the NFIP has also grown rapidly, selling millions of new policies in thousands of communities across the United States. Yet, again, problems have arisen through this development as insurance through the NFIP is concentrated in certain areas, especially coastal ones. 94% of NFIP policies are located in coastal states (NFIP 2022).[115] Under these circumstances, it is not surprising that the programme has not been financially viable. In order to meet its obligations over the years, the NFIP has borrowed enormously from the Treasury and stood at more than $20 billion in debt in 2022.[116] In addition, the private insurance sector is only involved in writing policies and processing claims on behalf of the NFIP but bears none of the risk and is also not involved in rate-setting.[117] Overall, the private flood insurance market remains small and usually only offers coverage above the NFIP cap in the lender-placed market.[118]

In summary, the NFIP remains inefficient and faces continuous increases in losses because it is not in a position to adapt its flood-plain maps to new risks and is furthermore unable to enforce comprehensive building codes and flood-plain management reforms which are central to the programme's original intent. The NFIP is criticised for keeping rates as low as possible which results in the inability to fund revisions in flood insurance maps in a timely and updated manner. Additionally, there is no mapping and requirement for local governments to regulate development in order to address future risks such as rising sea levels, for instance. Finally, the maps that exist are often not detailed enough to integrate them into local land-use planning for flood hazards.[119] This leads to the false reflection of hazards in insurance premiums, which contributes to losses. When "insurance premiums are too low and do not reflect the actual risk of loss, the resulting subsidy seems likely to encourage floodplain development and could account for some portion of the rising flood losses."[120] One could argue that under NFIP rules, the programme actually insures moral hazard: instead of promoting risk-averse behaviour, the programme actually entices people to take risks, for example, by moving to and living in a flood-prone area along the coasts. The programme then "rewards" them with relief measures once their luck runs out.[121] The moral hazard critique is complex, but data show that there continues to be low and slow take-up of insurance even when it is available. This does not only have to do with moral hazard but also with factors such as cognitive bias that affect risk-related decisions, continuous inadequacy of flood maps and management advice, lack of enforcement of flood-plain policies in the communities, a lack of adequate information for home buyers, and local political pressure from homeowners and the construction industry. Flood mapping is crucial but actually remains controversial, as many property owners petition locally to be excluded from the maps in order not to be obligated to buy insurance. As indicated before, it is quite expensive to re-evaluate regions for flood hazards every five years, and the mapping is thus behind in many regions, or the NFIP has decided not to map certain areas that are at risk from storm-water drainage or coastal erosion, for instance. Thus, the programme continues to suffer because the ambitions of the policy are not underpinned by the knowledge and funds to carry it out successfully.[122]

5.7 Conclusion

After several reauthorisations over the years, the 2012 Biggert-Waters Flood Insurance Reform Act was passed to systematically overhaul the NFIP and address inadequacies.[123] Among the most important reform ideas was the implementation of the findings of disaster science such as accurate flood-plain maps and risk calculations for public policy and flood protection. One of the most important parts of this reform was to remove insurance subsidies for certain properties and move towards premiums that realistically reflect calculated flood risks, which remains an extremely difficult endeavour. Soon after the 2012 law was introduced, however, homeowners complained that new FEMA maps overestimated their exposure and that they could not afford the rise in premiums. The reform of 2012 turned out to be extremely unpopular politically,[124] and with a new 2014 law (Menendez-Grimm Homeowner Insurance Affordability Act), the strongest new provisions of the 2012 NFIP reform—namely the movement towards actuarial insurance premium rates—were rolled back.[125] This shows that it still remains challenging to advocate effectively for a sound insurance program in the court of public opinion today. The political appetite to fund disaster mitigation policy when no crisis is at hand is often weak. Instead, post-disaster relief and recovery spending is popular[126]—a vicious cycle that proves to be disastrous in the medium and long run, as the historical development of the NFIP shows.

Even though the United States should have had the capacity to improve disaster mitigation effectively over the twentieth century due to its economic and political position, this was not the case. In comparison with the rest of the world, disaster mitigation has been stagnant (death toll due to disasters) or worsened (number of affected/injured as well as economic losses due to disasters), whereas other regions in the world, especially in Asia, have been able to effectively improve mitigation. One aspect in this regard is effective policy measures and government agencies that coordinate mitigation policies—something that the United States has not achieved. Even though, as this chapter has shown, in the late 1960s it was remarkable that it was even possible to pass a federally backed flood insurance system such as the NFIP in the United States against the background of various societal, cultural, and political factors, the policy has not been effective almost since its inception. Since then, any consensus on environmental or disaster policy has vanished in the U.S. political context. Historical records show that very early on, and even before the programme was passed, there had been concerns about the effectiveness of the programme by disaster policy researchers—which were eventually proven right. Especially the fact that the programme was not mandatory but also that it involved perpetually subsidised rates, made it ill-suited to reflect the actual risk of residing in flood-hazard zones. Until now, the political goals for the NFIP and the technocratic goals seem to diverge.

What was the role of the insurance industry in this development? Historically, insuring flooding and related disasters has never been an attractive business avenue for insurers in the United States and was consequently not pursued. Additionally, when the NFIP was drafted, the insurance industry advocated for a limited role of the government, opposing a mandatory scheme and one that extends beyond flood risks. To this day, private insurers' commitment to insuring flooding is limited and will possibly remain so in the face of climate change and the additional risks that are coming with it.

Thus, the United States maintains a system that does

> not only fail to meet its prescribed goals, it also systematically benefitted some groups at the expense of others; specifically, it subsidized the cost of living and doing business in areas at high risk of flooding at the expense of taxpayers everywhere.[127]

At the same time, the system of subsidies for insurance premiums has not incentivised large numbers of people to purchase insurance. Yet, in the case of the NFIP's failures, a mere technocratic adjustment might not be enough in today's America to fix the problem as could be seen in the quick reversal of NFIP reforms after 2012.[128] With a Congress that already serves particular interests more than the public at large[129] and is polarised by ideologically entrenched positions on climate change, it is questionable if reform of the NFIP will be possible in the foreseeable future. Moreover, bringing in expertise from academic research has proven to be difficult in U.S. economic policies, and especially so in the case of disaster policy; as we have shown, the "disaster science-policy action gap"[130] has developed over a long period—and might be here to stay. An example to the contrary, however, is the history of fire insurance where research successfully informed policy in the nineteenth and twentieth centuries.[131]

Despite some adjustments to the programme over the years, it is questionable if the NFIP can truly be reformed, given the path-dependent nature of its political and historical context. Nevertheless, researchers are suggesting specific areas of reform that could make the programme more effective. Such suggestions include encouraging all-hazard policies as a standard offering to provide more financial protection for households, extending mandatory purchase requirements for flood insurance to a broader area (which would reduce the flood insurance gap and possibly spur the involvement of the private insurance market with appropriate indicators and pricing of risk), providing more information and effective communication to home buyers, encouraging public-private partnerships to close the flood insurance gap, and finally supporting community investment in infrastructure for flood protection.[132]

Notes

1 Swiss Re Institute, *Sigma 2/2019*, 5.
2 The term *resilience* is used across many disciplines, including physics, ecology, material science, social science, economics, engineering, psychology, sociology, anthropology, public health, and disaster management (Bergstrand et al., "Assessing the Relationship," 393; Rus, Kilar, and Koren, "Resilience Assessment of Complex Urban Systems to Natural Disasters," 312). Generally,

socio-ecological definitions of resilience refer to adaptive capacity, meaning how society responds to, recovers from, and adapts to new conditions (Rus, Kilar, and Koren, "Resilience Assessment of Complex Urban Systems to Natural Disasters," 312). Resilient communities display a certain level of preparation and capacity in disaster prevention, which includes factors such as technological development, education levels, economic status, quality of the natural environment and resource management, stable political structures, adequate infrastructure, and flows of knowledge between political levels and inside communities (UNDRR, "Understanding Disaster Risk."). Improving resilience levels includes both the capacity to avoid the loss of life due to disaster as well as high levels of economic losses. In a developed country like the U.S., higher economic losses rather than loss of life can be expected, whereas it is the other way around for developing countries (Mochizuik et al., "Revisiting the 'Disaster and Development' Debate," 41f; UNDRR, "Disaster Risk Reduction in Least Developed Countries.").

3 Ritchie and Roser, "Our World in Data."
4 Ritchie, "A Century of Global Deaths from Disasters."
5 Ritchie, "The World Has Become More Resilient."
6 Ritchie, "The World Has Become Resilient."
7 Ritchie and Roser, "Our World in Data."
8 Roser, "What Is Economic Growth?"
9 Werner, "Natural Catastrophes and Their Effects on Reinsurance," 260.
10 Weinkle et al., "Normalized Hurricane Damage in the Continental United States, 1900–2017," 808–13.
11 Mener, "Disaster Response."
12 Hallegatte et al., "Future Flood Losses in Major Coastal Cities," 802–6; Gibbs, "Climate Change Impact on the U.S. Real Estate Market"; Jones, "U.S. Cities Most Impacted."
13 Yorke, "House of Cards."
14 NOAA, *What Percentage of the American Population Lives Near the Coast?*
15 Yorke, "House of Cards."
16 Jones, "U.S. Cities Most Impacted."
17 Bachaud, "Home Buyers Are Paying More Attention."
18 Yorke, "House of Cards."
19 Bachaud, "Home Buyers Are Paying More Attention."
20 Kousky et al., "Flood Risk and the U.S. Housing Market," 3–24.
21 Bernstein, Gustafson, and Lewis, "Disaster on the Horizon," 253–72.
22 Roser, "What Is Economic Growth?"; FRED, "Gross Domestic Product, USA, 1947–2024."
23 United States Census Bureau, "Coastline County Population," 2012.
24 United States Census Bureau, *New Florida Estimates Show Nation's Third-Largest State Reaching Historic Milestone*.
25 Strader, "We Keep Moving"; Jones, "U.S. Cities Most Impacted."
26 Weinkle et al., "Normalized Hurricane Damage."
27 Swiss Re Institute, *Sigma 2/2019*, 5.
28 Swiss Re Institute, *Sigma 1/2016*, 4.
29 Swiss Re Institute, *Sigma 1/2023*, 29.
30 Frank, Gesick, and Victor, "Inviting Danger."
31 Jones, "U.S. Cities Most Impacted."
32 Strader, "We Keep Moving."
33 Werner, "Natural Catastrophes."
34 Rohland, "Adapting to Hurricanes"; Birkland and Waterman, "Is Federalism the Reason?" 697–714; Shugart II, "Disaster Relief as Bad Public Policy," 519–39; Schuster, "Political Disasters"; Gasper, *Make It Rain*.
35 Greener, "Path Dependence."
36 Castaldi and Dosi, "The Grip of History and the Scope for Novelty."
37 Path dependence in the context of flood risk management has also been studied for other regions, for example, in alpine mountain regions in Austria (Seebauer et al., "How Path Dependency Manifests in Flood Risk Management"; Frost, "Path Dependent"; Tier, "Overcoming Contemporary Reform").
38 Thaler et al., "Addressing Path Dependencies."
39 Cutter and Finch, "Temporal and Spatial Changes," 2301–6.
40 Rauscher, "American Philanthropy in the Age of Political Polarization."
41 Knowles and Kunreuther, "Troubled Waters," 329.
42 Mener, "Disaster Response."

43 Knowles and Kunreuther, "Troubled Waters," 330.
44 Knowles and Kunreuther, "Troubled Waters," 330.
45 Tier, "Overcoming Contemporary Reform."
46 Knowles and Kunreuther, "Troubled Waters," 335; Rohland, "Adapting to Hurricanes," 11.
47 Knowles and Kunreuther, "Troubled Waters," 335.
48 Knowles and Kunreuther, "Troubled Waters," 335.
49 Strother, "The National Flood Insurance Program," 462.
50 Taurus Flood, "What We Can Learn from Flood Control in the Netherlands"; Shimizu, "Policy Lessons for Japanese Disaster Responses"; Hallegatte et al., "Future Flood Losses," 802–6.
51 Strother, "The National Flood," 468.
52 Strother, "The National Flood," 468; Rauscher, "In the Face of Disaster."
53 American Institutes for Research, *A Chronology of Major Events*, 332.
54 Knowles and Kunreuther, "Troubled Waters," 332.
55 American Institutes for Research, *A Chronology of Major Events*, 2.
56 American Institutes for Research, *A Chronology of Major Events*, 2.
57 Knowles and Kunreuther, "Troubled Waters," 332.
58 American Institutes for Research, *A Chronology of Major Events*, 3.
59 American Institutes for Research, *A Chronology of Major Events*, 4.
60 American Institutes for Research, *A Chronology of Major Events*.
61 American Institutes for Research, *A Chronology of Major Events*, 5.
62 American Institutes for Research, *A Chronology of Major Events*, 7.
63 Knowles and Kunreuther, "Troubled Waters," 332.
64 Rohland, "Adapting to Hurricanes."
65 American Institutes for Research, *A Chronology of Major Events*, 7.
66 Strother, "The National Flood," 462.
67 American Institutes for Research, *A Chronology of Major Events*.
68 Rohland, "Adapting to Hurricanes," 10.
69 Hurricane Science, "1965—Hurricane Betsy."
70 Knowles and Kunreuther, "Troubled Waters," 332.
71 American Institutes for Research, *A Chronology of Major Events*, 9.
72 Knowles and Kunreuther, "Troubled Waters," 333.
73 American Institutes for Research, *A Chronology of Major Events*, 10.
74 American Institutes for Research, *A Chronology of Major Events*, 10.
75 Knowles and Kunreuther, "Troubled Waters," 332.
76 Knowles and Kunreuther, "Troubled Waters," 336.
77 American Institutes for Research, *A Chronology of Major Events*, 12.
78 Knowles and Kunreuther, "Troubled Waters," 339.
79 Knowles and Kunreuther, "Troubled Waters," 342.
80 Knowles and Kunreuther, "Troubled Waters," 340.
81 Knowles and Kunreuther, "Troubled Waters," 340.
82 Knowles and Kunreuther, "Troubled Waters," 341.
83 Knowles and Kunreuther, "Troubled Waters," 341.
84 FEMA, "NFIP Debt."
85 Knowles and Kunreuther, "Troubled Waters," 332.
86 Tier, "Overcoming Contemporary Reform."
87 Knowles and Kunreuther, "Troubled Waters," 334.
88 Knowles and Kunreuther, "Troubled Waters," 334.
89 Knowles and Kunreuther, "Troubled Waters," 333.
90 Knowles and Kunreuther, "Troubled Waters," 334.
91 Knowles and Kunreuther, "Troubled Waters," 335.
92 Knowles and Kunreuther, "Troubled Waters," 335.
93 Rohland, "Adapting to Hurricanes," 11.
94 Rohland, "Adapting to Hurricanes," 11.
95 Rohland, "Adapting to Hurricanes," 11.
96 Rohland, "Adapting to Hurricanes," 11.
97 Rohland, "Adapting to Hurricanes," 11.
98 Rohland, "Adapting to Hurricanes," 11.
99 Kousky et al., "Flood Risk," 11.

100 Rohland, "Adapting to Hurricanes," 12.
101 Knowles and Kunreuther, "Troubled Waters," 337; Yorke, "House of Cards."
102 Strother, "The National Flood," 462.
103 Tier, "Overcoming Contemporary Reform."
104 Tier, "Overcoming Contemporary Reform."
105 Greener, "Path Dependence."
106 Knowles and Kunreuther, "Troubled Waters," 336.
107 Knowles and Kunreuther, "Troubled Waters," 337.
108 Knowles and Kunreuther, "Troubled Waters," 337.
109 Knowles and Kunreuther, "Troubled Waters," 338.
110 Knowles and Kunreuther, "Troubled Waters," 338.
111 Burby, "Flood insurance," 113.
112 Knowles and Kunreuther, "Troubled Waters," 339.
113 Knowles and Kunreuther, "Troubled Waters," 339.
114 United States Census Bureau, "Coastline County"; Gabriel and Kunreuther, "Troubled Waters," 328.
115 NFIP, "National Flood Insurance Program (NFIP) & Risk Rating 2.0."
116 FEMA, "NFIP Debt."
117 Kousky, "Financing Flood Losses," 15.
118 Kousky, "Financing Flood Losses," 16.
119 Burby, "Flood Insurance," 113–14.
120 Burby, "Flood Insurance," 114.
121 Knowles and Kunreuther, "Troubled Waters," 343.
122 Knowles and Kunreuther," Troubled Waters," 344.; Kousky et al., "Flood Risk," 18.
123 Knowles and Kunreuther," Troubled Waters," 329.
124 Strother, "The National Flood," 452–80.
125 Knowles and Kunreuther, "Troubled Waters," 329; Strother, "The National Flood," 452–80.
126 Knowles and Kunreuther, "Troubled Waters," 345.
127 Strother, "The National Flood," 454.
128 Strother, "The National Flood," 454.
129 Gilens and Page, "Testing Theories of American Politics," 564–81.
130 Knowles and Kunreuther, "Troubled Waters," 344.
131 Knowles and Kunreuther, "Troubled Waters," 344; Werner, "Natural Catastrophes."
132 Kousky et al., "Flood Risk," 19.

Bibliography

American Institutes for Research. *A Chronology of Major Events Affecting the National Flood Insurance Program*. Washington, DC: Federal Emergency Management Agency, 2005.

Bachaud, Nicole. "Home Buyers Are Paying More Attention to Flood Risk, but Not Other Climate Events." https://www.zillow.com/research/home-buyers-climate-events-31497.

Bergstrand, Kelly, Brian Mayer, Babette Brumback, and Yi Zhang. "Assessing the Relationship Between Social Vulnerability and Community Resilience to Hazards." *Social Indicators Research* 122, no. 2 (2015): 391–409.

Bernstein, Asaf, Matthew T. Gustafson, and Ryan Lewis. "Disaster on the Horizon: The Price Effect of Sea Level Rise." *Journal of Financial Economics* 134, no. 2 (2019): 253–72.

Birkland, Thomas, and Sarah Waterman. "Is Federalism the Reason for Policy Failure in Hurricane Katrina?" *The Journal of Federalism* 38, no. 4 (2008): 697–714.

Burby, Raymond J. "Flood Insurance and Floodplain Management: The U.S. Experience." *Environmental Hazards* 3, no. 3 (2001): 111–22.

Castaldi, Carolina, and Giovanni Dosi. "The Grip of History and the Scope for Novelty: Some Results and Open Questions on Path Dependence in Economic Processes." In *Understanding Change: Models, Methodologies and Metaphors*, edited by Andreas Wimmer and Reinhart Kössler, 99–128. Palgrave Macmillan, 2006.

Cutter, Susan L., and Christina Finch. "Temporal and Spatial Changes in Social Vulnerability to Natural Hazards." *Proceedings of the National Academy of Sciences of the United States of America* 105 (2008): 2301–6.

Dinan, Terry, Perry Beider, and David Wylie. "The National Flood Insurance Program: Is It Financially Sound?" *Risk Management and Insurance Review* 22, no. 1 (2019).

FEMA. "Risk Rating 2.0 Is Equity in Action." https://www.fema.gov/sites/default/files/documents/fema_rr-2.0-equity-action 0.pdf#:~:text=Risk%20Rating%202.0%20is%20equity%20in%20action.%20FEMA%2C,than%20their%20fair%20share%20in%20flood%20insurance%20premiums.

FEMA. "NFIP Debt." https://www.fema.gov/case-study/nfip-debt.

Frank, Sadie, Eric Gesick, and David G. Victor. "Inviting Danger: How Federal Disaster, Insurance and Infrastructure Policies Are Magnifying the Harm of Climate Change." https://www.brookings.edu/articles/inviting-danger-how-federal-disaster-insurance-and-infrastructure-policies-are-magnifying-the-harm-of-climate-change/.

Frank, Thomas. "On This Flood Policy, Biden and Trump Administrations Agree." June 16, 2022. https://www.eenews.net/articles/on-this-flood-policy-biden-and-trump-administrations-agree/.

FRED. "Gross Domestic Product, USA 1947–2024." https://fred.stlouisfed.org/series/GDP#0.

Friedman, Lisa, and Jonathan Weisman. "Delay as the New Denial: The Latest Republican Tactic to Block Climate Action." *The New York Times*, July 21, 2022. https://www.nytimes.com/2022/07/20/us/politics/climate-change-republicans-delay.html.

Frost, Rebecca. "Path Dependent: Lessons from America's Ill-Equipped Flood Insurance System." *The Observer*, September 17, 2018. https://theobserver-qiaa.org/path-dependent-lessons-from-americas-ill-equipped-flood-insurance-system.

Gasper, John. "Make It Rain: How Politics Impacts Disaster Relief." https://www.cmu.edu/tepper/news/stories/2018/march/politics-disaster-relief.html.

Gibbs, David. "Climate Change Impact on the U.S. Real Estate Market; with Particular Examination of the Sea Level Rise." Bachelor Thesis, Modul University, Vienna, 2023.

Gilens, Martin, and Benjamin Page. "Testing Theories of American Politics: Elites, Interest Groups, and Average Citizens." *Perspectives on Politics* 12, no. 3 (2014): 564–81.

Greener, Ian. "Path Dependence." In *Encyclopedia Britannica*. https://www.britannica.com/topic/path-dependence.

Hallegatte, S., C. Green, R.J. Nicholls, and J. Corfee-Morlot. "Future Flood Losses in Major Coastal Cities." *Nature Climate Change* 3 (2013): 802–6.

Hurricane Science. "1965—Hurricane Betsy." https://hurricanescience.org/history/storms/1960s/betsy/index.html.

Jones, Jonathan. "U.S. Cities Most Impacted by Sea Level Rise." https://constructioncoverage.com/research/us-cities-most-impacted-by-sea-level-rise.

Knowles, Scott Gabriel, and Howard C. Kunreuther. "Troubled Waters: The National Flood Insurance Program in Historical Perspective." *The Journal of Policy History* 26, no. 3 (2014): 327–53.

Kousky, Carolyn. "Financing Flood Losses: A Discussion of the National Flood Insurance Program." *Risk Management and Insurance Review* 21, no. 1 (2018): 11–32.

Kousky, Carolyn, Howard C. Kunreuther, Michael LaCour-Little, and Susan Wachter. "Flood Risk and the U.S. Housing Market." *Journal of Housing Research* 29, no. 1 (2020): 3–24.

Kusumastuti, Ratih Dyah, Zaafri Ananto Husodo Viverita, Lenny Suardi, and Dwi Nastiti Danarsari. "Developing a Resilience Index Towards Natural Disaster in Indonesia." *International Journal of Disaster Risk Reduction* 10 (2014): 327–40.

Mener, Andrew S. "Disaster Response in the United States of America: An Analysis of the Bureaucratic and Political History of a Failing System." *CUREJ: College Undergraduate Research Electronic Journal, University of Pennsylvania* (2007). http://repository.upenn.edu/curej/63.

Mochizuik, Junko, Reinhard Mechler, Stefan Hochrainer-Stigler, Adriana Keating, and Keith Williges. "Revisiting the 'Disaster and Development' Debate—Toward a Broader Understanding of Macroeconomic Risk and Resilience." *Climate Risk Management* 3 (2014): 39–54.

NFIP. "National Flood Insurance Program (NFIP) & Risk Rating 2.0." https://www.monroecounty-fl.gov/628/Flood-Insurance-NFIP#:~:text=With%20more%20policyholders%20in%20NFIP,greatest%20beneficiary%20of%20the%20NFIP.

NOAA. *What Percentage of the American Population Lives Near the Coast?* National Ocean Service, National Oceanic and Atmospheric Administration, 2024.

Rauscher, Natalie. "American Philanthropy in the Age of Political Polarization: Conservative Megadonors and Foundations and Their Role in Spreading Climate Skepticism." *Politische Vierteljahresschrift* (2023).

Rauscher, Natalie. "In the Face of Disaster: Narratives of Community Vulnerability and Resilience in Media Coverage of Natural Catastrophes in the USA." In *Real, Imagined, and Displayed Fragility in Language and Society*, edited by Silvia Bonacchi. V&R Unipress, 2024.

Ritchie, Hannah. "A Century of Global Deaths from Disasters." https://ourworldindata.org/century-disaster-deaths.

Ritchie, Hannah. "The World Has Become More Resilienct to Disasters, but Investment Is Needed to Save More Lives." https://ourworldindata.org/the-world-has-become-more-resilient-to-disasters-but-investment-is-needed-to-save-more-lives.

Ritchie, Hannah, and Max Roser. 2014. "Our World in Data: Natural Disasters." https://ourworldindata.org/natural-disasters.

Rohland, Eleonora. "Adapting to Hurricanes: A Historical Perspective on New Orleans from Its Foundation to Hurricane Katrina. 1718–2005." *WIREs Climate Change* 9, no. 1 (2018).

Roser, Max. 2021. "What Is Economic Growth? And Why Is It so Important?" https://ourworldindata.org/what-is-economic-growth.

Rus, Katarina, Vojko Kilar, and David Koren. "Resilience Assessment of Complex Urban Systems to Natural Disasters: A New Literature Review." *International Journal of Disaster Risk Reduction* 31 (2018): 311–30.

Schuster, Natalie M. "Political Disasters: The Politics of U.S. Disaster Relief, 1927–2005." Doctor of Philosophy, University of Houston, 2014.

Seebauer, Sebastian, Thomas Thaler, Susanne Hanger-Kopp, and Thomas Schinko. "How Path Dependency Manifests in Flood Risk Management: Observations from Four Decades in the Ennstal and Aist Catchments in Austria." *Regional Environmental Change* 23, no. 31 (2023).

Shimizu, Mika. "Policy Lessons for Japanese Disaster Responses." *Asia Pacific Bulletin*, no. 121 (2011).

Shugart II, William F. "Disaster Relief as Bad Public Policy." *The Independent Review* 15, no. 4 (2011): 519–39.

Strader, Stephen M. "We Keep Moving to the Bull's Eye." *The New York Times*, September 16, 2018.

Strother, Logan. "The National Flood Insurance Program: A Case Study in Policy Failure, Reform, and Retrenchment." *Policy Studies Journal* 46, no. 2 (2018): 452–80.

Swiss Re Institute. *Sigma* 1/2016.

Swiss Re Institute. *Sigma* 1/2018.

Swiss Re Institute. *Sigma* 2/2019.

Swiss Re Institute. *Sigma* 1/2023.

Taurus Flood. "What We Can Learn from Flood Control in the Netherlands." https://nationalfloodservices.com/blog/what-we-can-learn-from-flood-control-in-the-netherlands/.

Thaler, Thomas, Susanne Hanger-Kopp, Thomas Schinko, and Ralf Nordbeck. "Addressing Path Dependencies in Decision-Making Processes for Operationalizing Compound Climate-Risk Management." *iScience* 26 (2023).

Tier, Melissa. "Overcoming Contemporary Reform Failure of the National Flood Insurance Program to Accelerate Just Climate Transition." *Journal of Public & International Affairs* (2021). https://jpia.princeton.edu/news/overcoming-contemporary-reform-failure-national-flood-insurance-program-accelerate-just-climate.

UNDRR. "Understanding Disaster Risk." https://www.preventionweb.net/understanding-disaster-risk/key-concepts/resilience.

UNDRR. "Disaster Risk Reduction in Least Developed Countries." https://www.undrr.org/disaster-risk-reduction-least-developed-countries.

United States Census Bureau. *Coastline County Population*. United States Census Bureau, 2012. https://www.census.gov/dataviz/visualizations/039/.

United States Census Bureau. "New Florida Estimates Show Nation's Third-Largest State Reaching Historic Milestone." https://www.census.gov/library/stories/2022/12/florida-fastest-growing-state.html.

Weinkle, Jessica, Christopher W. Landsea, Douglas Collins, Rade Musulin, Ryan P. Crompton, Philip J. Klotzbach, and Roger Jr Pielke. "Normalized Hurricane Damage in the Continental United States 1900–2017." *Nature Sustainability* 1 (2018): 808–13.

Werner, Welf. "Natural Catastrophes and Their Effects on Reinsurance." In *Managing Risk in Reinsurance: From City Fires to Global Warming*, edited by Niels Viggo Haueter and Geoffrey Jones. Oxford University Press, 2017.

Yorke, Oliver. "House of Cards: The Intersection of Climate Risk and the U.S. Real Estate Bubble." https://www.climate-x.com/articles/industry/house-of-cards-the-intersection-of-climate-risk-and-the-us-real-estate-bubble.

6

RAINFALL SEASONALITY, DROUGHTS, AND BUSINESS

Tirthankar Roy and Chinmay Tumbe

6.1 Introduction

Human civilisation requires water. Tropical climatic regions require not only a lot of water, because of the high evaporation rates they experience, but also special systems of recycling and storage. The green revolution depends on it.[1] So do urban water systems, which sustain industry and services. There is, however, no linear or straightforward relationship between climate and economic growth, partly because water sources are not always shaped by local climatic conditions. For example, Egypt receives little rainfall (less than 20 mm annually) but thrives due to the Nile River.[2] The Indian subcontinent receives heavy monsoonal rains (about 1,000 mm of rainfall annually), but most of it falls in a two-month window (mid-June to mid-August). The rest of the year can be quite dry, and the summer exceptionally so because of high levels of solar radiation.[3] This combination of monsoon and aridity is of particular interest in this chapter. We explore how business decisions adapt to these conditions and speculate on how climate change might alter seasonality. Nearly all illustrations come from India, a region probably more intensively researched than most other countries of the tropical and emerging world, and a region both authors know best. However, given the frequent references we make to the climatic conditions, many of our general claims may apply to other world regions that have similar climates.

The monsoon is a keyword for this chapter. The monsoon refers to the moist summer circulations that provide most of the yearly rainfall in the tropics and subtropics. The season is short, usually a few weeks, though in India, where one of the stronger monsoons occur, the season can last for two months. The pathway and strength of the monsoon are annually variable. Given the extreme tropical aridity, the variability causes frequent droughts and floods. Before the twentieth century, droughts in India inevitably translated into famines and epidemics with mass mortality.[4] In a normal year, heavy rains and extreme aridity occur at different times of the year.

Seasonality in this sense has historically had important implications for the economy of the Indian subcontinent, causing extreme events in abnormal years, and much effort to conserve water in normal ones. As for extreme events, studies are documenting even greater variability of rainfall in 1950–2015, owing to "climate change."[5] Much less is known about the future patterns of seasonality in normal years.

DOI: 10.4324/9781003478089-7

Business historians of India have yet to address the interface between climate and investment patterns in the long run. This oversight stands in contrast with the counterpart literature for Europe and North America.[6] There is a rich scholarship on Indian environmental history, but little of it engages with questions of interest to the economic historian—questions about growth, inequality, and business decisions.[7] There are potentially many ways that this relationship could be explored.

To keep the task manageable, we use the concept of "seasonality" to restrict the discussion to two large fields of business response to climate. These are financial market responses to seasonality, or the sequence of a busy season followed by a slack one, common in most rain-fed agricultural systems, and private investment in water systems that, in turn, sustain both rural and urban economic growth. There is a rough chronological logic to the selection of these two fields. The financial market response was more visible during colonial rule in India, not least because private banks and governments collected a lot of data on seasonality. In post-independence India, the markets were regulated, and the effect became harder to observe, though it almost certainly did not disappear.

The socialist ideological leanings that shaped economic thought in 1940s India worked against leaving financial markets free to set credit terms when many borrowers were poor farmers. The same instinct led to a significant rise in public investment in water infrastructure, mainly by building large multipurpose river valley projects to store and recycle water. The roughly 2,000 dams that were built between 1950 and 1980 generated electricity, supplied water to cities, controlled monsoon flooding, and fed irrigation canals. By the 1980s, however, public and expert opinion turned against dams for the environmental and economic damage that they entailed.[8] The problems of water supply and recycling that the dams were trying to address remained.

From the end of the millennia, private sector investment in water took off. Since then, as a field of investment, it has grown at a high speed, sustaining migration to the cities and generating enormous spillovers on engineering industries in several small towns. In a sense, borewells, essentially pipes used to extract water from holes created underground by narrow-drilling methods, have sustained India's economic emergence and urbanisation since the 1980s in crucial ways. We ask why it happened, how it happened, and with what effect.

The two examples—financial markets and money-lending activities, and direct private investment in water—both illustrate, in different ways, how private investment responds and adapts to climatic seasonality. While these two cases apply to India, the latter case of private investment in bore well technology has widespread resonance, and we also briefly discuss issues beyond India. Before we move to these examples, the concept of climatic seasonality requires a brief discussion in Section 6.2. Sections 6.3 and 6.4 of the chapter deal with the two sets of illustrations. Finally, we provide a conclusion with an overview of the challenges associated with business and climate in the context of seasonality.

6.2 Climate and Seasonality

What does climate in the Indian context mean? Much of India falls in the tropical monsoon climatic zone. The tropical lands experience extreme heat that dries up surface water for a long stretch of the year. Parts of the tropics also experience seasonal floods and rains owing to the actions of the intertropical convergence zone, commonly known as the monsoon. The southwest monsoon, so called because of the winds that blow over the Indian subcontinent from the southwest, occurs for four months, from June to September, and supplies most of the rainfall

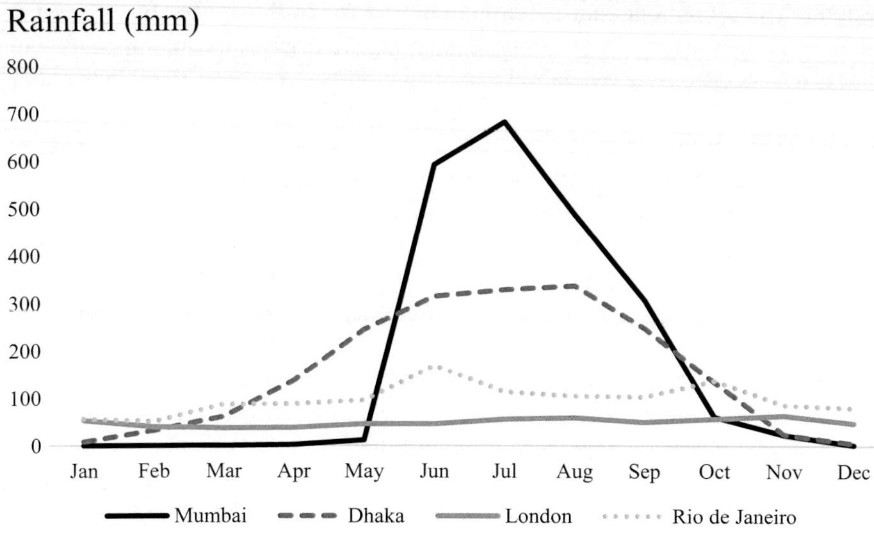

FIGURE 6.1 Seasonality of rainfall around the world, 2009–2024

Source: Average monthly rainfall data from Climatemps.com

in the year for large parts of the Indian subcontinent. As Figure 6.1 shows, the seasonality of rainfall in Mumbai on the western side of the Indian subcontinent and Dhaka on the eastern side, in any given year on average, is much more pronounced than observed in London or Rio de Janeiro. Following September, temperatures tend to cool off, settling into a winter that lasts until February or March and then a scorching summer that runs through April and May. The onset of the monsoon in June then again lowers temperatures.

Variations in rainfall, the only security against aridity, meant the risk of floods and droughts. These conditions did not influence equally all parts of the large region. Some seaside places could access the stable flows of a river and, by trading with the rest of the world, reduce their dependence on agriculture. The Himalayan Mountain range created a river system that carried water throughout the year due to snowmelt. Some of the strongest empires of the ancient and modern worlds ruled in the Indo-Gangetic Basin. But these were not the norm, and floods and famines elsewhere could also upset the less vulnerable regions. For example, even when famines occurred in the arid tracts, affected persons would migrate to the more well-stocked regions or to towns in search of food and cause shortages elsewhere. "Migration," Tim Dyson writes, "is manifestly a major feature of most famines."[9]

Indian history suggests diverse patterns of adaptation to aridity and drought: animal herding, seasonal migration, and the construction of small and local storages of water.[10] Evidence of ancient waterworks can be found all over India, from the stepwells of Rajasthan to the human-made lakes of Karnataka. Despite these actions, the population growth rate was low for centuries, suggesting that if these worked well during milder crises, they failed during the severe ones.[11]

The agricultural seasons were short everywhere because of the region's tropical monsoon climate. The tropical climate and attendant water scarcity depressed the possibility of production and thus limited the prospects of agricultural commercialisation to only a few regions.

The monsoon rains ensured one rain-fed crop in most parts of the region. But extension of the cultivation season, the raising of water-intensive crops like sugarcane, and raising the intensity of cultivation depended on the availability of irrigation water from rivers, lakes, and wells. Groundwater offered an alternative. But in the best of times, the mechanisms to extract and recycle groundwater were exceedingly expensive, well beyond the means of most peasants. This is so because in two-thirds of the South Asian landmass, underground water occurs in hard-rock aquifers that until recently were costly to drill and difficult to map.[12] The risk that a well, created with a relatively large investment, would dry up or fail to supply water was relatively high. In the Gangetic plains, by contrast, groundwater occurs in alluvial aquifers, which the peasants exploited more fully. Borewell technology, studied in depth in Section 6.4, was not affordable until the 1970s.

In the late nineteenth century, severe drought conditions triggered famines in peninsular India. Droughts and famines, directly via shortages of food and water and indirectly by unleashing epidemic diseases, killed many people. But they were not just demographic events. Economic history observes that disasters activated markets by creating sudden shortages. For instance, the Black Death of the fourteenth century is considered to have stimulated markets in labour and land in Europe for generations to follow.[13] That said, the markets that were activated after disasters tended to be anarchic, unregulated, and predatory. These are markets without rules, whether made by the state or by members of society.[14] Studies on livelihoods in conflict-ridden societies show a variety of ways in which emergencies lead to fierce competition over necessities. Communities caught up in civil war suffer or use opportunities for exploitative gains. Political power is used to peddle security and food. Acute emergencies create gainers as well as losers, and the power of the gainers can account for the persistence of dysfunctional situations. The syndrome of enhanced competition goes hand in hand with the syndrome of weak cooperation mentioned earlier. Indeed, one problem aggravates the other.[15]

In the twentieth century, expansion of the railway network, large river basin projects to supply irrigation water, development of urban water supply, and high-yield seeds of the 1970s and 1980s reduced the incidence and impact of droughts. The measurement of rainfall and forecasts improved vastly, helping farmers and businesses make better decisions, as has been documented in other contexts as well.[16] The economy's long-standing dependence on seasonality and aridity was reduced, but not enough. Areas receiving irrigation formed less than 20% of cultivable land in 1940.[17] In most places, agriculture was monsoon-dependent.

One of the early effective steps in this direction began around 1900. The British engineers built, and in some cases revived, a few large river-based canals.[18] These canals came up in areas where perennial rivers existed and were large enough in size. The most important such areas were Punjab, where the Himalayan snowmelt rivers were used effectively; the Godavari and Krishna deltas on the south-eastern coast; and the Ganges plains in north India. If the agricultural village was located in one of the few canal-irrigated zones, the peasants benefited by producing more crops for the export market. But if it was located in a rain-fed and arid zone, it was more than likely that the land would produce too little, even for a comfortable subsistence.[19]

After independence in 1947, the country stared at a rapidly growing population and shortages of food and jobs. Politicians responded with plans for hundreds of river valley projects, which would produce cheap electricity to sustain industrialisation and irrigation to grow more food. In the late twentieth century, another instrument joined the battle: deep drilling for groundwater. Per head water usage increased from 200 to 800 cubic metres from 1885 to 2015.[20] These projects sustained India's impressive green revolution.

The risk of extreme rainfall seasonality could in theory be covered by insurance markets. The insurance firm Lloyd's in England was known to be insuring against rainfall-related risks since the late nineteenth century, with even cricket matches being claimants, but there is little evidence of its implementation in India.[21] An early crop insurance scheme to protect against a downturn in yields was in place by 1949 in south India, but that did not last too long because of low literacy rates and poor awareness by farmers of the policy.[22] Since then, rainfall insurance in India has made little progress, with failures usually blamed on low awareness and high pricing. For instance, in 2006, these were the reasons specifically cited for the low uptake of the Agricultural Insurance Company's *Varsha Bima* (rainfall insurance) policy.[23] With no real recourse to insurance against rainfall seasonality risks, businesses reacted in other ways.

With this overview of seasonality in India, we now turn our attention to how this rainfall seasonality influenced businesses in manufacturing and the financial sector in the colonial era and then present a case study on the evolution of the business of borewells in India to adapt to water stress and droughts.

6.3 Climate and Business in Colonial India

6.3.1 Manufacturing and Commerce

Most manufacturing industries before independence processed natural resources. The supply of raw materials was highly seasonal, and therefore, their operations were also often highly seasonal. The exceptions were the large urban textile factories, which could store their materials, or resource them from different regions where the harvest dates did not exactly coincide. On the other hand, nearly all industrial occupations in small towns, in artisanal industries, had work intensity that rose sharply in the dry winter months (or after major harvests) and fell during the monsoons. Most food products—sugarcane production in north India, edible oils, cotton gins in western India and jute presses in the east, brickmaking in the Indo-Gangetic basin, and tile-making in Malabar (and, in turn, construction activity)—followed this pattern.[24]

It was well known to historians of trade in India that the monsoon influenced shipping seasons in the era of sailing ships and also regulated overland trading seasons because rivers were often dangerous or too dry to navigate, and roads were flooded. Steam technology applied to both types of transport, via ships and railways, reduced the seasonal effect considerably. There is rather little research available on just how much these did matter. One recent work on railways in North India is an exception. A flat and well-watered land, much of the United Provinces was already well connected by roads and boats before the railways came.[25] The railways changed trade by increasing the speed of cargo, reducing the cost of freight per ton per mile, and reducing the effect of seasonality upon transportation. Railways were less susceptible to the effects of the rains and, therefore, could smooth consumption within the year. However, railway passenger traffic itself could be seasonal, as seasonal migrant workers went back to their native places during the sowing season just before the monsoon.[26]

6.3.2 Seasonality and Financial Markets

The growth in agricultural trade needed bigger flows of credit. Between 1870 and 1930, the volume of credit linked to agricultural trade increased by 200-300% in real terms.[27] The indigenous bankers, operating as family firms, met the need for credit well enough. Most of these firms dealt with the local merchants, and some with the peasants. On the other side, corporate banks and bigger indigenous firms based in the cities dealt with the small-scale bankers' needs for money.[28]

As the banking business became more tied to commodity trade, the seasonal influence on the financial market increased. One of the first systematic histories of Indian banking in 1864 did not mention the words "season" and "rain."[29] In contrast, the evidence collected during the Banking Enquiry Commission (1929–30) used these words hundreds of times. A large amount of the cash reserves of banks was converted into trade credit during the winter months. The apex bank, the Imperial Bank, increased its interest rate to protect its reserves. During the primary harvest season (January–March), demand for money was at its peak, and loan rates between traders and peasants could be as high as 50–100%, though rates were a more moderate 12–24%, among known clients. By April, the market started cooling. In the rains (June–August), the rates crashed, and money markets disintegrated as movements of capital between small and large centres reduced.[30]

The Imperial Bank's rate influenced other rates. It saw big fluctuations between seasons. The rate ranged between 4 and 10% in the 1870s. The real interest rate (interest rate divided by a price index) fluctuation was higher because the price of food grain fell during harvest when money rates were high and increased in the rains when money rates were low. As corporate banking and the bill market expanded, there was some moderation in the seasonal effect. The peaks and troughs came closer, and the average monthly interest rate fell (1870–1920). There was more money coming into the system, and it was circulating between seasons better than before, but still not quite enough to remove the seasonality.[31]

Beyond the general features of the financial market, the market was quite complex internally. For example, it had several actors who did not face seasonality quite in the same way nor respond to it in the same way. Who were these actors? Corporate banking began in the early nineteenth century. Corporate banks included the foreign-owned exchange banks, licensed to deal in foreign currency transactions, the partly government-owned Presidency banks, and other banks. There was also a mixed set of cooperative credit societies, loan offices, and land mortgage banks. The Presidency Banks were the Bank of Bengal (est. 1806), Bank of Bombay (est. 1840), and the Bank of Madras (est. 1843). These functioned as the government's bankers and were amalgamated to form the Imperial Bank in 1921. These banks handled domestic trade and remittance and accepted bills issued by the indigenous bankers. Most were based in the port cities that rarely faced droughts.[32]

Indian merchants and landlords frequently floated banks. Only a few of these survived the worst business cycles. Most small banks depending on local savings suffered from a high closure rate. There was a high incidence of insider lending in these businesses. That is, bankers lent money to firms that they owned, or their friends did. Corporate banks dealt in other people's money. A weak regulatory system encouraged fraud and waste. When a merchant group floated a bank, invariably a lot of the money went to the same trades that these owners were interested in.[33]

Extreme seasonality was a major cause of high firm mortality. There was a disparity between the assets and liabilities of banks. Most of their loans went to businesses affected by seasonality. In a bad harvest year, a lot of the loans went bad. The obligation to pay interest on deposits remained intact. The outflow was not a seasonal one; the inflow was seasonal. If a bank suffered a loss in one year due to a harvest failure and unpaid debts, the rumour that it might sink would cause a run on the deposits and ensure it did sink.[34]

Non-corporate family firms were not so vulnerable because they did business based on their own savings and profits and did not rely too much on the banking system. Indeed, most such bankers avoided getting involved in the deposit business and thus take on a non-seasonal liability. Indigenous banking firms belonged to the families that had been doing the business for generations. Many had earlier been based in towns situated on the overland and river-borne trade routes such as Benares, Mathura, Delhi, or Jaipur. In the eighteenth and nineteenth centuries, many of these families relocated to the port cities of Bombay, Calcutta, and Madras. The bankers belonged to certain castes and communities, the prominent among these being the Multani and the Marwari, the Bengali Saha, Nattukottai Chettiar, Kallidaikurichi Brahmins in South India, the Jains and Gujaratis in Bombay, and Rohilla Afghans. For some of these communities, participating in business, droughts, and rainfall risk also led to their outmigration in search of greener pastures.[35] The biggest of these firms funded export trade and other bankers, sometimes did some deposit business, and were somewhat less susceptible to seasonality. However, the dealings between the lender and the borrower closer to agricultural land were less diversified and fiercely seasonal, in the same manner as described earlier.

There were two types of risk in the business: not knowing the debtor well enough, or having asymmetric information, as economists call it, and not being able to predict the quality of the harvest season. Economists do not have a name for the second risk because they have not studied climatic risk in credit markets enough. Asymmetric information was easier to address. Every member of every set selected their clients based on personal information. For example, the indigenous bankers charged lower rates for transactions among members of their own caste or community.[36]

However, a lot of the risk did not stem from a lack of information but was seasonal and thus linked to the climate. No amount of information on the client's accounts or trustworthiness could mitigate the risk of a bad season, which would bring the debtor down, and the lender with him or her. Nobody could predict the season. Most local loans were unsecured and not mortgage-backed and hence very risky.[37]

The big city bankers had a diversified portfolio to mitigate such risks. Their clients included landlords, warehouse owners, merchants with personal reputation, agents of outstation trading firms, tea estates, traders buying jute, tobacco, and chili, crops where these profitable cash crops were grown, and urban services like cinema and theatre industries. They discounted bills of exchange and issued remittance instruments like commercial papers when they were branches of Bombay or Calcutta firms. The Chettiar bankers transferred funds in the slack season to their overseas operations in Rangoon, Malaya, and the Straits Settlements.

The local bankers, by contrast, had little defence against seasonal risk. Bills failed to circulate below the banker-to-banker level, leading contemporary critics of Indian banking to complain of "inelasticity" of the supply of money in response to trade demand.[38] The financial system was not well equipped to spread credit between seasons and types of borrowers. Such a season-bound financial market excluded many types of borrowers whose investments would have sustained businesses outside agriculture.

This section has detailed how climatic factors affected manufacturing, commerce, and banking in colonial India. These factors became less significant for the economy in the second half of the twentieth century as the Indian economy diversified away from rain-fed agriculture and also because the government intervened heavily in financial markets by nationalising banks in 1969 and then expanding rural credit to overcome aspects of seasonality. While seasonality was a core issue for financial markets in colonial India, discussions on "climate change" were generally absent. Financial markets themselves deepened in the second half of the twentieth century with a wider participation from people and institutions.

The next section deals with the second of the two cases, illustrating again a link between climatic seasonality and business growth. This is a more direct kind of link than the one mediated by financial markets. We present a case study of the business of borewells in the late twentieth century as a drought-resistant mechanism and the challenges the adoption of bore well technology, in turn, raised.

6.4 The Business of Borewells

By the early 2020s, India was home to over 20 million borewells and was the largest user of this technology to extract groundwater.[39] As Table 6.1 indicates, India extracted more groundwater than the United States and China combined.[40] Borewells were the principal means of water supply in most cities and served 60% of the irrigated area in contemporary rural India.

This had not always been the case. Private businesses and government intervention played an important role in the evolution of the groundwater extraction industry since the late nineteenth century. Then the technology did not look very promising, as is evident by this quote by W. B. Muloch, a British administrator in western India:

> The pump portion of the well is most defective and weak. I found two of the pumps broken, bent, and cracked, from the strain required to keep up anything like a flow of water. I am of [the] opinion that wells such as Norton's will never be much used by the natives of this country; and I have been most disappointed in the results of my experiments. The exertion, required to use them is too great, the water too muddy, and the supply too meagre.[41]

In hindsight, Muloch clearly misjudged the revolutionary potential of new technologies of groundwater extraction that were emerging in the late nineteenth century. The first of these technologies, credited to A. L. Norton in 1868, was an invention of British military engineers in the Ethiopian desert during the Abyssinian War as an effort to quickly pump water for drinking purposes.[42] Initially its adoption appears to have been low in India, even as the art of drilling

TABLE 6.1 Groundwater withdrawals in 2017 by region

Region	*Groundwater withdrawal (KM cube/year)*	*% of total freshwater withdrawal*
North America	156	26
South America	27	13
Europe	65	23
Africa	45	20
Australia & Oceania	8	31
Asia	657	26
South Asia	*401*	*39*
East Asia	*132*	*18*
Western Asia	*64*	*39*
World	959	25

Source: UNESCO. *Groundwater: Making the invisible visible.* Paris: UNESCO, 2022, p. 20.

and extraction, for water and oil, was being perfected in America. These technologies, first called "tube-wells" and much later refined to "borewells" (with slight differences between them), gained further traction in India for irrigation purposes in the 1910s and municipalities in the 1920s. From then on, a series of files in the Local Self Government Public Health archival records across the Indian subcontinent record hundreds of petitions for sinking tube-wells in rural and urban neighbourhoods to address water deficiencies.[43] It was slowly gaining traction as a potential substitute to canal irrigation, tanks, and aqueducts.

While the government was roped in to start several of them, a lot of the effort was put in by private owners employing the services of private firms. Empire Engineering Ltd., based in Cawnpore (now Kanpur) in northern India, in 1913–14, used patented technology and advertised the multiple benefits of its technology to extract groundwater. It conveyed the message that it was being adopted by the public works department, municipalities, railways, private companies, and landowners. It marketed the "Brownlie's convoluted tube well" with a peak capacity of pumping 45,000 gallons of water per hour. It also clearly depicted the process of extraction and its efficiency over standard open wells which did not use the drilling principle. The city of Kanpur itself became a place where several such firms were being set up to cater to the growing need for tube wells across the subcontinent. Companies with "Tube Wells" in their names began to proliferate and advertise more and more in newspapers. However, the overall adoption of the technology was still low at the time of Indian independence in 1947, as the costs were seen to be prohibitive.[44]

In the 1950s and 60s, the Nehru government's thrust on water resources management was principally through hundreds of dams and canals for irrigation purposes.[45] Since the 1970s, there has been a distinct tilt of policies to promote borewell usage and eventually an explosion of using borewells in rural and urban areas such that it eventually became the principal source of water for both irrigation and drinking.[46] One reason for this switch was the preference of farmers to use decentralised borewells which were far more portable than canals, where farm location mattered. As one researcher has eloquently described:

> The tubewell is the great liberator of the topographic surface and symbol of economic liberalization. It is a fantastic technology, precisely because of its differences to a canal: it can be sunk almost anywhere, it is portable, it provides water on demand, it is managed independently, and it is not a monument.[47]

Borewells were as important in aiding India's Green Revolution as high-yield seeds and fertilisers, that collectively raised agricultural productivity and enabled food self-sufficiency. Subsidised and often free electricity given to farmers to run the borewells further incentivised the switch to borewells.[48] In the 1980s, a Free Boring Scheme (FBS) was launched in the populous state of Uttar Pradesh in northern India, marking robust government intervention in promoting the technology. By 2001, there were around two hundred thousand private borewells operating in the national capital, Delhi, reflecting extensive inroads in urban localities.[49]

Another reason for the dramatic uptake was that supply responded to the latent demand by increasing capacity and driving down costs. In the 1960s, drilling technologies were brought to India from the United States, and assembly workshops were initiated in the cities of Bangalore and Hyderabad in south India.[50] Because the heavy rigs were often attached to trucks, the industry gravitated towards truck manufacturing sites which in India were located in the southern state of Tamil Nadu. The truck bodybuilding cluster that eventually became India's bore well capital was a small town called Tiruchengode in that state.

The pioneers in this business had originally moved to Tiruchengode in the 1940s and were blacksmiths specialised in repairing vehicles and manufacturing ox-carts. The region itself receives little rainfall and is subject to high seasonality, causing people to look for other work in the lean seasons. In the 1950s, the servicing of trucks and lorries became an important activity in Tiruchengode since they were manufactured close by in the town of Namakkal (Leyland being a prominent manufacturer). The blacksmiths began a new line of work called lorry body building, similar in principle to the work of ox-cart making. A pioneer, Ponusamy employed 30 members of his extended family to run the largest workshop in the 1960s.[51] Another pioneering enterprise, called Paranthaman Engineering Works, was established in 1972 (later the PRD Rigs group), which innovated in improving the load-bearing capacity of the lorries. These entrepreneurs observed the new drilling machines for borewells brought into their area by the local government for boring operations and servicing, and they purchased a few rigs to experiment with the machines. The machines themselves were imported from Sweden (Atlas Corpo), the United States (Chicago Pneumatic), and Germany (Uraca) and assembled in Hyderabad. With an acute water shortage in Tiruchengode, the entrepreneurs spotted an opportunity to make the rigs themselves and immersed themselves in a learning process, making frequent trips to Hyderabad. By 1985, after several years of experimentation and raising capital from family and friends (banks did not consider this business line bankable), a local prototype was developed and marketed. It had a relatively low assembly cost of Rs. 100,000–500,000 per rig. In the 1990s, a technology shift led to pneumatic drills being taken over by hydraulic drills, as they required half the amount of diesel to do the same work, and these developed in a distinctive manner in Tiruchengode. Demand shot up, and thousands of truck rigs began their march out of Tiruchengode. Trucks would leave the town with the rigs attached to them, and workers would spend several months away from home during the year.[52] Peak business was before and after the monsoon, from mid-January to June, and was hence seasonal. During the lean season, or the monsoon season, from July to October, the trucks would return for repair and maintenance.[53]

From ox-carts to lorry body building to truck rigs, the Tiruchengode entrepreneurs adapted in the span of a couple of decades, faced with a water crisis in their own area and also elsewhere. As with several economic clusters in India, this was a family- and caste-based entrepreneurial cluster where workers could bootstrap themselves into being owners within a few years of operating in the business due to relatively lower setup costs. Caste networks of the dominant agrarian castes of the region provided trust and credit within the community.[54] The Tiruchengode Rig Owners Association or Tiruchengode Rig Urimaiyalargal Sangam (TRUS) was set up in 1994 and by the early 2020s consisted of hundreds of firms. A few of the more prominent were Giri Tubewells started by N. Ponnuswamy, considered to be a pioneer in the industry, PRD Rigs, Three Star Bore Well, Royal Rock Drillers, J K Bore Wells, Sanjay Rock Drills, KM Rock Drills, Vijaya Borewells, Sri Mayilsamy Borewells, Ratna Borewells, Sri Dhanalaxmi Borewell Services, Sri Murugan Rig, Raasi Borewells, Sri Kurunchi Borewells, Sri Amman Rig Services, Paari Rock Drills, Bharani Borewells, Srinivasa Borewells, Sathya Bore Wells, R V Raja Drillers, Reena Power Drill, Vigneshwar Bore Wells, Kongu Power Drillers, Ashok Bore Well, Sun Rock Drills, KCS Rig Service and MRM Agro Service.[55] By the 2010s, the Tiruchengode borewell cluster directly or indirectly employed over a hundred thousand people, which is considered to be large even by Indian standards.[56] Many of these workers were migrant workers working across the length and breadth of the country. Each sortie involved a truck rig and a trailing lorry employing around 15 people (drivers, managers, helpers, operators, cooks) and managed to drill 200–500 borewells in a year, sometimes as deep as 1,400 feet below the ground.

The life of a rig truck was about ten years, and the payback period of the investment was three to four years. Three-fourths of all drilling was for agricultural purposes, with the remaining for other purposes, often in urban areas. By 2020, there were around 10,000 rigs.[57]

The success of the industry led to more credit flow from formal banking channels. In 1999, the PRD Rigs group was approached by a customer in Kenya and that led its foray into exports. Africa was receptive to the borewell products from India because they offered quality products that were far cheaper than those from Europe and United States.[58] PRD's success inspired other firms from Tiruchengode to explore international markets since the early 2000s, enjoying varied levels of success. The scale of operations was further intensified when a special economic zone (SEZ) was situated a little outside Tiruchengode around 2010 to promote exports. With international success came diversification for some firms into drilling operations beyond the search of water, towards mining. Thus, in the relatively short time span of four decades, a group of humble blacksmiths in Tiruchengode paved the path towards a groundwater revolution in India and beyond by reverse engineering foreign technologies to local conditions. The family firm and caste networks were central in sustaining the early momentum and eventually its transition to second-generation leadership. That many of the pioneers were not qualified engineers but self-taught is also a testimony to ingenuity and adaptability in the face of adverse climatic considerations.[59]

Borewells were attractive because they helped communities deal with droughts, solving local co-ordination issues that were required for canal irrigation. Virtually any private property holder could commission a borewell rigging operation and strike water. Its rapid adoption across India greatly alleviated water stress and the worst impacts of deficient rainfall and helped in boosting agricultural productivity and reducing poverty.[60] However, the revolution in groundwater extraction and the success of the borewell business led to a new concern by the 1980s itself: that of depletion of groundwater tables.[61] State governments began monitoring the alarming dip in groundwater tables as drilling continued unabated. Drilling would go on until water was found, often going more than 1,500 feet under.[62]

In 1970, a Central Ground Water Board was established in India as a regulatory body to oversee groundwater management. This was a couple of decades before "climate change" became a buzzword in Indian policy circles. The board set up thousands of stations to monitor groundwater levels and provide alerts to areas where rapid depletions take place every year. A good monsoon recharges groundwater tables, and a drought year pushes them down. In addition, rainwater harvesting and other means are used to artificially charge groundwater tables. Despite the existence of a regulatory body, lax enforcement of rules has allowed the borewell business to grow unrestrained until the early 2020s. Apart from registered borewell drilling, unauthorised borewell drilling in areas with stressed water tables was also rampantly flouting government rules by the early 2020s.[63]

Demand-side management of water was by the early 2020s ineffective, and thus groundwater tables across India continued to fall. The availability of borewells led to the replacement of drought-resistant crops with water-intensive crops in some regions, leading to greater demand for water for agricultural purposes than before. If groundwater was priced correctly (by making farmers pay the price of electricity to extract it), extraction could be more efficient. But energy subsidies have also been an important part of India's political economy since the 1970s and have hugely incentivised groundwater extraction.[64] In urban areas, when water has run out, water tankers met, by the early 2020s, the demand by extracting from neighbouring areas and transporting the water to the city. But that, in turn, pushed down the groundwater tables from the neighbouring areas.[65]

India's case presents a classic vicious cycle where businesses found a timely solution to a climate-induced crisis, but the solution itself led to another problem which by the early 2020s had no clear solutions in sight yet. Recent research in India has also established a clear link between climate change, temperature changes, and groundwater storage capacities, accentuating the problem.[66] Further, increased local rainfall variability associated with climate change is also affecting bore well adoption rates to detrimental effect.[67]

India is the biggest consumer of groundwater in the world, but it is not the only country with concerns related to its management.[68] As Table 6.1 shows, outside South Asia, western Asia also has a high groundwater extraction rate, as measured by the share of total freshwater withdrawals. Saudi Arabia, famed otherwise for oil extraction, also followed India's path in rapidly extracting groundwater resources. With mass immigration and rapid economic growth between 1960 and 2022, Saudi Arabia's population grew ninefold from 4 million to 36 million people, greatly increasing its water demand in a desert landscape.[69] Groundwater was extracted in large quantities, serving over 90% of agricultural needs and over 30% of urban requirements.[70] While groundwater tables dipped precipitously over the decades, in recent years there has been stabilisation and even recharge in some aquifers due to better water conservation and the emergence of desalination as an important substitute for urban areas. By the early 2020s, it was the largest producer of desalinated water in the world, and this diversification has been part of its overall water management strategy.[71] Whether this proved to be a long-term solution for Saudi Arabia and indeed many other countries of the world facing climate change remains to be seen.

6.5 Conclusion

The chapter tries to connect climate and business history via the concept of seasonality, which entails the risk of droughts and floods. We studied the connection using two sets of examples: one of these comes from financial markets in colonial India, and the second, private investment in water systems since the late twentieth century. In both cases, climatic factors led to market activity, institutions, information sets, and technologies that supplemented the main livelihoods—agriculture, industry, and trade. In one case, it was banks, and in another, the industry supplying borewells. The latter also encouraged the diversification of the economy away from the agricultural sector by sustaining India's rapid urbanisation. The bore well case shows how fraught this pathway has been, especially in terms of environmental impact. Governments and businesses must constantly seek newer and different responses to climate change. Solutions that have worked well for a while had to be re-looked or expanded for more sustained development as contexts changed. These contexts include the political, social, economic, and environmental. The environmental context here, that is seasonality, itself is becoming more unpredictable due to climate change.

Will climate change make these challenges harder to meet? To answer this question, we need first of all to distinguish between the risk of extreme events (droughts and floods) rising due to global warming and the prospect of the average influx of water changing due to, say, a change in the recharge of underground sources. There is some evidence so far of the risk of extreme events rising in India, though the evidence is uneven at best. The evidence of a shift in the average influx of water is even more patchy because of two issues: there is a general lack of research on the prospect, and the average influx depends on the counteracting forces of falling recharge and rising snowmelt in Himalayan rivers.[72] With the prospect still open to interpretation, it is premature to discuss how businesses might respond.

No matter the general prospect, recent history does show certain patterns of entrepreneurial responses to climate change that are quite distinctive. For example, these responses have come not from big corporations and multinationals, but mainly from small-scale, family-based businesses working in industrial districts, such as those in the bore well industry of Tiruchengode. The pattern suggests that often the most effective recognition of a climatic constraint happens within a local or regional context. The growing international success of the entrepreneurs of Tiruchengode suggests that local solutions can also have a global market to address concerns of climate change.

Notes

1 The green revolution refers to the increase in agricultural productivity in much of the tropical world that began in the late twentieth century due to high-yield variety seeds and the intensive use of fertilisers.
2 https://climateknowledgeportal.worldbank.org/country/egypt/climate-data-historical.
3 https://climateknowledgeportal.worldbank.org/country/india/climate-data-historical.
4 Roy, *Natural Disasters and Indian History*; Tumbe, *The Age of Pandemics, 1817–1920*.
5 Roxy et al., "A Threefold Rise," 1–11; Zolin and Rodrigues, *Impact of Climate Change*.
6 Bansal and Hoffman, *The Oxford Handbook of Business and the Natural Environment*; Jones, *Profits and Sustainability*; Haueter and Jones, *Managing Risk in Reinsurance*; Rosen and Sellers, "The Nature of the Firm," 577–600; Smith and Greer, "Uniting Business History," 987–1009.
7 Fisher, *An Environmental History of India*.
8 Roy, *Monsoon Economies*.
9 Dyson, "Demographic Responses to Famines in South Asia," 17–26.
10 Discussed in Roy, *Monsoon Economies*.
11 Praveen and Visaria, "Population," 463–532, cite estimates of pre-Census population growth (1600–1872), which almost uniformly indicate an average annual rate of less than 0.5% in the Indian subcontinent.
12 Roy, *Monsoon Economies*.
13 See the discussion of the literature in Clark, "Microbes and Markets," 139–65.
14 Roy, *Natural Disasters*.
15 Roy, *Natural Disasters*, for citations of some relevant works on this theme.
16 Colpitts, "Knowing Nature," 1054–80; MacFadyen, "Long-Range Forecasts," 1010–33; Roy, *Natural Disasters*.
17 Roy, *The Economic History of India, 1857–2010*.
18 Amrith, *Unruly Waters*.
19 Amrith, *Unruly Waters*.
20 Roy, *Monsoon Economies*.
21 Times of India (TOI), "Insurance Against Rain."
22 Times of India (TOI), "Crop Insurance to Cover Agricultural Risks."
23 Times of India (TOI), "No Takers for Weather Insurance."
24 In the interwar years, the major source of industrial employment, the Inspectorate of Factories' data compiled and published in *Large Industrial Establishments in India*, designated these factories as seasonal.
25 Derbyshire, *Railways' Economic Impact*.
26 Roy, *Monsoon Economies*.
27 Roy, *Monsoon Economies*.
28 Roy, "Monsoon and the Market," 324–57.
29 Cooke, *The Rise, Progress, and Present Condition of Banking in India*.
30 Roy, "Monsoon and the Market."
31 Roy, "Monsoon and the Market."
32 Roy, *Economic History of India*, for a general description of the banking system in pre-independence India.
33 A systematic history of bank failures in pre-independence India, based on balance sheets, especially investment patterns, does not yet exist. The available accounts suggest the presence of a common syndrome: "discretion was used by the agents to give liberal loans to themselves," which were likely loans recycled into trading in which the directors were interested. Most small Indian banks were established by merchants. The citation comes from Muranjan, *Modern Banking in India*, 300.

34 Muranjan, *Modern Banking in India*, on banking panics in interwar India.
35 Tumbe, *The Age of Pandemics, 1817–1920*.
36 Roy, "Monsoon and the Market," discusses instances of these.
37 Roy, "Monsoon and the Market."
38 Rau, *Present-Day Banking in India*.
39 Shah, *Taming the Anarchy*; United Nations UNU-EHS, *Interconnected Disaster Risks 2023*.
40 Siebert et al., "Groundwater Use for Irrigation," 1863–80.
41 Muloch, "Norton's Tube Wells," 3.
42 Acciavatti, "Re-Imagining."
43 For example, the National Archives of Bangladesh in Dhaka contain a series of files on this issue.
44 Roy, *Monsoon Economies*.
45 Saravanan, *Water and the Environmental History of Modern India*.
46 Dhawan, *Development of Tubewell Irrigation in India*; UNESCO, *Groundwater: Making the Invisible Visible*.
47 Acciavatti, "Re-Imagining," 207.
48 Subramanian, "The Quest for Groundwater in Postcolonial India."
49 Subramanian, "The Quest for Groundwater in Postcolonial India."
50 Tastevin, "From Ox-Carts to Borewell Rigs," 497–525.
51 Tastevin, "From Ox-Carts to Borewell Rigs," 505.
52 Ananth, "Drilling Deep to Make a Living."
53 Ananth, "Drilling Deep to Make a Living."
54 Damodaran, *India's New Capitalists*.
55 From the website of the rig association in Tiruchengode https://trus.co.in/.
56 Tiruchengode https://trus.co.in/.
57 Ramnath, "How Tiruchengode Really Became the Borewell Rigs Capital of India."
58 Ramnath, "How Tiruchengode Really Became the Borewell Rigs Capital of India."
59 Tastevin, "From Ox-Carts to Borewell Rigs."
60 Sekhri, "Wells, Water, and Welfare," 76–102; Jain, "Long-Run Impacts of Tube Well Irrigation in India."
61 Shah, *Taming the Anarchy*.
62 Shah, *Taming the Anarchy*.
63 Shah, *Taming the Anarchy*.
64 Shah, *Groundwater Markets and Irrigation Development*; UNESCO, *Groundwater*.
65 Shah, *Groundwater Markets and Irrigation Development*; UNESCO, *Groundwater*.
66 Asoka et al., "Relative Contribution of Monsoon Precipitation," 109–17.
67 Steinhübel, Wegmann, and Mußhoff, "Digging Deep and Running Dry," 685–706.
68 UNESCO, *Groundwater*.
69 Rebecca Proctor, "How Saudi Arabia Is Acting to Stabilize."
70 Rebecca Proctor, "How Saudi Arabia Is Acting to Stabilize."
71 Rebecca Proctor, "How Saudi Arabia Is Acting to Stabilize."
72 Rani et al., "Spatio-Temporal Assessment," 1–10 (online journal).

Bibliography

Acciavatti, Anthony. "Re-Imagining the Indian Underground: A Biography of the Tubewell." In *Places of Nature in Ecologies of Urbanism*, edited by Anne Rademacher and K. Sivaramakrishnan, 206–37. Hong Kong: University of Hong Kong Press, 2017.

Amrith, Sunil. *Unruly Waters: How Mountain Rivers and Monsoons Have Shaped South Asia's History*. London: Penguin, 2020.

Ananth, M.K. "Drilling Deep to Make a Living." *The Hindu*, July 7, 2012.

Asoka, Akarsh, Tom Gleeson, Yoshihide Wada, and Vimal Mishra. "Relative Contribution of Monsoon Precipitation and Pumping to Changes in Groundwater Storage in India." *Nature Geoscience* 10 (2017): 109–17.

Bansal, Pratima, and Andrew J. Hoffman, eds. *The Oxford Handbook of Business and the Natural Environment*. Oxford: Oxford University Press, 2012.

Clark, Gregory. "Microbes and Markets: Was the Black Death an Economic Revolution?" *Journal of Demographic Economics* 82, no. 2 (2016): 139–65.

Colpitts, George. "Knowing Nature in the Business Records of the Hudson's Bay Company, 1670–1840." *Business History* 59, no. 7 (2017): 1054–80.

Cooke, C.N. *The Rise, Progress, and Present Condition of Banking in India*. Calcutta: Unknown Publisher, 1864.

Damodaran, Harish. *India's New Capitalists*. Ranikhet: Permanent Black, 2008.

Derbyshire, Ian D. *Railways' Economic Impact on Uttar Pradesh and Colonial North India (1860–1914): The Iron Raj*. Cambridge: Cambridge Scholars Publishing, 2022.

Dhawan, B.D. *Development of Tubewell Irrigation in India*. New Delhi: Agricole Publishing Academy, 1982.

Dyson, Tim. "Demographic Responses to Famines in South Asia." *IDS Bulletin* 24, no. 4 (1993): 17–26.

Fisher, Michael. *An Environmental History of India: From Earliest Times to the Twenty-First Century*. New Delhi: Cambridge University Press, 2018.

Haueter, Niels Viggo, and Geoffrey Jones, eds. *Managing Risk in Reinsurance: From City Fires to Global Warming*. Oxford: Oxford University Press, 2017.

Jain, Parul. "Long-Run Impacts of Tube Well Irrigation in India." *Ideas for India*, February 17, 2023.

Jones, Geoffrey. *Profits and Sustainability: A History of Green Entrepreneurship*. Oxford: Oxford University Press, 2017.

MacFadyen, J. "Long-Range Forecasts: Linseed Oil and the Hemispheric Movement of Market and Climate Data, 1890–1939." *Business History* 59, no. 7 (2017): 1010–33.

Muloch, W.B. "Norton's Tube Wells." *Times of India*, February 11, 1871.

Muranjan, S.K. *Modern Banking in India*. Bombay: New Book, 1940, 300.

Proctor, Rebecca. "How Saudi Arabia Is Acting to Stabilize and Replenish Its Groundwater Reserves." *Arab News*, March 22, 2024.

Ramnath, N.S. "How Tiruchengode Really Became the Borewell Rigs Capital of India." *Founding Fuel*, June 19, 2020.

Rani, Akanksha, Devesh Sharma, Mukand S. Babel, and Aditya Sharma. "Spatio-Temporal Assessment of Agro-Climatic Indices and the Monsoon Pattern in the Banas River Basin, India." *Environmental Challenges* 7 (2022): 1–10. (Online Journal).

Rau, B.R. *Present-Day Banking in India*. Calcutta: University of Calcutta, 1929.

Rosen, Christine Meisner, and Christopher C. Sellers. "The Nature of the Firm: Towards an Ecocultural History of Business." *The Business History Review* 73, no. 4 (1999): 577–600.

Roxy, M.K., S. Ghosh, A. Pathak, R. Athulya, M. Mujumdar, R. Murtugudde, P. Terray, and M. Rajeevan. "A Threefold Rise in Widespread Extreme Rain Events Over Central India." *Nature Communications* 8, no. 708 (2017): 1–11.

Roy, Tirthankar. *Natural Disasters and Indian History*. New Delhi: Oxford University Press, 2012.

Roy, Tirthankar. "The Monsoon and the Market for Money in Late-Colonial India." *Enterprise & Society* 17, no. 2 (2016): 324–57.

Roy, Tirthankar. *The Economic History of India 1857–2010*. New Delhi: Oxford University Press, 2020.

Roy, Tirthankar. *Monsoon Economies: India's History in a Changing Climate*. Cambridge: MIT Press, 2022.

Saravanan, Velayutham. *Water and the Environmental History of Modern India*. London: Bloomsbury, 2020.

Sekhri, Sheetal. "Wells, Water, and Welfare: The Impact of Access to Groundwater on Rural Poverty and Conflict." *American Economic Journal: Applied Economics* 6, no. 3 (2014): 76–102.

Shah, Tushaar. *Ground Water Markets and Irrigation Development: Political Economy and Practical Policy*. Bombay: Oxford University Press, 1993.

Shah, Tushaar. *Taming the Anarchy: Groundwater Governance in South Asia*. Washington, DC: Resources for the Future, 2009.

Siebert, S., J. Burke, J.M. Faures, K. Frenken, J. Hoogeveen, P. Döll, and F.T. Portmann. "Groundwater Use for Irrigation—A Global Inventory." *Hydrology and Earth System Sciences* 14 (2010): 1863–80.

Smith, Andrew, and Kirsten Greer. "Uniting Business History and Global Environmental History." *Business History* 59, no. 7 (2017): 987–1009.

Steinhübel, Linda, Wegmann, Johannes, and Oliver Mußhoff. "Digging Deep and Running Dry—The Adoption of Borewell Technology in the Face of Climate Change and Urbanization." *Agricultural Economics* 51, no. 5 (2020): 685–706.

Subramanian, K. "The Quest for Groundwater in Post-Colonial India." *The Hindu*, November 8, 2017.

Tastevin, Yann. "From Ox-Carts to Borewell Rigs: Maintenance, Manufacture and Innovation in Tiruchengode." In *Subaltern Urbanization in India: An Introduction to the Dynamics of Ordinary Towns*, edited by Eric Denis and Marie-Hélène Zérah, 497–525. New Delhi: Springer, 2017.

Times of India (TOI). "Insurance Against Rain." *The Times of India*, April 8, 1911.
Times of India (TOI). "Crop Insurance to Cover Agricultural Risks." *The Times of India*, April 26, 1949.
Times of India (TOI). "No Takers for Weather Insurance." *The Times of India*, May 31, 2006.
Tumbe, Chinmay. *India Moving: A History of Migration*. Gurugram: Penguin India, 2018.
Tumbe, Chinmay. *The Age of Pandemics, 1817–1920*. Gurugram: Harper Collins India, 2020.
UNESCO. *Groundwater: Making the Invisible Visible.* Paris: UNESCO, 2022.
United Nations. *UNU-EHS Interconnected Disaster Risks 2023: Risk Tipping Points*. New York: United Nations, 2023.
Visaria, Praveen, and Leela Visaria. "Population." In *Cambridge Economic History of India, Vol. 2: 1757–1970*, edited by Dharma Kumar, 463–532. Cambridge: Cambridge University Press.
Zolin, Cornelio, and Renato de A.R. Rodrigues, eds. *Impact of Climate Change on Water Resources in Agriculture*. Boca Raton: CRC Press, 2016.

7

FROM PASTORAL TO INDUSTRIAL AGRICULTURE

Global Food Chains and the Drivers of Climate Change

Simon Mowatt

7.1 Introduction

Agriculture includes the raising of crops, livestock, and fish, which all human life is dependent upon, and is synonymous with human civilisation itself.[1] Mankind has remade the environment to favour cropping and farming, with around 40% of the Earth's usable landmass bound to its activities.[2] The expansion of the global cropland area continues to accelerate, and increased by 9% from 2003 to 2019.[3]

From its inception, agriculture has affected the environment—changing patterns of land use, re-creating local climates through land clearance for crops and pasture, and diverting and using water resources. The Columbian Exchange, which took place in 1492 following Christopher Columbus's voyages to the Americas, led to the widespread transfer of plants, animals, culture, human populations, and also technology and ideas between the Americas and the Old World. This movement of animals and crops to new regions and continents profoundly altered the natural world.[4] However, *global* anthropometric climate change has been exacerbated by the industrialisation of agriculture, which intensified from the start of the use of inorganic fertilisers in the nineteenth century and accelerated dramatically following the "green revolution" from the late 1950s. Then, the introduction of engineered seeds favoured the creation of large-scale monocultures, intensive animal farming, and grazing, which delivered enormous improvements in productivity, enabling the sector to feed exponentially growing populations and reduce world food insecurity, while dramatically decreasing the amount of labour required.[5]

Despite increased productivity, 821 million people were still undernourished in 2019, and 25–30% of the total food produced was lost or wasted.[6] Moreover, modern agricultural practices, which are highly resource-intensive, were increasingly recognised for their negative environmental impacts. Engineered crops, often reliant on pesticides and inorganic fertilisers, locked farmers into specific crops and dependency on powerful firms, excluding local or traditional crops.[7] New farmland has often come at the expense of old-growth forests, which acted as carbon sinks and fostered biodiversity. The cultivation of marginal land such as sloping lands which were initially made productive by chemical fertilisers has left many areas exhausted and vulnerable to erosion and flooding.

DOI: 10.4324/9781003478089-8

The preproduction, production, and postproduction processes related to food production all contribute directly and indirectly to carbon emissions.[8] The highest-impact segments are growing rapidly. Global meat consumption is predicted to increase by 15% by 2031,[9] and along with dairy farming, is a significant driver of climate change.[10] In 2016, methane emitted by livestock and nitrous oxide from the use of synthetic fertilisers had 25 and 300 times respectively the impact on the climate that CO_2 had.[11] In 2019, the overall agricultural sector was responsible for around 22% of total global greenhouse gas emissions, almost the same amount as both industrial sources and the production of electricity and heat.[12] In the sobering assessment of the Food System Economics Commission,

> the economic value of this human suffering and planetary harm was well above 10 trillion USD a year, more than food systems contributed to global GDP. In short, by 2024, our food systems had been destroying more value than they had created.[13]

Climate change's impact on agriculture has also been exacerbating global food insecurity and inequality.[14] The potential disruptions have significant impacts for business and systemic geopolitical implications, but policy responses to addressing these issues have been slow, with agricultural emissions omitted from key climate accords such as the Kyoto Protocol and the Paris Agreement on Climate Change.[15] This absence of policy has arisen because agriculture and the food it produces have been at the fulcrum of highly politicised issues between powerful stakeholders holding divergent views about what sustainable food systems look like.[16] These issues have included food's role in national identity and culture, food security, controversies around technology and genetic modification, ultra-processed foods (UPFs) and obesity, employment patterns and globalisation, and agriculture's relationship to urbanisation.[17]

It is therefore important to examine the historical drivers in which agricultural practices are embedded and to explore why and how private enterprises decided to commit to environmental strategies and what has driven or restrained them. In 1957, Davis and Goldberg proposed the concept of "agribusiness" to acknowledge the impossibility of disentangling agriculture and the production of food from the complex web of large, vertically integrated corporations which had grown to control the supply chain, from inputs to the shaping of consumer markets.[18] Seed, fertiliser, and grain markets and the large agrichemical firms and retailers who drive activities in the food chain have become increasingly concentrated, resulting in oligopolies capable of influencing agricultural practices and climate policies.[19]

This chapter follows Horowitz in using the term "food chain"[20] to describe these value chains, and the literature review loosely adopts the "from farmyard to shopping cart" structure employed by Belasco and Horowitz.[21] First, the chapter provides an overview of the historical research that has analysed environmental and agricultural issues in the light of business. Then it looks at research into the firm and industry activity in the upstream activities related to agricultural inputs and farming, and the emergence of alternative agriculture. The following section reviews research in the downstream activities of processing, marketing, retail, and consumption, and then food chains themselves. Next, it explores the responses from social movements developing "alternative" agricultural systems. Then, the chapter reviews examples of how entrepreneurs, policymakers, and business leaders in New Zealand attempted to "green" processes in the food chain in the 1980s and 1990s to mitigate its perceived negative aspects, and the obstacles they encountered to realise these changes. It notes how certain companies ("lead firms") in the chain can have significant influence and control over the entire supply chain in the food industry, influencing the co-ordination and integration of activities in the supply chain and the introduction of (or resistance to) new innovations and standards. The chapter concludes by arguing that policymakers and business leaders could gain valuable insights by integrating business history scholarship into current economic and scientific discourse concerning industry's adoption of more sustainable practices mitigating the climate impacts of the food system.

7.2 Environmental Business History and Agriculture

Discussions on climate change have been notably absent from the research on the history of agro-business. While the dairy industry is a significant producer of greenhouse gases, for example, there is no mention of the industry's wider impact on climate change in Smith-Howard's expansive *Pure and Modern Milk: An Environmental History Since 1900*.[22] A contextual understanding of the origins of the path-dependent structural features of the sector that are now understood to have impacts on the climate can provide valuable insights into the extremely complex and contested discourses around alternatives, mitigation, and amelioration. First, it is worth acknowledging that environmental business history is still an emerging field, as until the end of the twentieth century, business history had largely been silent on both the environmental impacts of the agricultural sector and climate change more generally, as discussed elsewhere in this volume.[23] However, a body of work slowly developed, with special issues of the *Business History Review* (1999) and *Enterprise & Society* (2007) focused on environmental issues.[24] Unsurprisingly, although business historians often touched on environmental issues and food, many focused on industrial firms, the upstream activities of processing and market shaping, managerial innovations, and issues such as pollution control, regulation, corporate responses to social movements, developments in certification systems, and the shaping of consumer markets.[25] Mostly, the primary activities of agriculture were left to environmental and economic historians. Delving into the decision-making processes of managers and entrepreneurs through archival and interview-based cases, Jones in *Deeply Responsible Business* emphasised how ethical and environmental decisions are often highly personal, driven by a complex raft of enmeshed issues, that motivations matter, and he highlighted the specific circumstances by which corporate strategy in some locations embraced sustainable strategies while others did not.[26] Bergquist (2019), in a special issue of *Business and Environment Revisited*, highlighted and added to the growing literature on the topic, still mostly firm-based. This new vibrancy was reinforced by Berghoff and Rome's (2017) *Green Capitalism*, a collection of papers attempting to further link business history to environmental history, but once again agriculture was mostly neglected.[27] Despite this, a body of work has slowly been emerging, covering many elements of the evolution of food chains. More recently business historical research around this topic has also expanded its scope to explore global governance issues, focusing on the influence that social movements, NGOs, and global institutions have on corporate strategies in response to climate change. These areas of study are central to a better understanding of complex transboundary challenges such as agricultural transformation.[28]

7.3 Upstream: Agricultural Inputs, Farming, and the Creation of Global Commodity Chains

By the mid-nineteenth century, the agricultural practices of the first "agrarian revolution," including enclosure, early mechanisation, crop rotation, and selective breeding, had dramatically increased agricultural productivity in Britain, leading to the development of international agri-commodity markets.[29] During this period of growth, the global grain trade became one of the most enduring global oligopolies. By the 1850s, the London Corn Trade Association had established itself as the global hub of the international grain trade. By the 1880s, half of Britain's wheat was imported from the United States, but the U.S. share fell as global commodity chains developed, and Russia, Argentina, India, Australia, and New Zealand also became major global exporters.[30] At the same time Asian countries expanded their rice exports. Powerful, vertically integrated trading companies emerged alongside innovations in grain handling, from purchase and storage systems such as grain elevators to factories for processing flour, and sales and distribution systems such as the railways.[31] At this stage, Peruvian guano and Chilean nitrates were the primary nitrogenous fertilisers employed and were not manufactured on a large scale. However, the globalisation of grain markets was rapidly underpinned by biological innovations such as new varieties of seeds adapted to different soils and climates, the increasing use of chemical fertilisers (nitrogen, phosphorus, and potassium), and the beginning of the mechanisation process in agriculture.[32] This second "agrarian revolution," combined with the second industrial revolution, intertwined chemical and agricultural industries and markets.[33]

British urbanisation and the growth of the middle class increased the demand for fruit and meat, and these new patterns of consumption were spread through trade to the peripheries, which literally fed the Empire's new hungers in return with exports. The British Empire was very important in the spread of agricultural systems, achieved largely through "geographies of connection" such as networks of people, commodities, ships, flora, fauna, and ideas.

The imperial networks formed by Britain and other empires had long-term environmental consequences, locking territories across the globe into industrialised agricultural systems.[34] Countries such as New Zealand became highly integrated into European, mostly British, food production systems and chronically dependent on chemical fertiliser imports. Ian Hunter (2005) examined how, following the introduction of refrigeration in the 1880s, meat, and then, from the turn of the century, dairy, emerged as major export sectors driven primarily by British demand.[35] These sectors had major ecological impacts on the environment, driving the conversion of much of the native landscape to pasture and introduced grasslands, and, from them, high long-term carbon emissions.[36]

Elsewhere, large-scale wheat farms, especially in the United States and Canada, became prevalent long-term economic features, as the high sunk costs required for converting land and introducing mechanisation favoured large-scale monocultures and intensive grazing. The collapse of the first wave of globalisation helped embed these features, as the U.S. government supported wheat farmers with protectionist agricultural and trade policies, resulting in continued increases in production despite falling consumption, ensuring that overproduction was to become a perennial feature of the market, despite resulting in the ecological dust bowl disaster of the 1930s.[37] A solution for soil exhaustion was to be found in an increased dependence on inorganic fertilisers, as Olmstead and Rhode noted in their examination of biological and chemical technologies in the industrialisation of agriculture in the United States between 1800 and 1940.[38] In the early post–World War II period, commodity chains evolved fully into vertically integrated food chains dominated by "agribusiness."[39]

From the 1950s, scientific research led to genetic improvements in major staple crops, creating high-yield strains of wheat, rice, and maize. Increasingly these crops required pesticides and large-scale land clearance to thrive. Public investments in seed research were supported by infrastructure, market development, and policy support leading to a "green revolution."[40] While the critical importance of this success in feeding the world cannot be underestimated, these developments also created large climate externalities, as outlined in the introduction to this chapter, and food systems also continued to experience shocks.[41] Ferguson-Cradler's work on the collapse of the Peruvian fishery emphasised how climate change can have catastrophic impacts on agriculture, leading to widespread consequences that affect the structure and organisation of the economy, in that case resulting in the nationalisation of key enterprises.[42]

Focused on increasing yields, many of the actors in the food chain felt limited concern for the consequences of industrial farming and the chemicalisation of agriculture upon the environment. The drivers behind this transformation were complex, as shown by Hamilton who highlights the significance of the "farms race" in Cold War politics. He argues that the American industrial food system was as much the product of governmental subsidy and intervention as it was the result of market and technological forces.[43] But throughout, the scientific focus on "feeding the world's hungry," aligned with the earlier moral ethos of pastoral "improvements," which were seen as virtuous and godly, bringing land into "productive use," has helped to justify the "second green revolution." This ethos has supported the development of genetic engineering, biotechnology, and related uses of mobile information technologies, big data, and artificial intelligence, resulting, among other things, in the creation of closed markets where specific seeds are dependent on bespoke fertilisers and pesticides.[44]

The scramble to acquire the property rights for genetically modified seeds in the late 1990s and early 2000s resulted in a series of mergers and acquisitions, which further consolidated the industry. According to the Food and Agriculture Organization (FAO), the top three seed companies—Monsanto (acquired by Bayer in 2018), DuPont Pioneer, and Syngenta—controlled roughly 43% of the global commercial seed market in 2020. Furthermore, the seed trade had become closely integrated into the chemical industry, and over 60% of the global agrochemical market was controlled by just four firms.

According to FAO, the unprecedented level of market concentration throughout the global agri-food system—spanning crop seeds, agricultural chemicals, veterinary pharmaceuticals, agricultural machinery, fertilisers, livestock genetics, fishing rights, finance, food processing, and commodity trading—represents a challenge for the resilience, equitability, and sustainability of agri-food systems, had the risk of causing farmers, and even nations, to become dependent on a handful of powerful firms.[45]

Since the 2000s vertically integrated agribusiness firms have also been major players in biofuel production, which has also increased drives to convert more land for production, perhaps at the expense of food crops. Boon highlighted the sustainability issues arising from the use of food crops for biofuels, noting that rising CO_2 emissions from land use and changes related to fuel crop production have prompted petrochemical firms to shift away from this sector towards non-food crops.[46]

7.4 Downstream: Processing, Marketing, Retail, and Consumption

The downstream activities of meat processing and packing created noxious urban environmental problems. Rosen has examined the growing importance of regulation and the influence of wider social stakeholders in addressing these issues.[47] This contribution was followed by a growing body of fine-grained firm-based and industry studies integrating environmental issues, examining topics such as waste and pollution.[48] Marketing and the development of markets were another growing focus of scholarship.[49] Silva Lopes reviewed the important role standards could have in driving more ethical global food chains, examining the evolution of fair trade certification in chocolate. Competitors such as Cadbury and Rowntree collaborated together with local governments and producers, bringing new locations and techniques into the food system, resulting in more equitable outcomes.[50]

A major area of research in this field originated from projects at the Hagley Museum and Library. This interdisciplinary work explored the cultural and business aspects of food, resulting in two important edited volumes. In the first, Belasco and Scranton edited a volume exploring food's role from marketing and consumer culture perspectives, as well as its central and politicised position in the creation of, sometimes imagined, national identities.[51] Belasco summarised debates around the "sustainable food movement" and the issues such as "food miles" and the sector's ecological footprint. The rich studies also unpack the influence of wider social contexts such as how constructed national identities and cultures influence firm responses and the trajectory of markets.[52] Elsewhere, Jones and Mowatt presented comparative studies attempting to explain national divergences in the mainstreaming of the organic retail market, identifying factors which supported and frustrated entrepreneurial success.[53]

7.4.1 Food Chains

The "agribusiness" lens proposed by Davis and Goldberg gained new prominence through the rise of the "new political economy of agriculture" and global value-chain analysis.[54] Much of the business history literature prior to this dealt with the processing and downstream stages of food production systems, rather than the upstream primary activities, but this began to change as environmental and business historians sought to connect with other disciplines interested in food and embrace value-chain perspectives. Smith and Greer presented a special issue of *Business History* which moved beyond a firm-level analysis by examining the increasingly internationalised knowledge flows within agricultural commodity chains, attempting to unite business history and *global* environmental history, and forge deeper connections with environmental historians.[55] This issue presented papers by Goodchild and MacFadyen concerned with agriculture.[56] The new research agenda in environmental history, embracing global value chain (GVC) research, also had a parallel in agricultural history. Until its turn to "the new agricultural history," agricultural historians generally viewed the environment as "natural resources" for humans to utilise in the process of production in the economist's sense. Recent work, however, has drawn on the commodity-chain/supply-chain literature and a growing body of work examining the relationship between climate and agriculture. Business historians have so far been slow to seek collaborations with scholars in this field.[57]

In the second volume originating from the Hagley Museum, Roger Horowitz usefully adopted a global value-chain framework to analyse agricultural "food chains."[58] The edited book offered papers covering all parts of the process, from farms to consumers, and offered some insights into key drivers in the sector.[59] Godley and Williams, for example, highlighted how retailers can drive the growth and practices of upstream activities and shape market demand.[60] Elsewhere, the importance of supermarkets in driving the creation and growth of food markets and through the supply chain was also analysed by Cox, Mowatt, and Prevezer, who explore the growth of chilled ready meals in the UK, and by Jones and Mowatt in the creation of organic retailers.[61]

Hamilton argued that if eating is essentially an "agricultural act," then this has become part of a highly industrialised system in which economic structures are "continually shaped and reshaped by business and political decisions and by shifting cultural patterns of diet, agrarian ideology, environmental concern, and scientific knowledge."[62] A GVC perspective lends itself to analysing this complexity, and also the exhortation to business historians by Francesca Carnevali to consider the intertwined complexities of production, distribution, and consumption at once.[63] This lens is critical as food manufacturers and retailers have also become highly concentrated and shape the food chain. They have shifted food chains from being producer- to buyer-driven, potentially locking them into patterns that prioritise high-carbon and energy-intensive production. In 2020, the top U.S. market-concentration ratios in key food manufacturing segments were between 54% and 83%: 83% in cold cereal, 82% in soft drinks, 77% in beer, 63% in salty snacks, 58% in bread, and 54% in ice cream.[64] The consequences of all the recent changes driving global, rather than national, food chains have so far received less attention from a business history perspective, as many of the changes came after 1985.[65]

7.5 The Development and Commercialisation of Alternative Agriculture

Not everyone was convinced that the chemicalisation of agriculture was healthy for humans or the environment, and in response the idea of "alternative agriculture" developed. Jones has examined how key evangelists such as the Austrian philosopher Rudolf Steiner and the Life Reform movement, which championed his concept of biodynamic agriculture in the 1920s, provided the genesis of what in time evolved into the organic agriculture (OA) movement and the eventual slow growth of organic markets.[66]

Certification was central to the eventual growth of this sector. In their chapter in this volume on "Business, Institutions, and Climate Change," Bergquist and Jones review the establishment of some of the key international institutions and standards that helped legitimise organic agriculture such as the International Federation of Organic Agriculture Movements (IFOAM) in the 1970s and 1980s. Initially, however, many of the acolytes of alternative agriculture consciously rejected capitalist structures and were suspicious of dealing with the conventional food chain, instead drawing inspiration from Steiner, agriculturists studying traditional Asian farming systems, and forms of communal living outside of conventional society.

It was only during the 1980s, and even more so during the 1990s, that organic agriculture began to transition from the marginal to the more mainstream, as entrepreneurs sought to "green" food chains by attempting to mainstream organic agriculture. Jones and Mowatt questioned why this was more successful in some countries rather than others, and, alongside certification systems and supportive policies, a major driver was purposeful market shaping by retailers such as Whole Foods in the United States and the cooperative Super Brugsen chain in Denmark, alongside the awareness of ecological disasters and food scares in the public consciousness.[67]

While the early proselytisers of OA required a commitment to its wider principles, later adherents were often more pragmatic, embracing integrative and regenerative agriculture along with conventional retail channels in order to compete in the mainstream food chain. Integrated farming has been defined by the Integrated Arable Crop Alliance as

> a whole farm policy aiming to provide the basis for efficient and profitable production which is economically viable and environmentally responsible. It integrates beneficial natural processes into modern farming practice using advanced technology and aims to minimize the environmental risks while conserving, enhancing and recreating that which is of environmental importance.[68]

Regenerative agriculture stemmed from OA farmers who departed from its strict requirements to include elements of conventional practices where such inclusion was thought to be necessary. It does not have a formal definition but is based on the premise that "the health of soil, plants, animals, people, and the environment is one and indivisible," and there is a strong focus on "soil health, biodiversity, and [reducing] socioeconomic disparities."[69]

By 2007 worldwide sales of organic products had developed into a US$28 billion market. However, this growth remained highly uneven and prone to decline in economic downturns, as consumers switched back to cheaper conventional products, with supply vulnerabilities as producers often left the industry or reverted to conventional production, exacerbated by OA returning around 19–25% lower crop yields.[70] While the development of large organic retail chains acted to foster OA production, firms could also change strategy, leave the market, or be acquired, as Whole Foods Market was by Amazon in 2017.

OA has significantly lower impacts on climate change by rejecting inorganic fertilisers, creating healthier soils which sequester higher levels of carbon, and many secondary benefits such as healthier and more stable water systems. By 2024 the share of land under OA production reached 2% globally and more than 10% across the European Union, for the first time. Although these represent impressive milestones, they underscore that OA is still but a small part of the food system.[71] Despite the limitations of OA, it is increasingly being positioned by policymakers as a cornerstone for transitioning to a more efficient, lower-impact, more inclusive, resilient, and sustainable food chain. FAO and the European Union (EU) both have ambitious plans to champion OA. The FAO sees OA as being a strong contributor to key UN Sustainable Development Goal (SDG) achievements.[72] For OA to achieve this potential, it will have to expand rapidly and overcome the obstacles which restrain conversion in the food chain. However, entrepreneurs have faced many challenges to mainstream OA markets.

7.5.1 The State of Knowledge

The literature highlights the role of firms in shaping markets and consumer demand, emphasising the significant influence large companies have in controlling practices throughout value chains, including those in the food and agricultural sectors. It also examines their motivations for doing so. Business history, with its diverse methodologies for exploring the decision-making processes of business owners, can provide valuable insights into why and how firms commit to environmental and climate action.[73]

The U.S. meatpacking industry[74] and the German chemical industry[75] illustrate how managers and firms have historically addressed increasing regulatory pressures and rising public environmental consciousness, and how they have developed strategies to deal with emerging environmental and legal regulations. In these processes, leadership within companies was critical in shifting corporate culture towards valuing sustainability and recognising that environmental responsibility can enhance competitiveness and profitability in the long term. These insights can inform policy debates concerning agriculture and climate change.

Business historical research on climate change can offer valuable insights into the concept of corporate agency, highlighting how industry actions are ultimately driven by decision-makers and the outcomes of deliberate business strategies.[76] The next section, therefore, provides two cases, giving some insight into examples where firms have been unsuccessful in greening the food chain and a short review of a firm attempting to remake the coffee food chain to combat climate change impacts more successfully.

7.6 Driving Change in Agricultural Food Chains

7.6.1 *The Example of New Zealand Organic Mainstreaming*

This section examines the question of why OA production in New Zealand was unexpectedly low in comparison with other countries with similar agricultural profiles such as Denmark after the take-off period in the 1980s and 1990s. Revisiting the study of New Zealand agriculture presented by Jones and Mowatt, it questions whether the situation has changed by providing a longitudinal update to the 2012 study, conducted with follow-up interviews with the key informants closest to organic food production in late 2023.[77]

The New Zealand organic community had always maintained strong connections with international movements, and there was a New Zealand representative on the board of IFOAM from its foundation until 1999. And the New Zealand Biological Producers Council, created as an independent body formed by the Soil Association, Bio-dynamic Farmers & Gardeners Association, and the Henry Doubleday Association, had established the BioGro standard in 1982 without any input from New Zealand policymakers.[78] A new influx of European immigrants and returning migrants to New Zealand from the 1970s brought a new enthusiasm to organic production, but initially many of these advocates rejected participating in the mainstream food system.[79] Entrepreneurs such as Malcolm Rands, who founded ecostore, selling personal and homecare products (initially as a mail order and later as a conventional retailer and wholesaler to supermarkets), were influenced by biodynamic advocates such as the Austrian artist and environmentalist Friedensreich Hundertwasser, who had settled in Northland. Rands, with his family and associates, set up a village based on permaculture, the practice of designing long-term sustainable organic subsistence farming, integrating mixed crops, fruits, and trees. Rands decided to support this way of life by using ecostore as a vehicle to finance the Fairground Foundation, created to finance organic, ecological, and social causes, rather than mainstream food production directly.[80]

Established in 1984, Millton Vineyards & Winery was New Zealand's first organic biodynamic winegrower. James and Annie Millton had witnessed organic production on a trip to Europe, which led them to question the "recommended sprays of herbicide insecticide and systemic fungicide every 14 days" dictated by the chemical companies as normal practice in New Zealand viticulture. Although they embraced biodynamic production and Biogro certification, they chose not to market it as organic, as they considered it likely to put consumers off.[81]

By the late 1980s, however, entrepreneurs were making concerted efforts to mainstream OA, and several such as Common Sense Organics and Ceres Wholefoods had come to the conclusion that integrating forwards into retail and processing, and dealing with conventional supermarkets, would be required to develop the market. With impetus in the market, some organic produce wholesalers such as Fresh Direct were set up to capitalise on the expected growth opportunities mainstreaming presented. By the early 2000s these players could rely on supportive organic institutions and had a presence in all stages of the food chain, selling not only in organic retail chains but also in conventional supermarkets.[82]

This approach pushed against a totally unsupportive policy framework which did not incentivise organic conversion, and government policy which failed to adopt organic standards to support the market, despite being aware of New Zealand's potential to be an early adopter in the development of commercial organic markets. Without stimulus, New Zealand's largest agricultural producers, focused on serving export markets, were uninterested in OA. Fonterra, the effective monopoly dairy cooperative, had no real commitment. Neither did Wattie Frozen Food Ltd, the dominant processor, which had even less interest when purchased by H.J. Heinz Co. Ltd in 1992. The largest crop producer, Zespri (formerly the kiwifruit New Zealand Marketing Board), focused on satisfying the EurepGap/GlobalGap standards required for export to European supermarkets rather than OA.

The government and export sectors were heavily invested in maintaining and selling New Zealand's reputation as a "100% pure," "clean and green" environment, as it had been positioned since the introduction of refrigeration and beef, lamb, and dairy exports a century earlier. The supermarket duopoly also had little interest in promoting organic food in a country where the public perception was that New Zealand food was already healthy and green. Against this backdrop, New Zealand entrepreneurs were unable to make the breakthrough to grow OA significantly compared to the take-off in leading countries such as Denmark, Sweden, and the United States.[83]

By the 2020s it was becoming increasingly evident that climate change would pose a serious existential threat to New Zealand's position as a major food exporter, affecting which crops could thrive in different regions as temperatures increased.[84] Although the New Zealand organic market had continued to grow, the barriers identified in 2012 remained in place and ensured that systemic transformation has not occurred, and in some ways policymakers continued sending mixed signals to the sector. Only in March 2023 was legislation passed aimed with establishing a government-approved standard, motivated by the requirements to meet the same or equivalent standards as organic products produced in the EU to maintain access to export markets and to continue to supply European and British supermarkets.[85]

In 2022, New Zealand was one of the countries in Oceania with the smallest percentage of land under OA, 0.8%. And New Zealand was the only country in the world which exports the majority of its organic production, mostly accounted for by apple and kiwifruit exports.[86] In 2024 the new National party government in New Zealand introduced legislation to remove agriculture from the carbon Emissions Trading Scheme.[87] These factors continue to restrain the mainstreaming of OA in the domestic market, and in 2023 founding entrepreneurs of key firms such as Ceres Organics and Commonsense Organics still perceived that the situation had remained mostly unchanged.[88] In 2012, global warming, as climate change was generally referred to at the time, was one of many interwoven factors around attempting to ameliorate the externalities of conventional agriculture, promote social and economic justice, and create fairer, healthier lives and environments for workers, suppliers, and consumers in the food chain. Activists saw addressing climate change as inseparable from tackling these issues.

7.6.2 *Lead Firm Impacts in Producer-Driven Food Chains*

While New Zealand firms have been unable or unwilling to systemically change the food system, there are examples of multinationals attempting to use their leading position in food chains to make them more resilient and sustainable. Climate change presents major risks for food producers, in both raising costs and affecting the quality and security of supply.[89] But it also offers opportunities for firms, especially lead firms in buyer chains, to create shared social value by confronting these issues.[90] A good example of this is provided by Nestlé and their strategies for the premium speciality coffee brand, Nespresso, which allows them to derive higher profits from commodity products.[91] This strategy is at risk, as climate change will jeopardise the locations of production, quality, and yield of the high-value but temperature-sensitive Arabica bean.

Climate change also promotes coffee diseases and pests, extreme weather events, and changing rainfall patterns. At a time of growing demand for coffee, it was predicted that between 65 and almost 100% of current coffee-growing areas would be unsuitable for production by 2080. More than 70% of the 25 million coffee producers in 2020 were small-scale farmers with neither the training nor the resources to adapt to these challenges, although certification systems covering around 24% of production in 2019 (such as Rainforest Alliance and Fairtrade) filled some of these voids.[92]

Nestle's new approach began in 2009 with the Nespresso AAA sustainability programme, then further elaborated through the Positive Cup Initiative in 2020, setting higher sustainability goals such as securing 100% of beans from Rainforest Alliance-certified sources by 2020. Nestlé attempted to secure a constant and consistent quality supply of premium Arabica beans, including attempting to shift to a carbon-neutral supply chain by 2050, reduce its supply-chain costs, and mitigate climate change by farmer education and moving to more renewable energy throughout the system, as well as scientific investment in developing climate-resistant Arabica strains. The benefit for farmers is a claimed potential 90% increase in profits, as well as investments in their communities and farms.[93] Nestle's approach led to competitors competing with this strategy, and by 2012, the share of sustainably produced coffee reached 40% across the sector.[94] Despite concerns about the transparency of some aspects of the reporting, and that the firm's other interests as a UPF manufacturer may offset or negate the overall positive impacts,[95] Nestlé presents an example of a firm making long-term commitments to make a more climate-resilient and neutral food chain through purposeful business strategy.

7.7 Conclusion

The agricultural system faces an unprecedented challenge. It must find a way to feed nearly 50% more people as the global population is projected to reach 9 billion by 2050. This requires increasing food production by between 25 and 70% while dramatically reducing its climate and environmental impacts.[96] These conflicting demands may explain why sectoral reform has largely been overlooked in climate accords, despite agriculture being the second largest contributor to carbon emissions after industry. According to the FAO, it is still feasible to transition the food chain in time to keep global warming below the 1.5 degree Celsius threshold and achieve numerous social benefits, but this would require a major change in the behaviour of many actors in the system.[97]

Despite this complexity, the adoption of sustainable farming practices is one of the key steps that needs to be developed at large scale to support long-term economic development without negatively impacting future social and environmental conditions.[98] The choice between adopting either nature-based farm management or agri-technological solutions has profound socio-ecological considerations and implications for future sustainable landscapes, biodiversity, and the balance of ecosystem services and disservices these provide.[99] The European Union's European Green Deal has the goal of 25% of the EU's agricultural land under organic production by 2030.[100] Although policies may be enacted to drive change, the research examined in this chapter reminds us that it is business owners and decision-makers in the food chain who will ultimately make these choices.

The food system will continue to develop along two pathways—a highly intensive industrial path, seeking answers from genetic engineering and artificial intelligence to mitigate its climate emissions and risks, and some continued transition towards more regenerative, lower-impact systems. The critical question is where the balance will lie.

Business historians are particularly apt at investigating the critical "turning points" which help explain change processes, and this is the importance of bringing a business history perspective to this debate.[101] The New Zealand example shows that powerful actors anchoring food chains can have an influence over food chain practices, and therefore climate consequences, but that the desire to prioritise ecological strategies is precarious when left to the personal inclinations of individual entrepreneurs without supporting policy frameworks and incentives.[102] In that example, leading players in the market such as Fonterra and the national supermarket duopoly failed to drive change, but European retailers exerted a strong influence on steering the agricultural sector towards prioritising its EurepGap/GlobalGap standards.

It is therefore important to examine the drivers by which private enterprises decide to commit to environmental and climate strategies to enact change. Private regulatory systems, including voluntary efforts by firms and visionary entrepreneurs to restrain their own behaviour or take leadership positions, have been historically the primary form of global climate change governance, and the decisions to develop, support, and drive these and the certification schemes that scaffold them have been mainly firm-level business decisions.[103] Lead firms in food chains have the potential to drive positive actions towards climate change through the supply chain, as the Nespresso example demonstrates.

As agricultural markets have faced increasing challenges by climate change, firms have increasingly considered its impacts. Their responses have included making their supply chains more resilient and remaking them towards more positive climate outcomes. Firm-level research, especially when contextualised into historical institutional, social, regulatory, legal, and governance contexts, can help us understand the specific circumstances in which companies change their strategies so as to protect the environment and suggest which factors conditioned their success or failure in particular contexts.

Notes

1 Belasco gives a succinct review of how agriculture led to the development of bureaucracies and trade. See Belasco et al., *Food Nations*, Chapter 1.
2 Lorenz et al., *Organic Agriculture and Climate Change*, 2.
3 Lorenz et al., *Organic Agriculture and Climate Change*, especially Chapter 5–6.
4 Sidhoum et al., "Agriculture and Sustainability."
5 The global population has grown six-fold since 1800. Federico, *Feeding the World.*
6 ICCC, "Special Report on Climate Change and Land."
7 Pingali, "Green Revolution."
8 Preproduction emissions include fertiliser manufacture, energy use in feed production, and pesticide production. Production emissions include the direct and indirect emissions from livestock and crop production, and the indirect effects from land use change. And postproduction emissions include emissions from primary and secondary processing; storage, packaging, and transport refrigeration; retail activities; catering and domestic food management; and waste disposal. Vermeulen et al., "Climate Change and Food Systems," section 2, 198–202.
9 ICCC, "Special Report on Climate Change and Land"; FAO, "Strategic Framework 2022–31."
10 Sidhoum et al., "Agriculture and Sustainability."
11 Verschuuren, "The Paris Agreement on Climate Change," 54.
12 IPCC, *Climate Change 2022.*
13 Ruggeri et al., *The Economics of the Food System Transformation,* 7.
14 FAO, "Strategic Framework 2022–31."
15 Verschuuren, "The Paris Agreement on Climate Change."
16 Ruggeri et al., "The Economics of the Food System Transformation."
17 Hamilton argues that food production systems can never truly be examined by logic, or reduced to patterns of market forces because food chains are so complex and deeply saturated with cultural meaning, "Analyzing Commodity Chains," 25.

18 Davis and Goldberg, *A Concept of Agribusiness*.
19 The global food chain is witnessing increasing retail concentration, with supermarket chains becoming dominant in food sales in many territories, increasing from 5–10% across developing regions in 1990 to more than 20–50% in Mexico, Central America, and Southeast Asia and 50–60% in South America and South Africa by 2007. Aside from the high concentration in seed markets, some global markets such as coffee are also highly concentrated. Vermeulen et al., "Climate Change and Food Systems," 197.
20 Horowitz, "Making Food Chains." Vermeulen et al., describe the food chain as the manufacturing and distribution of inputs (seeds, animal feed, fertilisers, pest control); agricultural production (crops, livestock, fisheries, wild foods); primary and secondary processing, packaging, storage, transport and distribution; marketing and retail; catering; domestic food management; and waste disposal. In some cases, supply is linked through a "cold chain" in which continuous refrigeration is used to extend and ensure the shelf life of fresh and processed foods. They note that food systems also encompass outcomes of these activities and their governance. "Climate Change and Food Systems," 197.
21 Belasco et al., *Food Nations*.
22 Smith-Howard, *Pure and Modern Milk*.
23 The start of research streams moving towards remedying this deficit was initially stimulated by Christine Rosen and Christopher Sellers, who in a special issue of *Business History Review* on "Business and the Environment" in 1999 called for an "eco-cultural history of business" and the integration of business history and environmental history. Rosen et al., "The Nature of the Firm." Then in 2007, Rosen criticized "a generation of business historians who abstracted the rise of businesses from the environmental as well as the regulatory and legal contexts in which they arose." Rosen, "The Role of Pollution Regulation and Litigation," 302.
24 Rosen, "The Role of Pollution Regulation and Litigation."
25 Uekoetter examined issues around pollution control in "The Age of Smoke," 20–42; Rosen looked at groups of stakeholders seeking to mitigate agricultural waste issues, "Businessmen against Pollution." Other research such as Hoffman, "From Heresy," examined corporate responses to environmental campaigners and regulators. Jones and Lubinski highlighted the evolution of how regional context led to US and German firms developing different strategic responses to environment issues, Jones et al., "Making 'Green Giants.'"
26 Jones, *Deeply Responsible Business*; Jones, *Varieties of Green Business*; Jones, and Mowatt, "National Image as a Competitive Disadvantage"; Jones and Lubinski, "Making 'Green Giants'"; Jones, *Profits and Sustainability*.
27 Berghoff and Rome, *Green Capitalism?* This collection builds on Berghoff and Mutz's earlier investigation into the need to forge links with environmental historians, "Missing Links."
28 Ballor and Pitteloud, "Introduction: Capitalism and Global Governance" provide a good analysis of this topic.
29 Prior to this, internationally traded grain (wheat, maize, rice, and cereals) accounted for only a small fraction of grain consumption. See Mercier, "The Evolution of World Grain Trade."
30 Drivers of these changes included political factors such as the repeal of the corn laws (1846) and resulting fall in tariffs; technological innovations leading to decreased transaction and transport costs and changes in farming practices; and institutional changes such as increased market efficiency and a more stable international monetary system. See Aparicio and Pinilla, "International Trade in Wheat"; Ejrnæs, Persson, and Rich, "Feeding the British"; Rothstein, "Multinationals in the Grain Trade."
31 The nineteenth century saw the growth of major trading family firm dynasties such as The Bunge (Buenos Aires), Andre, Louis-Dreyfus and Fribourgs/Continental Grain (France). In the United States the Cargill, Peavey, and Pillsbury families were key in the integration of the global grain market. See Morgan, *Merchants of Grain;* Mercier, "The Evolution of World Grain Trade."
32 Aparicio and Pinilla, "International Trade in Wheat."
33 Dixon, "Chemical Fertilizer in Transformations."
34 Smith and Geer, "Uniting Business History and Global Environmental History."
35 Hunter, "Commodity Chains and Networks in Emerging Markets."
36 Pawson and Brooking, "Empires of Grass."
37 Aparicio and Panilla, "International Trade in Wheat."
38 Olmstead and Rhode, *Creating Abundance*.
39 Davis and Goldberg, *A Concept of Agribusiness*.
40 Pingali, "Green Revolution."
41 Mansel Blackford, for example, examined the consequences of overfishing and urged business historians to consider environmental factors in their research. Blackford, "Fishers, Fishing, and Overfishing."

42 Ferguson-Cradler, "Coping with Crisis."
43 Hamilton, *Supermarket USA*.
44 Pawson and Brooking, "Empires of Grass," 101–3. Elsewhere, Pawson notes that "If growth, like 'improvement', is a modern morality tale, New Zealand—along with Australia—added an extra twist in the 1950s. This was the urge to enhance levels of primary production in order to 'feed the world' and stave off famine in war-torn Europe, and latterly in countries experiencing rapid population growth. This was the justification for ongoing land clearance, supported by the state through the Lands and Survey Department, and the enthusiastic acceptance of chemical farming that was the subject of Rachel Carson's *Silent Spring* in 1962." Pawson, "Into the Anthropocene," 62.
45 FAO, "Strategic Framework 2022–31," 35.
46 Boon, "A Climate of Change?"
47 Rosen, "The Role of Pollution Regulation and Litigation."
48 Elmore, *Citizen Coke*; Stokes, Roman, and Sambrook, *The Business of Waste*.
49 For instance, Barnes and Higgins examine the growth of global brands such as Anchor. Barnes and Higgins, "Brand Image, Cultural Association and Marketing."
50 Silva Lopes, "Building Brand Reputation." See also Milov, "Promoting Agriculture."
51 Belasco and Scranton, *Food Nations*. In this volume Wilk, for example, explores the highly politicised role food has not only in the creation of national identity but in threats to it from globalisation. Wilk, "Food and Nationalism."
52 Key papers examining these issues include Godley and Williams, "The Chicken, the Factory Farm, and the Supermarket," Godley and Williams, "Democratizing Luxury"; Barnes and Higgins, "Brand Image, Cultural Association and Marketing"; Jones and Mowatt, "National Image as a Competitive Disadvantage."
53 Jones and Mowatt, "National Image as a Competitive Disadvantage," and "Organic Food and National Image."
54 Davis and Goldberg coined the term in *A Concept of Agribusiness*. See also Hamilton, "Analyzing Commodity Chains," 23 and "Revisiting the History of Agribusiness" for an appraisal and limitations of this concept.
55 Smith and Geer, "Uniting Business History and Global Environmental History."
56 Goodchild, "The Problem of Milk"; MacFadyen, "Long-Range Forecasts."
57 Coclanis gives a succinct review of this literature in "Field Notes: Agricultural History's New Plot."
58 Horowitz, "Making Food Chains."
59 Belasco and Horowitz, *Food Chains*.
60 Godley and Williams, "The Chicken, the Factory Farm"; see also Godley and Williams, "Democratizing Luxury."
61 Cox, Mowatt, and Prevzer, "New Product Development"; Jones and Mowatt, "National Image as a Competitive Disadvantage."
62 Hamilton, "Analyzing Commodity Chains." [quotation needs page number; I don't have access to de Gruyter]
63 Di Martino et al., "People, Places and Business Cultures."
64 Hendrickson et al., "The Food System," Figure 4, p. 9.
65 Smith and Geer, "Uniting Business History and Global Environmental History."
66 Jones, *Deeply Responsible Business*, Chapter 9; Jones, *Profits and Sustainability*, Chapter 1; Jones and Mowatt, "National Image as a Competitive Disadvantage"; Jones and Mowatt, "Organic Food and National Image." For broad historical survey of the history of organic food, see Lockeretz, *Organic Farming. An International History*.
67 Jones and Mowatt, "National Image as a Competitive Disadvantage."
68 Quoted in Leake, "The Development of Integrated Crop Management in Agricultural Crops," 952.
69 Lorenz and Lal, "Organic Agriculture and Climate Change," sections 5.4 and 5.6.
70 Lorenz and Lal., "Organic Agriculture and Climate Change," 202–16, 220–30.
71 The global market in 2022 was judged to be around €135 billion (roughly US$144 billion.) Willer, Trávníček, and Schlatter, "The World of Organic Agriculture."
72 Its most notable contributions are claimed to be to SDG 1 (no poverty), SDG 2 (zero hunger), and SDG 10 (reduced inequalities), as well as to supporting the broader SDG agenda. FAO, "Strategic Framework 2022–31."
73 Pawson considers the question of what environmental history can teach about the future. Initially sceptical, he considers that this is a "compelling moral issue, one that requires honesty, consistency, and a

willingness both to look and to see. Environmental historians would seem well equipped on each of these grounds," as would business historians I would suggest. Pawson, "Into the Anthropocene," 73.
74 Rosen, "Businessmen Against Pollution."
75 Jones and Lubinski, "Making 'Green Giants'."
76 Smith and Greer note that "none of the participants in the recent methodological debates about how to do business history has mentioned either research collaboration with natural scientists or the integration of scientific research findings into single author papers by business historians." Smith and Geer, "Uniting Business History and Global Environmental History." Perhaps this is because science-based literature often works better with economists and their generally abstracted conceptualisation of business, See Sörlin and Lane, "Historicizing Climate Change."
77 This included the senior managers/directors of organic food retail groups, produce wholesalers and food processing companies interviewed in 2012 by Jones and Mowatt for "National Image as a Competitive Disadvantage."
78 Campbell and Leipins, "Naming Organics," 29.
79 Coombes and Campbell noted the importance of European immigrants with experience in organics in stimulating the organic community market garden culture in the Nelson region but noted many left the industry after a few years. Coombes and Campbell, "Recent Developments in Organic Food," 7.
80 Malcolm Rands interview, January 24, 2012, Auckland; Rands, *Ecoman*, 17.
81 New Zealand Winegrowers, "Millton Vineyards."
82 Jones and Mowatt, "National Image as a Competitive Disadvantage."
83 Jones and Mowatt, "National Image as a Competitive Disadvantage."
84 Lilburne et al., "Progress in Understanding Challenges and Opportunities for Agriculture Under Climate Change."
85 Ministry of Primary Industries, "Changes to Organic Products Law."
86 46% apples and over 32% kiwis. Willer, Trávníček, and Schlatter, "The World of Organic Agriculture," section: Organic Agriculture in Oceania: Key Facts and Figures, 257–62.
87 Daniels, "Agriculture Minister Todd McClay Defends Removing Farmers from Emissions Trading Scheme."
88 Noel Josephson (co-founder, Ceres Organics), interview conducted November 20, 2023. Marion Woods (co-founder, Commonsense Organics), interview conducted December 11, 2023.
89 Bager and Lambin, "Sustainability Strategies by Companies in the Global Coffee Sector."
90 Porter suggest improving environmental outcomes is in firms' interest and can lead to competitive advantages and higher returns. For an introduction to the "Porterian hypothesis" see Porter and Kramer, "Creating Shared Value." Nicholas Stern and Joseph Stiglitz have lent recent support to the conception that tackling climate change can drive future economic growth with far-reaching benefits to society in terms of wellbeing. Stern and Stiglitz, "Climate Change and Growth."
91 Initially competitive advantage was further enhanced by consumer lock-in to a pod deliver system, but the patent protection for this ended in 2006 as competitors offered compatible pods, putting pressure on the firm to find other ways to maintain their position.
92 Bianco, "Climate Change Adaptation, Coffee, and Corporate Social Responsibility"; Verburg et al., "An Innovation Perspective to Climate Change Adaptation in Coffee Systems."
93 Gonçalves and Zylbersztain note that the shift away from commodity supply of robusta beans to focusing on arabica, and working directly with international firms such as Nestlé and Starbucks, is a major factor in increased profitability. This is discussed in their work "Business History and Challenges for Coffee Cooperatives in Brazil."
94 Levy, Reinecke, and Manning, "The Political Dynamics of Sustainable Coffee."
95 The use of aluminium for the coffee pods for example is problematic, and the firm's success in recycling them has been questioned. Gunther, "The Good, the Bad and the Ugly." Levy, Reinecke, and Manning give a nuanced view of how the motivations of coffee producers change the relative power dynamics between NGOs and producers and can lead to shifts towards multinational firm dominance, yet present evidence of the success of these programmes in changing practices. Levy, Reinecke, and Manning, "The Political Dynamics of Sustainable Coffee."
96 Lorenz and Lal, *Organic Agriculture and Climate Change*, 220.
97 Ruggeri et al., *The Economics of the Food System Transformation*, 7.
98 Sidhoum and Vrachioli, "Agriculture and Sustainability."
99 Vanbergen et al., "Transformation of Agricultural Landscapes in the Anthropocene."

100 European Commission, "Action Plan for Organic Production in the EU."
101 Rollings, "Business and Global Capitalism."
102 Hamilton, "Analyzing Commodity Chains," 22.
103 Bergquist and Jones, "Private Regulation, Institutional Entrepreneurship, and Climate Change." Specific examples of how rich historical cases shed light this are well posed by Geoffrey Jones 'paradoxes'—such as why states without the best natural endowments became leaders in wind power generation, or why those with factor endowments failed to become leaders such as in the New Zealand organic market for example. Berghoff et al., *Green Capitalism?* Is an excellent collection of contributions in this vein.

Bibliography

Aparicio, Gema, and Vicente Pinilla. "International Trade in Wheat and Other Cereals and the Collapse of the First Wave of Globalization, 1900–1938." *Journal of Global History* 14, no. 1 (2019): 44–67.
Ballor, Grace, and Sabine Pitteloud. "Introduction: Capitalism and Global Governance in Business History." *Business History Review* 97, no. 3 (2023): 459–79. https://doi.org.ezproxy.aut.ac.nz/10.1017/S0007680523000855.
Bager, Simon L., and Eric F. Lambin. "Sustainability Strategies by Companies in the Global Coffee Sector." *Business Strategy and the Environment* 29, no. 8 (2020): 3555–70.
Barnes, Felicity, and David M. Higgins. "Brand Image, Cultural Association and Marketing: 'New Zealand' Butter and Lamb Exports to Britain, c.1920–1938." In *The Brand and Its History*. Routledge, 2022.
Belasco, Warren, and Roger Horowitz, eds. *Food Chains: From Farmyard to Shopping Cart*. University of Pennsylvania Press, 2009.
Belasco, Warren, and Philip Scranton, eds. *Food Nations: Selling Taste in Consumer Societies*. Routledge, 2002.
Berghoff, Hartmutt, and Mathias Mutz. "Missing Links? Business History and Environmental Change." *Jahrbuch für Wirtschaftsgeschichte* 2 (2009): 9–21.
Berghoff, Hartmutt, and Adam Rome, eds. *Green Capitalism? Business and the Environment in the Twentieth Century*. University of Pennsylvania Press, 2017.
Bergquist, Ann-Kristin, and Geoffrey Jones. "Private Regulation, Institutional Entrepreneurship and Climate Change: A Business History Perspective." Harvard Business School Working Paper, No. 24-041, January 2024.
Bianco, Gino B. "Climate Change Adaptation, Coffee, and Corporate Social Responsibility: Challenges and Opportunities." *International Journal of Corporate Social Responsibility* 5, no. 1 (2020): 3.
Blackford, Mansel. "Fishers, Fishing, and Overfishing: American Experiences in Global Perspective, 1976–2006." *Business History Review* 83, no. 2 (2009): 239–66. https://doi.org/10.1017/S0007680500000519.
Boon, Marten. "A Climate of Change? The Oil Industry and Decarbonization in Historical Perspective." *Business History Review* 93, no. 1 (2019): 101–25. https://doi.org.ezproxy.aut.ac.nz/10.1017/S0007680519000321.
Campbell, Hugh, and Ruth Liepins. "Naming Organics: Understanding Organic Standards in New Zealand as a Discursive Field." *Sociologia Ruralis* 41, no. 1 (2001): 22–39.
Coclanis, Peter A. "Field Notes: Agricultural History's New Plot." *The Journal of Interdisciplinary History* 50, no. 2 (2019): 187–212. https://doi.org/10.1162/jinh_a_01409.
Cox, Howard, Simon Mowatt, and Martha Prevezer. "New Product Development within a Network Setting: The Case of the Chilled Ready-Meals Industry in the UK." *Industry and Innovation* 10, no. 2 (2003): 197–217.
Cronon, William. "The Trouble with Wilderness: Or, Getting Back to the Wrong Nature." In *Uncommon Ground: Rethinking the Human Place in Nature*, edited by William Cronon. W.W. Norton & Company, 1995.
Daniels, Chelsea. "Agriculture Minister Todd McClay Defends Removing Farmers from Emissions Trading Scheme." *New Zealand Herald*, June 12, 2024.
Di Martino, Paolo, Andrew Popp, Peter Scott, and Francesca Carnevali. *People, Places and Business Cultures: Essays in Honour of Francesca Carnevali*. Boydell Press, 2017.
Dixon, Marion W. "Chemical Fertilizer in Transformations in World Agriculture and the State System, 1870 to Interwar Period." *Journal of Agrarian Change* 18, no. 4 (2018): 768–86.
Ejrnæs, Mette, Karl Gunnar Persson, and Søren Rich. "Feeding the British: Convergence and Market Efficiency in the Nineteenth-Century Grain Trade." *The Economic History Review* 61 (2008): 140–71.
Elmore, Bartow. *Citizen Coke: The Making of Coca-Cola Capitalism*. Norton, 2014.

European Commission. "Action Plan for Organic Production in the EU." 2021. https://agriculture.ec.europa.eu/farming/organic-farming/organic-action-plan_en#organicsintheeu.

Federico, Giovanni. *Feeding the World: An Economic History of Agriculture, 1800–2000*. Princeton University Press, 2005.

Ferguson-Cradler, Gregory. "Coping with Crisis: The Peruvian State-Owned Fishing Enterprise Pesca Perú, 1973–1998." *Business History* (January 2024): 1–24.

Food and Agriculture Organization of the United Nations (FAO). *Strategic Framework 2022–31*. FAO, October 2021.

Godley, Andrew, and Bridget Williams. "The Chicken, the Factory Farm, and the Supermarket: The Emergence of the Modern Poultry Industry in Britain." In *Food Nations: Selling Taste in Consumer Societies*, edited by Warren Belasco and Philip Scranton. Routledge, 2002.

Godley, Andrew, and Bridget Williams. "Democratizing Luxury and the Contentious 'Invention of the Technological Chicken' in Britain." *Business History Review* 83, no. 2 (2009): 267–90. https://doi.org/10.1017/S0007680500000520.

Goodchild, Hayley. "The Problem of Milk in the Nineteenth-Century Ontario Cheese Industry: An Envirotechnical Approach to Business History." *Business History* 59, no. 7 (2017): 1081–110.

Gonçalves, Caroline, and Decio Zylbersztajn. "Business History and Challenges for Coffee Cooperatives in Brazil: The Case of Cooxupé Cooperative." In *Coffee Consumption and Industry Strategies in Brazil*, 203–28. Woodhead Publishing, 2020.

Gunther, Marc. "The Good, the Bad and the Ugly: Sustainability at Nespresso." *The Guardian*, May 27, 2015.

Hamilton, Shane. "Analyzing Commodity Chains: Linkages or Restraints?" In *Food Chains: From Farmyard to Shopping Cart*, edited by Belasco Warren and Roger Horowitz, 16–26. University of Pennsylvania Press, 2009. https://doi.org/10.9783/9780812204445.16.

Hamilton, Shane. "Revisiting the History of Agribusiness." *Business History Review* 90, no. 3 (2016): 541–45. https://doi.org/10.1017/S000768051600074X.

Hamilton, Shane. *Supermarket USA: Food and Power in the Cold War Farms Race*. Yale University Press, 2018.

Hendrickson, Mary K., Philip H. Howard, Emily M. Miller, and Douglas H. Constance. "The Food System: Concentration and Its Impacts." *A Special Report to the Family Farm Action Alliance* 26 (2020).

Horowitz, Roger. "Making Food Chains." In *Food Chains: From Farmyard to Shopping Cart*, edited by Belasco Warren and Roger Horowitz, 2–5. University of Pennsylvania Press, 2009.

Hunter, Ian. "Commodity Chains and Networks in Emerging Markets: New Zealand, 1880–1910." *Business History Review* 79, no. 2 (2005): 275–304.

The Intergovernmental Panel on Climate Change (ICCC). *Special Report on Climate Change and Land*. IPCC, 2019.

Jones, Geoffrey. *Profits and Sustainability: A History of Green Entrepreneurship*. Oxford University Press, 2017.

Jones, Geoffrey. *Varieties of Green Business: Industries, Nations and Time*. Edward Elgar Publishing, 2018.

Jones, Geoffrey. *Deeply Responsible Business: A Global History of Values-Driven Leadership*. Harvard University Press, 2023.

Jones, Geoffrey, and Christina Lubinski. "Making 'Green Giants': Environment Sustainability in the German Chemical Industry, 1950s–1980s." *Business History* 56, no. 4 (July 2014): 623–49.

Jones, Geoffrey, and Simon Mowatt. "National Image as a Competitive Disadvantage: The Case of the New Zealand Organic Food Industry." *Business History* 58, no. 8 (2016): 1262–88.

Jones, Geoffrey, and Simon Mowatt. "Organic Food and National Image: The Paradox of New Zealand." In *Varieties of Green Business: Industries, Nations and Time*, edited by Geoffrey Jones. Edwin Elgar, 2018.

Levy, David, Juliane Reinecke, and Stephan Manning. "The Political Dynamics of Sustainable Coffee: Contested Value Regimes and the Transformation of Sustainability." *Journal of Management Studies* 53, no. 3 (2016): 364–401.

Lilburne, Linda, A.G. Ausseil, T. van der Weerden, J. Guo, E. Teixeira, A. Sood, H. Smith, A. Neverman, D. Mitchell, and A. Parker. "Progress in Understanding Challenges and Opportunities for Agriculture Under Climate Change." Technical Report, Adaptive Strategies for Future Farming. Occasional Report No. 34. Farmed Landscapes Research Centre, Massey University, 2022.

Lockeretz, William, ed. *Organic Farming. An International History*. Cromwell Press, 2007.

Lorenz, Klaus, and Rattan Lal. *Organic Agriculture and Climate Change*. Springer, 2023. https://doi.org/10.1007/978-3-031-17215-1.

MacFadyen, Joshua. “Long-Range Forecasts: Linseed Oil and the Hemispheric Movement of Market and Climate Data, 1890–1939.” *Business History* 59, no. 7 (2017): 1010–33.

Mercier, Stephanie. “The Evolution of World Grain Trade.” *Applied Economic Perspectives and Policy* 21, no. 1 (1999): 225–36.

Milov, Sarah. “Promoting Agriculture: Farmers, the State, and Checkoff Marketing, 1935–2005.” *Business History Review* 90, no. 3 (2016): 505–36. https://doi.org/10.1017/S0007680516000714.

Ministry of Primary Industries. “Changes to Organic Products Law.” www.mpi.govt.nz.

Morgan, Dan. *Merchants of Grain*. Viking Press, 1979.

New Zealand Winegrowers. “Millton Vineyards: Organic Trailblazers.” September 20, 2019. https://www.nzwine.com/en/media/our-people/millton/.

OECD. *Concentration in Seed Markets: Potential Effects and Policy Responses*. Paris: OECD Publishing, 2018. https://doi.org.ezproxy.aut.ac.nz/10.1787/9789264308367-en.

Olmstead, Alan L., and Paul W. Rhode. *Creating Abundance: Biological Innovation and American Agricultural Development*. Cambridge University Press, 2008.

Pawson, Eric. “Into the Anthropocene: Environmental History and the Morality of Climate Change.” *International Review of Environmental History* 6, no. 2 (2020): 57–73.

Pingali, Prabhu L. “Green Revolution: Impacts, Limits, and the Path Ahead.” *Proceedings of the National Academy of Sciences* 109, no. 31 (2012): 12302–8.

Poore, Joseph, and Thomas Nemecek. “Reducing Food’s Environmental Impacts Through Producers and Consumers.” *Science* 360, no. 6392 (2018): 987–92.

Porter, Michael E., and Mark R. Kramer. “Creating Shared Value: How to Reinvent Capitalism—and Unleash a Wave of Innovation and Growth.” In *Managing Sustainable Business: An Executive Education Case and Textbook*, edited by Gilbert G. Lenssen and Craig Smith. Springer, 2019.

Rands, Malcolm. *Ecoman: From a Garage in Northland to a Pioneering Global Brand*. Random House, 2013.

Rollings, Neil. “Business and Global Capitalism: Continuities and Change.” *Business History Review* 97, no. 3 (2023): 620–26. https://doi.org.ezproxy.aut.ac.nz/10.1017/S0007680523000429.

Rosen, Christine Meisner. “Doing Business History in the Age of Global Climate Change.” *Enterprise & Society* 8, no. 2 (2007): 221–26.

Rosen, Christine Meisner. “The Role of Pollution Regulation and Litigation in the Development of the U.S. Meatpacking Industry, 1865–1880.” *Enterprise & Society* 8, no. 2 (2007): 297–347. http://www.jstor.org/stable/23700979.

Rosen, Christine Meisner, and Christopher C. Sellers. “The Nature of the Firm: Towards an Ecocultural History of Business.” *Business History Review* 73, no. 4 (1999): 577–600.

Rothstein, Morton. “Multinationals in the Grain Trade, 1850–1914.” *Business and Economic History* (1983): 85–93.

Ruggeri Laderchi, C., H. Lotze-Campen, F. DeClerck, Benjamin Bodirsky, Quitterie Collignon, Michael Crawford, Simon Dietz, Lukas Fesenfeld, Claudia Hunecke, Debbora Leip, Steven Lord, Sarah Lowder, Sebastian Nagenborg, Toby Pilditch, Alexander Popp, and Isabella Wedl. *The Economics of the Food System Transformation*. Food System Economics Commission (FSEC), Global Policy Report, 2024. http://www.fao.org/family-farming/detail/en/c/1696203/.

Sharp, Paul, Sofia Henriques, Eoin McLaughlin, Xanthi Tsoukli, and Christian Vedel. “A Microlevel Analysis of Danish Dairy Cooperatives: Opportunities for Large Data in Business History.” *Enterprise & Society* 11 (2023): 1–29. https://doi.org/10.1017/eso.2023.5.

Sidhoum, Amer Ait, and Maria Vrachioli. “Agriculture and Sustainability.” In *The Palgrave Handbook of Global Sustainability*, edited by Robert Brinkmann. Cambridge University Press, 2023. https://doi.org/10.1007/978-3-031-01949-4_44.

Silva Lopes, Teresa da. “Building Brand Reputation Through Third-Party Endorsement: Fair Trade in British Chocolate.” *Business History Review* 90, no. 3 (2016): 457–82. https://doi.org/10.1017/S0007680516000738.

Smith, Andrew, and Kirsten Greer. “Uniting Business History and Global Environmental History.” *Business History* 59, no. 7 (2017): 987–1009. https://doi.org/10.1080/00076791.2017.1338688.

Smith-Howard, Kendra. *Pure and Modern Milk: An Environmental History Since 1900*. New York: Oxford University Press, 2013.

Sörlin, S., and M. Lane. “Historicizing Climate Change—Engaging New Approaches to Climate and History.” *Climatic Change* 151 (2018): 1–13. https://doi.org/10.1007/s10584-018-2285-0.

Stern, Nicholas, and Joseph Stiglitz. "Climate Change and Growth." *Industrial and Corporate Change* 32, no. 2 (2023): 277–303. https://doi.org/10.1093/icc/dtad008

Stokes, Raymond, Roman Köster, and Stephen Sambrook. *The Business of Waste: Great Britain and Germany, 1945 to the Present*. Cambridge University Press, 2013. https://doi.org/10.1017/CBO9781139225663.

Vanbergen, Adam, Marcelo Aizen, Stephane Cordeau, L.A. Garibaldi, M.P. Garratt, A. Kovács-Hostyánszki, L. Lecuyer, H.T. Ngo, S.G. Potts, J. Settele, and E. Skrimizea. "Transformation of Agricultural Landscapes in the Anthropocene: Nature's Contributions to People, Agriculture and Food Security." In Bohan, David, and Vanbergen, Adam, eds., *Advances in Ecological Research*. Vol. 63, 193–225. Elsevier, 2020. https://doi.org/10.1016/bs.aecr.2020.08.002.

Verburg, René, Eric Rahn, Pita Verweij, Marijke Van Kuijk, and Jaboury Ghazoul. "An Innovation Perspective to Climate Change Adaptation in Coffee Systems." *Environmental Science & Policy* 97 (2019): 16–24.

Vermeulen, Sonja, Bruce Campbell, and John Ingram. "Climate Change and Food Systems." *Annual Review of Environment and Resources* 37 (2012): 195–222. https://doi.org/10.1146/annurev-environ-020411-130608.

Verschuuren, Jonathan. "The Paris Agreement on Climate Change: Agriculture and Food Security." *European Journal of Risk Regulation* 7, no. 1 (2016): 54–57. http://www.jstor.org/stable/24769997.

White, Richard. *The Organic Machine: The Remaking of the Columbia River.* Hill and Wang, 1995.

White, Sam, Christian Pfister, and Franz Mauelshagen, eds. *The Palgrave Handbook of Climate History*. Palgrave Macmillan, 2018.

Wilk, Richard. "Food and Nationalism: The Origins of 'Belizean Food'." In *Food Nations: Selling Taste in Consumer Societies*, edited by Warren Belasco and Philip Scranton, 67–89. Routledge, 2002.

Willer, Helga Jan Trávníček, and Bernhard Schlatter, eds. *The World of Organic Agriculture: Statistics and Emerging Trends 2024*. Research Institute of Organic Agriculture FiBL, Frick, and IFOAM-Organics International, 2024.

8

CLIMATE CHANGE AND THE BUSINESS (HISTORY) OF ENERGY

Marten Boon

8.1 Introduction

Fossil energy consumption is the primary cause of climate change. Around 75% of the greenhouse gas emissions from the human consumption of fossil fuels can be attributed to the birth of the modern energy industry in the late eighteenth century, with over half of those emissions occurring since the 1970s.[1] Of all the themes, sectors, and industries in this volume, perhaps none of them is so fundamentally and causally related to climate change as the energy industry.

The energy industry is vast. The fossil energy industry makes up a large part of it, but energy is also produced from other primary and secondary sources such as electricity and heat from nuclear energy sources or from renewable sources such as hydropower, wind, solar, biomass, and a few others. Over time, scholars have talked about energy in many different ways, depending on what seemed important at the time they were writing. Presently, our understanding of energy is mostly shaped by the dichotomy of fossil versus renewable. The fossil energy industry's undeniable contributions to climate change have made it the target of public vilification, while the renewables industry is often portrayed as the ultimate panacea to counteract climate change. Neither characterisation is obviously true. But as the producers of the primary cause of anthropogenic climate change, oil companies like Exxon and Shell have become the climate's worst enemies in the public eye since the 1990s, spurring politicians, NGOs, minority groups, concerned citizens, investors, and even the courts to call upon these companies to radically decarbonise their businesses.[2]

Conflict, opposition, and environmental concerns, however, have not only been present in fossil industries but also prevalent in non-fossil energy industries. The need to decarbonise the global energy system has dramatically increased the pressure on wind, solar, and other low- and zero-carbon energy alternatives to grow, resulting in various kinds of conflict over their environmental impact, and pressure on land and water resources as well as concerns to make for a sustainable and just energy transition that conforms to sustainable development goals and planetary boundaries.[3] As a consequence, there is arguably no other sector that will need to be so fundamentally transformed to mitigate climate change as the energy sector.[4]

DOI: 10.4324/9781003478089-9

Two research themes where business history can make a useful contribution take centre stage. The first centres on the ways in which the energy industry has responded to the climate change phenomenon, in particular to its scientific basis and the translation of climate science into policy. The second theme centres on the transition to a decarbonised energy system, addressing the question of how policy and business responses to climate change have shaped the ways in which energy is produced and consumed and how they have impacted the business of energy. Both themes have so far received modest, though growing, attention in the business historical literature. This is in part due to climate change's relatively late arrival on the scene of mainstream environmental issues but also because the discipline has been slow to incorporate an environmental perspective.[5] In addition, business historical approaches to energy, although relatively numerous, are fragmented. There are corporate histories of energy companies and there are many, often national, case studies of various energy sectors at varying points in time, leaving a blind spot for the longue durée and systemic character of historical patterns of energy consumption.[6] And when business historians address energy, they often excel in the history of winners: the major, long-lived companies, the great industries, the successful entrepreneurs, and the energy sources and technologies that have dominated modern history.[7] To better understand how the energy industry interacted with climate change, it is important, both from an intellectual point of view and for practice, to re-appraise the extant business historical literature on energy and raise new questions that address the industry's interaction with energy systems in a long-run perspective.

The emerging research agenda that seeks to recast the humanities in the Anthropocene has stimulated such a re-appraisal and renewal of the business historiography of the energy industry.[8] This chapter aims to adopt this new perspective by bringing together the business historiography on the energy industry and climate change within a global framework. That is, however, no small task within the limited space available. The chapter is therefore structured according to a periodisation that attempts to capture the major global turning points in the relationship between the energy industry and global climate concerns and attempts to draw in the most relevant and up-to-date literature to provide a state of the art of the business history literature available.

The periodisation proposed in this chapter follows the development stages of the global climate change governance regime to delineate distinct phases in the historical evolution of the energy industry's interaction with climate change. The reason for sticking to an essentially political periodisation is that greenhouse gas emissions are typical externalities—that is, unpriced and untaxed discharges into an ungoverned global commons, the atmosphere. The only thing that brought climate change onto the managerial horizon of the energy industry was the fact that growing scientific concern inspired political action that forced industries, one way or another, to account for their emissions. The ways in which this regime was negotiated and established produced varying degrees of opportunities and threats for fossil, low-carbon, and renewable energy industries alike. Moreover, the role of the state has been, and still is, central to the business history of climate change. The failure of the market to realise the eradication of an unpriced externality such as greenhouse gases required the state to step in and guide, force, or coax the technological, socio-cultural, and economic changes necessary for radical decarbonisation.

The chapter is structured into four sections. The first deals with the rise of climate change concerns in the 1970s and 1980s, culminating in the first global call for climate change mitigation by the World Commission on Environment and Development (WCED) in 1987. The second section straddles the period between 1987 and 1997, when a global system of scientific and intergovernmental cooperation was established under the auspices of the United Nations (UN), culminating in the 1997 Kyoto Protocol, aiming to curb global greenhouse gas emissions with fundamental consequences for the energy industry. The third section deals with the period between Kyoto and the 2015 Paris Agreement, during which scientific consensus and an established global climate governance regime began to make a deepening impact on energy systems, setting in motion a transition towards decarbonisation. The fourth section discusses the period after 2015, which will be based primarily on a case study of the Norwegian state-owned energy company Equinor (formerly Statoil until 2018).

Each section sketches the global context, traces the outlines of the development of the energy industry, and discusses the state of the art of the historiography of the period concerned. The chapter concludes that the energy industry has been slow to adapt to the reality of climate change. First, although the energy crises of the 1970s fostered a window of opportunity for renewable energy technologies to grow, the lack of environmental urgency ultimately choked off a potential breakthrough, rendering most of these technologies uncommercial by the time the world came to recognise that radical decarbonisation was required in the early 1990s. Second, it took the best part of the 1980s and 1990s to garner sufficient political and societal consensus about the causes and consequences of climate change. And even if consensus was by and large achieved by the 2000s, this did not extend to the ways in which climate change was to be avoided and mitigated, allowing the fossil energy industry to drag its feet and lobby for policy instruments that exerted limited direct pressure on GHG emissions from fossil fuels and left room for their proliferation. Many of the world's major oil and gas companies remained ambivalent about climate change and decarbonisation well into the 2020s. The case study of Equinor clearly portrays this ambivalence. Nonetheless, starting in the 1990s and accelerating in the 2000s, energy transition policies provided the protective niches that allowed the commercialisation of clean energy technologies, to the point where, by the late 2010s, wind and solar energy became cost-competitive with fossil energy. Business historians, meanwhile, are beginning to dig into this rather general narrative in search of the hitherto ignored business histories of non-fossil energy and the role of business in international governance regimes.

8.2 Back to the Future of Energy Transition, 1970s to Early 1980s

At the core of writing energy history in the Anthropocene is the notion that our past (and current) pattern of energy consumption is unsustainable and requires an energy transition to a clean, zero-carbon energy system. Whereas energy itself has a long historiography, the concept of energy transition is considerably younger and can be traced back to the 1970s.[9] For the historiography of energy, the 1970s had undeniable implications, but it was arguably also a decade lost, as it fostered neither a political nor an academic agenda to address the energy-climate nexus. It took the best part of the 1980s and 1990s for this awareness to emerge, even though some nations adopted climate change policies as early as the late 1970s such as under the Jimmy Carter presidency in the United States.[10]

The first half of the 1970s witnessed the culmination and coincidence of concerns over energy, development, and the environment. Among these concerns was the relationship between climate change and the greenhouse gas emissions from fossil fuel combustion, a relationship that had long been acknowledged but became more urgent in the context of the environmental discourse of the 1970s. For example, the famous *Limits to Growth* report, which appeared in 1972, noted in a brief but very concrete passage that the rate of fossil energy consumption was related to the possibility of climate change caused by rising concentrations of CO_2 in the atmosphere.[11] Likewise, the report of the UN Conference on the Human Environment in Stockholm in 1972 called for strengthening and expanding international scientific cooperation "to understand better the general circulation of the atmosphere and the causes of climatic changes whether these causes are natural or the result of man's [*sic*] activities," which reflected the emergence of the notion of human-induced climate change as a potential threat in the scientific and environmental community.[12] Although such international scientific cooperation indeed intensified after that, greenhouse gas emissions were not widely perceived as an urgent problem in the 1970s.

Within this discourse of limits to growth and environmental degradation, international business became apprehensive of potentially costly environmental regulations threatening to hamper trade and investment.[13] Apprehension gave rise to concerted action through transnational business interest associations, in particular, the International Chamber of Commerce.[14] The petroleum industry responded by establishing its own international environmental body, the International Petroleum Industry Environmental Conservation Association (IPIECA) in 1974, which shortly thereafter developed a partnership with the UN Environment Program, also established in 1974, to map the energy industry's environmental problems and develop concerted international action to address them. The resulting assessment report did not define carbon dioxide (CO_2) as a pollutant, ignored its potential climatic effects, and clearly stated that any pollution deriving from the consumption of fossil fuels was not the responsibility of the industry.[15]

Although environmental concerns became more urgent in the 1970s, the politics and business of energy were dominated by the decade's energy crises, in particular the oil price hikes of 1973–4 and 1979–80, in combination with fears about rapidly depleting oil reserves, the so-called "peak oil" theory, which were particularly pronounced in the United States.[16] The idea of an energy transition from oil to alternative energy sources was therefore framed primarily in terms of energy security, not energy sustainability, even though there were some that captured popular attention with their environmentally motivated pleas for a renewable energy future in the 1970s.[17] In addition, the energy crises focused attention on energy conservation and the notion of decoupling economic growth from energy consumption.[18] There was even a global moment for energy transition, culminating in the 1981 UN Conference on New and Renewable Sources of Energy in Nairobi, the first North-South dialogue on energy.[19] However, the conference followed shortly after the second oil price shock of 1979–80, engendering its focus on energy security. If the 1970s opened potential pathways for a transition to a non-fossil energy system, it was not because of concern for the environment, let alone climate change.[20]

Nonetheless, energy policies and public research programmes aimed at reducing oil dependence in many, mostly Western, countries also benefited the development of renewable energy.[21] Perhaps most spectacular was the boom in the solar photovoltaics (PV) and wind energy industries. PV panel technology developed primarily in the United States, from a publicly funded space technology in the 1950s and 1960s into a commercial, though still small-scale, industry for the consumer market in the 1980s. In particular, investments by oil and gas companies like Exxon, Amoco, Chevron, and Shell fuelled innovation in PV technologies.[22] These firms were attracted by government tax credits and subsidies, the potential of PV panels to power off-grid operations (offshore, for example), and a general trend in the industry towards diversification into other energy sources and technologies in response to the energy crises of the 1970s and early 1980s.[23]

Another stand-out history, one where the oil industry played no significant role, is that of the Danish, and to a lesser degree, the German, wind turbine industries. Like the United States, Denmark had a tradition of small-scale, home-spun wind turbine manufacturers, but whereas the United States deployed government-funded research programmes into utility-scale wind power, the Danish response to the energy crises of the 1970s was to support small-scale and incremental development through the introduction of a feed-in tariff, a guaranteed price that allows renewable energy to compete with other sources, and subsidy programmes for existing turbine developers and manufacturers in the late 1970s. Whereas the U.S. programme reached a technological and commercial dead end by the early 1980s, Danish policies fostered a thriving industry.[24] Although U.S. policies did not engender an American wind industry, the 1978 Public Utility Regulatory Policies Act, forcing utilities to purchase power from smaller-scale power producers, and tax credits for renewable power generation by consumers, created the world's first large-scale market for renewable energy. The state of California deployed these policy instruments particularly vigorously, fostering the so-called "wind rush" in the 1980s, from which, in particular, the Danish wind turbine manufacturers—with Vestas as the most striking example—but also American entrepreneurs, benefitted.

Environmental concerns, if not climate change, had a more negative effect on the prospects of nuclear energy. With some exceptions, nuclear power, once perceived as the backbone of a future energy system, lost most of its shine because of the environmental concerns about nuclear waste, to which health and security concerns were added following the Three Mile Island accident (1979) and the Chernobyl disaster (1986).[25] Nuclear power generation growth, high in the 1970s and early 1980s in countries like the United States, France, and Germany, levelled off by the late 1980s, with the main exceptions being major emerging economies like China and India.[26] In some industrialised countries, for example, Finland, nuclear power's low carbon emissions have sparked renewed business interest in nuclear power in the 2000s, which has also rekindled business historical interest in this sector.[27]

The energy crises of the 1970s and early 1980s did not entrench a transition towards a more environmentally sustainable energy system. To the contrary, despite examples of experimentation and government attempts at a managed energy transition to new or alternative energy, renewables remained far too expensive to compete with fossil energy, which became clearly apparent when tax credits and subsidy schemes were phased out in most developed countries from the mid-1980s onwards, stopping the growth of solar and wind energy in their tracks. From the early 1980s, and in particular after oil prices crashed in 1986, the scare of oil scarcity dissipated, and most governments in the Global North switched from government-led energy conservation and substitution programmes to neoliberal policies focused on strengthening the functioning of international energy markets.[28]

One of the most compelling examples of how the energy, and in particular oil, supply security concerns of the 1970s and 1980s helped entrench reliance on petroleum rather than break it is the emergence of Norway as Western Europe's largest oil and gas producing nation. The main company used in this chapter as a case study, Equinor, was established as a fully state-owned operating company by parliamentary decision in 1972, as part of the country's determination to use the offshore petroleum resources that were discovered in the late 1960s for national economic development.[29] Although Norway can be considered an early adopter of environmental regulation, also in the petroleum industry, the main purpose of Equinor was to capture resource rents, develop into a competitive oil company, and facilitate the growth of a Norwegian petroleum industry that would provide jobs and income for Norwegians.[30] Over time, Equinor did indeed meet these goals, providing a remarkably successful example of state-led hydrocarbon development that is unique in the world.[31] It helped propel Norway from one of the lesser developed economies in Western Europe in 1970 into one of the wealthiest nations on earth. However, this firmly established developmental purpose increasingly pitted both the country's and the company's focus on petroleum-led growth against the growing concerns over climate change in the 1990s, more on which later.

Even though the world emerged from the 1970s as addicted to hydrocarbons as before, leaving renewable energy at the fringes of the global energy mix, the growth rate of energy use did change considerably. Whereas global primary energy consumption grew at an approximate annual rate of 4% between 1950 and 1973, growth fell substantially to an average of 1.6% in the decades after the energy crises of the 1970s.[32] Energy conservation efforts and economic stagnation in the late 1970s and early 1980s even caused a temporary decline in CO_2 emissions, at least in developed economies.[33] But by the mid-1980s, CO_2 emissions were on the rise again, especially in China and other emerging non-OECD countries (see Figure 8.1 in Appendix).

8.3 Climate Change Emerges, 1987–97

Climate change remained primarily the object of scientific interest until the late 1980s, when the UN World Commission on Environment and Development (the Brundtland commission) gave human-induced climate change a prominent place in its final report, *Our Common Future*, in 1987. By then, an improved understanding of the causes and effects of climate change had emerged in the early 1980s, based on the work of UN bodies like the World Meteorological Organization (WMO), the UN Environment Program (UNEP), and related international research programs.[34] The Brundtland Commission report coincided with the 1987 Montreal Protocol that curbed global emissions of substances harmful to the ozone layer, which provided traction to the possibility of global action on climate change.[35] This spurred the UN into action, leading to the establishment of a global scientific climate assessment body, the Intergovernmental Panel on Climate Change (IPCC) in 1988 and the initiation of the process that would lead to the UN Framework Convention on Climate Change (UNFCCC). Once the IPCC and the UNFCCC had established climate change on the international political agenda and began to assemble the scientific proof for human-induced climate change, the urgency for a transition away from fossil fuels (re)emerged in earnest, which culminated in the first international climate agreement in Kyoto in 1997, the Kyoto Protocol.

However, as climate change climbed the political agenda, it soon hit numerous roadblocks that needed to be cleared to facilitate the decarbonisation of the global energy system. Three problems were particularly vexing. The first was to gain global acceptance of the science proving human-induced climate change, in particular from those sections of industry that depended on fossil energy. Recent research has established that major oil companies like ExxonMobil, Total Energies, and Shell had a well-developed understanding of climate science and the potential effects of climate change, and, in some cases, actively contributed to the public dissemination of their insights in the 1970s and 1980s.[36] However, by the time the IPCC and the preparations for a Framework Convention emerged in the late 1980s, signalling the potential for government intervention through emissions and energy policies, the oil industry, together with a number of other fossil energy-dependent industries, joined forces in undermining the validity of climate science and opposing government regulation.[37] U.S. corporations were particularly active, but U.S. industry bodies like the Global Climate Coalition, probably the most well-known example of the corporate climate change countermovement, active between 1989 and 2002, had some European members through the U.S. subsidiaries of BP, Shell, and the chemical company Hoechst, as well as close ties to more global industry bodies like the International Chamber of Commerce.[38] These private efforts to negate and undermine climate science and counteract public action on climate change also highlight the role of private business in global environmental governance (GEG) and its increasing role in the regulation of the energy industry.[39]

The role of business in GEG is also of growing interest to business historians in relation to the second problem facing climate action, that is, to determine how emissions reductions were to be realised and how the burden would be distributed among nations. This proved to be a very contentious issue indeed, as major industrialised economies like the United States and emerging economies like China opposed the idea of imposing emissions restrictions upon their economies, which came to bear on the negotiations for the world's first international climate agreement at Kyoto in 1997. Moreover, such concerns were shared and echoed by the business community, and strengthened business initiatives to gain a stronger position in GEG, a fact first noted by political scientists and international business scholars.[40] The gist of this body of work is that business, and in particular the energy industry, gained an influential position in GEG, but (business) historical analyses of how and why this came about are still sparse.[41]

Research from the 2020s argues that concerted efforts by organised business interests to gain a voice and presence in GEG have managed to direct environmental regulation towards voluntary and market-based instruments.[42] This is clearly reflected in the support of business for market-based instruments such as carbon credit trading and other so-called flexibility mechanisms that formed the core of the Kyoto Protocol, to ensure, ostensibly, flexible and cost-effective emissions reductions, while vehemently opposing non-market instruments such as a carbon tax. While a carbon tax imposes a direct and uniform price (cost) on carbon, fossil-dependent industries opposed it, arguing that it would reduce competitiveness and destroy jobs, economic growth, and development.[43]

The experiences of Norway and Equinor are a case in point, their climate change efforts in the 1990s (and beyond) highlighting the trade-offs and roadblocks involved in climate change mitigation. Gro Harlem Brundtland, who chaired the World Commission on Environment and Development that put climate change on the international agenda, was also prime minister of Norway in the 1980s and early 1990s. She was determined to turn Norway into a climate leader, introducing a carbon tax in 1991 and formulating ambitious greenhouse gas emission reduction targets.[44] Equinor outwardly accepted the new policy direction but was privately opposed to the tax and the reduction targets, although it did turn Equinor's attention to CO_2 capture and storage (CCS) technology in the 1990s.[45]

By the late 1990s, it became painfully clear that it was quite impossible to continue producing oil and gas and simultaneously realise ambitious emissions reduction—a problem that has continued into the 2020s. Norway's emissions related to petroleum extraction rose 63% between 1990 and 2020, constituting over 80% of Norway's emissions growth since 1990.[46] From this dilemma emerged a peculiar Norwegian policy mix, that on the one hand continued to extract oil and gas at a breakneck pace, while directing large public expenditure towards state support for carbon capture and storage projects, financing international forest conservation and afforestation, and one of the world's most encompassing subsidy packages for electrifying the country's fleet of passenger cars on the other.[47] In addition, both the national government and Equinor rallied behind a discourse that framed Norway as the greenest producer of petroleum, as a way of rationalising its continued petroleum extraction; a narrative that became both forceful and influential because Norwegian climate and energy policy was shaped by a close-knit network of policymakers, scholars, and business and labour leaders with a background in the petroleum industry.[48] The discourse has a grain of truth, as Norwegian petroleum has some of the world's lowest greenhouse gas emissions associated with its production, so-called "scope 1" emissions. However, these emissions represent only a sliver of the total emissions from petroleum combustion, around 10%.[49]

Norway's emissions paradox goes some way in explaining the outcome of the Kyoto climate conference in 1997, which could be construed as a victory of sorts for the fossil energy industry, including a major European petroleum producer like Norway. Kyoto's flexibility mechanisms allowed fossil energy-dependent industries and industrialised countries with hard-to-abate emissions to realise emissions reductions elsewhere—either in less developed countries through the Clean Development Mechanism or in developed countries through the Joint Implementation instrument—and claim these reductions as carbon credits to cover their own emissions, without having to implement costly reduction measures themselves. In addition, the EU Emissions Trading System, the world's first large-scale version of Kyoto's third flexibility mechanism, emissions trading, which started trading in 2005, initially failed to foster large-scale emissions reductions because flaws in the system's design kept carbon prices too low to make an impact.[50] Revisions of the trading system, undertaken in the early 2020s, have helped foster higher carbon credit prices and thus created a stronger incentive to reduce emissions for European industries.

Despite that fact that the Kyoto Protocol's market-based instruments enabled the proliferation of fossil fuels because they posed very little direct limits on fossil fuel consumption, the very fact that the developed world had concluded a climate agreement in the first place marked the significance of the work of the IPCC, and the Protocol itself set in motion further changes in political, policy, and market responses to climate change that pressured the energy industry to take action.[51] This focused the energy industry's attention on the third problem that emerged from international climate change mitigation efforts that led to the Kyoto Protocol: how to adapt corporate strategies to climate change and develop technologies and energy sources able to decarbonise the energy system? Many of these technologies and sources, which had emerged in the 1980s and 1990s, were still much more costly than fossil fuels, exemplified, for instance, by the state of wind power technology in the mid-1990s.[52] Kyoto required policymakers and business leaders to recast their thinking about how to turn them into viable alternatives.

8.4 From Kyoto to Paris, 1997–2015

Kyoto was a pivotal moment in global climate governance. However, its effect on global greenhouse gas emissions was divergent, the result of political fractures that the agreement failed to overcome. Among industrialised countries (known as Annex 1 and Annex 2 signatories to the UNFCCC), the United States and Europe, in particular the members of the EU, took radically opposing views of the Kyoto Protocol. The United States failed to ratify the protocol, while the EU proposed and implemented the world's most ambitious climate policy programme. Least developed and developing countries (known as LDCs and Non-Annex I signatories to the UNFCCC) were exempt from the Protocol, reflecting the principles of the *polluter pays* and the *common but differentiated responsibilities* that were established under the Framework Convention. Only with the Paris Agreement of 2015 did emerging and least-developed economies take on binding emission targets as well. Because of the divergent outcomes of Kyoto, global emissions developed along different pathways (see Appendix, Figure 8.1). CO_2 emissions from energy in present-day EU member states started to decline already in the 1990s, reaching 27% below 1990 levels in 2021 (see Table 8.1 in Appendix), albeit with large variation among countries. While Denmark and the UK managed 50 and 43% reductions respectively, the Netherlands managed only a 7% and Spain and Norway (although the latter is not an EU country) actually grew emissions by 15%. In the case of Norway, emissions growth can almost entirely be attributed to its petroleum industry, and in particular to the state-owned Equinor, which is responsible for 76% of the country's oil and gas production.[53] CO_2 emissions from energy in the United States continued to grow and only started declining in the late 2000s. In least developed and emerging economies, in particular China, CO_2 emissions from energy soared, especially from the turn of the millennium when China overtook the United States as the world's largest emitter, more than tripling its emissions from energy between 1990 and 2021.

At the corporate and industry level, Kyoto triggered a similarly fractured reaction in the fossil energy industry, which was soon noted by international business scholars. Particularly striking was an apparent split between the European and U.S. oil companies, as, for instance, BP and Shell diverged from their American counterparts by publicly acknowledging climate change, supporting the Kyoto Protocol, and developing ostensibly proactive strategies to tackle climate change.[54] Initial scholarly interest was apprehensive but generally upbeat about the potential of the oil industry to positively contribute to decarbonisation, as demonstrated by, among others, management scholars Ans Kolk and David Levy, two early students of corporate responses to climate change.[55] Contemporaries interpreted the divergence between European and American oil companies as a positive sign for substantive contributions towards decarbonisation on behalf of the European companies.[56]

Since then, both the management and business historical literature have grown increasingly critical (and normative) of fossil fuel industry responses to climate change. This includes the recent scholarship on what the oil industry knew about climate change and what it did with that knowledge prior to Kyoto (see Section 8.3), but there has also been increased attention to the inconsistencies and discrepancies between the green rhetoric and business-as-usual strategies and investments of the oil industry as time went by.[57]

Part of the problem for business—a problem business had helped create through its own insistence on voluntary action and markets—was that the actual goals and policies that came out lobbying and of Kyoto were vague and often politically charged, which created considerable uncertainty for corporate strategic adaptation to climate change.[58] This partially helps explain why the variety in individual company responses is so large, ranging from a company like ExxonMobil that maintained its negative stance on climate science and mitigation policies into the late 2000s, to Ørsted, the former Danish state-owned oil and gas company DONG, that reinvented itself as a renewable energy company by entering offshore wind power in 2008 and finally divesting all of its fossil energy assets in 2017.[59]

Part of the explanation for the divergence and heterogeneity observed in the corporate responses to the Kyoto Protocol appears to be the political, institutional, and technological environments that businesses operate in. This is particularly relevant for the extent and direction of energy industry investments into renewable energy and other technologies identified as clean energy technologies such as carbon capture and storage (CCS). For instance, explanations for the transition of Ørsted emphasise the importance of Danish policies stimulating the development of a domestic market for on- and offshore wind power, in which earlier Danish experiences with wind power technologies and policies proved beneficial.[60] However, the example of Ørsted, which established itself as the world's largest developer of offshore wind farms by the early 2020s, is an outlier in the petroleum industry. Judging by the industry's long history of involvement with renewable energy, in particular solar, oil and gas companies have not been particularly successful at incorporating renewable energy into their business models, even though oil company investments in solar in the 1970s and 1980s did enable solar PV commercialisation to a limited extent.[61]

Nonetheless, solar energy, specifically solar PV panel manufacturing, was a central part of BP and Shell's post-Kyoto climate strategies, and both companies developed dominant positions in the global solar PV panel market in the early 2000s.[62] In particular, BP's high-profile rebranding into *beyond petroleum* in 2000, with a prominent role for solar energy in both its corporate logo and its organisational structure, appeared to capture the turnaround from climate change opposition to support in the European petroleum industry. However, the Deepwater Horizon oil spill from the BP-operated Macondo well in the Gulf of Mexico in 2010 seemed to expose that BP's true concerns were not for the environment—or the safety of its workers, for that matter—but for its bottom line.[63]

BP and Shell's positions in solar were in part driven by strategic considerations to enhance their environmental and sustainable reputations, but they were mainly facilitated by renewed attempts at government intervention to foster the growth of renewable energy. Energy policy experiments with tax credits, subsidies, feed-in tariffs, and mandatory renewable energy off-take obligations (which obligated utilities to purchase and distribute power generated from non-fossil energy sources) in, for instance, Denmark and the United States in the 1970s and early 1980s had proven how successful government intervention could be. Setting a minimum tariff for energy generated from renewable sources provided the predictability for investors and manufacturers that was necessary for sustained innovation and growth. Growing international consensus on climate change provided the momentum for new policies that rekindled markets for renewables in Denmark and the United States and created new ones in other countries such as the UK, Germany, and Spain, where feed-in tariffs and other measures were introduced in the 1990s. These facilitated the creation of domestic wind turbine industries in the latter two countries that fostered internationally competitive companies that were able to compete in other major growth markets for wind that appeared in the 1990s and 2000s such as the United States and India.[64] The widespread adoption of feed-in tariffs and other public policy support for renewable energy only started in the 2000s.

Early adopters of renewable energy policies also boosted the development of solar energy, in particular solar PV panels in the 1990s. Once dominant, the U.S. solar industry had barely survived the Reagan years, and what was left ended up in the hands of BP, Siemens, and then Shell by the late 1990s and early 2000s. Renewed support for renewables proved a boon for Japanese electronics firms like Sharp, Kyocera, and other major Japanese conglomerates that had long been experimenting with solar PV technologies and quickly grew into global companies in the 1990s.[65] By the mid-2000s, Chinese manufacturers entered the international market with support from the state, which perceived renewable energy as both a strategic industry and a means of decarbonising China's rapidly rising energy consumption.

The solar and wind energy booms, unleashed by the growing number of countries offering public subsidies for renewables after 2000, fostered intense competition that caused an accelerating decrease in the cost of renewable energy technologies.[66] As lower costs and public subsidies fuelled growth and competition, the composition of the solar and wind energy industries changed remarkably in the 2000s, as low-cost production in China propelled Chinese firms onto the global market and forced consolidation in the United States and other established markets.[67] Table 8.2 (see Appendix) illustrates how limited renewable energy uptake still was around 2000, with pioneering countries such as the United States, Denmark, Germany, and, to a lesser extent, Spain, in the lead, whereas by 2023, renewables had grown five-fold globally, with China far ahead of the rest. Although dependence on public subsidies and other policy instruments remained a significant risk for renewable energy industries in the decades following Kyoto, the 2015 Paris Agreement consolidated and intensified the trend of an accelerating uptake in renewable energy capacity in most developed and developing countries (Figure 8.2 in Appendix) as both solar PV and on- and offshore wind became cost-competitive with fossil energy in the 2010s.[68]

With the accelerating uptake of renewable energy after 2000 also came (business) historical interest in renewable and clean energy technologies, producing a steadily rising range of academic and non-academic publications retracing the histories of these technologies and the firms and entrepreneurs that attempted to craft viable businesses and markets out of them.[69] The vast majority of this work is centred on the West, although work emerging in the early twenty-first century has begun to unearth the history of renewable energy experimentation and innovation in the Global South.[70]

8.5 Post Paris: The Case of Equinor

Until the 21st Conference of the Parties of the UNFCCC in Paris in 2015, most oil and gas companies remained hesitant about their own strategies towards climate change, despite markedly more proactive rhetoric after Kyoto. The Paris Agreement bound virtually all countries in the world to the obligation to reduce emissions to the extent of limiting global warming to 2 degrees Celsius, and to efforts to reach 1.5 degrees, for the first time. Paris seemed to mark a point of no return for the oil and gas industry as it changed the terms of the discourse considerably. Under pressure from activists, politicians, and even investors, a growing number of oil and gas companies have introduced climate change and energy transition strategies since Paris. This has also produced a growing body of contemporary scholarship scrutinising the substantiveness of these strategies and attempting to evaluate their adherence to decarbonisation pathways consistent with a 1.5-degree global warming scenario, also referred to as “net zero in 2050.”[71] Variation in companies’ strategic alignment with a net zero in 2050 scenario continues to be large, with U.S. firms and many oil-producing nations typically lagging behind.

In general, oil companies’ climate change strategies remained modest until the 2020s. In the early 2000s, those companies that proactively acknowledged climate change, like Shell, BP, and Equinor, had done little to substantively contribute to decarbonisation beyond the low-hanging fruit: ditching coal assets, promoting natural gas as a cleaner-burning fossil fuel, avoiding gas flaring, making token investments in renewable energy, directing R&D spending towards technologies that could prolong fossil fuel consumption such as CCS and formulating ambiguous and difficult-to-evaluate corporate emissions reduction targets.[72] Moreover, even token renewable energy investments remained secondary to and dependent on the companies’ oil and gas interests. For instance, the widely publicised renewables strategies of BP and Shell in the early 2000s were soon scaled back again as both companies divested most of their solar assets and paused wind energy investments in the late 2000s and early 2010s: Shell in response to its reserve booking crisis of 2004, while BP was forced to divest around 25 billion U.S. dollars in assets following its Deepwater Horizon disaster in 2010.[73] Nonetheless, the Paris Agreement marked an important shift, forcing increasing numbers of companies to take into account their so-called "scope 3" emissions from the consumption of their products, which most companies had long regarded as not their responsibility. Much of the contemporary literature, however, highlights the ambiguous, opaque, and uncommitted character of many oil company strategies to deal with these emissions.[74]

A look at Equinor's climate change and energy transition strategies is instructive to gain a sense of the conflicted and ambiguous character of how the industry has attempted to keep abreast of an increasingly critical and demanding external environment in the past decade.[75] Because Equinor is the operator of three-quarters of the country's oil and gas production and provides jobs, investments, and R&D to numerous local communities in Norway, it is considered a strategic and vital cornerstone business, which explains why the state has held on to its majority ownership in the company. Therefore, any discussion of Equinor must take into account its close relations to the state and its discursive and material influence on Norwegian energy and climate policies. As mentioned earlier, Norway's inability to reconcile its climate ambitions with its petroleum extraction in the 1990s gave rise to a particular rhetoric that framed Norway as a green or sustainable producer of oil and gas in order to continue extracting petroleum for as long as possible. Norway's continued exploration and extraction of petroleum became increasingly inconsistent with the Paris Agreement, most clearly illustrated by the International Energy Agency's call for an immediate stop to investments in new fossil fuel extraction beyond already producing fields in 2021.[76] Despite such calls, Norway persists in its sustainable extraction rhetoric, continuing to award new exploration licenses.[77]

Perhaps reflecting the important role of Gro Harlem Brundtland for spurring global action on climate change, Equinor came out in support of climate science and taking precautionary action since the early 1990s.[78] It could be argued, hence, that Equinor was a front-runner among its peers when it came to climate change. However, despite generous lip service, the company made little material contributions to decarbonisation. For example, unlike other major European oil companies, Equinor had not made any investments in renewable energy around the turn of the millennium. A possible reason for this seeming paradox might be Norway's, and Equinor's, early insistence on flexibility mechanisms to realise global greenhouse gas emissions reductions. The rationale was that Norway's high-cost, high-productivity economy made emissions reductions in Norway relatively expensive, especially in the petroleum sector. Flexible, market-based instruments would make it possible for reductions to be realised in low-cost countries and sectors, leaving Norway to concentrate on extracting petroleum as efficiently and with as low scope 1 emissions as possible. As mentioned earlier, this logic developed into a national mantra and goes some way in explaining why Equinor was a vocal supporter of climate change mitigation, but without much substantive action. The only area where Equinor did make an effort was in carbon capture and storage. This was part of a long-standing Norwegian debate on the use of Norwegian natural gas for electricity generation, which was favoured by Labour Party proponents of industrial growth but opposed by the environmental movement for the associated CO_2 emissions.[79]

In 2007, as part of a new and ambitious climate change policy of Jens Stoltenberg's red-green coalition (which governed from 2006 to 2013), a political compromise allowed the development of gas-fired power plants but only if carbon capture and storage were integrated. Because the technology to capture CO_2 emissions from power plant exhaust gases was still unproven at scale at the time, Equinor ended up managing the country's only gas-fired power plant as one of very few Norwegian enterprises with the industrial size and capacity to potentially realise such an encompassing project. The promise of CCS as a silver bullet to Norway's climate paradox led Stoltenberg to call the project Norway's moon landing, in reference to U.S. President John F. Kennedy's 1961 speech promising American boots on the moon within a decade. The project failed in 2013 because the technical feasibility at the time did not live up to the high political expectations, but it established CCS as a key technology for Equinor and Norway because it had the potential to prolong the life of oil and gas extraction without leading to higher CO_2 emissions. Indeed, consecutive governments and Equinor have continued to support the development of CCS as a strategic technology for Norway.[80] Otherwise, beyond CCS, Equinor's environmental and climate change efforts prior to 2007 had been very modest. This changed in 2007, in no small part because of growing external pressure, including from Stoltenberg's renewed efforts to reduce Norway's greenhouse gas emissions.[81]

A particular flashpoint for Equinor was its investment in Canadian oil sands, also in 2007. Oil sands—also known as tar or bituminous sands—are a thick type of crude oil that is prolifically available, particularly in Canada and Venezuela, but producing it is costly and emits more CO_2 than most other petroleum resources. Equinor's oil sands investment elicited public condemnation in Norway and abroad because it clashed violently with Equinor's proactive climate rhetoric and with Norway's renewed climate ambitions. The company's feeble defence was that the world desperately needed new oil and gas resources—the investment was made in the context of skyrocketing oil and gas prices that left the world in the grip of a peak oil scare between 2005 and 2014—and that Equinor would revolutionise oil sands production by dramatically cutting emissions through technological innovations.[82] This quickly turned out to be a pipe dream, and Equinor decided to exit oil sands assets in 2013, divesting its assets completely in 2017. The public outcry that the oil sands investment caused in 2007 was the first wake-up call for the company to engage more substantively with decarbonisation. However, the question was what shape its decarbonisation strategy should take?

Until 2014, Equinor's primary strategic priority was to find new oil and gas reserves. As part of its growth strategy, Equinor had merged with the oil and gas division of Norsk Hydro, a major Norwegian aluminium smelter that had also been active in fertiliser and energy, in 2007.[83] Although Equinor was primarily interested in Hydro's oil and gas assets to boost its own clout in the global race for new resources, the merger, rather fortuitously, also included Hydro's wind power assets, which consisted of stakes in onshore wind farms in Norway, a stake in an offshore wind farm under development in the UK (Sheringham Shoal), and a floating wind turbine prototype called Hywind (short for Hydro wind). In response to the rising external pressure, Equinor's top management soon identified offshore wind as the company's key strategic gambit for a more credible climate change strategy, along with CCS. Offshore wind was close to the company's core business (large offshore projects), and the possibility to further develop in-house floating wind technology appeared to be a competitive advantage for the future direction of the global offshore wind energy market.[84]

But this New Energy strategy was more symbolic than substantive at first. Whereas BP and Shell had already established separate divisions for new energy ventures around the turn of the millennium, Equinor's wind power assets were placed in a small New Energy unit in the Technology & Projects division, without direct representation in the corporate executive committee. The relatively low status of the new unit and the organisation's unfamiliarity with the different financial and operational characteristics of wind power projects were initially significant roadblocks to the development of the unit. Although support from a few key managers ensured that New Energy could develop its first offshore wind farm in the UK and acquire licenses for additional offshore wind projects in the UK between 2010 and 2012, investments in new energy made up a mere 3% of the company's capital expenditure before 2014. Yet, despite significant internal limitations, New Energy did manage to build experience that would later prove vital for Equinor's growth in offshore wind, even though outside observers at the time interpreted Equinor's slow pace as an expression of greenwashing and window-dressing. Nonetheless, as a new CEO took the helm in 2015, in the wake of the 2014 oil price crash and the 2015 Paris Agreement, these first forays into offshore wind provided a springboard for a new strategic direction that unfolded in the following years, entailing a separate New Energy Solutions division (later renamed Renewables), with the aim of growing investments in renewable energy from 3 to 50% of capital expenditure by 2030, and increasingly concrete and elaborate decarbonisation plans pledging the company's leadership in the energy transition and alignment with net-zero emissions in 2050.

To what extent did this strategic pivot contribute to decarbonisation? There are several ways in which the energy transition and decarbonisation strategies of the oil and gas industry continue to be flawed, and Equinor is no exception. First of all, Equinor is not only committed to offshore wind (and some solar energy) but has also built its so-called low-carbon solutions business, consisting of carbon capture and storage and hydrogen projects. Throughout the twentieth century, and in particular since the 1970s, hydrogen has been hyped by scientists and techno-futurists as the solution for a sustainable and secure energy future. Stubbornly high costs and low availability of pure hydrogen, however, plagued such future prospects until the late 2010s, as the necessity to reduce emissions from hard-to-abate sectors like steelmaking and heavy transport has become increasingly urgent, rekindling old dreams of a "hydrogen economy."[85] Hydrogen, which could substitute for natural gas and other fossil fuels in industry and power fuel cells in vehicles, can be derived in several ways, but the most common are from natural gas and from water using electricity (electrolysis). Hydrogen is considered a clean fuel if it is produced from water using renewable energy (so-called green hydrogen) or from natural gas with the associated CO_2 emissions captured and stored (so-called blue hydrogen).[86] With Equinor's enormous gas reserves in Norway, its wind power investments, and long-running engagement with CCS, hydrogen has clear attractions for the company, not least the potential for blue hydrogen, which would allow Equinor and Norway to continue to produce natural gas but sell it as clean hydrogen using CCS.

Second, as other oil companies do, Equinor relies on carbon offsets and relative metrics to measure and report its progress towards net zero. The problem for any oil company is that it has no direct control over scope 3 emissions from the consumption of its products, which represent the bulk of total fossil fuel emissions. One way of dealing with those emissions is to simply divest oil and gas assets, but this only displaces emissions, as other investors with fewer environmental qualms might buy those assets and most likely continue producing oil and gas. Another option is to write off oil and gas assets, which would reduce the supply of oil and gas, but this would ultimately be at the expense of shareholders, who are unlikely to support such a measure, irrespective of whether they are private or state investors. Consequently, most oil and gas companies opt for offsets and relative emission metrics to account for scope 3 emissions. However, these instruments have attracted increasing criticism. For instance, the quality and effectiveness of carbon offset instruments that oil company decarbonisation strategies are relying on to reduce scope 3 emissions such as afforestation and forest conservation projects are not actually reducing emissions but merely offsetting them. Scholarly and media research suggests that several offsetting programmes are creating "phantom [carbon] credits" that fail to contribute to decarbonisation and might even lead to higher emissions.[87] Likewise, relative emission metrics serve to mask or offset absolute emissions. Equinor, for instance, reports its emissions per unit of energy supplied, a so-called carbon intensity metric.[88] As long as the share of non-emitting energy in its total energy production grows, which, in the case of Equinor, comes predominantly from its wind power projects, absolute CO_2 emissions from oil and gas may rise even when emissions appear to go down in relative terms.

Although offsets and relative emissions can be construed as misleading or shifting the problem, dealing with scope 3 emissions leaves oil companies with few alternatives. In the case of Equinor, the use of relative emission metrics in dealing with the problem of scope 3 emissions points out the connection between the various parts of their climate and energy transition strategies. Renewable energy investments increase the share of green energy, while carbon offsets and CCS reduce emissions, improving the relative emissions metric at both ends of the fraction. Because such tactics create the room for manoeuvre to continue investing in oil and gas extraction, Equinor's approach is not challenged by the Norwegian state, as it serves the state's own interest in prolonging the life of petroleum revenues and the Norwegian petroleum sector. Even though Equinor heralds from a nation that has spearheaded international climate action for decades, it is not markedly different in its approach to climate change than its privately owned peers due to the paradoxical character of Norwegian climate and energy policy.[89]

The example of Equinor points to a wider problem with approaches to decarbonisation of the energy industry as practiced over the past 40 years. Policy has emphasised the innovation of technologies to substitute fossil energy while eschewing imposing direct limits on fossil fuel consumption. Although this has facilitated a remarkable acceleration of renewable energy production, which in turn helped reduce greenhouse gas emissions, the voluntary and market-based instruments aimed at the direct emissions from fossil energy consumption have proven much less effective, allowing ample room for the proliferation of fossil fuels.[90] This became of increasing concern to policymakers, experts, and scholars from the 2010s onwards, leading to a growing interest in the possibility of a phase-out, that is, imposing direct limits on the use of fossil fuels, for example by deliberately winding down fossil energy technologies.[91] Although phase-out of fossil fuels has developed into a scholarly interest in particular, there are ample historical examples of managed phase-outs such as the case of the ozone-degrading chlorofluorocarbon gases that were phased out by means of the 1987 Montreal Protocol or the complex history of eradicating polluting vehicle emissions such as lead compounds and nitrous oxides.[92] With regard to decarbonisation, prominent examples are the phasing out of unabated coal-fired power plants or fossil energy subsidies, which became an intensely debated topic at the 26th UNFCCC Conference of the Parties in Glasgow in 2021 and a focal point of public activism by groups such as Extinction Rebellion.[93] For business historians, the concept of phase-out is an interesting but underexplored topic—one that will undoubtedly grow more urgent and relevant as the goal of net zero in 2050 draws closer.

8.6 Conclusion

The energy industry has been slow to adapt to the reality of climate change. First, although the energy crises of the 1970s fostered a window of opportunity for renewable energy technologies to grow, the lack of environmental urgency in the policy responses to the decade's energy crises ultimately choked off a potential breakthrough. By the time the world came to recognise that radical decarbonisation was required in the late 1980s, most of these technologies were still far from being commercially competitive with fossil energy, while energy policies in most industrialised countries stimulated the proliferation of fossil fuels.

Second, it took the best part of the 1980s and 1990s to garner sufficient political and societal consensus about the causes and consequences of climate change. And even if consensus was, by and large, achieved by the 2000s, this did not extend to the ways in which climate change was to be avoided and mitigated, allowing the fossil energy industry to drag its feet concerning the decarbonisation of the energy system. The extant and emerging business historical literature on business regulation and global environmental governance suggests that organised business interests were successful at gaining policy influence, in particular from the late 1980s onwards, lobbying for flexible, voluntary, and market-based instruments that imposed no or very limited direct restrictions on the production and consumption of fossil fuels. Moreover, the threat of direct emissions limits choking off competitiveness, jobs, and economic growth proved a powerful ally to business as two of the largest emitters, the United States and China, resisted imposing national emissions reduction targets until the 2015 Paris Agreement. Within this ambivalent international environment, the oil industry enjoyed ample room for manoeuvre to continue business as usual. Nonetheless, by the 2000s, energy transition policies aiming to decarbonise the energy system provided the protective niches that allowed the commercialisation of clean energy technologies, to the point where, by the late 2010s, wind and solar energy became cost-competitive with fossil energy, leading to a rapid growth of renewable energy capacity in most industrialised and emerging economies.

In terms of historiography, much of the cited literature in this chapter includes many contributions by political scientists and management and international business scholars from the early twenty-first century, reflecting how business historians are still catching up to the fact that the environment in general, and climate change in particular, have too long been treated as external to the object of their studies, leaving open several questions that have been only fleetingly touched upon in this chapter. For instance, how can we account for the role of business in international governance regimes? How do individual corporate responses to climate change relate to international and transnational corporate collective action? What explains the large variety in individual corporate responses to climate change? What role did various stakeholders play in shaping those responses? What is the role of energy companies in the clean energy transition? And what does the concept of phase-out have to offer for the business history of climate change? In addition, this chapter has shown the significance of bottom-up, niche experimentation with alternative energy technologies in the late 1970s and 1980s for the subsequent uptake of renewable energy since the early 2000s. Although these earlier entrepreneurial experiments did seldom lead to large-scale, durable enterprises, collectively, their efforts were vital to the decarbonisation transition that began to take form several decades later. A better understanding of the history of this early phase of experimentation (and often failure)—a key stage in any transition[94]—is vital for a better understanding of the business history of the energy industry in the era of climate change. Our historical understanding of these questions is still limited, but this chapter has attempted to highlight the contours of this growing research agenda.

Notes

1 Griffin, "CDP Carbon Majors Report 2017"; Pachauri and Meyer, "Climate Change 2014 Synthesis Report," 4–6; Lee et al., *Climate Change 2023: Synthesis Report*, 4–5.

2 Boffey, "Court Orders Royal Dutch Shell"; Cheon and Urpelainen, *Activism and the Fossil Fuel Industry*; Krauss and Eavis, "Climate Activists"; McGreal, "Big Oil and Gas."

3 Temper et al., "Movements Shaping Climate Futures"; Child et al., "Sustainability Guardrails."
4 IEA, "Net Zero by 2050," 13.
5 Rosen, "Greening of Business History"; Rosen, "Doing Business History"; Rosen, "What Is Business History?"; Jones, van Leeuwen, and Broadberry, "The Future of Economic, Business, and Social History," 234; Smith and Greer, "Uniting Business History."
6 Massard-Guilbaud, "From the History."
7 Fridenson, "Business Failure and the Agenda of Business History," 562–63.
8 Bergquist, "Renewing Business History"; Crutzen and Stoermer, "The Anthropocene"; McNeill and Engelke, *The Great Acceleration*.
9 Massard-Guilbaud, "From the History"; Graf, *Oil and Sovereignty*, 395.
10 Henderson, "Governing the Hazards of Climate."
11 Meadows et al., *The Limits to Growth*, 71–72.
12 UN, "Report of the United Nations Conference on the Human Environment," 21; Ward and Dubos, *Only One Earth*; Bolin, *A History of the Science and Politics of Climate Change*, 28.
13 Bergquist and David, "Beyond Planetary Limits!," 7.
14 Bergquist and David, "Business In (Action)."
15 UNEP, "Environmental Aspects of the Petroleum Industry," 8.
16 Araújo, "The Emerging Field of Energy Transitions."
17 Lovins, "Energy Strategy"; Commoner, *The Poverty of Power*.
18 Gross, "Reimagining Energy."
19 Basosi, "Lost in Transition."
20 Bonneuil and Fressoz, *The Shock of the Anthropocene*, 187; Basosi, "A Small Window," 340–41.
21 Basosi, "A Small Window."
22 Boon, "A Climate of Change?"; Jones and Bouamane, "'Power from Sunshine'."; Varadi, *Sun Above the Horizon*; Perlin, *From Space to Earth*.
23 Mody, "Surveying the Landscape."
24 Jones and Bouamane, "Historical Trajectories"; Jones, "Entrepreneurship, Policy, and the Geography of Wind Energy."
25 Melosi, "Nuclear Energy."
26 BP Statistical Review of World Energy 2022, "Nuclear: Generation."
27 Jensen-Eriksen, "Looking for Cheap and Abundant Power"; Rubio-Varas, De la Torre, and Connors, "The Atomic Business."
28 Basosi, "A Small Window"; Türk, "Reducing Dependence."
29 Thomassen, *Statoil and Equinor 1972–2001*.
30 Nissen, "A Greener Shade of Black?," 412.
31 Lie, "The Norwegian State and the Oil Companies"; Ryggvik, "A Short History of the Norwegian Oil Industry."
32 Original data based on: Energy Institute—Statistical Review of World Energy (2023); Smil (2017)—with major processing by Our World in Data. "Direct primary energy from biofuels" [dataset]. Energy Institute, "Statistical Review of World Energy"; Smil, "Energy Transitions: Global and National Perspectives," https://ourworldindata.org/grapher/global-primary-energy.
33 Based on BP Statistical Review of World Energy 2022, "CO2 Emissions from Energy, 1965–2021."
34 Bolin, *A History of the Science and Politics of Climate Change*, 31–32.
35 Bolin, *A History of the Science and Politics of Climate Change*, 45–46.
36 Supran, Rahmstorf, and Oreskes, "Assessing ExxonMobil's Global Warming Projections"; Bonneuil, Choquet, and Franta, "Early Warnings."
37 Oreskes and Conway, *Merchants of Doubt*; Brulle, "Institutionalizing Delay"; Brulle, "Advocating Inaction"; Bonneuil, Choquet, and Franta, "Early Warnings"; Franta, "Early Oil Industry Disinformation"; Franta, "Weaponizing Economics."
38 Bergquist and David, "Beyond Planetary Limits!."; Bergquist and David, "Business in (Action)"; Climate Files, "1989 GCC Membership List," "1996 GCC Progress Report on U.S. Industry Voluntary Actions to Curb Greenhouse Gases" and Membership List," https://climateinvestigations.org/global-climate-coalition-documents-index/.
39 Bergquist and Jones, "Private Regulation"; Pitteloud, "Let's Coordinate!"; Ballor and Pitteloud, "Introduction"; Pitteloud et al., "Capitalism and Global Governance."
40 McNeill and Engelke, *The Great Acceleration*, 77–79. On the role of business in GEG in general, see for example Falkner, "Business and Global Climate Governance"; Falkner, "Global Environmental Politics and Energy"; Wright and Nyberg, *Climate Change*.

41 Vormedal, "The Influence of Business"; Bergquist and David, "Business In (Action)."
42 Bergquist and David, "Beyond Planetary Limits!"; Bergquist and Jones, "Private Regulation."
43 Vormedal, "The Influence of Business"; Meckling, "The Globalization of Carbon Trading"; Cartel, Aggeri, and Caneill, "L'histoire méconnue du marché européen du carbone"; Newell and Paterson, "A Climate for Business."
44 Nissen, "A Greener Shade of Black?"
45 Nissen, "A Greener Shade of Black?" 419.
46 Gavenas, Rosendahl, and Skjerpen, "CO2-Emissions from Norwegian Oil and Gas Extraction."
47 Tjernshaugen, "The Growth of Political Support for CO2 Capture and Storage in Norway"; Roettereng, "Carbon Capture and Storage"; Roettereng, "The Foreign Policy of Carbon Sinks"; Park, Choi, and Youn, "REDD+."
48 Nissen, "A Greener Shade of Black?" 418.
49 Masnadi et al., "Global Carbon Intensity of Crude Oil Production."
50 Sato et al., "Allocation, Allocation, Allocation!"
51 Vormedal and Meckling, "How Foes Become Allies."
52 Milborrow, "Economics of Wind Power and Comparisons with Conventional Thermal Plant."
53 Based on Statistics Norway, Statbank, table 08940: Greenhouse gases, by source, energy product and pollutant, GWP values due to Kyoto protocol (AR4) (closed series) 1990–2022 [Klimagasser, etter statistikkvariabel, kilde (aktivitet), energiprodukt, komponent og år], https://www.ssb.no/en/statbank/table/08940; Norwegian Offshore Directorate, "Production Per Operator in 2023", https://www.norskpetroleum.no/en/facts/historical-production/.
54 Boon, "A Climate of Change?"
55 Kolk and Levy, "Winds of Change"; Levy and Kolk, "Strategic Responses to Global Climate Change"; Levy and Kolk, "Multinationals and Global Climate Change"; van de Wateringen, "The Petroleum Industry and Reputation"; Kolk, Levy, and Pinkse, "Corporate Responses in an Emerging Climate Regime"; Kolk, "Developments in Corporate Responses"; Pinkse and Kolk, *International Business and Global Climate Change*; Davis, "The Changing World of Oil"; Levy and Newell, "Oceans Apart?"
56 Rowlands, "Beauty and the Beast?"
57 Beale and Fernando, "Short-Termism"; Bell and Lundblad, "A Comparison of ExxonMobil's Sustainability Reporting to Outcomes"; Skjærseth and Skodvin, *Climate Change and the Oil Industry*; Skjærseth, *ExxonMobil*; Skjærseth, "Major Oil Companies in Climate Policy"; Saeverud and Skjaerseth, "Oil Companies and Climate Change"; Li, Trencher, and Asuka, "The Clean Energy Claims"; Le Menestrel, van den Hove, and de Bettignies, "Processes and Consequences"; van Halderen et al., "Managing Impressions."
58 Woerd, Levy, and Begg, *The Business of Climate Change*.
59 Dzhengiz, Henry, and Malik, "The Role of Partnership Portfolios"; Supran and Oreskes, "Rhetoric and Frame Analysis."
60 Voldsgaard and Rüdiger, "Innovative Enterprise."
61 Lüdeke-Freund, "BP's Solar Business Model"; Miller, "Why the Oil Companies Lost Solar"; Pinkse and van den Buuse, "The Development and Commercialization of Solar PV"; Backer, "When Oil and Wind Turbine Companies Make Green Sense Together"; Mody, "Surveying the Landscape."
62 Sluyterman, "Royal Dutch Shell"; Lowe and Harris, "Taking Climate Change Seriously"; Browne, *Beyond Business*; Jones and Bouamane, "'Power from Sunshine'," 52.
63 Balmer, Powell, and Greyser, "Explicating Ethical Corporate Marketing"; Gendron et al., "Rôle et Responsabilités."
64 Jones, "Entrepreneurship, Policy, and the Geography of Wind Energy," 223–28.
65 Jones and Bouamane, "'Power from Sunshine'," 42–43.
66 Elia et al., "Wind Turbine Cost Reduction"; Elia et al., "Impacts of Innovation."
67 Jones and Bouamane, "Historical Trajectories," 51–58; Jones and Bouamane, "'Power from Sunshine'," 69.
68 IRENA, *Renewable Power Generation Costs in 2022* (Abu Dhabi: International Renewable Energy Agency, 2023), https://www.irena.org/Data/View-data-by-topic/Costs/Global-LCOE-and-Auction-values.
69 Besides the already cited work by Geoffrey Jones, other examples include Madrigal, *Powering the Dream*; Eisler, *Overpotential*; Eisler, *Age of Auto Electric*; Turner, *Charged*.
70 Escobar Andrae and Arellano Escudero, "Green Innovation from the Global South."
71 Vieira, Longo, and Mura, "From Carbon Dependence to Renewables"; Dietz et al., "How Ambitious Are Oil and Gas Companies' Climate Goals?"; Rekker et al., "Evaluating Fossil Fuel Companies' Alignment with 1.5 °C Climate Pathways"; Trencher, Blondeel, and Asuka, "Do All Roads Lead to Paris?"

72 Skjærseth and Skodvin, *Climate Change and the Oil Industry*, 52–73.
73 Boon, "A Climate of Change?" 117–23; Taylor, "Shell Shock."
74 Trencher, Blondeel, and Asuka, "Do All Roads Lead to Paris?"
75 The following is based on Boon, *National Champion*, 321–61.
76 IEA, "Net Zero by 2050," 21.
77 Ministry of Energy, "62 New Production Licenses."
78 Skjærseth and Skodvin, *Climate Change and the Oil Industry*, 62.
79 Boon, *National Champion*, 347.
80 Boon, *National Champion*, 352.
81 Boon, *National Champion*, 253.
82 Boon, *National Champion*, 251–54.
83 Boon, *National Champion*, Chapter 5.
84 The remainder of this section is based on Boon, *National Champion*, Chapter 9.
85 Eisler, *Overpotential*, 99–100.
86 Depending on method and source there are many more hydrogen colours but grey (from natural gas without CCS), blue and green are the most common types.
87 Lakhani, "'Worthless'"; Greenfield, "Revealed"; Jones and Lewis, "Forest Carbon Offsets Are Failing."
88 Equinor, "Energy Transition Plan," 12–14.
89 Trencher, Blondeel, and Asuka, "Do All Roads Lead to Paris?"
90 Trencher et al., "The Rise of Phase-Out," 2.
91 For an extensive review of this literature, see Trencher et al., "The Rise of Phase-Out."
92 Näsman, "The Political Economy of Emission Standards."
93 Sparrow, "As Resistance Grows."
94 Kanger and Schot, "Deep Transitions," 8.

Bibliography

Araújo, Kathleen. "The Emerging Field of Energy Transitions: Progress, Challenges, and Opportunities." *Energy Research & Social Science* 1 (March 1, 2014): 112–21. https://doi.org/10.1016/j.erss.2014.03.002.

Backer, L. "When Oil and Wind Turbine Companies Make Green Sense Together." *Business Strategy and the Environment* 18, no. 1 (January 2009): 43–52. https://doi.org/10.1002/bse.570.

Ballor, Grace, and Sabine Pitteloud. "Introduction: Capitalism and Global Governance in Business History." *Business History Review* 97, no. 3 (2023): 459–79. https://doi.org/10.1017/S0007680523000855.

Balmer, John, Shaun Powell, and Stephen Greyser. "Explicating Ethical Corporate Marketing. Insights from the BP Deepwater Horizon Catastrophe: The Ethical Brand That Exploded and then Imploded." *Journal of Business Ethics* 102, no. 1 (2011): 1–14. https://doi.org/10.1007/s10551-011-0902-1.

Basosi, Duccio. "A Small Window: The Opportunities for Renewable Energies from Shock to Counter-Shock." In *Counter-Shock: The Oil Counter-Revolution of the 1980s*, edited by Duccio Basosi, Giuliano Garavini, and Massimiliano Trentin, 336–56. London: I. B. Tauris, 2018.

Basosi, Duccio. "Lost in Transition: The World's Energy Past, Present and Future at the 1981 United Nations Conference on New and Renewable Sources of Energy." *Journal of Energy History/Revue d'Histoire de l'Énergie* 4 (2020): 1–16.

Beale, F., and M. Fernando. "Short-Termism and Genuineness in Environmental Initiatives: A Comparative Case Study of Two Oil Companies." *European Management Journal* 27, no. 1 (February 2009): 26–35. https://doi.org/10.1016/j.emj.2008.05.002.

Bell, Jan, and Heidemarie Lundblad. "A Comparison of ExxonMobil's Sustainability Reporting to Outcomes." *The Journal of Applied Business and Economics* 12, no. 1 (2011): 17–29.

Bergquist, Ann-Kristin. "Renewing Business History in the Era of the Anthropocene." *Business History Review* 93, no. 1 (2019): 3–24. https://doi.org/10.1017/S0007680519000369.

Bergquist, Ann-Kristin, and Thomas David. "Beyond Planetary Limits! The International Chamber of Commerce, the United Nations, and the Invention of Sustainable Development." *Business History Review* 97, no. 3 (2023): 481–511. https://doi.org/10.1017/S0007680522001076.

Bergquist, Ann-Kristin, and Thomas David. "Business in (Action): The International Chamber of Commerce and Climate Change from Stockholm to Rio." *Uppsala Papers in Economic History* 3 (2024): 1–26.

Bergquist, Ann-Kristin, and Geoffrey Jones. *Private Regulation, Institutional Entrepreneurship and Climate Change: A Business History Perspective*. Cambridge, MA: Harvard Business School, 2024.

Boffey, Daniel. "Court Orders Royal Dutch Shell to Cut Carbon Emissions by 45% by 2030." *The Guardian*, May 26, 2021.

Bolin, Bert. *A History of the Science and Politics of Climate Change: The Role of the Intergovernmental Panel on Climate Change*. Cambridge: Cambridge University Press, 2007.

Bonneuil, Christophe, Pierre-Louis Choquet, and Benjamin Franta. "Early Warnings and Emerging Accountability: Total's Responses to Global Warming, 1971–2021." *Global Environmental Change* 71 (2021). Online. https://doi.org/10.1016/j.gloenvcha.2021.102386.

Bonneuil, Christophe, and Jean-Baptiste Fressoz. *The Shock of the Anthropocene: The Earth, History and Us*. Paperback ed. London: Verso, 2017.

Boon, Marten. "A Climate of Change? The Oil Industry and Decarbonization in Historical Perspective." *Business History Review* 93, no. 1 (2019): 101–25.

Boon, Marten. *National Champion: Statoil and Equinor After 2001*. Oslo: Scandinavia University Press, Universitetsforlaget, 2022.

Browne, John. *Beyond Business: An Inspirational Memoir From a Visionary Leader*. London: Weidenfeld & Nicolson, 2010.

Brulle, Robert J. "Institutionalizing Delay: Foundation Funding and the Creation of U.S. Climate Change Counter-Movement Organizations." *An Interdisciplinary, International Journal Devoted to the Description, Causes and Implications of Climatic Change* 122, no. 4 (2014): 681–94. https://doi.org/10.1007/s10584-013-1018-7.

Brulle, Robert J. "Advocating Inaction: A Historical Analysis of the Global Climate Coalition." *Environmental Politics* 32, no. 2 (February 23, 2023): 185–206. https://doi.org/10.1080/09644016.2022.2058815.

Cartel, Mélodie, Franck Aggeri, and Jean-Yves Caneill. "L'histoire méconnue du marché européen du carbone: archéologie du secteur électrique." [In Fr]. *Entreprises et histoire* 86, no. 1 (2017): 54–70. https://doi.org/10.3917/eh.086.0054.

Cheon, Andrew, and Johannes Urpelainen. *Activism and the Fossil Fuel Industry*. Abingdon, Oxon: Routledge, 2018.

Child, Michael, Otto Koskinen, Lassi Linnanen, and Christian Breyer. "Sustainability Guardrails for Energy Scenarios of the Global Energy Transition." *Renewable and Sustainable Energy Reviews* 91 (August 1, 2018): 321–34. https://doi.org/10.1016/j.rser.2018.03.079.

Commoner, Barry. *The Poverty of Power: Energy and the Economic Crisis*. New York: Alfred A. Knopf, 1976.

Crutzen, Paul J., and Eugene F. Stoermer. "The Anthropocene, with Commentary by Will Steffen." In *The Future of Nature: Documents of Global Change*, edited by Libby Robin, Sverker Sörlin, and Paul Warde, 483–90. New Haven: Yale University Press, 2013.

Davis, Jerome. "'Why Don't Oil Companies Commit More to 'Renewables'? A Corporate Constraints Perspective." In *The Changing World of Oil: An Analysis of Corporate Change and Adaptation*, edited by Jerome Davis, 169–89. Abingdon: Routledge, 2006.

Dietz, Simon, Dan Gardiner, Valentin Jahn, and Jolien Noels. "How Ambitious Are Oil and Gas Companies' Climate Goals?" *Science* 374, no. 6566 (2021): 405–8. https://doi.org/10.1126/science.abh0687.

Dzhengiz, Tulin, Leona A. Henry, and Khaleel Malik. "The Role of Partnership Portfolios for Sustainability in Addressing the Stability-Change Paradox: Dong/Orsted's Transition from Fossil Fuels to Renewables." *Business & Society* (2023). https://doi.org/10.1177/00076503231211214.

Eisler, Matthew N. *Overpotential: Fuel Cells, Futurism, and the Making of a Power Panacea*. New Brunswick, NJ: Rutgers University Press, 2012.

Eisler, Matthew N. *Age of Auto Electric: Environment, Energy, and the Quest for the Sustainable Car*. Cambridge, MA: The MIT Press, 2022.

Elia, A., M. Kamidelivand, F. Rogan, and B. Ó Gallachóir. "Impacts of Innovation on Renewable Energy Technology Cost Reductions." *Renewable and Sustainable Energy Reviews* 138 (March 1, 2021): 110488. https://doi.org/10.1016/j.rser.2020.110488.

Elia, A., M. Taylor, B. Ó Gallachóir, and F. Rogan. "Wind Turbine Cost Reduction: A Detailed Bottom-Up Analysis of Innovation Drivers." *Energy Policy* 147 (December 1, 2020): 111912. https://doi.org/10.1016/j.enpol.2020.111912.

Equinor. *Energy Transition Plan*. Stavanger: Equinor, 2022.

Escobar Andrae, Bernardita, and Nelson Arellano Escudero. "Green Innovation from the Global South: Renewable Energy Patents in Chile, 1877–1910." *Business History Review* 93, no. 2 (2019): 379–95. https://doi.org/10.1017/S000768051900062X.

Falkner, Robert. "Business and Global Climate Governance: A Neo-Pluralist Perspective." In *Business and Global Governance*, edited by Morten Ougaard and Anna Leander. Warwick Studies in Globalisation, 99–117. London: Routledge, 2010.

Falkner, Robert. "Global Environmental Politics and Energy: Mapping the Research Agenda." *Energy Research & Social Science* 1 (2014): 188–97. https://doi.org/10.1016/j.erss.2014.03.008.

Franta, Benjamin. "Early Oil Industry Disinformation on Global Warming." *Environmental Politics* 30, no. 4 (2021): 663–68.

Franta, Benjamin. "Weaponizing Economics: Big Oil, Economic Consultants, and Climate Policy Delay." *Environmental Politics* 31, no. 4 (June 7, 2022): 555–75. https://doi.org/10.1080/09644016.2021.1947636.

Fridenson, Patrick. "Business Failure and the Agenda of Business History." *Enterprise & Society* 5, no. 4 (2004): 562–82.

Gavenas, Ekaterina, Knut Einar Rosendahl, and Terje Skjerpen. *CO2-Emissions from Norwegian Oil and Gas Extraction*. Oslo: Statistics Norway, 2015.

Gendron, Corinne, Bernard Girard, Silvester Ivanaj, Vera Ivanaj, and Alice Friser. "Rôle et responsabilités des hauts dirigeants face aux changements climatiques: Réflexions à partir du cas de BP." *Entreprises et Histoire* 86 (2017): 34–53.

Graf, Rüdiger. *Oil and Sovereignty: Petro-Knowledge and Energy Policy in the United States and Western Europe in the 1970s*. 1st ed. Online Resource Vols. New York: Berghahn Books, 2018.

Greenfield, Patrick. "Revealed: More Than 90% of Rainforest Carbon Offsets by Biggest Certifier Are Worthless, Analysis Shows." *The Guardian*, 2023. https://www.theguardian.com/environment/2023/jan/18/revealed-forest-carbon-offsets-biggest-provider-worthless-verra-aoe.

Griffin, P. *The Carbon Majors Database—CDP Carbon Majors Report 2017*. London: Carbon Disclosure Project, 2017.

Gross, Stephen G. "Reimagining Energy and Growth Decoupling and the Rise of a New Energy Paradigm in West Germany, 1973–1986." *Central European History* 50, no. 4 (2017): 514–46.

Henderson, Gabriel. "Governing the Hazards of Climate: The Development of the National Climate Program Act, 1977–1981." *Historical Studies in the Natural Sciences* 46, no. 2 (2016): 207–42. https://doi.org/10.1525/hsns.2016.46.2.207.

IEA. *Net Zero by 2050*. Paris: International Energy Agency, 2021.

Jensen-Eriksen, Niklas. "Looking for Cheap and Abundant Power: Business, Government and Nuclear Energy in Finland." *Business History* 64, no. 8 (November 17, 2022): 1413–34. https://doi.org/10.1080/00076791.2020.1772761.

Jones, Geoffrey. "Entrepreneurship, Policy, and the Geography of Wind Energy." In *Green Capitalism? Business and the Environment in the Twentieth Century*, edited by Hartmut Berghoff and Adam Rome, 206–30. Philadelphia: University of Pennsylvania Press, 2017.

Jones, Geoffrey, and Loubna Bouamane. *Historical Trajectories and Corporate Competences in Wind Energy*. Cambridge, MA: Harvard Business School, 2011.

Jones, Geoffrey, and Loubna Bouamane. *"Power from Sunshine": A Business History of Solar Energy*. Cambridge, MA: Harvard Business School, 2012.

Jones, Geoffrey, Marco H.D. van Leeuwen, and Stephen Broadberry. "The Future of Economic, Business, and Social History." *Scandinavian Economic History Review* 60, no. 3 (November 1, 2012): 225–53. https://doi.org/10.1080/03585522.2012.727766.

Jones, Julia P.G., and Simon L. Lewis. "Forest Carbon Offsets Are Failing." *Science (New York, N.Y.)* 381, no. 6660 (2023): 830–31. https://doi.org/10.1126/science.adj6951.

Kanger, Laur, and Johan Schot. "Deep Transitions: Theorizing the Long-Term Patterns of Socio-Technical Change." *Environmental Innovation and Societal Transitions* 32 (September 1, 2019): 7–21. https://doi.org/10.1016/j.eist.2018.07.006.

Kolk, Ans. "Developments in Corporate Responses to Climate Change Within the Past Decade." In *Economics and Management of Climate Change: Risks, Mitigation and Adaptation*, edited by B. Hansjürgens and R. Antes, 221–30. New York, NY: Springer, 2008.

Kolk, Ans, and David Levy. "Winds of Change: Corporate Strategy, Climate change and Oil Multinationals." *European Management Journal* 19, no. 5 (2001): 501–9. https://doi.org/10.1016/S0263-2373(01)00064-0.

Kolk, Ans, David Levy, and Jonatan Pinkse. "Corporate Responses in an Emerging Climate Regime: The Institutionalization and Commensuration of Carbon Disclosure." *European Accounting Review* 17, no. 4 (2008): 719–45. https://doi.org/10.1080/09638180802489121.

Krauss, Clifford, and Peter Eavis. "Climate Activists Defeat Exxon in Push for Clean Energy." *The New York Times*, May 26, 2021.

Lakhani, Nina. "'Worthless': Chevron's Carbon Offsets Are Mostly Junk and Some May Harm, Research Says." *The Guardian*, May 24, 2023. https://www.theguardian.com/environment/2023/may/24/chevron-carbon-offset-climate-crisis?CMP=Share_iOSApp_Other.

Lee, Hoesung, José Romero, and Core Writing Team. *Climate Change 2023: Synthesis Report. Summary for Policymakers*. Geneva: IPCC, 2023.

Le Menestrel, M., S. van den Hove, and H.C. de Bettignies. "Processes and Consequences in Business Ethical Dilemmas: The Oil Industry and Climate Change." *Journal of Business Ethics* 41, no. 3 (December 2002): 251–66. https://doi.org/10.1023/a:1021237629938.

Levy, David, and Ans Kolk. "Strategic Responses to Global Climate Change: Conflicting Pressures on Multinationals in the Oil Industry." *Business and Politics* 3 (2002): 275–300.

Levy, David, and Ans Kolk. "Multinationals and Global Climate Change: Issues for the Automotive and Oil Industries." In *Multinationals, Environment and Global Competition*, edited by M.S. Lundan, 171–93. Bingley: Emerald Group Publishing Limited, 2003.

Levy, David, and Peter Newell. "Oceans Apart? Business Responses to the Environment in Europe and North America." *Environment* 42, no. 9 (2000): 8–20.

Li, Mei, Gregory Trencher, and Jusen Asuka. "The Clean Energy Claims of BP, Chevron, ExxonMobil and Shell: A Mismatch Between Discourse, Actions and Investments." *PLOS One* 17, no. 2 (2022): e0263596. https://doi.org/10.1371/journal.pone.0263596.

Lie, Einar. "Context and Contingency: Explaining State Ownership in Norway." *Enterprise & Society* 17, no. 4 (2016): 904–30. https://doi.org/10.1017/eso.2016.18.

Lie, Einar. "The Norwegian State and the Oil Companies." In *Oil Producing Countries and Oil Companies: From the Nineteenth Century to the Twenty-First Century*, 267–86. Brussels: Peter Lang, 2011.

Lovins, Amory. "Energy Strategy: The Road Not Taken?" *Foreign Affairs* 55, no. 1 (1976): 65–96.

Lowe, Ernest A., and Robert J. Harris. "Taking Climate Change Seriously: British Petroleum's Business Strategy." *Corporate Environmental Strategy* 5, no. 2 (December 1, 1998): 22–31. https://doi.org/10.1016/S1066-7938(00)80063-9.

Lüdeke-Freund, Florian. "BP's Solar Business Model: A Case Study on BP's Solar Business Case and Its Drivers." *International Journal of Business Environment* 6, no. 3 (2014): 300–28. https://doi.org/10.1504/ijbe.2014.063775.

Madrigal, Alexis. *Powering the Dream: The History and Promise of Green Technology*. Cambridge, MA: Da Capo Press, 2011.

Masnadi, Mohammad S., Hassan M. El-Houjeiri, Dominik Schunack, Yunpo Li, Jacob G. Englander, Alhassan Badahdah, Jean-Christophe Monfort, J.E. Anderson, T.J. Wallington, J.A. Bergerson, and D. Gordon. "Global Carbon Intensity of Crude Oil Production." *Science* 361, no. 6405 (2018): 851–53. https://doi.org/10.1126/science.aar6859.

Massard-Guilbaud, Geneviève. "From the History of Sources and Sectors to the History of Systems and Transitions: How the History of Energy Has Been Written in France and Beyond." *Journal of Energy History/Revue d'Histoire de l'Énergie* 1, no. 1 (December 4, 2018): 1–37.

McGreal, Chris. "Big Oil and Gas Kept a Dirty Secret for Decades: Now They May Pay the Price." *The Guardian*, June 30, 2021.

McNeill, John R., and Peter Engelke. *The Great Acceleration: An Environmental History of the Anthropocene Since 1945*. Cambridge, MA: Belknap Press, 2014.

Meadows, Donella H., Dennis L. Meadows, Jørgen Randers, and William W. Behrens. *The Limits to Growth: A Report for the Club of Rome's Project on the Predicament of Mankind*. New York: Potomac Associates, 1972.

Meckling, Jonas. "The Globalization of Carbon Trading: Transnational Business Coalitions in Climate Politics." *Global Environmental Politics* 11, no. 2 (2011): 26–50. https://doi.org/10.1162/GLEP_a_00052.

Melosi, Martin V. "Nuclear Energy and the Rise of Environmentalism in the United States." In *Oil Shock: The 1973 Crisis and Its Economic Legacy*, edited by Elisabetta Bini, Giuliano Garavini, and Federico Romero, 246–60. London: I.B. Tauris, 2016.

Milborrow, D. "Economics of Wind Power and Comparisons with Conventional Thermal Plant." *Renewable Energy* 5, no. 1–4 (1994): 692–99. https://doi.org/info:doi/10.1016/0960-1481(94)90454-5.

Miller, D. "Why the Oil Companies Lost Solar." *Energy Policy* 60 (2013): 52–80. https://doi.org/10.1016/j.enpol.2013.05.043.

Ministry of Energy. "62 New Production Licenses Offered in the APA 2023 Licensing Round." News Release, 2024. https://www.regjeringen.no/en/aktuelt/62-new-production-licences-offered-in-the-apa-2023-licensing-round/id3021963/.

Mody, Cyrus C.M. "Surveying the Landscape: The Oil Industry and Alternative Energy in the 1970s." In *Electrical Conquest: New Approaches to the History of Electrification*, edited by W. Bernard Carlson and Erik M. Conway, 51–79. Cham: Springer, 2024.

Näsman, Mattias. "The Political Economy of Emission Standards: Politics, Business and the Making of Vehicle Emission Regulations in Sweden and Europe, 1960–1980s."Doctoral thesis, monograph, Umeå universitet, 2021. http://urn.kb.se/resolve?urn=urn:nbn:se:umu:diva-186828(51).

Newell, Peter, and Matthew Paterson. "A Climate for Business: Global Warming, the State and Capital." *Review of International Political Economy* 5, no. 4 (1998): 679–703.

Nissen, Ada. "A Greener Shade of Black? Statoil, the Norwegian Government and Climate Change, 1990–2005." *Scandinavian journal of history* 46, no. 3 (2021): 408–29. https://doi.org/10.1080/03468755.2021.1876757.

Oreskes, Naomi, and Erik M. Conway. *Merchants of Doubt: How a Handful of Scientists Obscured the Truth on Issues from Tobacco Smoke to Global Warming*. New York: Bloomsbury Press, 2010.

Pachauri, R.K., and L.A. Meyer. *Climate Change 2014 Synthesis Report*. Geneva: IPCC, 2014.

Park, Mi Sun, Esther Sekyoung Choi, and Yeo-Chang Youn. "REDD+ as an International Cooperation Strategy Under the Global Climate Change Regime." *Forest Science and Technology* 9, no. 4 (December 1, 2013): 213–24. https://doi.org/10.1080/21580103.2013.846875.

Perlin, John. *From Space to Earth: The Story of Solar Electricity.* Cambridge, MA: Harvard University Press, 2002.

Pinkse, Jonatan, and Ans Kolk. *International Business and Global Climate Change.* Abingdon: Routledge, 2009.

Pinkse, Jonatan, and Daniel van den Buuse. "The Development and Commercialization of Solar PV Technology in the Oil Industry." *Energy Policy* 40 (January 1, 2012): 11–20. https://doi.org/10.1016/j.enpol.2010.09.029.

Pitteloud, Sabine. "Let's Coordinate! The Reinforcement of a 'Liberal Bastion' Within European Industrial Federations, 1978–1987." *Business History* 65, no. 2 (February 17, 2023): 345–65. https://doi.org/10.1080/00076791.2021.1905797.

Pitteloud, Sabine, Grace Ballor, Patricia Clavin, Nicolás M. Perrone, Neil Rollings, and Quinn Slobodian. *Capitalism and Global Governance in Business History: A Roundtable Discussion*. Cambridge, MA: Harvard Business School, 2022.

Pitteloud, Sabine, Grace Ballor, Patricia Clavin, Nicolás M. Perrone, Neil Rollings, and Quinn Slobodian. *Capitalism and Global Governance in Business History: A Roundtable Discussion*. Cambridge, MA: Harvard Business School, 2023.

Rekker, Saphira, Guangwu Chen, Richard Heede, Matthew C. Ives, Belinda Wade, and Chris Greig. "Evaluating Fossil Fuel Companies' Alignment with 1.5 °c Climate Pathways." *Nature Climate Change* 13, no. 9 (September 1, 2023): 927–34. https://doi.org/10.1038/s41558-023-01734-0.

Roettereng, Jo-Kristian S. "Carbon Capture and Storage in Climate Policy Perspective: The Case of Norway." M.Sc., Norwegian University of Science and Technology NTNU, 2011.

Roettereng, Jo-Kristian S. "The Foreign Policy of Carbon Sinks: Carbon Capture and Storage as Foreign Policy in Norway." *Energy Procedia* 63 (January 1, 2014): 6927–44. https://doi.org/10.1016/j.egypro.2014.11.727.

Rosen, Christine Meisner. "Industrial Ecology and the Greening of Business History." *Business and Economic History* 26, no. 1 (1997): 123–37.

Rosen, Christine Meisner. "Guest Editors' Introduction: Doing Business History in the Age of Global Climate Change." *Enterprise & Society* 8, no. 2 (2007): 221–26.

Rosen, Christine Meisner. "What Is Business History?." *Enterprise & Society* 14, no. 3 (2013): 475–85.

Rowlands, I.H. "Beauty and the Beast? BP's and Exxon's Positions on Global Climate Change." Article; Proceedings Paper. *Environment and Planning C-Government and Policy* 18, no. 3 (June 2000): 339–54. https://doi.org/10.1068/c9752.

Rubio-Varas, M., J. De la Torre, and D.P. Connors. "The atomic business: structures and strategies." *Business History* 64, no. 8 (2022/10/18 2022): 1395–412. https://doi.org/10.1080/00076791.2020.1856080.

Ryggvik, Helge. "A Short History of the Norwegian Oil Industry: From Protected National Champions to Internationally Competitive Multinationals." *Business History Review* 89, no. 1 (2015): 3–41. https://doi.org/10.1017/S0007680515000045.

Saeverud, Ingvild Andreassen, and Jon Birger Skjærseth. "Oil Companies and Climate Change: Inconsistencies Between Strategy Formulation and Implementation?" Article. *Global Environmental Politics* 7, no. 3 (August 2007): 42. https://doi.org/10.1162/glep.2007.7.3.42.

Sato, Misato, Ryan Rafaty, Raphael Calel, and Michael Grubb. "Allocation, Allocation, Allocation! The Political Economy of the Development of the European Union Emissions Trading System." *WIREs Climate Change* 13, no. 5 (2022): e796. https://doi.org/10.1002/wcc.796.

Skjærseth, Jon Birger. *ExxonMobil: Tiger or Turtle on Social Responsibility?* FNI Report. Trykt Utg. Vol. 7. Lysaker: Fridtjof Nansens institutt, 2003.

Skjærseth, Jon Birger. "Major Oil Companies in Climate Policy: Strategies and Compliance." In *Implementing the Climate Regime: International Compliance*, edited by Jon Hovi, Olav Schram Stokke, and Geir Ulfstein, 187–208. London: Earthscan, 2005.

Skjærseth, Jon Birger, and Tora Skodvin. *Climate Change and the Oil Industry: Common Problem, Varying Strategies*. Issues in Environmental Politics. Manchester: Manchester University Press, 2003.

Sluyterman, Keetie. "Royal Dutch Shell: Company Strategies for Dealing with Environmental Issues." *The Business History Review* 84, no. 2 (2010): 203–26.

Smith, Andrew, and Kirsten Greer. "Uniting Business History and Global Environmental History." *Business History* 59, no. 7 (October 3, 2017): 987–1009. https://doi.org/10.1080/00076791.2017.1338688.

Sparrow, Jeff. "As Resistance Grows to the Fossil Fuel Regime, Laws Are Springing Up Everywhere to Suppress Climate Activists." *The Guardian* (London), September 18, 2022. https://www.theguardian.com/commentisfree/2022/sep/19/as-resistance-grows-to-the-fossil-fuel-regime-laws-are-springing-up-everywhere-to-suppress-climate-activists.

Supran, Geoffrey, and Naomi Oreskes. "Rhetoric and Frame Analysis of ExxonMobil's Climate Change Communications." *One Earth* 4, no. 5 (May 21, 2021): 696–719. https://doi.org/10.1016/j.oneear.2021.04.014.

Supran, Geoffrey, S. Rahmstorf, and Naomi Oreskes. "Assessing ExxonMobil's Global Warming Projections." *Science* 379, no. 6628 (2023): eabk0063. https://doi.org/10.1126/science.abk0063.

Taylor, Bernard. "Shell Shock: Why Do Good Companies Do Bad Things?*." *Corporate Governance: An International Review* 14, no. 3 (2006): 181–93. https://doi.org/10.1111/j.1467-8683.2006.00498.x.

Temper, Leah, Sofia Avila, Daniela Del Bene, Jennifer Gobby, Nicolas Kosoy, Philippe Le Billon, Joan Martinez-Alier, P. Perkins, B. Roy, A. Scheidel, and M. Walter. "Movements Shaping Climate Futures: A Systematic Mapping of Protests Against Fossil Fuel and Low-Carbon Energy Projects." *Environmental Research Letters* 15, no. 12 (2020): 123004. https://doi.org/10.1088/1748-9326/abc197.

Thomassen, Eivind. *Statoil and Equinor 1972–2001: Commerce and Politics*. Oslo: Universitetsforlaget, 2022.

Tjernshaugen, Andreas. "The Growth of Political Support for CO2 Capture and Storage in Norway." *Environmental Politics* 20, no. 2 (March 1, 2011): 227–45. https://doi.org/10.1080/09644016.2011.551029.

Trencher, Gregory, Mathieu Blondeel, and Jusen Asuka. "Do All Roads Lad to Paris?" *Climatic Change* 176, no. 7 (June 20, 2023): 83. https://doi.org/10.1007/s10584-023-03564-7.

Trencher, Gregory, Adrian Rinscheid, Daniel Rosenbloom, and Nhi Truong. "The Rise of Phase-Out as a Critical Decarbonisation Approach: A Systematic Review." *Environmental Research Letters* 17, no. 12 (November 25, 2022): 123002. https://doi.org/10.1088/1748-9326/ac9fe3.

Türk, Henning. "Reducing Dependence on OPEC Oil: The IEA s Energy Strategy Between 1976 and the Mid-1980s." In *Counter-Shock: The Oil Counter-Revolution of the 1980s*, edited by Duccio Basosi, Giuliano Garavini, and Massimiliano Trentin, 241–58. London: I. B. Tauris & Company, Limited, 2018.

Turner, James Morton. *Charged: A History of Batteries and Lessons for a Clean Energy Future*. Seattle: University of Washington Press, 2022.

UN. *Report of the United Nations Conference on the Human Environment*. New York: United Nations, 1973.

UNEP. *Environmental Aspects of the Petroleum Industry—an Overview*. Paris: United Nations Environment Programme, 1978.

van de Wateringen, S. "The Petroleum Industry and Reputation: Developments in Corporate Reputation Over the Period 1990–2002." In *Perspectives on International Corporate Responsibility*, edited by J. Hooker, A. Kolk, and P. Madsen. International Corporate Responsibility Series, 119–44. Philosophy Documentation Center, 2005.

van Halderen, M.D., M. Bhatt, Gajm Berens, T.J. Brown, and C.B.M. van Riel. "Managing Impressions in the Face of Rising Stakeholder Pressures: Examining Oil Companies' Shifting Stances in the Climate Change Debate." Article. *Journal of Business Ethics* 133, no. 3 (February 2016): 567–82. https://doi.org/10.1007/s10551-014-2400-8.

Varadi, Peter F. *Sun Above the Horizon: Meteoric Rise of the Solar Industry.* Vol. 5. Singapore: Pan Stanford Publishing, 2014.

Vieira, Leticia Canal, Mariolina Longo, and Matteo Mura. "From Carbon Dependence to Renewables: The European Oil Majors' Strategies to Face Climate Change." *Business Strategy and the Environment* 32, no. 4 (2023): 1248–59. https://doi.org/10.1002/bse.3185.

Voldsgaard, Asker, and Mogens Rüdiger. "Innovative Enterprise, Industrial Ecosystems, and Sustainable Transition: The Case of Transforming DONG Energy to Ørsted." In *Handbook of Climate Change Mitigation and Adaptation*, edited by Maximilian Lackner, Baharak Sajjadi, and Wei-Yin Chen, 3633–84. Cham: Springer International Publishing, 2022.

Vormedal, Irja. "The Influence of Business and Industry NGOs in the Negotiation of the Kyoto Mechanisms: The Case of Carbon Capture and Storage in the CDM." *Global Environmental Politics* 8, no. 4 (2008): 36–65. https://doi.org/10.1162/glep.2008.8.4.36.

Vormedal, Irja, and Jonas Meckling. "How Foes Become Allies: The Shifting Role of Business in Climate Politics." *Policy Sciences* 57, no. 1 (March 1, 2024): 101–24. https://doi.org/10.1007/s11077-023-09517-2.

Ward, Barbara, and René J. Dubos. *Only One Earth: The Care and Maintenance of a Small Planet.* New York: W.W. Norton & Company, Inc., 1972.

Woerd, Frans van der, David L. Levy, and Katie Begg, eds. *The Business of Climate Change: Corporate Responses to Kyoto*. Sheffield: Greenleaf Pub., 2005.

Woerd, Frans van der, David L. Levy, and Katie Begg. "Introduction." In *The Business of Climate Change: Corporate Responses to Kyoto*, edited by Frans van der Woerd, David L. Levy, and Katie Begg, 11–20. Sheffield: Greenleaf Pub., 2005.

Wright, Christopher, and Daniel Nyberg. *Climate Change, Capitalism, and Corporations: Processes of Creative Self-Destruction.* Cambridge: Cambridge University Press, 2015.

APPENDIX

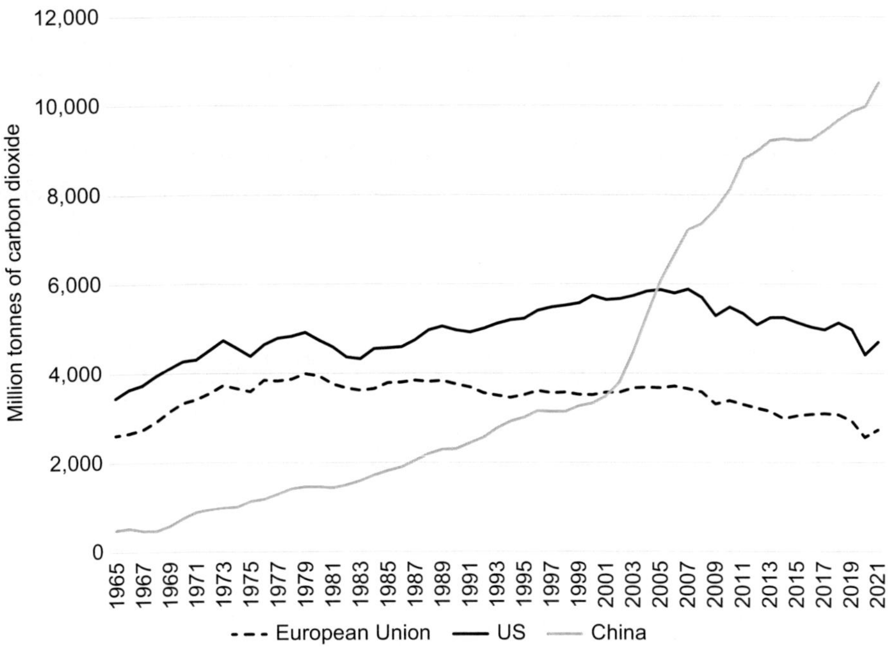

FIGURE 8.1 CO_2 emission from energy, 1965–2021

Source: Energy Institute, "EI Statistical Review of World Energy 2024." London: Energy Institute, 2024, 16.

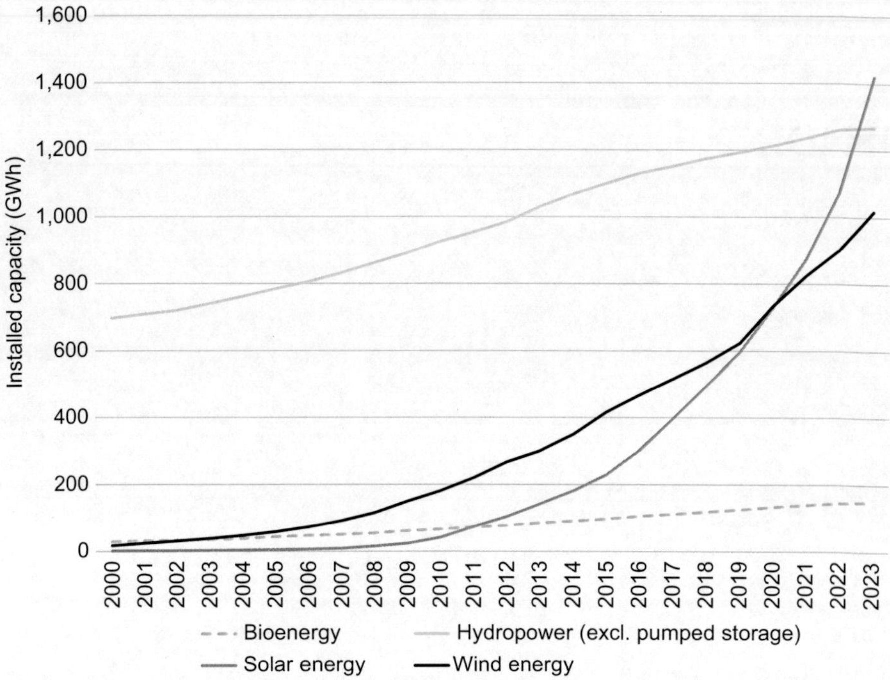

FIGURE 8.2 Global installed renewable energy capacity, in GWh, 2000–2023

Source: International Renewable Energy Agency, "Renewable Capacity Statistics, 2024." Abu Dhabi: IRENA, 2024.

TABLE 8.1 National CO_2 emissions from energy in 2022, percentage change from 1990

Country	*% change since 1990*
Denmark	–50
United Kingdom	–43
Czech Republic	–41
Sweden	–40
Germany	–38
Russian Federation	–29
France	–26
Italy	–23
Venezuela	–21
Poland	–17
Netherlands	–7
US	–5
Japan	–3
Spain	15
Norway	15

(*Continued*)

TABLE 8.1 (Continued)

Country	*% change since 1990*
Australia	34
South Africa	35
Argentina	81
Algeria	114
Brazil	122
Egypt	154
South Korea	156
Iraq	157
Chile	174
Saudi Arabia	184
Turkey	196
Morocco	223
Iran	263
India	324
Indonesia	332
China	356

Source: Energy Institute, "EI Statistical Review of World Energy 2024." London: Energy Institute, 2024, 16.

TABLE 8.2 Top ten countries by installed renewable energy capacity, in GWh

	2000		*2023*
USA	30.0	China	1051.8
Germany	7.4	USA	289.9
Denmark	2.5	Germany	151.2
Spain	2.3	India	117.8
Philippines	1.8	Japan	92.7
Japan	1.5	Brazil	66.6
Italy	1.3	Spain	62.0
India	0.9	UK	45.9
Mexico	0.9	Australia	45.0
Netherlands	0.7	Italy	42.9

Source: International Renewable Energy Agency, "Renewable Capacity Statistics, 2024." Abu Dhabi: IRENA, 2024.

9

CLIMATE CHANGE AND CORPORATE ADAPTATION

The Evolution of the Wine Industry in the Andes[1]

Teresa da Silva Lopes and Andrea Lluch

9.1 Introduction

The last quarter of the twentieth century witnessed an acceleration of global warming, driven by increased greenhouse gas emissions, reduced precipitation, and rising cloud bases. Global warming led to glacier ice loss, one of the earliest and most visible indicators of human-induced climate change.[2] The Andes region,[3] which boasts the world's longest above-water mountain chain, has experienced some of the highest rates of glacier loss globally.[4]

Since the 1970s, the accelerated melting of these glaciers has been impacting not just the region's natural environment, but also the society and the economy of the Andes. Human lives and cultural and religious practices of communities that have historically regarded mountain glaciers as sacred places,[5] as well as local plant and animal life, the built environment, and industries, have all been affected in multiple ways.[6]

Several industries in the Andes both have been impacted and have impacted climate change. The mining industry is considered to be the industry that caused most environmental, social, and climate impacts.[7] On the opposite side, an industry that has been impacted is mountain tourism. For instance, the Chacaltayla glacier was once a major tourist destination in the Bolivian Andes, famous for its high-altitude ski slopes. However, this ceased to be a skiing resort due to glacier loss, and by 2009, the glacier had completely disappeared.[8] Hydropower is another important industry in the Andes, providing energy for populations. In 2021, hydropower contributed to more than 70% of the energy produced in Colombia and Ecuador, and more than 80% of the energy produced in Peru. This is one of the industries that has essentially been impacted by climate change[9] as it relies on glacial runoff to sustain the necessary water flows used in energy production.[10] Other industries that rely on glacier water resources from the Andes, and which have also been affected by climate change, are agriculture and viticulture.[11] In all these cases, businesses have had to adapt their strategies.

This chapter examines the impact of climate change and glacier ice melting on the wine industry in the Andean region. Analysing this topic is important for three reasons. First, the wine industry has historically held significant economic importance in the Andes region, especially in Argentina and Chile.[12] Second, climate change and the melting of glacier ice have significantly impacted the dynamics of the industry, requiring corporate adaptation and re-configuration of

DOI: 10.4324/9781003478089-10

international supply chains in which Andean wine firms participate. Third, the evidence collected here offers a historical approach and draws on a geographical region and industry through the lenses of climate change, all of which have not yet been researched from the point of view of business history.

The chapter has three main aims. The first is to provide a historical overview of the impact of climate change, particularly the melting of glacier ice on the wine industry. It focuses specifically on Argentina and Chile, which house two main wine-producing regions in the Andes. The second aim is to identify trends and patterns in the international wine supply chain configuration, comparing the strategies carried out by wine companies operating in the region to deal with such environmental challenges. The third objective is to contribute to the literature on the history of the wine industry by explaining how climate change has shaped the evolution of business strategies and models aimed at managing environmentally related risks. These strategies are distinct from traditional market-based strategies and business models, which do not take climate change into account. In doing so, the chapter explores other aspects of climate change beyond the melting of glacier ice, to also analyse other natural risks such as droughts, and CO_2 emissions and carbon footprint created by business.

Collaboration in business has long been acknowledged by business historians in the growth and success of both large and small enterprises.[13] These studies have often emphasised how collaborative efforts within and between businesses have driven innovation, efficiency, and competitive advantage. This case on the wine industry in the Andes identifies the significance of more recent collaborative initiatives as a preventive risk management strategy for addressing climate change, as it enables the sharing of innovative practices, knowledge, and resources and reduces costs. Given the significance of the problem, collaboration is shown in this study to be a unique way forward for growth and survival, and must involve not just business competitors, but multiple stakeholders within supply chains at local, regional, and global levels.

The research relies extensively on original sources, particularly oral interviews with different stakeholders associated with the wine industry. These include industry associations, civil society, and networks of experts, including academics, the state, and policymakers. The actors analysed here include domestic and foreign wine businesses operating in the region such as Bodegas Catena Zapata, Bodegas Salentein, and Bodegas Casarena from Argentina, Viña Concha y Toro from Chile, and Sogrape a leading wine producer from Portugal, which acquired Finca Flichman (Argentina) and Chateau Los Boldos (Chile). It also analyses the role of other institutions such as Bodegas de Argentina, Wines of Argentina, Instituto Nacional de Vitivinicultura (INV) in Argentina, Wines of Chile, and CORFO Innova in Chile,[14] among others, and industry specialists and academics.

The concepts used here are risk prevention and risk mitigation. Risk prevention is used to explain ex-ante strategies to reduce greenhouse emissions and minimise further melting of glacier ice. Risk mitigation (or adaptation) is used here to refer to measures or actions to increase the ability to adapt to the new contextual environment, as a result of climate change. Some actions aim to reduce vulnerabilities and/or benefit from opportunities resulting from current or future climate changes.[15] International supply chains focus on the trajectories of goods from producers to consumers, addressing the national competitive advantages of nations and industries. The international supply chain in the wine industry includes a range of activities—from soil management and vineyard cultivation to wine production and ageing, bottling, labelling, and distribution and logistics, before reaching the final consumer. International supply chain governance is shaped by actors beyond those involved in procuring, producing, marketing, and distributing products and services.

The chapter is structured into five sections. Following the introduction, Section 9.2 provides an overview of the evolution of the wine industry in the Andes, in particular in Chile and Argentina. Section 9.3 analyses how climate change, including glacier ice melting, has affected the configuration of the global supply chain of wines from the Andes. It also examines changes in risk management strategies followed by key wine firms with multinational activity as well as small and medium enterprises, and analyses the support role provided by public and private institutions in the Andes region to address such challenges. Section 9.4 analyses the various collaborative business models used to address climate change at different institutional levels, including the national, industry, region, and firm levels. Finally, Section 9.5 provides some conclusions.

9.2 The Wine Industry in the Andes

9.2.1 A Brief Historical Overview

Chile and Argentina are the most important producers of wines in South America. In 2023, both ranked among the top 10 countries in the world in terms of vineyard acreage, wine production, and exports.[16] The internationalisation of the Andean wine industry started essentially from the 1990s, first in Chile and then in Argentina.[17] The share of Chilean wines sold in foreign markets increased from 7% in 1989 to 63% in 2002, and to more than 80% in the 2020s, making Chile the country with the most globalised wine industry in the world.[18] This export growth was driven by several factors, in particular technological, organisational, and marketing innovations, of which product standardisation through the use of more easily identifiable varieties was central.[19] A pioneer in this innovation process was the Spanish winemaker Miguel Torres, who invested in Chilean vineyards in the late 1970s to start producing high-quality wines, for which he imported state-of-the-art technology.[20] Following Torres lead, other Chilean wineries started contracting international winemakers, who contributed to the modernization of the industry, revolutionising it.[21] This was achieved not only through strong entrepreneurship and innovation by firms, but also through strong collaboration between wineries and other stakeholders, such as governmental agencies and local research institutes and universities.[22]

By 2023, Chilean vineyards were quite fragmented and were located in the southern part of the country, between the Andes mountain range and the Pacific Ocean.[23] In that year, the Chilean wine industry comprised 800 active, registered wineries, 11,697 producers and 394 export companies, 76% of which were SMEs (small and medium-sized enterprises).[24] Even though the leading exporting companies were the largest (albeit relatively small when compared to the United States and Australian large wine producers), they remained family controlled or were owned by national economic groups. In 2023, Concha y Toro, founded in 1883, was Latin America's largest winery, with sales of around US$1 billion and an empire that extended beyond its native Chile to organic vineyards in California and Argentina.[25]

Argentina was the leading wine-producing country in the Americas until the second half of the 1980s, when it was surpassed by the United States.[26] Historically, Argentina's wine producers prioritised quantity over quality. This approach became unsustainable when a structural crisis hit the industry in the 1980s. This was due to a combination of external factors, including economic instability, decreasing per capita domestic consumption, and increasing demand for higher-quality wines. At industry level there was also a lack of technological and viticultural innovations, which led to lower productivity and an inability to produce higher-quality wines.[27] This crisis eventually led to a transformation in the Argentine wine industry, which shifted towards the production of higher-quality wines and developing a stronger focus on exports.

Argentina presents a marked diversity of soils and climates throughout the country that measures 3,800 kilometres from north to south. In 2023, Argentina's vineyards covered 204,847 hectares, distributed in 18 provinces and 23,027 winegrowers.[28] Most Argentinean viticulture occurs in the Andes valleys. Over 70% of Argentina's wines are grown in Mendoza.[29] In contrast to Chile, where the foreign market accounts for most of the sales, by 2023 about 70% of Argentine wine sales were still domestic.[30] Nonetheless, from the 1990s, the industry experimented with a process of restructuring, and technological modernisation, and concentration.[31] Between 1992 and 2008 the Argentine wine industry received an estimated foreign direct investment (FDI) of US$550 million. The origin of this FDI was the United States, but other important countries included France, Portugal, Spain, the UK, and, later on, Chile. Many local wineries sold their businesses to foreign investors during this period.

In further contrast to the Chilean wine industry, the Argentinian wine industry was by 2023 quite concentrated: 35 wine producers accounted for 80% of exports, and just 8 exporting companies were responsible for half of the bottled wine exports.[32] Most of these leading wine producers were companies that had carried out major organizational and technological transformations from the 1990s. Among these was Catena Zapata Winery. The decision to undertake such transformations was not based on climate change debates. It relied on detailed economic analysis of potential markets, and on a desire to develop high-quality wines to be exported to developed countries.[33] Catena Zapata introduced several important innovations in the industry from the 1980s. These included using a vinification style named "Californian French" to produce high-quality wines, planting from 1992 vines in high-altitude lands with access to water, and discovering that the alluvial soils in Mendoza are not homogenous, resulting in vineyard with lots with unique characteristics.[34]

9.2.2 *Climate Change and the Andes Wine Industry*

Since the 1990s, the global wine sector has been addressing "sustainability initiatives" such as organic farming, carbon footprint reduction, waste reduction, and sustainable certification programmes. These preventive and mitigating risk management strategies have aimed to deal with climate change impacts, market pressures, and new environmental policies. Media pressure, the growing interest for green products, retailers' concerns, and specific export-market requests, were decisive in driving the industry's commitment to sustainability practices.[35] Institutional pressure from international organisations, such as the International Organization of Vine and Wine (OIV), also started to become a significant driver for the development of sustainability initiatives by policymakers.[36] Producers from the New World wine-producing countries such as Australia, New Zealand, and South Africa were at the forefront of these initiatives from the early 1990s and relied on these sustainable environmental practices as a prerequisite for entering wine export markets.[37] However, it was not until 2004 that the OIV applied the definition and general principles of sustainable development as given by the 1987 Brundtland report to viticulture.[38] This absence of a common and global definition of sustainable viticulture for over a decade was a major drawback for the wine industry,[39] leading potentially to concerns about "greenwashing," as companies might falsely advertise their practices as environmentally friendly without adhering to rigorous standards. It is only from the early 2000s that the adoption of sustainable management methods in the wine industry became more relevant.[40] This was a result of the establishment of more clear guidelines and certifications to ensure sustainability efforts.

Historical analysis of Argentina and Chile indicates that initially, the development and implementation of initiatives to address climate change and sustainability have been more influenced by global trends (external demands) than local conditions. In comparative terms, climate change concerns in the wine industry appeared slightly earlier in Chile than in Argentina and followed different agendas. However, both countries approved self-assessment sustainability codes between 2010 and 2011. In Chile, previous initiatives related to carbon emissions, energy, and other social responsibility issues converged in the implementation of the first national sustainability code of the Chilean wine industry. It was published in December 2010 and implemented in 2011.[41] The first attempts to address carbon emissions in Chilean wineries focused on sales in international markets.[42] Several pioneers, including Concha y Toro and Viña De Martino, measured their carbon emissions starting in 2007. Two years later, De Martino introduced the first carbon-neutral wine in Latin America (Nuevo Mundo) and became the first carbon-neutral winery in South America.[43]

The Argentinian wine industry has been seen as a latecomer to environmental sustainability practices, and yet it was a pioneer in social issues. For example, Finca La Celia implemented an environmental management system and ISO 14,000 certification, which involved social assistance programmes for the communities living near the farms, as well as other social and educational assistance programmes.[44]

Since the early 2000s, sustainability issues in Argentina's wine industry have been associated with an increasing interest in promoting corporate social responsibility.[45] Discussions on its first sustainability protocol began in 2009. It was finally published in December 2011 by Bodegas de Argentina with the involvement of several institutions, including Bodegas Catena Zapata, which played a leading role and was the first to gain certification. By August 2023, 181 wineries were certified in sustainability, while 50 had been re-certified.[46]

9.3 Business Strategies of the Wine Industry in Argentina and Chile

Academic research on the impacts of climate change and the melting of glacier ice on the wine industry started to gain momentum in the early 1990s and focused on the management of both vineyards and the wine-producing business. Some studies have placed emphasis on the negative effects of climate change. Other studies have identified new opportunities for winemakers traditionally from cooler grape-growing regions, which, because of rising temperatures, have been able to develop new and more marketable wines.[47] Several studies emphasise the adaptation strategies required in the wine sector at the level of procurement of grapes to address and mitigate the effects of climate change.[48] Discussions on viticulture practices such as changes in plant material, viticultural techniques, and the wine-making process have received more attention from researchers than other activities in the supply chain, such as marketing and distribution.[49] This section takes into account the main agents in the global supply chain and, based on interviews with wine producers and other stakeholders, identifies some of the prevention and mitigation strategies they developed to deal with these challenges.

9.3.1 Procurement

Rising temperatures, extreme climate events, and a lack of water from the Andes Mountains, which have historically been a key source of fertility in agriculture and viticulture, have been posing a very significant threat to the more traditional viticulture areas in the Andes region. Previously predictable seasonal patterns have been disturbed.[50] The weather parameters traditionally used to help manage climatic incidents have become increasingly difficult to anticipate and foresee.[51] This is considered to be the most visible impact of climate change on the actual characteristics of wines.[52]

Procurement in the context of wine production refers to the process of sourcing and acquiring all the necessary goods, services, and raw materials for the entire winemaking process. This includes planting, dealing with irrigation and water management, changing grape varietal characteristics, and addressing the challenges posed by new pests and diseases.

9.3.1.1 Planting

Planting and vineyard management can generate a footprint through the use of fertilisers and pesticides, the use of systems of irrigation that consume high levels of energy, and soil management such as tiling which can release stored carbon into the atmosphere. Seventy percent of the world's winemaking regions could become unsuitable for growing wine grapes if global temperatures exceed 2°C above the preindustrial average.[53] Globally, the most threatened regions are already hot and dry regions such as Eastern Mendoza, which is Argentina's largest wine-producing region, and which has been experiencing increasingly serious problems associated with climate change.[54]

Long-term strategies have included site-specific planting choices, which allow an increase in the viticultural area under a particular climate. For example, grape cultivation in the upper part of the Uco Valley, a high-altitude region in Mendoza, started in the 1990s. As mentioned, Catena Zapata pioneered the strategy of moving vineyards to higher altitudes in the mountains with cooler temperatures and better access to water. This strategy was radical then, aiming to improve quality rather than prevent climate change-related challenges. This transition involved a learning process because drip irrigation was primarily used for fruit trees in Argentina, and applying it to wine production presented a challenge.[55]

In the early 2000s, there was a decisive shift in the industry as other companies followed this strategy. The proportion of vineyards in the Uco Valley increased from 30% in 2000 to 70% in 2010.[56] While this strategy proved effective in many ways, there were potential risks such as the need for costly irrigation systems, UV radiation intensity, and frosts, which can damage the vineyards.[57] An alternative strategy for wine growers who had not moved their vineyards to high altitudes was to move production to the southern areas, particularly to Patagonia. Some of these wine projects, initially considered to be risky and experimental, were undertaken in the 2010s by Montes in Chiloé (Chile) and Grupo Avinea in Chubut (Argentina). They marked a new southern frontier for grape growing and wine production in South America.[58]

In Chile, as most of the planted areas face the sea, the main changes have been associated with short-term variability of temperature and rain. This meant that vineyards' growth cycles shortened, making some grape varietals inadequate. As a result, some producers moved south, while others moved in altitude.[59] In Chile, changing vineyard altitude is not considered to be viable.[60] This is because Chilean wines are of comparatively lower prices, requiring mechanised processes in the management of vineyards, and these processes are harder to use for vineyards in the mountains. Despite such hurdles, the expansion continued as more self-assured winemakers took greater risks and sought out more extreme vineyard sites.[61]

Meanwhile, short-term strategies have included more flexible management practices that enable vine productivity to be adjusted to specific local climatic conditions such as water management, soil management, nutrient management, and harvest and post-harvest management.[62] For example, some wine growers in Argentina have decided to leave more leaves on the vines or use hail netting (usually utilised for a different purpose) to reduce the exposure of the vines to the sun, making them cooler. These preventive strategies have reinforced the need for total control and monitoring, addressing the gap and the diverging interests between grape growers and wine producers, and the need for stronger control by producers in the supply chain.

9.3.1.2 Irrigation and Water Management Challenges

Water is a critical resource for environmental and economic sustainability in wine regions, and inaccessibility to freshwater resources in wine-producing regions worldwide is expected to increase.[63] Glacier water has historically played a particularly significant role in the evolution of the wine industry in South America. For Chile's biggest wine producer, Concha y Toro, the challenge of climate change has not been rising temperatures but decreasing amounts of water.[64] In Argentina the concern has been similar, considering that water shortage is an existential threat.[65]

Circumstances vary throughout each country. Chile's central and northern vineyards depend on irrigation, but this water supply depends in turn on wintertime precipitation, which falls as snow in the Andes and then feeds agriculture in the country when it melts and fills up the rivers and dams. Several historical sources have pointed out how, over the past 50 years, northern wine regions have had less rain, explaining the growing tendency to move to the country's southern areas.[66] Other firms, particularly from quality producers, have opted for drip irrigation systems. They have also added technology into the decision-making process, using tools such as soil moisture sensors and meteorological stations. Since 2007, Viña Montes has gradually implemented a series of water-saving initiatives at its Marchigüe vineyard in Colchagua. Also, after several cover crop studies, it started mulching the vineyard floor with pine bark to reduce evaporation, leading to water savings of 19%.[67]

Other new systems involved the development of predictive models and the use of satellites to avoid wasting water beyond what the plant needs.[68] Once again, this system was mainly used by large firms. Concha y Toro became the first winery in the world to measure its water footprint in 2010. Measuring water use in more than 9,000 hectares spread over many states in different regions of the country was a laborious task at the time. But the system allowed the company to adapt their agriculture to the different types of grapes and land and, as a result, to develop a complete portfolio of varieties and climates expressed in their wines.

However, canals, the traditional system of irrigation, were, by the early 2020s, still used for 50% of agricultural land, even though they waste 50% of the water compared to a 5% loss with drip irrigation systems. Another element that has attracted renewed attention is the control of water resources by using the Chilean Water Model, which is now considered the world's leading example of the free-market approach to water law.[69] The private ownership of water was established during the Pinochet dictatorship in 1973 and enshrined in the 1980 Constitution. The number of water-related disputes that reach courts in Chile has increased in the last decades.[70] Some voices in the wine industry have claimed that the Chilean government should rise to the challenge of adapting to this complex new reality by reviewing water regulations and investing in water infrastructure. Viña Montes's chief winemaker, Aurelio Montes Jr., considered that none of the past governments had invested enough in water infrastructure.[71]

In the case of Argentina, water problems vary between high elevations in the mountains and lower altitudes in the plains. Periods of water scarcity that impacted the territories of the province of Mendoza have been observed from the second half of the nineteenth century,[72] and have critically impacted a semi-arid region, home to most Argentinean wine production, which relies on surface streamflow from several snow-fed rivers.[73] The wine industry has been the dominant user of streamflow for over a century, thanks to a complex irrigation system.[74] A major change that occurred because of climate change relates to the irrigation technique used. The most common irrigation practice is extremely resource intensive. Businesses have implemented alternative solutions for dealing with the lack of water, and drip irrigation has become more widespread. The first shift from flood systems to drip systems took place in the 1990s. Other sustainable technologies have been high-performance water treatment plants, which reuse water from the winemaking process, or constructing vegetative or artificial structures to shade the vines. However, climate change and ice melting are not the only factors influencing water availability. Weak implementation of existing water regulations and the use of illegal wells make it difficult for authorities to monitor the amount of water extracted. This situation has also caused debates on water governance to rise.[75] Industry specialists noted the need for new infrastructure and a more active role of public policies.[76] Nevertheless, by 2023, about 70 to 80% of the vineyards in Argentina were still irrigated by flood systems, and the waterproofing of canals (to avoid water loss) remains inadequate.[77]

9.3.1.3 The Changing Characteristics of Grape Varietals

An important impact of climate change has been related to the actual characteristics of the grape varietals in terms of flavour and aroma. Hotter weather conditions have been creating challenges for grape varietals traditionally planted in some areas of the Andes. Certain grape varietals like Cabernet Sauvignon, which favours a specific temperature range, or Pinot Noir, which grows in cool climates, are being challenged. Higher temperatures have resulted in riper grapes with elevated sugar concentrations in the must, higher potential alcohol levels, lower acidity, and cooked flavours.[78] Such modifications, even when small, can make a big difference between a poor, a good, and an excellent crop.

Mendoza's signature wine is Malbec. Recent research has proposed that climate change would negatively impact the aromatic composition of Malbec coming from conventional vineyards.[79] Others have a different view, considering Malbec to be a highly versatile variety. Its versatility has enabled it to thrive in the majority of Argentina's terroirs, making it a wine with potential to withstand the challenges of climate change.[80] These challenges and their impacts on grape quality have led major wineries like Catena Zapata and Concha y Toro to establish internal research centres. These research centres have concentrated on rapidly accumulating knowledge to make timely decisions in response to current climate dynamics.[81] However, in the long run, these responses to climate change, often involving the cultivation of different types of grape varietals, may challenge traditional associations between wine regions and their established reputations for distinctive flavours.

9.3.1.4 Dealing with Pests and Diseases

The spread from the nineteenth century of the disease phylloxera required growers to graft their vines onto new, disease-resistant roots. Chile is one of the areas of the world where phylloxera has not penetrated; consequently, most of the national vines are planted on their own, original roots. Therefore, in Chile, pressure from pests and diseases has historically been relatively low when compared to many locations worldwide, and biodiversity is quite high. However, climate change has led to an increase of pests and diseases that have been affecting vineyards in the Andes. For instance, Catena Zapata Winery in Argentina witnessed a spread of powdery mildew related to water scarcity and increased difficulty in controlling it. In the past, flood irrigation controlled the root form of phylloxera in Argentina by asphyxiating the pest with a strong flow of water. However, these traditional systems have been challenged recently as they waste scarce water. While the drip system is effective in water management, it is not in dealing with the negative impacts of pests and diseases. Therefore, pests and diseases have become harder to control, raising the risks of lower crops. As a prevention strategy, wine growers have instead been replanting vineyards that rely on more resistant types of roots.[82]

9.3.2 Wine Production

Although the wine industry generates a comparatively small carbon footprint compared to other industries, companies have pursued different prevention and mitigation strategies to measure and reduce greenhouse gas emissions since the 2000s.[83] Wine production can contribute to a carbon footprint through several processes involving the use of fuel (Scope 1 emissions, with direct release of greenhouse gases) and energy purchases (Scope 2, indirect emissions).[84] Concha y Toro started measuring its carbon footprint in 2007 to comply with foreign buyers' requirements, as a response to the market.[85] By 2011, Concha y Toro carried out the first version of its sustainability strategy, and managed to analyse the footprint of all its production processes, in accordance with the Corporate Standards for Accounting and Reporting of the GHG Protocol, whose application is certified by international auditing firms such as Deloitte, PwC, KPMG, and Ernst & Young.[86] As result, the firm was able to reduce its direct carbon emissions by 49% between 2007 and 2017.[87] It became the first winery to apply the Science Based Targets initiative to commit to the measurement of CO_2 in 2018.[88] Other companies have also been trying to address this issue. Sogrape, for instance, created an R&D department in 2003 and invested in different technologies with a view to monitoring climate change.[89]

An important source of carbon footprint in wine production relates to energy use. As Concha y Toro recognised, this was a challenge but also an opportunity linked to the specific context of Chile. They were buying and privileging renewable energy suppliers, However, that has not been something companies have been able to procure easily in Latin America.[90] Argentine wineries, on the contrary, have expressed concerns about the high costs of implementing solar energy projects and the lack of institutional incentives due to long-term macroeconomic instability faced by the local economy.

The first wines in Argentina to obtain carbon footprint certification were produced by Bodegas Salentein, based Uco Valley. In July 2014, the wine company Portillo announced the certification of its entire line of wines: Carbernet Sauvignon, Syrah, Merlot, Tempranillo, Rosé Malbec, Chardonnay and Sauvignon Blanc. To achieve that certification the company had hired Ernst & Young to carry out a study of the full evolution of its production processes. The Carbon Trust Certification of the UK issued the certification. The calculations and certification were carried out in accordance with the PAS 2050:2008 regulations and the International Code of Good Practice for Greenhouse Gas Emissions and Reduction Claims.[91]

However, these aims have not always been easy to achieve. There has been a great discrepancy between the strategies of larger and smaller vineyards, particularly those not exposed to foreign markets, which started much later measuring their carbon footprint and had difficulty in developing mitigation strategies and applying for certification. Larger wineries have been able to comply with life-cycle assessments, measure their carbon quality, or access certain certifications.[92]

Meanwhile, smaller businesses have encountered challenges in accessing capital and implementing certain initiatives, and the need for specific public policies to deal with these pressing challenges has been highlighted by some wine specialists.[93] In the case of Argentina, these difficulties have been, once again, exacerbated by the complex macroeconomic landscape of recent years, characterised by increasing inflation, exchange controls, and local currency devaluation.

9.3.3 *Bottling, Packaging, and Ageing*

The bottling of wines and their international transport are stages that generate significant greenhouse gas emissions. By the end of 2007, wine producers in Chile began to receive requests from customers in the developed world to produce lighter bottles and track greenhouse gas emissions per bottle. For example, Cristal Chile, a glass manufacturer, announced in 2009 that they used the UK market as a reference for the design and subsequent production of bottles. This led to the development of new types of bottle 10 to 15% lighter.[94] Another firm, Cristalerías Toro, applied for certifications such as the SO14.064 Standard to address the concerns of developed countries. However, despite sustainability concerns, premium wines continued to use traditional heavy bottles and wooden cases. International surveys have shown that in 2023 for premium wines it was still important for consumers to have a traditional bottle.[95]

In Argentina, as part of their risk prevention strategies, wine producers began selecting suppliers based on sustainability evaluations of their activities. This assessment reflected their practices, protocols, and certification goals. However, firms identified several obstacles to the selection process. One significant challenge was the difficulty in choosing glass suppliers, as the glass industry operates as a dual monopoly. Despite this, Argentina's main glass supplier, Veralia, of European origin, demonstrated a commitment to sustainability principles and certification processes. For instance, in 2023, they introduced a new 300-gram model for bottles. Importing bottles was becoming a non-viable alternative due to its additional environmental impact.[96]

Another option to reduce greenhouse gas emissions has become to sell wine in bulk. For years, wine companies have been resisting selling in bulk. While it makes sense from an environmental point of view, from a business point of view, it poses several challenges.[97] First, prices tend to be lower, creating difficulties for Argentinian and Chilean wines to become more competitive in an environment where consumers are favouring higher-quality wines. Countries like Chile, which rely on low costs and sell at low prices, have been open to the possibility of selling in bulk, and in recent years, the country became one of the world's top five bulk wine exporters, shipping to around 30 different international markets. A second challenge is associated with the removal of jobs from the Andean region. A third challenge is related to the need to create flexible mechanisms to keep the reputation of the denominations of origin in such circumstance. A possibility considered has been the use of third parties to oversee the bottling and certification of wines exported in bulk, in the countries of destination.[98] In Argentina these challenges have led to efforts by wine producers of using other means for climate change mitigation and compensation to enable the continuation of exports in bottles.[99]

There have also been changes in the corks used to seal the wine bottles, and barrels used for aging the wines. Companies have been trying to source cork and oak-barrel near suppliers whose materials retain relevant tannin levels. Changes in weather patterns and carbon dioxide levels affected the development and quality of oak, the primary wood used to age wine in barrels.[100] Studies of several oak species have indicated that increasing atmospheric CO_2 has accelerated the production of "tree mass" to levels twice the rate as levels previously observed. For instance, Tuscan oak forests of Quercusilex, American forests of Quercus alba, and European forests of Quercus robur have been observed to exhibit increasing growth rates. The result of this quickening growth has been that the size and number of conducting vessels in the stems has increased, creating expanded passages in the wood, making it more porous and vulnerable to damage and failure.[101]

In terms of packaging, companies have been trying to deal with packaging suppliers committed to reducing emissions. For instance, by the early 2020s, more than 50% of Concha y Toro's key suppliers were committed to generating less waste, for instance, by using boxes as standard as possible for the packaging of the bottles of wine.[102] Similarly, to other firms such as the subsidiaries of multinational enterprises in Argentina and Chile were working with suppliers to ensure that the paper packaging used was recyclable and relied on minimum amounts of ecologically damaging ink.[103]

9.3.4 Distribution

The distribution of wines to the destination is the stage in the global supply chain that generates the greatest emissions (Scope 3, which companies' are indirectly responsible for). Companies in Chile have developed mitigation strategies with the aim of offsetting their emissions in the distribution of wines since the early 2000s. One group of wineries has chosen to offset their emissions through their own CO_2 absorption and capture projects, using their own plantations or forests. Other companies have opted instead to acquire carbon credits to neutralise the international transportation of wine.[104] Within this group, some vineyards have focused on reducing emissions during transport. Companies often obtain certifications such as the CarboNeutral Delivery Certification, which covers the entire process from storage to the final destination, for being carbon neutral.[105] When asked why this certification was sought, some wineries acknowledged that it has been a marketing strategy since, in developed markets, there has been a growing awareness of the environmental impact of international transport, putting countries such as Chile at a potential disadvantage. By neutralising the emissions from shipping, this disadvantage can be mitigated. While important to kick-start the combat on climate change, certifications like the CarboNeutral Delivery could be regarded as greenwashing. Even though legislation such as EU directives prohibits companies from making sustainability claims about their wines without data to support them, many producers have been reluctant to discuss sustainability due to the fear of being perceived as insincere and greenwashers.[106]

9.3.5 Consumer

Wine drinking patterns depend on various factors, consumer preferences being a key one.[107] From the early 2000s, consumers, particularly younger generations, have become more concerned with sustainability issues and climate change. According to data collected for the International Wines and Spirits Record (IWSR) Covid Tracker 2021, nearly half of American adult alcohol drinkers (48%) and 70% of Chinese alcohol drinkers said they were "positively influenced" to buy brands that had demonstrable environmental or sustainability credentials.[108] This led retailers to request wines that addressed environmental challenges. The alcohol monopolies of the Nordic countries (Sweden, Finland, Norway), as well as large retailers from developed countries, were also key in placing such demands. As mentioned before, in 2007, in the UK, a key market for Chilean wines, supermarkets started requesting that wine producers provided information about the carbon footprint they generated.[109] In order to address these risks, and as a preventive measure, companies incorporated sustainability considerations into their marketing practices.[110] In 2007, the British supermarket chain Tesco, announced the creation of a "carbon rating."[111] In wine producing, emerging markets, such as Chile studies have also shown that wine consumers are willing to pay a premium for eco-certified wines.[112] From the producer's perspective, these local and international tendencies create a potential differentiation in sales because both importers and new consumers increasingly focus on the winery and the production practices behind the product they consume.

9.4 Collaboration

A constant pattern emerged from the 2000s in terms of the risk management strategies developed by businesses in the global supply chain for wines, which involved establishing various forms of horizontal collaboration. These interactions have occurred among competing wine producers, suppliers, clients, employees, governments, universities, industry associations, environmental organisations, and other stakeholders.

While larger wine companies have been more proactive in forming partnerships, industry associations have become increasingly crucial in driving other collaborative initiatives. They have collaborated with academia and government agencies to help small and medium-sized enterprises participate in climate change adaptation efforts and develop mitigation strategies. Historical research highlights that pre-existing conflicts in the corporate wine sector have diminished due to mergers that consolidated corporate collective action, particularly since the early 2000s, which, as a result, promoted further collaboration.[113]

In Chile, reducing energy consumption has been a key tool in fighting climate change. In 2003, the National Council for Clean Production (CPL) signed and promoted the first Clean Production Agreement in partnership with the Chilean Wine Corporation (CCV). Around 2007 and 2008, the largest wineries took the first steps to measure their GHG emissions. It was in 2009 that the first cross-cutting associative projects on the carbon footprint of Chilean wines were launched to involve smaller wineries. These included the "Energy and Climate Change" programme, promoted by Vinos de Chile,[114] and the study "Carbon Footprint of Agricultural Products for Export," conducted by the Agricultural Research Institute (INIA). The project promoted by Vinos de Chile was financed by Innova Chile-CORFO and focused on energy efficiency and reducing carbon emissions. This programme included training and agreements with consulting companies specialised in these areas. Chile also pioneered the establishment of the Sustainability Code, which began as a collaboration between the Association of Colchagua Wineries and the Universidad de Talca, and later expanded due to the support of the "Consorcio I+D de Vinos de Chile." In 2024, 90 wineries received certification.[115]

A case of private-public collaboration was promoted by Concha y Toro with their suppliers, small and medium enterprises, to measure carbon footprint. As a result, an agreement was signed with the Center for Sustainability of the Universidad Central de Chile. Under this agreement, students conducted carbon-free footprint measurements for different suppliers. This alliance operated successfully for about five years, enabling numerous suppliers to adopt systematic carbon footprint measurement practices. This experience was considered to be a win-win collaboration because it involved a university, a territory, and its suppliers.[116] Overall, in Chile, it was this combination of different collective efforts that allowed small and medium-sized wineries to get involved in environmental issues in general, and carbon emissions in particular.

Collaborative initiatives started later in Argentina. In 2009, in what seems to have been the first official collaborative climate change initiatives, the National Institute of Viticulture of Argentina

(INV), through a grant provided by the OIV, evaluated the current knowledge and perceptions of the industry regarding the subject and its potential impact on viticulture.[117] This research provided a first contact with different actors in the sector, including winegrowers, winemakers, teachers, and politicians, making it possible for these actors to learn more about expectations, levels of credibility and scepticism, degree of knowledge, demands for information, and potential for adaptation, with respect to climate change at that time.[118] Nonetheless, there existed already several institutions which studied climate change in Argentina, such as some national universities (like Universidad Nacional de Cuyo, Facultad Regional de Mendoza, Universidad de Buenos Aires), research institutions (such as CONICET and the National Institute of Agricultural Technology (INTA)), and government entities (such as the National Institute of Viticulture (INV)).

In the horizontal associative path, a transcendent milestone was the formation of the Sustainability Commission in Bodegas de Argentina in 2010.[119] The first Wine Sustainability Self-Assessment Protocol in Argentina was developed initially with people associated with the firm Catena Zapata such as agronomists, geologists, and sales and foreign-trade specialists. The collaboration of the Faculty of Agrarian Sciences of the National Universidad de Cuyo, the INV, and INTA was considered to be a virtuous circle enriched by the interaction with multiple institutions and the sharing of knowledge.[120] Collaborations such as these have been increasing in the wine sector, being considered the way forward, given the very high the costs associated with tackling climate change.[121]

As a result, small and medium sized firms have followed strategies that rely on institutional support and various forms of collaboration, which have provided them with better access to capital, knowledge and technologies. A recent initiative to support smaller companies was the launch of a new carbon footprint measurement tool by Corporación Vitivinícola Argentina (COVIAR) that integrates various elements rather than generating them separately. The new calculator was financed by the National Ministry of Economy and was developed by the CLIOPE research group of the Universidad Tecnológica Nacional, Facultad Regional de Mendoza, and CONICET. Other key institutions such as Wines of Argentina (WoA)[122] have redefined their strategies and started forming partnerships with both the public and private sectors to assist small and medium enterprises, by giving them the tools and guidance to apply for funding to make investments and develop sustainability practices.[123] WoA has also connected social and gender issues with sustainability, with the idea of increasing the visibility of women in the wine industry.[124]

In both countries, in these types of collaborative experiences, academic institutions such as CONICET, INTA, the Universidad de Cuyo, and the Universidad de Mendoza have played a crucial role in developing new professionals and promoting knowledge exchange between the industry and the academy. In the case of Chile, the academic sphere included the Centre for Vine and Wine of the University of Talca (CTVV) and the Centre for Viticulture of the Catholic University of Chile (CEVIUC). In the first decades of the 2000s, many of these institutions have been actively involved in vine and wine research-related activities, financed mainly by public funds managed by CONICYT (The National Commission for Research on Science and Technology).[125]

Large firms have also created alliances in viticulture. For example, in Chile for wine producers that rely on independent farmers for grape production, the vulnerability and uncertainty created by climate change have led to changes in their contractual relationships to move towards stronger forms of collaboration.[126] In Argentina, private collaborations in viticulture focused more on soil studies, grape varietal suitability to different soil types, and pest management. For example, in 2007 the Catena Institute of Wine (dedicated to the advancement and promotion of wine knowledge for the benefit of wine technicians and the wine community), began collaborating with local Institutions and the Universidad de Cuyo to focus on Malbec and sustainability research. Another illustration is Finca Flichman, which formed alliances with the Universidad de Cuyo and CONICET in Argentina, as well as with Copernicus in Europe, National Oceanic and Atmospheric Administration (NOAA) in the United States, and other observatories such as the Royal Observatory of Belgium. This collaboration led to important improvements on the mapping of climate change projections, and involved domestic and foreign researchers. As a result, the firm developed preventive strategies to address climate change, informing decisions such as where to plant in new geographical regions, how to combat pests and diseases, how to manage new humidity conditions, and which grape varietals to plant. For example, in Chile, climate change has led Finca Flichman to replace Chardonnay and Sauvignon Blanc with grapes such as Riesling and Viognier.[127]

9.5 Conclusion

This chapter contributes to emerging research in business history on climate change by examining the impact of climate change on the prevention and mitigation risk management strategies of businesses in Andean wine industry, in particular in Argentina and Chile. It argues that traditional corporate strategies for building competitiveness have become insufficient. Beyond altering their corporate boundaries by investing in vineyards at higher altitudes or different geographical regions, wine firms have also developed prevention and mitigation strategies, which have involved greener capabilities, with a global supply chain impact. These innovations encompass both technological and organisational advancements.

Large wine firms with international operations have successfully implemented risk management strategies, many of which have involved substantial investments in innovations, which relied on their financial capacity. In contrast, small and medium-sized enterprises, particularly those not involved in international business, have faced challenges in implementing such strategies due to their more limited financial resources. Adaptation has been costly, and as illustrated in this chapter, producers' unequal access to capital, knowledge, and technologies has further impeded these investments. To overcome these difficulties, smaller firms have sought support from other stakeholders such as governmental agencies and industry associations, forming various collaborations. Associations like Vinos de Chile, Bodegas de Argentina, and Wines of Argentina have played significant roles in assisting businesses, particularly from the 2020s.

Large firms have also engaged in individual collaborations to mitigate the impact of extreme risks in the medium and long term across the supply chain. These alliances have often included competitors and other stakeholders such as suppliers, universities, research institutes, and governmental agencies. The business community in both countries, particularly in Argentina, has historically faced many significant challenges and more uncertain contexts. The response to such exposure has increased the communities' flexibility, considred to be potentially an advantage in the new contextual environment.

Overall, climate change has been transforming business models and practices in the wine industry. Strategic risk management has impacted the industry's configuration, leading to an environment where collaboration is crucial for survival and competitiveness. This collaboration has been possible due to incremental shifts in mindsets. From high executives to field workers and suppliers of bottles and oak casks, the rising awareness regarding the significance of sustainability practices, carbon footprint reduction, and social responsibility, has been central in ensuring growth and survival in an industry where climate change has been so impactful.

Notes

1 We are solely responsible for interpreting these data obtained through interviews. However, we are very grateful to all those who participated in the interviews, particularly Ms. Carla Aruani, Coordinator of R&D on Sustainability at Instituto Nacional de Vitivinicultura (Argentina); Mr. Fernando Buscema from the Catena Institute of Wine; Dr. Laura Catena, CEO of Catena Zapata; Professor Barbara Civit, Researcher at CONICET (Argentina); Mr. José Galante, Winemaker and CEO of Saleintein Wines; Ms. Martina Galeano former winemaker at Casarena Bodega and Viñedos; Mr. António Graça, Director of R&D at Sogrape; Mr. Filipe Neves, New World Production Business Development Manager at Sogrape; Ms. Mafalda Guedes, Head of Corporate Communications and Sustainability at Sogrape; Ms. Valentina Lira, Sustainability Director of Concha y Toro; Ms. Magdalena Pesce, Director of Wines of Argentina; Mr. Rogelio Rabino, Director of Viticulture and Enology, Finca Flichman; Dr. Luis Romito, Expert on Climate Change and first Coordinator of the Commission for Sustainability of Bodegas de Argentina; Ms. Gabriela Vergara, Coordinator of Viticulture at Finca Flichman; and Mr. Diego Vergara Eneros, Director of Enology from Chateau Los Boldos. We want to thank Professor Facundo Rojas, a researcher at CONICET (Argentina), and Professor Ximena Rueda, from the Universidad de los Andes (Colombia), for their suggestions and feedback at the beginning of the research. We also thank the participants in the Workshop held at the Centre for Global Dialogue, SwissRe in 2023; and the panel sessions on climate change at the Annual European Business History Congress in Oslo (2023) and Lisbon (2024) for their value comments and suggestions.
2 Castellanos et al., "Central and South America."
3 The Andes region covers a contiguous mountain area spreading across Argentina, Bolivia, Chile, Colombia, Ecuador, Peru, and Venezuela.
4 The glaciers in the Cordillera Blanca in the Andes, including the Palcacocha area in northern Peru, began retreating in the late nineteenth century, but this process accelerated and became particularly visible from the late 1970s. By 2000, 98% of all the glaciers in the Andes were showing some level of shrinkage Huggel et al., "Anthropogenic Climate."
5 Allison, "The Spiritual Significance"; Interview with Diego Vergara Enero, Manager at Chateau Los Boldos; May 12, 2023.
6 Schoolmeester, "Outlook on Climate Change."
7 Odell et al., "Mining and Climate."
8 Kaenzig et al., "Climate Change Adaptation."
9 Schoolmeester et al., "The Andean Glacier"; Castellanos et al., "Central and South America."
10 IEA, "Climate Impacts."
11 IPCC, *Climate Change 2022.*
12 Unwin, "The Wine"; Lopes et al., "The Changing."
13 See for example Wilkins, *The History of Foreign Investment in the United States*; Silva Lopes, *Global Brands*, 87–106; Galambos, *Competition and Collaboration*; Silva Lopes, "Trademarks as 'Merchants of Skill'."
14 CORFO is a public-sector organisation dedicated to promoting entrepreneurship, innovation and growth in Chile.
15 Definition based on Casson and Silva Lopes, "Foreign Direct Investment in High-Risk Environments."
16 Thach, "Climate Change Propels."
17 Anderson and Pinilla, "Wine's Belated."
18 Mora, "The Chilean Wine Industry."
19 Felzensztein, "The Chilean."
20 Bas and Kunc, "National Systems."
21 Knowles and Sharples, "The History."
22 Fernandez-Stark and Bamber, "Wine Industry in Chile."
23 Wines of Chile, "Winegrowing Regions."
24 Wines of Chile, "Winegrowing Regions."
25 Mora, "The Chilean Wine Industry."
26 Medina-Albaladejo et al., "El mercado mundial de vino."
27 Richard-Jorba, "Crisis y transformaciones."
28 *INV, Informe* Anual de Superficie 2023.
29 Richard-Jorba, "Crisis y Transformaciones."
30 Vinodata, "Informe El Vino Argentino en los Mercados Globales."

31 Artopoulos et al., "Export Emergence."
32 Vinodata, "Informe El Vino Argentino en los Mercados Globales."
33 Interview with Laura Catena, CEO of Catena Zapata, April 24, 2023.
34 Interview with Laura Catena, CEO of Catena Zapata, April 24, 2023.
35 Corbo et al., "Functional Beverages"; Merli et al., "Sustainability Experiences."
36 Pomarici and Vecchio, "Will Sustainability Shape."
37 Moggi et al., "From Sustainability."
38 OIV, "Resolution Cst 1/2004."
39 Trigo et al., "Sustainability Awareness."
40 Rugani and Lamastra, "A Common Framework."
41 Moscovici and Reed, "Comparing Wine."
42 Olmos, "La huella de carbono."
43 Interview with Valentina Lira, General Manager of Sustainability of Concha y Toro, June 6, 2023.
44 No es posible florecer en un ambiente deteriorado, *Revista de Vinos y Viñas* 996 (August 2005): 68.
45 McDermott, "The Politics of Institutional"; McDermott et al., "Public–Private."
46 Devia, "More Wineries Certified in Sustainability."
47 Gladstones, "Viticulture and the Environment," is considered the first highly cited publication that discussed viticulture about climate change, specifically global warming; Jones and Web, "Climate Change."
48 Sacchelli et al., "Climate Change"; van Leeuwen and Darriet, "The Impact of Climate Change"; Gonen et al., "Adapting and Thriving."
49 Ashenfelter and Storchmann, "Climate Change."
50 Interview with José Galante, former enologist at Catena Zapata and Bodegas Salentein's Chief Winemaker, May 17, 2023.
51 Interview with Magdalena Pesce, CEO of Wines of Argentina, June 29, 2023.
52 Interview with António Graça, Director of R&D at Sogrape, June 26, 2023, and interview with Martina Galeano, winemaker of Bodegas Casarena, July 11, 2023.
53 Van Leeuwen and Darriet. "Climate Change Impacts."
54 Interview with Magdalena Pesce, CEO of Wines of Argentina, June 29, 2023.
55 Interview with José Galante, former enologist at Catena Zapata and Bodegas Salentein's Chief Winemaker, May 17, 2023.
56 Interview with Laura Catena, CEO of Catena Zapata, April 24, 2023.
57 Interview with Filipe Neves, New World Production Business Development Manager at Sogrape, June 26, 2023.
58 Schachner, "Extreme Conditions and a Changing Climate."
59 Interview with António Graça, Head of R&D at Sogrape, June 26, 2023; Interview with Filipe Neves, New World Production Business Development Manager at Sogrape, June 26, 2023.
60 Press Interview with Rodrigo Barría, Agricultural Manager of Viña Montes. https://www.france24.com/es/minuto-a-minuto/20230831-productores-de-vinos-chilenos-rescatan-viejas-Strategiestécnicas-para-afrontar-cambio-climático.
61 Interview with Diego Vergara Eneros, Manager at Chateau Los Boldos, May 12, 2023.
62 Naulleau et al., "Evaluating Strategies."
63 Upton and Nielsen-Pincus, "Climate Change."
64 Concha y Toro technical director Marcelo Papa expressed in 2023 "Chile is a country where we are getting less rain, so the biggest issue in Chile will not be the temperature, but the amount of water." Quoted in Schmitt, "Why It's Drought, Not Heat." add age number.
65 Interview with Laura Catena, CEO of Catena Zapata, April 24, 2023.
66 Bauer, "Against the Current."
67 "Chile's Approach to Water," https://www.meiningers-international.com/wine/styles-regions/chiles-approach-water.
68 Interview with Valentina Lira, General Manager Sustainability of Concha y Toro, June 6, 2023.
69 Herrera et al., "Understanding Water."
70 Blanco et al., "Water Conflicts."
71 Quoted in Chile's Approach to Water, https://www.meiningers-international.com/wine/styles-regions/chiles-approach-water.
72 Rojas et al., "Sequías en Mendoza."
73 Rivera et al., "A Decade of Hydrological."

74 Castex et al. "Water availability."
75 Interview with António Graça, Director of R&D at Sogrape, June 26, 2023; Interview with Luis Romito, Expert on Climate Change and first Coordinator of the Commission for Sustainability of Bodegas de Argentina, September 6, 2023.
76 Mussetta and Hurbert, "Creating Resilient Water."
77 Interview with Magdalena Pesce, CEO of Wines of Argentina, June 29, 2023.
78 Van Leeuwen and Darriet, "The Impact of Climate Change."
79 Martinez et al., "Understanding the Impact."
80 Interview with Martina Galeano, winemaker Bodegas Casarena, July 11, 2023.
81 Interview with Fernando Buscema, Executive Director of the Catena Institute of Wine, June 5, 2023. In 2014, the most extensive Malbec study that had been carried out to date was published in *Food Chemistry* and *The American Journal of Enology and Viticulture*, as a result of joint research between the CIW and UC Davis.
82 Interview with Laura Catena, CEO of Catena Zapata, April 24, 2023; Interview with Rogelio Rabino, Director of Viticulture and Enology, Finca Flichman, and Gabriela Vergara, Coordinator of Viticulture—Finca Flichman, May 12, 2023.
83 Olmos, "La Huella."
84 According to international Wineries for Climate Action, Scope 1 represents 10% of a winery's activities, while Scope 2 added a further 5%. This leaves 85% under Scope 3, which covers all of the other most significant climate impacts in the wine life cycle—from grape buying to dry goods and business travel.
85 Interview with Valentina Lira, General Manager of Sustainability at Concha y Toro, June 6, 2023.
86 See the chapter by Stroehle in this volume for a discussion of the reliability of these measurements.
87 Interview with Valentina Lira, General Manager of Sustainability at Concha y Toro, June 6, 2023.
88 Case Study—Viña Concha y Toro, https://sciencebasedtargets.org/companies-taking-action/case-studies/case-study-viña-concha-y-toro.
89 Interview with António Graça, Director of R&D at Sogrape, June 26, 2023.
90 Interview with Valentina Lira, General Manager of Sustainability at Concha y Toro, June 6, 2023.
91 Press Release from Salentein Winery, https://www.bodegasalentein.com/es/detalle-de-portillo-el-primer-vino-argentino-en-certificar-su-huella-de-carbono-en-todos-sus-varietales-52/index.html.
92 Interview with Carla Aruani, Instituto Nacional de Vitivinicultura (INV), Argentina, July 4, 2023. Interview with Barbara Civit, Researcher at CONICET, May 15, 2023.
93 Interview with Magdalena Pesce, CEO if Wines of Argentina, June 29, 2023.
94 Olmos, "La huella de carbono."
95 Interview with Laura Catena, CEO of Catena Zapata, April 24, 2023.
96 Interview with Martina Galeano, winemaker Bodegas Casarena, July 11, 2023.
97 Interview with Luis Romito, Sustainability and Commission Expert on Climate Change at Bodegas de Argentina, August 30, 2023; Interview with Mafalda Guedes, Head of Corporate Communications and Sustainability at Sogrape, 12 May 2023; Interview with Laura Catena, CEO of Catena Zapata, April 24, 2023.
98 Interview with António Graça, Head of R&D at Sogrape; and Filipe Neves, New World Production Business Development Manager, Sogrape, June 26, 2023.
99 Interview with Magdalena Pesce, CEO of Wines of Argentina, June 29, 2023.
100 Mira de Orduña, "Climate change."
101 Interview with Rogelio Rabino, Director of Viticulture and Enology, Finca Flichman, May 12, 2023. Tate, "Global Warming's Impact," 104.
102 Así es Como Viña Concha Y Toro Disminuye Sus Emisiones De Co2, https://vinacyt.com/noticia/sustentabilidad/asi-es-como-vina-concha-y-toro-disminuye-sus-emisiones-de-co2/.
103 Interview with Filipe Neves, New World Production Business Development Manager at Sogrape, June 26, 2023.
104 Olmos, "La huella."
105 Olmos, "La huella."
106 Interview with Laura Catena, CEO of Catena Zapata, April 24, 2023.
107 Lopes, *Global Brands.*
108 When Will Sustainability Matter to Wine Consumers?, https://www.wineintelligence.com/when-will-sustainability-matter-to-wine-consumers/.
109 Interview with Valentina Lira, General Manager of Sustainability at Concha y Toro, June 6, 2023.

110 Borsellino et al., "'Green' Wine."
111 Gwynne, "UK Retail Concentration."
112 Valenzuela et al., "Consumer Willingness to Pay."
113 Erices, "Disputas de poder"; Chazarreta, "Organizaciones Empresariales."
114 Wines of Chile (WOC)—Vinos de Chile in Spanish—is a non-profit, private organisation of Chilean wine producers dedicated to promoting the quality and image of Chilean wines. It is the oldest organisation in the New World, formed in 1919 as the Liga de Defensa del Vino.
115 "Sustainability Code of the Chilean Wine Industry," https://www.sustainable.cl/en/.
116 Interview with Valentina Lira, General Manager Sustainability of Concha y Toro, June 6, 2023.
117 Aruani, "Cambio Climático y Vitivinicultura."
118 Interview with Carla Aruani, Instituto Nacional de Vitivinicultura, Argentina, July 4, 2023.
119 Bodegas de Argentina is a business chamber created in 2001 as a result of the merger of the Centro de Bodegueros de Mendoza, based in Mendoza and founded in 1935, and the Asociación Vitivinícola Argentina, based in Buenos Aires and founded in 1904.
120 Interview with Luis Romito, Expert on Climate Change and first Coordinator of the Commission for Sustainability of Bodegas de Argentina, September 6, 2023.
121 Interview with Laura Catena, CEO of Catena Zapata, April 24, 2023.
122 Wines of Argentina is an organisation founded in 1993. Its main mission is to collaborate in consolidating Argentina as one of the main wine exporting countries.
123 In 2023 Wines of Argentina presented its sustainability programme for the sector, Sustenta-Vitis, made possible through AL-INVEST Verde, an initiative of the European Union (EU), developed jointly with Interco Nouvelle-Aquitaine. The intention of WoA is also supported for SMEs to apply to the Sustainability Protocols Certification of Bodegas de Argentina (BdA) and to encourage initiatives that facilitate a more sustainable, diverse, and inclusive viticultural industry.
124 Interview with Magdalena Pesce, CEO of Wines of Argentina, June 29, 2023; Interview with Filipe Neves, New World Production Business Development Manager at Sogrape, June 26, 2023.
125 Farinelli, "Innovation and Learning."
126 Interview with Valentina Lira, General Manager of Sustainability at Concha y Toro, June 6, 2023.
127 Interview with António Graça, Head of R&D at Sogrape, June 26, 2023.

Bibliography

Allison, Elizabeth A. "The Spiritual Significance of Glaciers in an Age of Climate Change." *Wiley Interdisciplinary Reviews: Climate Change* 6, no. 5 (2015): 493–508.

Anderson, Kym, and Vicente Pinilla. "Wine's Belated Globalization, 1845–2025." *Applied Economic Perspectives and Policy* 44, no. 2 (2022): 742–65.

Artopoulos, Alejandro, Daniel Friel, and Juan Carlos Hallak. "Export Emergence of Differentiated Goods from Developing Countries: Export Pioneers and Business Practices in Argentina." *Journal of Development Economics* 105 (2013): 19–35. https://doi.org/10.1016/j.jdeveco.2013.07.001.

Aruani, Carla. *Cambio Climático y Vitivinicultura en Argentina: Informe Final.* Beca de Investigación OIV 2009. Instituto Nacional de Vitivinicultura, 2011.

Ashenfelter, Orley, and Karl Storchmann. "Climate Change and Wine: A Review of the Economic Implications." *Journal of Wine Economics* 11, no. 1 (2016): 105–38.

Bauer, Carl J. *Against the Current: Privatization, Water Markets, and the State in Chile.* Kluwer Academic Publishers, 1998.

Blanco, Elisa, Guillermo Donoso, and Pablo Camus. 2023. "Water Conflicts in Chile: Have We Learned Anything from Colonial Times?" *Sustainability* 15 (19): 14205. https://doi.org/10.3390/su151914205.

Borsellino, Valeria, Giuseppina Migliore, Marcello D'Acquisto, Caterina Patrizia Di Franco, Antonio Asciuto, and Emanuele Schimmenti. "'Green' Wine Through a Responsible and Efficient Production: A Case Study of a Sustainable Sicilian Wine Producer." *Agriculture and Agricultural Science Procedia* 8 (2016): 186–92.

Brundtland, Gro Harlem. *Report of the World Commission on Environment and Development—Our Common Future.* United Nations, 1987.

Casson, Mark, and Teresa da Silva Lopes. "Foreign Direct Investment in High-Risk Environments." *Business History* 55, no. 3 (2013): 375–404.

Castellanos, Edwin J., Maria Fernanda Lemos, Laura Astigarraga, Noemí Chacón, Nicolás Cuvi, Christian Huggel, Liliana Miranda, Mariana Moncassim Vale, Jean Pierre Ometto, Pablo Luis Peri, Julio C. Postigo, Laura Ramajo, Lisandro Roco, and Matilde Rusticucci. "Central and South America." In *Climate Change 2022—Impacts, Adaptation and Vulnerability. Contribution of Working Group II to the Sixth Assessment Report of the Intergovernmental Panel on Climate Change*, edited by H.-O. Pörtner, D.C. Roberts, M. Tignor, E.S. Poloczanska, K. Mintenbeck, A. Alegría, M. Craig, S. Langsdorf, S. Löschke, V. Möller, A. Okem, and B. Rama, 1689–816. Cambridge University Press, 2023.

Castex, Victorine, Enrique Morán Tejeda, and Martin Beniston. "Water Availability, Use and Governance in the Wine Producing Region of Mendoza, Argentina." *Environmental Science & Policy* 48 (2015): 1–8.

Chazarreta, Adriana. "Organizaciones Empresariales y Conflictos en la Burguesía Vitivinícola Argentina." *Anuario IHES* 37, no. 2 (2015): 161–77.

Cohen, Zach. *Glacier Retreat in the Andes—A Triple Threat to Water, Energy, and Food Security*. Perspectives—A Development, Climate & Sustainability Publication. John Hopkins, School of Advanced International Studies, March 8, 2023.

Corbo, María Rosaria, Antonio Bevilacqua, Leonardo Petruzzi, Francesco Pio Casanova, and Milena Sinigaglia. "Functional Beverages: The Emerging Side of Functional Foods: Commercial Trends, Research, and Health Implications." *Comprehensive Reviews in Food Science and Food Safety* 13, no. 6 (2014): 1192–206.

Devia, Sol. "More Wineries Certified in Sustainability Through WofA and Sustenta-Vitis." *Wines of Argentina Blog*, October 9, 2023.

Erices Castro, José, and Miguel Escalona Ulloa. 2022. "Disputas de Poder en la Vitivinicultura Chilena: Valles del Biobío Bajo Las Sombras Del Desarrollo." *Economía, Sociedad y Territorio* 22, no. 69 (2022): 309–37. https://doi.org/10.22136/est20221713.

Farinelli, Fulvia. "Innovation and Learning Dynamics in the Chilean and Argentine Wine Industries." American Association of Wine Economics—Working Paper 145, 2013. https://doi.org/10.22004/ag.econ.164653.

Felzensztein, Christian. "The Chilean Wine Industry: New international Strategies for 2020." *Emerald Emerging Markets Case Studies* 4, no. 2 (2014): 1–12.

Fernandez-Stark, Karina, and Penny Bamber. "Wine Industry in Chile." In *Services in Global Value Chains: Manufacturing-Related Services*, 364–89. Singapore: APEC Policy Support Unit, 2016.

Galambos, Louis. *Competition and Collaboration: The Emergence of a National Trade Association*. Johns Hopkins University Press, 1966.

Gladstones, John. *Viticulture and the Environment: A Study of the Effects of Environment on Grape Growing and Wine Qualities with Emphasis on Present and Future Areas of Growing Winegrapes in Australia*. Wine Titles, 1992.

Gonen, Limor Dina, Tchai Tavor, and Uriel Spiegel. "Adapting and Thriving: Global Warming and the Wine Industry." *Sage Open* 14, no. 1 (2024). https://doi.org/10.1177/21582440241227750.

Grazia, Diego, Chiara Mazzocchi, Giordano Ruggeri, Stefano Corsi, and Kerry Wilkinson. "Grapes, Wines and Changing Times: A Bibliometric Analysis of Climate Change Influence." *Australian Journal of Grape and Wine Research* 2023, no. 1 (2023): 1–19.

Gwynne, Robert N. "UK Retail Concentration, Chilean Wine Producers and Value Chains." *Geographical Journal* 174, no. 2 (2008): 97–108.

Herrera, Mauricio, Cristian Candia, Diego Rivera, Douglas Aitken, Daniel Brieba, Camila Boettiger, Guillermo Donoso, and Alex Godoy-Faúndez. "Understanding Water Disputes in Chile with Text and Data Mining Tools." *Water International* 44, no. 3 (2019): 302–20. https://doi.org/10.1080/02508060.2019.1599774.

Huggel, C., M. Carey, A. Emmer, H. Frey, N. Walker-Crawford, and I. Wallimann-Helmer. "Anthropogenic Climate Change and Glacier Lake Outburst Flood Risk: Local and Global Drivers and Responsibilities for the Case of Lake Palcacocha, Peru." *Natural Hazards Earth System Sciences* 20, no. 8 (2020): 2175–93.

Hurlbert, Margot, and Paula Mussetta. "Creating Resilient Water Governance for Irrigated Producers in Mendoza, Argentina." *Environmental Science & Policy* 58 (2016): 83–94. https://doi.org/10.1016/j.envsci.2016.01.004.

IEA. *Climate Impacts on Latin American Hydropower*. IEA, 2020.

INV—Instituto Nacional de Vitivinicultura. *República Argentina: Informe Anual de Superficie 2023*. INV, 2023.

IPCC. "AR5 Climate Change 2013: The Physical Sciences." https://www.ipcc.ch/report/ar5/wg1/.
IPCC. "Climate Change 2021: The Physical Science Basis." https://www.ipcc.ch/report/sixth-assessment-report-working-group-i/.
IPCC. "Climate Change 2022—Impacts, Adaptation and Vulnerability." https://www.ipcc.ch/report/ar6/wg2/.
Jones, Gregory V., and Leanne B. Webb. "Climate Change, Viticulture, and Wine: Challenges and Opportunities." *Journal of Wine Research* 21, no. 2–3 (November 2010): 137–45.
Kaenzig, Raoul, Martine Rebetez, and Gaëlle Serquet. "Climate Change Adaptation of the Tourism Sector in the Bolivian Andes." *Tourism Geographies* 18, no. 2 (2016): 111–28.
Knowles, Tim, and Liz Sharples. "The History and Development of Chilean Wines." *International Journal of Wine Marketing* 14, no. 2 (2002): 7–16.
Liberman, Giselle Lara. "The Argentine Wine Industry: Creating New Spaces for Coordination?" *Prometheus* 31, no. 4 (2013): 305–18.
Lopes, Teresa da Silva. *Global Brands: The Evolution of Multinationals in Alcoholic Beverages*. Cambridge University Press, 2007.
Lopes, Teresa da Silva, Andrea Lluch, and Gaspar Martins Pereira. 2020. "The Changing and Flexible Nature of Imitation and Adulteration: The Case of the Global Wine Industry, 1850–1914." *Business History Review* 94, no. 2 (2020): 347–71. https://doi.org/10.1017/S0007680520000306.
Lopes, Teresa da Silva, and Shin Tomita. "Trademarks as 'Global Merchants of Skill': The Dynamics of the Japanese Match Industry, 1860s–1930s." *Business History Review* 96, no. 3 (2022): 559–88.
Martínez Liliana, Bianca Costa, Leonor Deis, Marta Dizy, and Mar Vilanova. "Understanding the Impact of Rising Temperatures due to Climate Change on Aromatic Compositions in Malbec Wines from Mendoza, Argentina." IVES Conference Series, ICGWS, 2023.
McDermott, G. "The Politics of Institutional Renovation and Economic Upgrading: Recombining the Vines That Bind in Argentina." *Politics & Society* 35, no. 1 (2007): 103–43.
McDermott, G., R. Corredoira, and G. Kruse. "Public-Private Institutions as Catalysts of Upgrading in Emerging Market Societies." *Academy of Management Journal* 52, no. 6 (2009): 1270–96.
Medina-Albaladejo, Francisco José, José Miguel Martínez-Carrión, and Josep Maria Ramon-Muñoz. "El Mercado Mundial de Vino y la Competitividad de los Países del Hemisferio Sur, 1961–2010." *Latin America in Economic History* 21, no. 2 (2014): 40–83.
Merli, Roberto, Michele Preziosi, and Alessia Acampora. "Sustainability Experiences in the Wine Sector: Toward the Development of an International Indicators System." *Journal of Cleaner Production* 172 (2018): 3791–805.
Mira de Orduña, R. "Climate Change Associated Effects on Grape and Wine Quality and Production." *Food Research International* 43, no. 7 (2010): 1844–55.
Moggi, Sara, Paul Pierce, and Nicole Bernardi. "From Sustainability to Thrivability: A Novel Framework for Entrepreneurial Ecosystems." *International Entrepreneurship and Management Journal* 18, no. 2 (2022): 829–53.
Mora, G. Marcos. "The Chilean Wine Industry." In *The Palgrave Handbook of Wine Industry Economics*, edited by A. Alonso Ugaglia, J.M. Cardebat, and A. Corsi, 177–200. Palgrave Macmillan.
Moscovici, Daniel, and Alastair Reed. "Comparing Wine Sustainability Certifications Around the World: History, Status and Opportunity." *Journal of Wine Research* 29, no. 1 (2018): 1–25. https://doi.org/10.1080/09571264.2018.1433138.
Naulleau, Audrey, Christian Gary, Laurent Prévot, and Laure Hossard. "Evaluating Strategies for Adaptation to Climate Change in Grapevine Production—A Systematic Review." *Frontiers in Plant Science* 11 (2021): 607859. https://doi.org/10.3389/fpls.2020.607859.
Odell, S.D., A. Bebbington, and K.E. Frey. "Mining and Climate Change: A Review and Framework for Analysis." *The Extractive Industries and Society* 5, no. 1 (2018): 201–14. https://doi.org/10.1016/j.exis.2017.12.004.
OIV. "Resolution Cst 1/2004 Development of Sustainable Vitiviniculture." 2014. https://www.oiv.int/public/medias/2074/cst-1-2004-en.pdf.
Olmos, Ximena. *La Huella de Carbono en el Comercio Internacional: El Caso de las Viñas Chilenas*. Universidad de Chile—Instituto de Estudios Internacionales, 2012. https://repositorio.uchile.cl/handle/2250/111175.
Pomarici, Eugenio, and Riccardo Vecchio. "Will Sustainability Shape the Future Wine Market?" *Wine Economics and Policy* 8, no. 1 (2019): 1–4.
Richard-Jorba, Rodolfo A. "Crisis y Transformaciones Recientes en la Región Vitivinícola Argentina: Mendoza y San Juan, 1970–2005." *Estudios Sociales (Hermosillo, Son.)* 16, no. 31 (2008): 81–123.

Rivera, Juan A., Sebastián Otta, Carolina Lauro, and Natalia Zazulie. "A Decade of Hydrological Drought in Central-Western Argentina." *Frontiers in Water* 3 (2021): 640544. https://doi.org/10.3389/frwa.2021.640544.

Rojas, Facundo, Osvaldo Sironi, and Facundo Martín. "Sequías en Mendoza (Argentina): Una Mirada Sociohistórica desde la Segunda Mitad del Siglo XIX." *Agua y Territorio = Water and Landscape* 22 (2023): 147–66.

Rugani, Benedetto, and Lucrezia Lamastra. "A Common Framework for Sustainability Indicators in the Wine Sector: Dream or Reality?" *Current Opinion in Environmental Science & Health* 31 (2023): 100408. https://doi.org/10.1016/j.coesh.2022.100408.

Sacchelli, Sandro, Sara Fabbrizzi, and Silvio Menghini. "Climate Change Effects and Adaptation Strategies in the Wine Sector: A Quantitative Literature Review." *Wine Economics and Policy* 5, no. 2 (2016): 114–26.

Santisi, Jennifer. "Warming Up the Wine Industry: Climate Change May Leave a Major Impact on Vineyards Across the Country." *E—The Environmental Magazine* 22, no. 6 (2011): 12.

Schachner, Michel. "Extreme Conditions and a Changing Climate." 2023. https://www.wineenthusiast.com/culture/wine/chile-argentina-patagonia-wine/.

Schmitt, Patrick. "Why It's Drought, Not Heat, That Threatens Chilean Viticulture." 2023. https://www.thedrinksbusiness.com/2023/08/why-its-drought-not-heat-that-threatens-chilean-viticulture/.

Schoolmeester, T., M. Saravia, M. Andresen, J. Postigo, A. Valverde, M. Jurek, B. Althan, and S. Giada. *Outlook on Climate Change Adaptation in the Tropical Andes Mountains.* Mountain Adaptation Outlook Series. United Nations Environment Programme, GRID-Arendal and CONDESAN, 2016.

"Sustainability Code of the Chilean Wine Industry." https://www.sustainable.cl/en/.

Tate, A.B. "Global Warming's Impact on Wine." *Journal of Wine Research* 12, no. 2 (2001): 95–109. https://doi.org/10.1080/09571260120095012.

Thach, Liz. "Climate Change Propels France to Number 1 Largest Global Wine Produce in 2023." *Forbes*, November 11, 2023. https://www.forbes.com/sites/lizthach/2023/11/11/climate-change-propels-france-to-1-largest-global-wine-producer-in-2023/.

Trigo, Ana, Rui Fragoso, and Ana Marta-Costa. "Sustainability Awareness in the Portuguese Wine Industry: A Grounded Theory Approach." *International Journal of Agricultural Sustainability* 20, no. 7 (2022): 1437–53. https://doi.org/10.1080/14735903.2022.2150377.

Unwin, Tim. *The Wine and the Vine: An Historical Geography of Viticulture and the Wine Trade*. Routledge, 2005.

Upton, Erin, and Max Nielsen-Pincus. "Climate Change and Water Governance: Decision Making for Individual Vineyard Owners in Global Wine Regions." *Frontiers in Climate* 3 (2021): 654953. https://doi.org/10.3389/fclim.2021.654953.

Valenzuela, Lionel, Rodrigo Ortega, Daniel Moscovici, Jeff Gow, Adeline Alonso Ugaglia, and Radu Mihailescu. "Consumer Willingness to Pay for Sustainable Wine—The Chilean Case." *Sustainability* 14, no. 17 (2022): 10910. https://doi.org/10.3390/su141710910.

Van Leeuwen, Cornelis, and Philippe Darriet. "The Impact of Climate Change on Viticulture and Wine Quality." *Journal of Wine Economics* 11, no. 1 (2016): 150–67.

Vinodata. "Informe El Vino Argentino en los Mercados Globales." IIEP, Universidad de Buenos Aires, 2023. https://iiep.economicas.uba.ar/vinodata/2023.

Wilkins, Mira. *The History of Foreign Investment in the United States to 1914*. Harvard University Press, 1989.

10

TOWARDS A NEW TYPE OF TOURISM?

From Conventional Tourism to Ecotourism in the Alps

Laurent Tissot

10.1 Introduction

It has become commonplace to write that "climate change" is overturning all the forecasts made in the field of tourism.[1] Since the middle of the nineteenth-century promoters, analysts, and other observers unhesitatingly developed strategies aimed at enabling the sector to welcome new flows of travellers. From the 1990s, the irruption of climate change has transformed the situation to such an extent that certainties have given way to hesitation and scepticism. The quantitative obsession that has underpinned nearly all prior development models is giving way to multidimensional approaches that give quality as much room as quantity. "It is now time to rethink tourism, the direction it is taking and its impact on humanity and the planet." Taken from the World Tourism Organisation's newsletter, this kind of phrase can be found on every website devoted to global tourism and its future.[2] The vast majority of experts from various governmental and non-governmental organisations have all come to the same conclusion. The future will be highly problematic for the tourism industry if nothing is done.[3]

The financial stakes are considerable. This can be captured simply by looking at traffic flows. Estimated at 435 million in 1990, the number of international tourists, that is, those who have crossed a border to satisfy their desire, had risen to over 1 billion 15 years later and reached almost 1.5 billion in 2019, one year before Covid-19. This figure does not include tourists who stay in their country of residence. The 2020 pandemic had a profound effect, with tourist numbers falling back to 1990 levels. However, the latest figures show an impressive recovery, with 968 million tourists in 2022, and all the signs pointing to a return to pre-Covid-19 levels by 2023.[4] Despite the comfort that these figures bring to many tourism promoters and stakeholders, the future remains overshadowed by the pressing issue of climate change.

To address this uncertainty, this chapter will focus on a particular area of tourism, the Alps.[5]

Four aspects of the effects of climate change will be raised. First, the chapter will look at ways climate has always shaped tourism in the Alps. A popular destination since the end of the eighteenth century under the influence of Romanticism, the Alps subsequently exerted a real fascination on countless tourists.[6] Over the years, the region has undergone a very high level of transformation in terms of the infrastructure needed to accommodate the growing number of

DOI: 10.4324/9781003478089-11

visitors. This has given rise to significant tensions between tourism promotion and climate protection. The second aspect will focus on the impact of climate change on Alpine regions. Specifically, what are the effects of climate, biology, and nature on mountain environments? And what are their effects on the tourist activities that rely on the mountains for leisure purposes? The third part will use a number of examples to show how global and local actors are adapting. The adaptation measures relate to technical aspects (the use of snow cannons) or strategic aspects (the tendency to give the summer season back its predominance). Far from being a sticking plaster on a bad leg, these measures are designed to give the illusion that anything is still possible for tourism to adjust to a changing context. The final section will analyse the activities now on offer that have taken into account the impact of climate change on the Alpine environment. These are new practices linked to new visions of the mountains. The focus will be on what is now known as "ecotourism" or "sustainable tourism" which is becoming increasingly popular with customers looking for something different from what conventional tourism can offer.

In addition to the financial issues that have until now governed the use of the mountains, there are now more fundamental choices to be made about the relationship that human beings want to have with their environment. This article should help to grasp the most important of these.

10.2 Alpine Tourism and its Features

From a historical perspective, it is now recognised that the tourism industry is a very vulnerable sector. As the manager of the Beau-Rivage Palace hotel in Lausanne put it in March 1939, "The slightest rumour from a political, health, economic, or simply fashion source makes travellers change their plans, and a place that one year was packed is empty the following year."[7] This is certainly one of its important characteristics: tourism is defined as an activity that is extremely exposed to any conjunctural variation, be it economic, political, meteorological, cultural, or social. For example, World War I, the economic crisis of the 1930s, the war in Iraq in the 1990s, the terrorist attacks in Tunisia or Paris in 2015, the social demonstrations of the *gilets jaunes* in France in 2018–2019, the loss of attraction for a resort that does not follow cultural fashions or is locked into an outdated model or more recently the Covid-19 pandemic reveal this extreme fragility that can prove disastrous for a particular destination. The effects are immediate: quickly, the flow of travellers stops, hotels empty, beaches are deserted, and museums close.[8]

In most of these historical examples, the actors acted in the same way and each time showed an unsuspected capacity for resilience whatever their function: promoters, local population, consumers, and observers. Let's take the case of winter sports. From the 1880s onwards and under the influence of the hygiene movements, exposure to the sun became a medical and therapeutic fact that many doctors proposed. These new recommendations had unprecedented consequences for tourism activities. Until this period, the sun was shunned by the elites who could afford to go on holiday or tourist trips, especially by upper- and middle-class women for whom a tanned complexion was inadvisable for social reasons. Hence, for these social classes winter was spent near the sea—especially on the Côte d'Azur—where tourists did not suffer from the heat. In summer they moved to the mountain resorts where they could enjoy the coolness of the altitude and avoid suffocating from the intense temperatures.[9]

Based on healthy towns and countryside, good eating habits, and physical exercise and sport, the emergence of hygienic theories led to a change in the perception of the body. Hygienists claimed that, far from being a factor of social distinction—a tanned complexion proves your origin—nature was a therapy. The sun not only helped the ecosystem—nature, plants, and so on—to live and grow, it also contained the best ingredients to make people stronger, healthier, and more resistant. It healed and contributed to the health of the body.[10] Combined with the freshness of the sea or the dry air of the mountains, it helped to make bodies stronger. This craze had a considerable impact on the seasonality of activities. Contrary to the habits of the previous period, from the 1880s summer became the season for the seaside and winter for the mountains. This reversal should be considered as a determining factor in understanding the tourism as well as the sporting expansions that marked the whole of the twentieth century and the transformation of these spaces into open places of well-being.

The emergence of winter sports reflects how mountain venues came to be presented as desirable destinations in the cold season. Skating, curling, sliding on sledges, bobsleds, or skis became characteristic activities of the Alpine areas. Leisure activities for the rich, certainly, but these carefree activities—not always pleasant in the cold or beneficial in the face of numerous falls—hide the more general transformations of winter tourism. Beyond the social and financial restriction, these activities more broadly delimited a new dynamic that gave a configuration to tourism that was not as visible before. A transition took place at the turn of the nineteenth to twentieth centuries in which multiple factors became newly significant.[11] So many were significant, that I will simply list them here, but their contribution will be clarified as the chapter progresses.

Cultural factors: new perception of the body: hedonism, tanning, self-representation which required new clothes and outfits for winter;

Technical factors: motorisation/mechanisation of mobility: bicycles, motorbikes, cars, aeroplanes, and so on, facilitating individual and collective travel to the resorts, these tools complemented the infrastructure providing access to the snowfields: mountain railways, cable cars, ski lifts, and so on;

Political factors: politicisation/ideologisation: state support (for various reasons—democratisation, popular tourism, social control, nationalism);

Medical and sanitary factors: hygiene, emergence of sanatoriums: "the good air of the Alps";

Educational factors: boarding schools, public schools, new pedagogical doctrines; media factors: coverage/spectacularisation—Olympic Games, world championships, and so on;

Economic factors: impetus in the marketing of activities in transport, accommodation (hotels, hostels, and so on), services (banking), the construction sector, equipment manufacturers, and so on;

Entrepreneurial factors: extension of seasonality, increase in profitability.

This is not to say that Alpine tourism has not changed since the beginning of the twentieth century.[12] As a number of studies have shown, the *laissez faire-laissez aller* has to be set against a changing context. So-called "modernity" has often been seen, when associated with the mountains, as forms of transgression.[13] Economically and socially, the mountains suffer from an indisputable judgement. A place of isolation, backwardness, ignorance, and conservatism, of exile, it would only accumulate the handicaps likely to leave it far behind the progress of the world below. Making it affordable means imposing standards from elsewhere. On the other hand, it is a forerunner in experimenting with novel forms of collective management of common goods and practices of governance commonly described as democratic, anticipating evocations that make it properly "magical." The emergence of sanatoria and winter sports also responds, as if in echo, to this valorisation of mountain areas. The impact of the various environmental "aggressions" that are multiplying with the physical transformation of the landscape and the densification of tourism, leading today to over-exploitation, is transforming the frames of reference of the Alpine populations. Often not content to unload the heavy burden represented of tradition and authenticity, they use new terms to make people believe that, despite all the shifts, nothing has changed.[14]

But what has unquestioningly changed is what is required of the mountain. Research clearly shows this reversal. Pascal Gutknecht and Jon Mathieu reveal the slow but gradual disappearance of the term *sublime* (a term that in English can be traced back to Alpine crossing by the poet William Wordsworth in the 1790s) in the language of mountain lovers." This decline corresponded to the emergence of what other researchers have called the "utilitarian" phase of the Alps, which have become solely objects of pleasure and leisure since the 1990s.[15] In order to better market these uses of the mountains, advice and practical proposals expelled the traditional emotional expressions that had previously made this space an object of wonder.

Another indicator of change is provided by the important issue of land management. Building a tourist system historically implied detecting when and how the land moved from agro-pastoral management to touristic real estate management. This transition is more precisely linked, in the long run, to the construction of secondary residences. For example, in Champéry in the Valais Alps in Switzerland, from the middle of the nineteenth century onwards, this tourism resort underwent a classic redevelopment with an increasing number of hotels. Further redeployment is also visible in the second half of the twentieth century, especially in connection with this new type of accommodation and the osmosis between the economic and political spheres of the locality. This characteristic of the tourist start-up tended to disappear in the fifties. The local tourism model gave way to a market phase, with land and property management "acquiring a higher exchange value than its use value." This is not to say that the transformation was driven by *laissez-faire laissez-aller* alone. On the contrary, the dynamics of the transactions reveal an attachment to certain heritage values regarding the landscape. As Luigi Lorenzetti and Daniela Delmenico point out, "the sales choices of the group of the ten most active surnames on the land market tend to limit the transfer of land (as well as real estate) ownership into the hands of non-domiciled buyers, without however preventing the transition of the village to residential tourism."[16] Thus, neither fatalism nor conservatism can be observed in the modes of regulation tested in Champéry. In other words, openness and lock-in are intertwined in the way the land issue is managed.

This example cannot, of course, be generalised. Yet the selling of land has been particularly intense in other regions of the Swiss Alps. The takeover of several Swiss resorts by the American multinational Vail Resorts shed a harsh light on the ambitions of its promoters to encourage mountain sports again and again, particularly in winter. With sales set to reach $2.9 billion by 2022, and ownership of three dozen ski resorts in the United States, one in Canada and three in Australia, the American company has spent 118.5 million Swiss francs on the Crans-Montana facilities, buying the 24 ski lifts, chairlifts, and gondolas that provide access to the 140 kilometres of slopes, as well as four shops that rent out equipment, a ski school, and 11 restaurants. The purchase price for another ski resort in Andermatt-Sedrun was 149 million Swiss francs.[17] The climax was reached in October 2023 when a newspaper broadcast video showing how the construction of a new ski run was being carried out. Several of the shovels were seen attacking a glacier near Zermatt to enable the creation of this slope dedicated to a World Cup ski race. Described as an "ecological aberration" by the press and denounced by a number of competitors and observers of the sporting world, this initiative took on the air of a veritable denial of the problems raised by climate change.[18]

But it is important to avoid a one-sided view of these trends. The diversity of trajectories must be taken as a fact. Above all, considering diversity leads to a better contextualisation of theoretical notions forged in the 1980s by geographers, notably Richard Butler, who developed the concept of the life cycle of a tourist resort over the long term.[19] Much criticised for its latent determinism, this theory maintains that any tourist destination goes through "a phase of exploration and discovery, of local involvement and training, of development, of consolidation, of stagnation, and ends with a phase of decline or benefits from a phase of regeneration, also known as renewal."[20] While the biophysical, psychosocial, and equipment load capacities condition the evolution of the cycle, the regeneration phase can be marked by the arrival of investors from outside the resort who ensure a new economic profile, as we see with the example of Crans-Montana and Andermatt.

10.3 Alpine Tourism and Climate Change

It is in this broad context of Alpine tourism that the effects of climate change must be accounted for. It is not surprising that climate change is emerging as a very disturbing factor. But there is a big difference between what we can see in the long term and what is currently defined as climate change. While cyclical disturbances were considered to be periodic and accidental disorders, they did not challenge the basic principles: wait it out then carry on with business as usual. This sector can no longer be conceived in these terms. Climate change brings into question the very purpose of tourism. Skiing without a brick of snow, bathing in tornadoes, walking in an unsettled nature are enough to show the inanity of the models that previously made tourism successful. In a 2022 article Daniel Scott and Stefan Gössling show the huge impact of climate change on tourism research. Based on an analysis of 1,290 articles, they provide a chronological overview of the evolution of the literature on climate change and tourism over the 35 years since the first journal publication on the topic in the mid-1980s. They also explore some of the key events that have influenced progress and areas of interest within the literature. The number of publications prior to 2000 is very limited and represents only 1.6% of the publications in this analysis. Tourism was not even mentioned in the First Assessment Report of the Intergovernmental Panel on Climate Change (IPCC) in 1990. The growth in publications began after the Third Assessment Report (AR3) in 2001 and the Djerba Declaration on Tourism and Climate Change in 2003, the first study of its kind in the tourism sector.[21]

As impressive as this scientific finding is, the critical question is whether the surge in publications has had a real impact on tourism policies. It is striking to note that, from the 1990s to the 2020s, tourist flows continued to grow. As mentioned earlier, it was only the Covid-19 pandemic that put an end to what was seen as a never-ending boom. The recovery from the pandemic took place under difficult conditions. Coupled with the war in Ukraine and the economic consequences that followed, it is not surprising that the cessation of tourist movements is no longer considered out of the ordinary. What is remarkable, however, is that climate change seems to have little impact on the evolution of the tourist industry. A smokescreen seems to be obscuring the reality of the issues raised by climate change, contrary to the warnings in the scientific literature. Are the developers and other sectors of tourism blinded to such an extent by the profit motive? Or do we need to adopt a different approach to identify possible shifts in these same policies? Could the impressive increase in tourist traffic be hiding more subtle changes in the capacity of the actors (promoters as well as consumers) to change habits? To answer these questions, we probably need to move away from heavy macroeconomic trends and refocus the analysis on specific and finer points of tourism activities.

The issue of tourism is part of a more general context: it is related to the assessment of the effects on the Alps themselves from a geological, biological, and environmental point of view as well as on their population from a social and demographic perspective. Climate change is a fact since the end of the twentieth century; however, the intensity of its consequences remains uncertain.[22] This uncertainty might be hiding the depth of the effects of climate change on Alpine regions. Tourism is only one element among others and is as much affected by climate change as it is by the existence of a very fragile ecosystem. Taking Alpine tourism as our case study, the reported consequences of the impact of climate change are clear. Far from hiding in denial, the Alpine resorts are facing fundamental changes in the reality of tourism.[23] Climate change means different things. In summer, the shrinking of the permafrost in conjunction with the disappearance of the glaciers presents serious dangers for walkers, whether they are mountaineers in search of summits or simple hikers on the prepared paths. In winter, the lack of snow makes it very difficult to maintain a season where sports, and especially skiing, can be practised over a long period. Such uncertainties lead winter sports enthusiasts to choose other destinations higher in altitude (see Section 10.4). But in invoking these consequences, we need to be more precise.

The consequences are clearly affecting the practicability of ski areas. Of course, temperature variations have different impacts depending on the altitude of the stations and the exposure of the slopes. Snow cover is not the same everywhere. But, on a global level, forecasts confirm the long-term fears about the viability of ski areas. The following table illustrates this situation (data from 2006).[24]

Based on figures available in 2007, Table 10.1 estimates how many resorts can still function with an average temperature increase of 1, 2 and 4 degrees Celsius in the different Alpine countries in the future. While an increase of 2 degrees does not affect most resorts—two-thirds of them are still viable—an increase of 4 degrees will cause a real disaster. Of the 666 resorts studied, only 202 would still be suitable for skiing. Switzerland is the country that still manages best with a reduced loss of just over half, while Austria can only count on the existence of 20% of the existing stations. The potential annual costs of climate change in Switzerland, considering losses in winter tourism, have been calculated to be 2.3 to 3.2 billion Swiss francs by 2050, that is 0.6% to 0.8% of the Swiss GDP for 1995.[25]

What do such figures mean in terms of attendance? Consider the number of people who take for the first time a ski lift in Switzerland and are transported to the summit in winter (see Figure 10.1).[26] This does not include people who stay in Switzerland but do not use these means of transport. However, it does give a better idea of the users, especially in winter, the

TABLE 10.1 Number of naturally snow-reliable ski areas in the European Alps under present and future climate conditions (national level)

Country	*Number of ski areas*	*Number of ski areas with current snow conditions.*	*+1° C (=2025) Number of ski areas*	*+2° C (=2050) Number of ski areas*	*+4° C (=2100) Number of ski areas*
Austria	228	199	153	115	47
France	148	143	123	96	55
Germany	39	27	11	5	1
Italy	87	81	71	59	21
Switzerland	164	159	142	129	78
Total	666	609	500	404	202

Source: Remontées mécaniques suisses (RMS), 14.

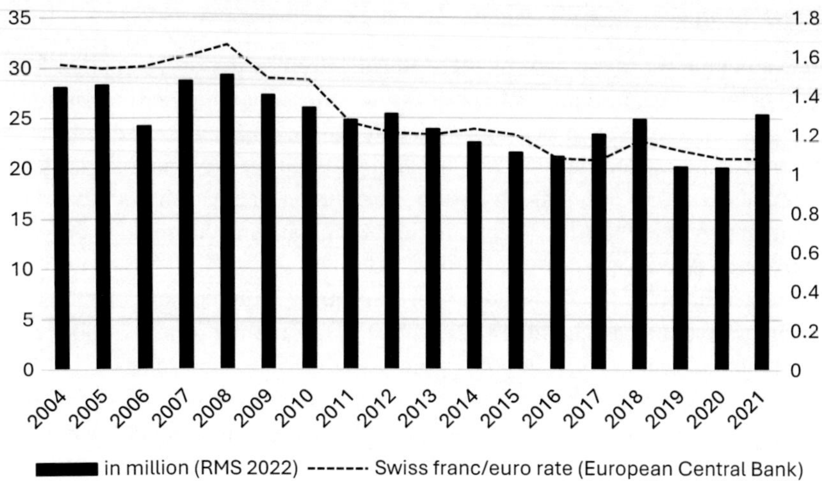

FIGURE 10.1 First winter passages in Switzerland

Source: Remontées mécaniques suisses (RMS),14.

vast majority of whom being skiers. In the winter of 2021–2022, the Swiss ski areas recorded around 25.4 million first-time visitors. The following graph relates these data to the exchange rate of the euro during the period under consideration. Several insights can be gained from this data. The first is the downward trend in the number of users of ski lifts. The second is that this downward trend is partly dependent on the euro exchange rate. For the Swiss clientele, staying in the country was certainly the cheapest thing to do, given the rate of the euro at the beginning of 2005. With the weakening of the euro, however, it became more advantageous, in the medium term, to go to neighbouring countries to benefit from very attractive prices. Consequently, it can be noted that during the same period, the number of customers coming from abroad tends to decrease. But does this mean that climate change has no impact? From the interests of the ski lifts, the erasure of any climatic impact and the insistence on the rate of the euro suggest reasons for the decrease in users. A third conclusion can be drawn: the difficulty for companies linked to Swiss tourism to admit the climate factor. This is an element to which I will return later.

10.4 Alpine Tourism and Adaptations

According to the literature, there are four main types of technological adaptations to climate change in the Alpine tourism industry: (1) landscaping and slope development, (2) moving to higher altitudes and north facing slopes, (3) glacier skiing and (4) artificial snowmaking. And there is a variety of different types of behavioural adaptations, ranging from operational practices and financial tools to new business models and a move towards the diversification of activities. A detailed description of the various adaptation measures can be found in various studies.[27] This section focuses on the use of artificial snow as an effective means of climate diversion, allowing a better understanding of the links between technology and tourism on the one hand and cost/benefit on the other. It will also look at new business models that aim to provide year-round tourism.

10.4.1 *Artificial Snow*

A snow gun is a device for making snow mechanically from water and air, all at low temperatures (0°C and below). The principle is to project a mixture of compressed air and water in sufficiently cold weather. While the processes are technically more or less efficient, artificial snow requires a lot of water and electricity, which has a negative impact on biodiversity. A study carried out by the International Commission for the Protection of the Alps has shown that the 3,000 or so snow cannons used in the Alps over a single season consume as much energy as a city the size of Basel over an entire year.[28]

In 1963, Switzerland was the first country in Europe to acquire snow cannons. They were installed at Chalet-à-Gobet in the heights of Lausanne. It was mainly from the 2000s onwards that the installations spread thanks to an improvement in the devices as we can see in Figure 10.2.[29] In winter 2021/22, Switzerland had around 22500 hectares of slopes, of which some 12100 could be technically snow-covered. This represented 54% of the total slope surface.

This phenomenon is, of course, not limited to the Swiss Alps. It is massively present and visible in all Alpine countries and even at intensities that go beyond the Swiss case. Hundreds of millions of euros have also been invested in snowmaking equipment throughout the Alpine countries. In France, for instance, almost half a billion euros were spent between 1990 and 2004.[30] Table 10.2 shows the share of the ski area using mechanical means for the production of snow in the Alpine countries at the end of the 2000s.[31]

If in Switzerland, artificial snow covered 56% of the ski area in the winter of 2022–23, in Austria, a major competitor in winter sports, the figure is 70%, and even 90% in South Tyrol. It seems that, for the near future, the resorts will also have to continue to equip themselves.

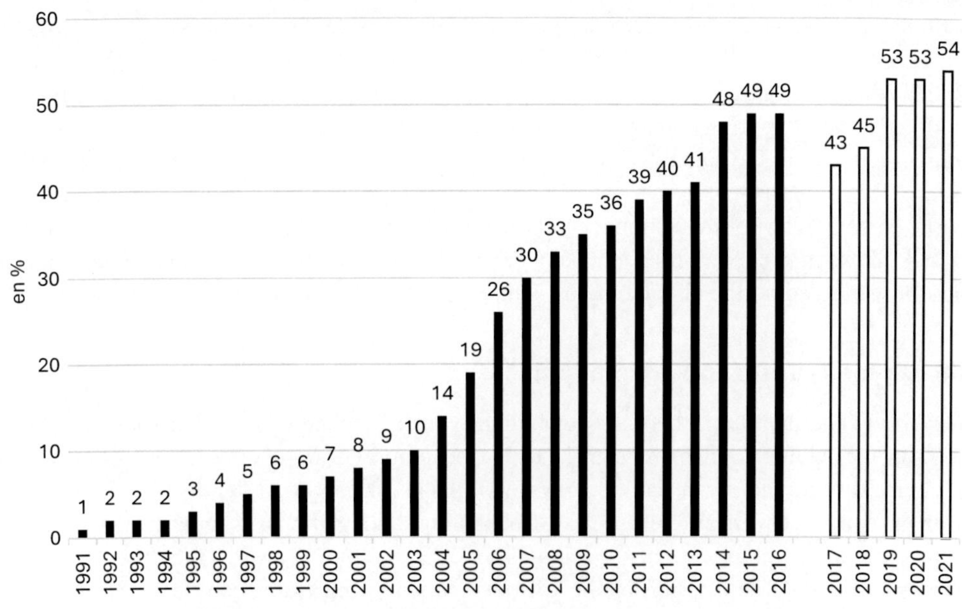

The proportion of the slope surface that could technically be snow-covered up to 2016 is shown with the black columns. A new calculation method has been adopted since 2017 based on a sample shown with the columns without the black fill.

FIGURE 10.2 Artificial snow cover in Switzerland (in % of available hectares of slopes)

Source: Remontées mécaniques suisses, 10.

TABLE 10.2 Extent of snowmaking in the Alpine countries (2008–2010)

	Slope surface	*Area suitable for artificial snow*	*In %*	*Sources*
Switzerland	22000 ha	7920 ha	36 %	Swiss ski lifts (2010)
Austria	25400 ha	16760 ha	66 %	Austrian Cable Car Association (2009)
Germany (Bavaria)	3700 ha	599 ha	16 %	Bavarian State Office for the Environment (2009)
Italy	22500 ha	15750 ha	70 %	Swiss ski lifts (2008)
France	25000 ha	5300 ha	21 %	Badré et alii (2009)
Slovenia	2100 ha	900 ha	75 %	Slovenian Tourist Board (2008)
Alpine countries	99938 ha	47289 ha	47 %	

Source: Badré M., Prime J.L., Ribière G., 2009.—" Neige de culture: Etat des lieux et impacts environnementaux." Note socio-économique. République Française. Ministère de l'Ecologie, de l'Energie, du Développement durable et de l'Aménagement du territoire. http://www.cpepesc.org/IMG/pdf/neige_de_culture_eta.pdf.

According to a poll, the industry wants to invest several tens of millions of francs in snowmaking systems in Switzerland from 2024–100 million in total. The Arosa-Lenzerheide ski lifts are at the top of the list. Thirty million francs are to be invested in the snowmaking infrastructure there over the next five years.[32] In the Leysin and Col des Mosses resorts in the canton of Vaud, more than 170 snow cannons are being planned at a total cost of CHF 30 million.[33]

Although this proposed installation has recently provoked violent reactions from environmental activists, the development of snowmaking systems is above all hampered by financial barriers. One kilometre of slope costs one million Swiss francs. To ensure a sufficient supply of water, the construction of new reservoirs, as in Lenzerheide, is particularly expensive. These massive investments require public aid. Under strict conditions, the Swiss Confederation could issue loans at reduced rates, not to mention the aid provided in certain tourist cantons such as Valais, Grisons, or Berne by their governments who consider the situation serious enough to intervene in a sector considered essential to their economy.[34]

The economic dimension does not stop at installation costs. It is certain that at higher temperatures, larger quantities of snow will be required, which will inevitably increase costs. Moreover, if ambient temperatures exceed a certain threshold, snowmaking will no longer be viable. In addition, the use and costs of snowmaking will not be evenly distributed throughout the season, as some periods such as the beginning of the season will require more snowmaking. Overall, the costs are considerable in terms of water consumption, energy demand, landscape, and ecology. The sum of the externalities raises the question of the real benefits of these facilities, which are not only environmentally but also economically questionable.[35]

These few developments suffice to demonstrate that artificial snowmaking is a technically, economically, and ecologically fragile response to climate change. Nevertheless, it remains the favoured response for many resorts ready to invest considerable sums to ensure ski conditions that will satisfy a clientele that is always eager for speed and snow.

10.4.2 A Year-round Season

Winter tourism is the main source of income for most mountain resorts.[36] The reliance on this economic model is fraught with dangers due to current climate variability that has already led to snow-poor winters and the potential further impacts of projected climate change. This perilous dependence has led developers and resort managers to look for alternatives, including year-round tourism. This comprises mainly summer tourism, including the intervening shoulder season, but also climate- and weather-independent services such as congress tourism, educational tourism, and health tourism. The objectives of this strategy of alternatives are to sustain the activity and to reduce the dependence of the resorts on snow.[37]

Alpine ski resorts will thus have to plan and implement adaptation strategies if they are to survive.[38] One of these strategies could be to make better use of the summer season. Gaëlle Serquet quantifies the observed impact of hot summer air temperature on tourism demand in the Swiss Alps by analysing the relationship between temperature and overnight stays at 40 Alpine resorts.[39]

Tests were carried out to establish whether there was a link with summer tourism. The results show that high temperature values have an impact at lower elevations on the number of overnight stays made by domestic tourists in the Swiss Alps: 24 resorts revealed a correlation between the number of hot days at lower elevations and the number of overnight stays in the mountains. Summer stays in Alpine resorts may be favoured by warmer temperatures at lower altitudes. If these observations are confirmed, mountain resorts could once again become attractive during the summer months.

In the Swiss Alps, there is a clear correlation between domestic tourist demand and higher summer temperatures, especially in June and, to a lesser extent, in August. The number of nights spent in the mountain resorts and the warm temperatures at lower altitudes, where most domestic tourists come from, appear to be interdependent. Alpine resorts closest to major cities were more popular and showed stronger correlations with high summer temperatures, which may be explained by the fact that responses to heat events occur in the short term and that nearby mountain resorts are chosen accordingly.

It seems likely that mountain hotels will welcome national tourists willing to react to hot days more frequently or for longer periods, especially if the frequency and intensity of heat waves increase.

This means that summer tourism in the Alps could benefit from climate change. Summers in the Mediterranean could become too hot, while Northern Europe, including the Alps, would offer a more pleasant climate. This would reverse the above-mentioned cyclical changes in seasonality. Philippe Bourdeau has clearly shown these changing configurations in the following diagram.[40]

As Figure 10.3 suggests, the Alps could be witnessing another historical shift in destinations. The Alpine arc was a favourite summer destination throughout the nineteenth century, but the attraction of the sea and seaside holidays, with the social and medical benefits of exposure to the sun, has had an adverse effect. The relative desertion of the mountains during the summer was reduced by the emergence of winter sports, which ensured the arrival of an increasingly large and regular clientele from the beginning of the twentieth century. Although winter has been the season for the mountains for a century, climatic changes have reignited interest in summer stays. Fleeing the stifling heat of the lower altitudes, fearing for their health, particularly in relation to the increase in skin cancers due to prolonged exposure to the sun at too high a temperature, tourists return to the altitudes in summer and take advantage of the milder temperatures. On the other hand, the melting of the snow due to high temperatures is jeopardising winter sports activities, at least for resorts below 2000 metres. As the argument given earlier suggests, the massive use of snow cannons in some places is not enough to make the ski slopes viable. The summer season is once again becoming the "queen" season of the mountains.

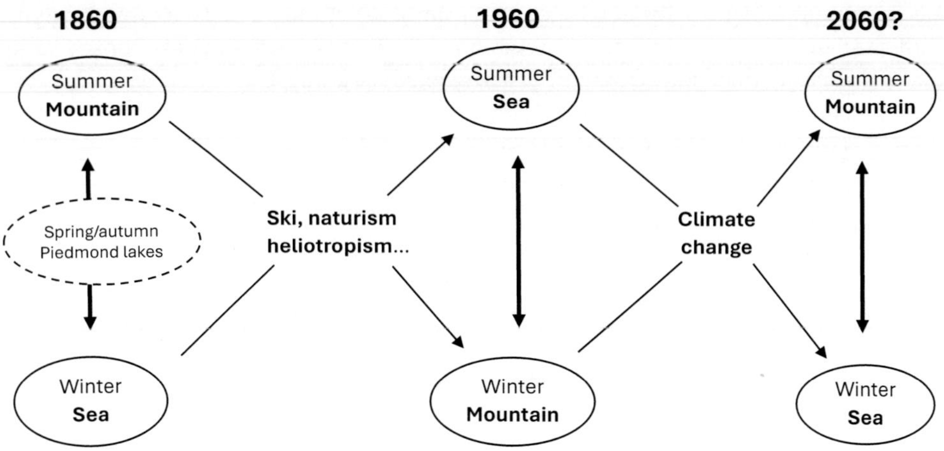

FIGURE 10.3 Towards a new seasonal reversal of tourism polarities?

Source: Bourdeau, Philippe. "De l'après-ski à l'après-tourisme, une figure de transition pour les Alpes?", *Journal of Alpine Research* | Revue de géographie alpine [En ligne], 97-3 | 2009—http://journals.openedition.org/rga/1049.

10.5 Ecotourism: A New Tourism?

The promotion of what is known as "soft tourism" or "ecotourism" is now present on all the websites of Alpine resorts. Many travel guides also devote special publications to these issues. The famous *Guide du Routard* and *Lonely Planet*, to take two examples of public authority in the world of tourism, have turned attention, in one way or another—specific volumes, extensive realisations on websites—to the possibility of travelling "correctly" with plenty of addresses for "responsible" or "sustainable" travel.[41] Presented as the ideal solution to the threats of climate change without damaging the economic equilibrium of the regions concerned, new activities—some old but brought up to date, others highly original in their conception—are rubbing shoulders with, but not necessarily wiping out, the so-called "polluting" activities.[42] If "greenwashing" is a weapon shamelessly used to avoid disappointing customers who are supposedly conservative in their choices and activities, and above all promoters who are sensitive to their remuneration, the substance of offers that respect ecosystems is based on a reality that is far from utopian. There is a great deal of serious thought being given to this subject, which needs to be analysed accordingly and without preconceived ideas. The scope of so-called "environmentally-friendly" measures has given rise to controversy, even among those who defend them. But all in all, they form a matrix of possibilities that have already been put into practice.

First of all, we need to define what we mean by these concepts, which are still highly adaptable. The concept of ecotourism has been the subject of numerous studies, resulting in definitions that vary from one area to another. Without going into all the details, we can use the definition given by the World Tourism Organisation as part of the preparations for the International Year of Ecotourism in 2002:[43]

While there is no universal definition of ecotourism, its general characteristics can be summarised as follows:

1. Ecotourism includes all forms of tourism close to nature, in which the tourists' main motivation is to observe and appreciate nature and traditional cultures in natural areas.
2. It includes educational and explanatory features.

3. The organisation of trips for small groups is mostly, but not always, the responsibility of small specialised local companies. Foreign organisers of varying sizes also organise, supervise or market ecotourism excursions, usually for small groups.
4. Ecotourism minimises negative impacts on the natural and socio-cultural environment.
5. It supports the protection of natural areas by:
   - providing an economic benefit to the local authorities concerned, to organisations and to authorities that administer natural areas for protection purposes,
   - creating jobs and alternative sources of income for local communities,
   - raising awareness among the local population and tourists of the need to protect natural and cultural assets.

During a survey carried out by Christophe Clivaz and Alexandre Savioz in Chamonix on the perception of climate change and the melting of the glaciers, one participant stated that "[p]roposing alternative or sustainable tourism is one thing but proposing an efficient and profitable model is another. Forcing us in our economic development when no one in any other sector is trying would be absurd for the region's economy and would be useless in view of the global situation. As we know today, the curves will probably never change or at least not for several hundred years." As we can see from these comments, the issues are complicated, with good intentions coming up against understandable socio-economic concerns.[44]

From a historical point of view, this perspective is not new. Protecting the environment and promoting the well-being of the local population go hand in hand with the idea that tourism should not be a bulldozer crushing everything in its path. In this sense, responsible travel in natural regions has been a perceptible trend since the end of the nineteenth century. The work of mountaineers, but also of the artists (painters, writers) not to mention the local population, has been significant in this respect. In the Swiss Alps, there are many examples of resolute opposition to the construction of mountain railways and funiculars, or hydroelectric power stations. The most emblematic case was the outcry following plans to build a railway from Zermatt to the summit of the Matterhorn.[45] While the preservation of a national identity is a key element in these oppositions at a time when societies are subject to major socio-technical transformations, it also involves the preservation of natural sites and, more broadly, the environment.[46] In this case, building a railway to the summit of the Matterhorn would have touched the very foundation of Switzerland, because it attacked the very essence of its mountainous identity. Strongly tinged with conservative, even reactionary, ideology, this argument was repeated many times and led to the abandonment of the project in 1905. The writer Gonzague de Reynold played a major role.[47] These outcries were also strongly echoed by a group of painters known from 1891 as the Ecole de Savièse, after the village in the canton of Valais where they met. But beneath this artistic and ideological argument lie convictions in which the protection of the environment from the damage caused by tourism is perceptible.

Another case highlights the challenge to the philosophy of tourism at all costs. From the 1970's, opposition to the construction of an altiport above the resort of Verbier in the canton of Valais sums up the rise of ecological awareness in the contemporary sense of the term against the economic ambitions of local elites shaped by the ideology of everything is possible everywhere. This event can be described as a turning point. The initiative was taken by Téléverbier, the company in charge of cable car and ski lift achievements in the resort of Verbier. Located at an altitude of 2,200 m, the installation was to be used for aircraft landing and taking off, the idea being to provide the Valais resort with a major promotional asset. It was opposed by a number of environmental groups led by a journalist, Franz Weber, who went on to become one of the leaders of ecological causes. From being a short-term issue, it became a structural one. Strictly speaking, it's not the impact of climate change that is driving the debate, but the event has attracted so much media attention that no further tourist adjustment is conceivable without analysing the environmental consequences it would generate.[48] Ecotourism is therefore part of a broad-based movement that is leading to the emergence of new attitudes to Alpine tourism and, as a result, new activities.

10.5.1 *A New Way of Exploring the Mountain*

What is meant by "sustainable activities"? Whilst hiking—in summer or winter with snow-shoes—mountain biking, and thermal baths are being stepped up considerably, a number of original activities are being achieved in parallel. New passions are emerging, especially for all that the mountains have to offer.[49] In this respect, the profession of "mountain leader" deserves a closer look. In France, the profession was officially founded in 1976 although the activity is already visible at the end of the 1960s.

The term "mountain leader" refers to an educational and cultural approach to the mountains under the guidance of a specialised instructor. The approach evolved without recourse to the techniques and tools of mountaineering. Rather, it allows tourists to discover the mountains using different methods of progression relying on activities that respect the ecosystems. The process wants to invest in a better knowledge of an area through geology, botany, natural sciences, zoology, history, heritage, botany, food, forestry, biology, and so on. It includes observation by walking, meditation, cooking with plants, organic or creating an herbarium. It seeks, among other things, to encourage wolf sightings by managing guard dogs. Overall, it seeks to promote a better understanding of the natural heritage and human environments of the mountains.

This approach quickly took on an international dimension, first in Europe and then worldwide. The Commission Européenne des Accompagnateurs en Montagne (CEMM) was created in 1989 on the initiative of representatives of national and professional organisations from a handful of countries. Fifteen years later, it led to the foundation of the Union of International Mountain Leader Associations (UIMLA), which is headquartered in Villars-sur-Ollon, Switzerland. Harmonisation of the content and level of training is the main objective. The UIMLA's mission is to:

> Make Mountain Leaders throughout the world aware of the existence of UIMLA and encourage the forming of National Associations.
>
> Work towards cultural, social and legal recognition of the profession of International Mountain Leader.
>
> Harmonise member countries' training and assessment in terms of minimum requirements and quality.
>
> Encourage and bring value to a sustainable development of mountainous regions.
>
> Encourage and agree the right of an IML to practise in all mountainous countries.
>
> Develop the visibility and professional reputation of the profession.[50]

In Switzerland, training was only launched in 1996 at the Gîte de St-Jean in the Val d'Anniviers, on the private initiative of a doctor. This delay was undoubtedly linked to resistance from mountaineering circles, particularly high-mountain guides and local people who did not see the point of such a specialisation. The success of the training programme broke down this initial resistance. Now, the course is aimed at future mountain professionals who wish to train for the Swiss Federal Mountain Leader Diploma. The course is based on the profession's charter of professional activities and the regulations and directives of a federal diploma. It is recognised by the State Secretariat for Education, Research and Innovation (SEFRI) as a preparatory course for the federal mountain leader certificate. It is certified by the UIMLA.[51] The training combines varied types of knowledge to equip future mountain professionals with the skills they need to bring the mountain to life and pass on all its "secrets" to others, without the need to climb it to the top at all costs. Apart from being a response to climate shift, this rise in the mid-range mountains is also understood to be a cure for modern ills: stress, noise, speed, agitation, hyperactivity, and so on. Highlighting the virtues of the mountains is also part of the "answer to everything" that they are supposed to provide, as some participants have pointed out.

10.5.2 What Economic Impacts?

Assessing the economic impact of all these adaptations to climate and in particular ecotourism remains a difficult exercise. Beyond the well-founded forecasts and euphoric projections of the companies involved in what is proving to be a new source of revenue, there is a lack of long-term data. As already noted, everyone involved in tourism now seems determined to move in a sustainable direction. Official and private annual reports appear as proof of their endeavours to implement the measures taken or to be taken. But it is hard to be sure what this really means. Between greenwashing and genuine commitment, the nuances blur and sometimes the two become confused. To take the example of the installation of new snow cannons in Arosa/Lenzerheide or Leysin/Les Mosses, the websites of these resorts make no distinction between their environmental commitments and their artificial snow initiatives. In many cases, economic profit is not the goal, with quality taking precedence over quantity. The approach also includes an anti-capitalist stance on a mountain used as a mere economic commodity.

The first problem is that the available statistics include all types of tourists. It is impossible to determine the proportion of those who have adopted a "sustainable" approach, conforming to environmentally friendly practices in their travel, accommodation, catering and entertainment, and those who, while praising the benefits of sustainable tourism, still continue their usual behaviours. This also means that it is impossible to measure the real impact on the economy in figures.

According to opinion surveys carried out among tourists, nearly 70% of those questioned indicated an interest in, if not a preference for, ecotourism-related activities.[52] This does not mean that these same people are actually physically involved in such activities. Based on surveys by American organisations, ecotourism generated $157.76 billion in 2021 and $185.43 billion in 2022. And according to the latest projections, this figure should approach $300 billion in 2026.[53]

All these figures are based on extrapolations that are difficult to prove. Yet, despite the lack of more precise statistical data, qualitative data does exist showing the ability of certain mountain resorts to firmly establish their tourism strategy in line with official recommendations. The example of Golm Silvretta Lünersee Tourismus GmbH is very relevant. In 2018, its resort was named Austria's first climate-neutral tourism region. It is a medium-sized tourism company in Vorarlberg Montafon, employing up to 250 people. It manages three areas—Golm, Silvretta, and Lünersee—with a total of 13 ski lifts, a ski sports school, four restaurants with 2,000 seats and one hotel, as well as the High Alpine Road. For a number of years now, the resort's achievement strategy has been based on paying particular attention to the quality of life of its residents and visitors through a series of concrete actions from regional products to an ingenious drainage system for snowmaking or a photovoltaic chairlift. The measures are outlined in a strategy document and verified each year.[54]

In a report on the challenges of sustainable tourism in the canton of Valais, where tourism plays a central role, the Valais Tourism Observatory also makes some interesting observations through a series of interviews with people directly involved in promoting ecotourism and case studies of ski lift companies and hotels. The measures range from making electric cars available to guests staying in the resort to concrete actions that hoteliers have undertaken to implement their goals—selling local foods in restaurants, returning the deposit on the ski pass to the guest, and selling ski passes whose price is automatically increased by a financial contribution, the amount of which is reinvested in a local project that is fairly climate-neutral. But some of those interviewed made no secret of the fact that it would be difficult, particularly for hoteliers, to be more proactive. The Swiss hotel industry's umbrella organisation, Hotelleriesuisse, admits that money is an obstacle to implementing more daring measures. "We have to show hoteliers that they can start with small things that don't cost too much. You don't have to start by changing the whole heating system. With food waste, for example, a lot can already be done," says Hotelleriesuisse's specialist in organisational achievement.[55]

The growth in these environmentally friendly activities is unquestionable, even if it is still difficult to measure their effects. But the balance remains fragile. In some regions, the success of winter hiking, whether on skis or snowshoes, is causing disturbance to wildlife. This is prompting the public authorities to set up quiet zones and regulate these leisure-related flows.[56] Switzerland now has almost 800 restricted quiet zones. You can only travel in these zones on designated winter routes. There are just under 300 zones recommended by the cantons under the Federal Law on Hunting and the Protection of Wild Mammals and Birds (LChP) which have no legal status but are intended to raise public awareness. Finally, 43 federal free zones protect threatened fauna and biotopes, where human activities are regulated at all times of year. In other words, ecotourism is not seen as an activity that can be developed endlessly. The profitability of economic aspect clashes head-on with the sustainability aspect.

10.6 Conclusion

It has become commonplace to accept the importance of climate change in daily lives. Tourism is no exception to what is now taken for granted. It is strongly dependent on climatic conditions, with "good" weather influencing the willingness to stay at the seaside and in the mountains. If the seasonal character has always marked tourist flows, the trends have not been constant, with seaside tourism and mountain tourism competing for preferences according to the relationship with the body, medical concerns, sporting aspirations, the enjoyment of well-being, self-image, and the like. In the Alps, these tourism models explain the long-term development of an activity that has shown, over the last two hundred years, a resilience that may seem surprising. Wars, economic crises, pandemics, accidents, and terrorist acts, however intense these often-tragic events may have been, have not overturned the desire of populations to escape the constraints of daily life by staying somewhere else.

With climate change visible since the end of the twentieth century, the rupture is deeper, and the transition is no longer based on waiting for the bad events to pass and for everything to return to the way it was before. The effects are dramatic from many points of view: ecological, geological, biological, environmental, but also social, economic, and cultural. The urgency appears to require concrete action based on new approaches.

From the point of view of tourism operators—promoters, investors, entrepreneurs, and the like—the evidence is less perceptible. Economic interests still monopolise actions, even if speeches are coated with an ecological varnish about what is happening. If the inevitability of climate change remains questionable, it is not obliterated. The revival of alpine summer tourism supports the idea that winter tourism, with the insufficiency and scarcity of snow cover, cannot be confined to the hopes placed in the creation of artificial snow. As technology, however sophisticated, costly, and massive it may be, has not provided an appropriate solution to these developments, a new paradigm has been imposed on the fields of action. The tourist industry has advocated a gradual adaptation, supported by a clientele addicted to the old models. On the other hand, the scientific community insists on considering new tourism models that can adapt to the new conditions in the world. Does this mean that mountain tourism is on the way out? It is perhaps premature to say so. There is plenty to choose from in some regions. More and more consumers are asking for it, and the tourist industry is aware that it has to listen to them. It is also up to the tourists and their ability to change their habits and choices to respond.

A balance between adaptation to climate change and economic profitability has yet to be found. The tourism operators are under no illusions as to the challenges they face. Whether they like it or not, mountain tourism must adapt to these new conditions.

Notes

1 For a convincing summary of the history of climate change and the impacts on tourism, see Pfister, *Climate and Society in Europe.*
2 United Nations, International tourism from January to July 2022 returns to 60% of its pre-pandemic level, posted on September 28, 2022, 2.
3 Becken, *Tourism and Climate Change*. The podcast that the World Economic Forum recently devoted to this issue is well worth a listen: Shin, *Rising Global Temperatures.*
4 Globe-Trotting, "Tourisme International 2022."
5 By Alps, we mean the western part of the Alps, which form an arc between the Mediterranean Sea and the Valais, and the central Alps, between eastern Switzerland and western Austria. We will not deal specifically with the eastern and southern Alps.
6 Reichler, *La découverte des Alpes.*
7 Archives Beau-Rivage Palace (ABR), "Report by the Board of Directors."
8 Tissot, "Storia del turismo."
9 Boyer, *L'hiver dans le Midi.*
10 Granger, *Les corps d'été*; Ory, *L'invention du bronzage.*
11 Quin, *Le ski en Suisse.*
12 Boyer, "Les Alpes et le tourisme," 19–20.
13 Tissot, "La montagne: territoire du moderne?," 8–9.
14 Crettaz, *La Beauté du reste*, 63–64.
15 Gutknecht und Mathieu, "Erhabene Berge."
16 For both quotations, Lorenzetti, "La transition touristique," 10.
17 Stern, "Globasierung des Wintertourismus," 4.
18 Imsand-Curty, "Descentes de Zermatt."
19 Butler, "The Origins of the Tourism Area Life Circle."
20 Suchet," Pour en finir," 8.
21 Scott, "A Review of Research into Tourism."
22 Einhorn, "Changements climatiques."
23 Clivaz, *Tourisme d'hiver.*
24 Abegg, "Climate Change and Tourism," 73–80.
25 Meier, *Sozioökonomische Aspekte.*

26 Remontées mécaniques suisses (RMS), *Faits et chiffres de*, 14.
27 Abegg, "Climate Change Impacts," 25–60. See also Njoroge, "Climate Change," 95–108.
28 CIPRA, *Künstliche Beschneiung.*
29 Remontées mécaniques suisses, *Faits et chiffres de*, 10.
30 Abegg, "Climate Change Impacts," 42.
31 CIPRA, *Le tourisme face au changement climatique.*
32 Schmidt, "Enneigement artificiel."
33 https://www.rts.ch.
34 https://www.admin.ch.
35 Rixen, "Does Artificial Snow Production," 219–30.
36 Humair, "Le tourisme comme moteur," 9–54 and Tissot, "La Suisse en tant que destination touristique," 585–99.
37 For a comparative approach, see Guex, *Tourisme, mobilités.*
38 Elsasser, "Climate Change," 253–57 and Elsasser, "The Vulnerability."
39 Serquet, *Changement climatique.*
40 Bourdeau, "De l'après-ski à l'après-tourisme."
41 Le guide du routard, *Tourisme durable* and Lonely Planet, "50 Years of Travel."
42 Lemaistre, "Année internationale de l'écotourisme."
43 Baumgartner, *De l'écotourisme au tourisme.*
44 Clivaz, "Recul des glaciers."
45 Dénoréaz, "Un funiculaire au sommet du Cervin."
46 Le Dinh, *Le Heimatschutz.*
47 Clavien, *Les Helvétistes.*
48 Tissot, *La Suisse se découvre*, 212–17.
49 For a convincing discussion, see Reichwein, "Nature Sport."
50 https://uimla.org/mission/.
51 Anniviers Formation, "Ecole de St. Jean."
52 *Médiaterre Website.*
53 *Ecotourism Global Market Report.*
54 *Golm der Erlebnisberg im Montafon.*
55 Observatoire valaisan du tourisme, "Tourisme Durable."
56 Terre et Nature, *L'Hebdo de la.*

Bibliography

Abegg, Bruno, Shardul Agrawala, F. Crick, and Anne de Montfalcon. "Climate Change Impacts and Adaptations in Winter Tourism." In *Climate Change in the European Alps*, edited by S. Agrawala. Paris: OECD Publishing, 2007.

Abegg, Bruno, Rolf Bürki, and Hans Elsasser. "Climate Change and Tourism in the Alps." In *Managing Alpine Future*, edited by A. Borsdorf, J. Stötter, and E. Veulliet. Wien: Verlag der Österreichischen Akademie der Wissenschaften, 2008. https://www.admin.ch/gov/fr/accueil/documentation/communiques.msg-id-2839.html.

Anniviers Formation. "Ecole de St. Jean." *Programme de la formation*, 2023. https://www.anniviersformation.ch/formation/accompagnateur-en-montagne/.

Archives Beau-Rivage Palace (ABR). "Report by the Board of Directors." March 13, 1939.

Baumgartner, Christian. *De l'écotourisme au tourisme durable dans les Alpes. Un rapport de synthèse.* Zurich: CIPRA, 2002. file:///D:/Colloque%20Zurich/Rapport_ecotourisme_F.pdf.

Becken, S., and J.E. Hay. *Tourism and Climate Change: Risks and Opportunities. Climate Change, Economies, and Society*. Clevedon, UK and Buffalo: Channel View Publications, 2007.

Bourdeau, Philippe. "De l'après-ski à l'après-tourisme, une figure de transition pour les Alpes?" *Journal of Alpine Research | Revue de géographie alpine* [En ligne] 97, no. 3 (2009). http://journals.openedition.org/rga/1049.

Boyer, Marc. *L'hiver dans le Midi*. La Tour d'Aigues: L'Aube, 2002.

Boyer, Marc. "Les Alpes et le tourisme." *Histoire des Alpes: Tourisme et changements culturels* 9 (2009): 19–30.

Butler, Richard W. "The Origins of the Tourism Area Life Circle." In *The Tourism Area Life Circle*, edited by R.W. Butler. Clevedon: Channel View Publications, 2006.

CIPRA. *Künstliche Beschneiung im Alpenraum: Ein Hintergrundbericht.* Schaan, LIE: CIPRA International, 2004. https://www.cipra.org/en/dossiers/11/dateien/454_de/@@download/file/Dossier_Kunstschnee_D.pdf?inline=true.

CIPRA. *Le tourisme face au changement climatique: Un rapport de synthèse de la CIPRA.* Schaan, LIE: CIPRA International, 2011. file:///D:/Colloque%20Zurich/110223f_compactTourismus.pdf.

Clavien, Alain. *Les Helvétistes: Intellectuels et politique en Suisse romande au début du siècle.* Lausanne: Ed. d'en bas, 1993.

Clivaz, Christophe, C. Gonseth, and C. Matasci. *Tourisme d'hiver: Le défi climatique.* Lausanne: Presses polytechniques et universitaires romandes, 2015.

Clivaz, Christophe, and Alexandre Savioz. " Recul des glaciers et appréhension des changements climatiques par les acteurs touristiques locaux: Le cas de Chamonix-Mont-Blanc dans les Alpes françaises." *Via* 18 (2020). http://journals.openedition.org/viatourism/6066.

Crettaz, Bernard. *La Beauté du reste: Confessions d'un conservateur de musée sur la perfection et l'enfermement de la Suisse et des Alpes.* Genève: Zoé, 1993.

Dénoréaz, Alice. " Un funiculaire au sommet du Cervin? Modernisation et identité nationale (1890–1914)." *Annales Valaisannes* (2012): 71–124.

Ecotourism Global Market Report 2022. https://www.businesswire.com/news/home/20221116005696/en/Ecotourism-Global-Market-Report-2022-ResearchAndMarkets.com#:~:text=The%20global%20ecotourism%20market%20is,at%20a%20CAGR%20of%2012.7%25.

Einhorn, B., N. Eckert, C. Chaix, L. Ravanel, P. Deline, M. Gardent, and P. Schoeneich. "Changements climatiques et risques naturels dans les Alpes: Impacts observés et potentiels sur les systèmes physiques et socio-économiques." *Revue de géographie alpine* 103, no. 2 (2015): 1–33.

Elsasser, Hans. "The Vulnerability of the Snow Industry in the Swiss Alps." *Mountain Research and Development* 21 (November 2001): 335–39. https://www.researchgate.net/publication/232679364_The_Vulnerability_of_the_Snow_Industry_in_the_Swiss_Alps.

Elsasser, Hans, and Rolf Bürki. "Climate Change as a Threat to Tourism in the Alps." *Climate Research* 20, no. 3 (2002): 253–57. https://www.researchgate.net/publication/242557516_Climate_change_as_a_threat_to_tourism_in_the_Alps.

Globe-Trotting. "Tourisme International 2022: Chiffres Clés, tendances et statistiques." 2023. https://www.globe-trotting.com/post/tourisme-international-2023.

Golm der Erlebnisberg im Montafon. 2023. https://www.golm.at/de/Nachhaltigkeit.

Granger, Christophe. *Les corps d'été: Naissance d'une variation saisonnière, XXe siècle.* Paris: Autrement, 2009.

Guex, Delphine. *Tourisme, mobilités et développement régional dans les Alpes Suisses: Mise en scène et valeur territoriale. Montreux, Finhaut et Zermatt du XIXe siècle à nos jours*. Neuchâtel: Alphil, 2016.

Gutknecht, Pascal, and Jon Mathieu. " 'Erhabene Berge': Eine korpuslinguistische Studie zu den Periodika des Schweizer Alpen-Clubs von 1864 bis 2014." *Histoire des Alpes* 20 (2020): 215–34.

Humair, Cédric. "Le tourisme comme moteur du développement socioéconomique et vecteur du rayonnement international de la Suisse (XIXe–XXe siècles)." In *Le tourisme suisse et son rayonnement international*, directed by Cédric Humair and Laurent Tissot, 9–54. Lausanne: Antipodes, 2011.

Imsand-Curty, Ugo, "Descentes de Zermatt: une aberration écologiques." *24Heures*, October 10, 2023. https://www.24heures.ch/glacier-creuse-a-la-pelleteuse-descentes-de-zermatt-une-aberration-ecologique-463440560288.html.

Le Dinh, Diana. *Le Heimatschutz, une ligue pour la beauté: Esthétique et conscience culturelle au début du siècle en Suisse*. Lausanne: Section d'histoire de la Faculté des Lettres, 1992.

Le guide du routard. *Tourisme durable: Plein d'adresses pour voyage responsable.* Paris: Hachette, 2010–2011.

Lemaistre, Philippe. "Année internationale de l'écotourisme: Un premier bilan." *Téoros* 21, no. 3 (2002). Connected January 28, 2024. http://journals.openedition.org/teoros/3274.

Lonely Planet. "50 Years of Travel." *Sustainable Travel*, 2023. https://www.lonelyplanet.com/articles/category/sustainable-travel.

Lorenzetti, Luigi, and Daniela Delmenico. "La transition touristique d'une 'Station-Village': Multipositionnalité et gestion foncière à Champéry (Valais), 1870–1970." *Journal of Alpine Research/Revue de géographie alpine* 105, no. 4 (2017). https://doi.org/10.4000/rga.3643.

Médiaterre Website at the World Summit on Sustainable Development or Johannesburg Summit, held from August 26–September 4, 2002. https://www.mediaterre.org/actu,20030425115721,1.html.

Meier, Ruedi. *Sozioökonomische Aspekte von Klimaänderungen und Naturkatastrophen in der Schweiz.* Zürich: Schlussbericht NFP 31, 1998.

Njoroge, Joseph M. "Climate Change and Tourism Adaptation: Literature Review." *Tourism and Hospitality Management* 21, no. 1 (2015): 95–108.

Observatoire valaisan du tourisme. "Tourisme Durable: Rapport sur les enjeux du tourisme durable dans le canton du Valais, automne-hiver 2021." file:///D:/Colloque%20Zurich/kona_tourobs_tourisme-durable_fr.pdf.

Ory, Pascal. *L'invention du bronzage. Essai d'une histoire culturelle.* Paris: Flammarion, 2018.

Pfister, Christian, and Heinz Wanner. *Climate and Society in Europe: The Last Thousands Years.* Berne: Haupt, 2021.

Quin, Grégory, Laurent Tissot, and Jean-Philippe Leresche, dirs. *Le ski en Suisse: Une histoire.* Orbe/Colombier: Château & Attinger, 2023.

Reichler, Claude. *La découverte des Alpes et la question du paysage.* Genève: Georg, 2002.

Reichwein, P.A., P.O. Schut, and G. Quin. "Nature Sport and Environmental History: Adulation or Alteration of Nature." *Sport History Review* 55 (2024): 1–12.

Remontées mécaniques suisses. *Faits et chiffres de la branche des remontées mécaniques suisses 2022.* Berne: Remontées Mécaniques Suisses, 2022. file:///D:/Colloque%20Zurich/221010_Fakten_Zahlen_2022_f.pdf.

Rixen, Christian, Veronika Stoeckli, and Walter Ammann. "Does Artificial Snow Production Affect Soil and Vegetation of Ski Pistes? A Review." *Perspectives in Plant Ecology, Evolution and Systematics* 5 (2003): 219–230. http://www.ekoarhiv.bg/system/files/documents/Rixen_2003_Perspectives-in-Plant-Ecology%2C-Evolution-and-Systematics_0.pdf.

RTS. https://www.rts.ch/info/regions/vaud/14181988-un-ambitieux-projet-de-canons-a-neige-a-leysin-et-aux-mosses-suscite-de-vives-oppositions.html.

Schmidt, Martin. "Enneigement artificiel: Les remontées mécaniques vont investir des millions." *Blick*, March 13, 2023. https://www.blick.ch/fr/news/suisse/pretes-a-tout-pour-lor-blanc-les-remontees-mecaniques-veulent-investir-100-millions-pour-senneiger-id18394188.html.

Scott, Daniel, and Stefan Gössling. "A Review of Research into Tourism and Climate Change—Launching the Annals of Tourism Research Curated Collection on Tourism and Climate Change." *Annals of Tourism Research* 95 (July 2022). https://www.sciencedirect.com/science/article/abs/pii/S0160738322000603?via%3Dihub.

Serquet, Gaëlle. "Changement climatique dans les régions de montagnes suisses: Une analyse de variables climatiques en lien avec le tourisme." Thèse de doctorat, Université de Neuchâtel, 2011. file:///D:/Colloque%20Zurich/Th%C3%A8se%20Serquet.pdf.

Shin Yan. "Rising Global Temperatures Are Already Affecting the Tourism Industry—Here's How." https://www.weforum.org/agenda/2023/08/temperatures-tourism-climate-impact/.

Stern, Daniel. "Globasierung des Wintertourismus: Kapitalakkumulation per Sessellift." *WOZ* 49 (December 7, 2023): 4.

Suchet, André. " Pour en finir avec le concept de Butler (1980) et son modèle d'évolution des destinations touristiques. Le cycle de vie comme un concept inadapté à l'étude d'une aire géographique." *Loisir et société/Leisure and Society* (2015): 7–19.

Terre et Nature. *L'Hebdo de la vie au vert.* Terre et Nature, 2024.

Tissot, Laurent. "Storia del turismo e storia economica: Considerazioni metodologiche ed epistemologiche." In *Tourismus und Entwicklung im Alpenraum, 18-20.Jh*, hg. von Andrea Leonardi und Hans Heiss, 23–41. Innsbruck: StudienVerlag, 2003.

Tissot, Laurent. "La Suisse en tant que destination touristique." In *Histoire économique de la Suisse au XXe siècle*, directed by P. Halbeisen, M. Müller, and B. Veyrassat, 585–99. Neuchâtel: Alphil, 2021.

Tissot, Laurent. *La Suisse se découvre: Trois siècles de tourisme en question (1730 à nos jours).* Neuchâtel: Alphil, 2023.

Tissot, Laurent, Christophe Jaccoud, and Grégory Quin. "La montagne: territoire du moderne?" *Les Sports Modernes* 1 (2023): 8–9.

Union of International Mountain Leader Associations. https://uimla.org/mission/.

United Nations—World Tourism Organisation (UNWTO). "Full Recovery of Tourism in Sight." *Barometer*, September 28, 2022: 2. https://www.unwto.org/fr/taxonomy/term/347#:~:text=Le%20moment%20est%20venu%20%C3%A0,humanit%C3%A9%20et%20pour%20la%20plan%C3%A8te.%20%C2%BB&text=On%20estime%20%C3%A0%20474%20millions,la%20m%C3%AAme%20p%C3%A9riode%20en%202021.

11

CLIMATE CHANGE AND THE INSURANCE INDUSTRY

The Long Road to the UNEP 1995 Insurance Statement[1]

Niels Viggo Haueter

11.1 Introduction

Climate change impacts insurance in four areas: underwriting, asset management, corporate citizenship, and insurers' own footprints. Underwriting is impacted by increasing losses from natural catastrophes such as storms, floods, droughts, and wildfires (referred to as "natcats" in the industry) which are well documented.[2] While the increase in financial losses is mainly due to continuously increasing concentration of values with rising population and industrialisation density, climate change exacerbates this trend. Climate change also impacts life and health business through, among other factors, "extreme heat, air pollution, and increased exposure to infectious diseases spread by non-human vectors."[3] Furthermore, the transition to renewable energies has brought about a new class of risks that still need to be better understood to design appropriate insurance products.

In addition to underwritten risks, insurers are exposed to transition risks on invested assets.[4] On the other hand, insurers as institutional investors can facilitate the transition to renewable energies. Such investments not only support renewables financially but, as insurers are seen as risk-conscious investors, can also serve as incentives for other investors.

Insurance companies' direct environmental impact is felt in areas such as business travel and facilities management. Measures to reduce carbon footprints may extend to incentivising staff to be environmentally conscious with, for example, public transport subsidies or sponsoring staff activities in green projects. Other areas that link the insurance industry with climate change are seen in the industry promoting risk management and adaptive measures,[5] raising awareness through public debates,[6] and political action through lobbying.[7]

Responses from the industry vary. Haufler identified European insurers as being more responsive to climate change than U.S. insurers.[8] Haueter and Jones see mostly large insurance companies actively communicating on climate change issues since the 1990s.[9] According to Mills, business responses to climate change vary across the sector and can include increasing premiums, reducing exposure to climate-related risks or entirely withdrawing from the market, buying more reinsurance, increasing reliance on the public sector, investing in risk monitoring and prevention, participating in climate research, supporting shifts towards renewable energies, and

DOI: 10.4324/9781003478089-12

becoming politically active.[10] Insurers can further choose to reduce or divest their investment in and underwriting of environmentally detrimental business, a trend that has only recently seen some momentum.[11] Investment strategies differ considerably. Brogi et al, for example, find that in the United States, "larger, more profitable, and more solvent insurance companies show the highest level of ESG awareness."[12]

While fewer today debate the fact that climate change is happening, some uncertainty remains as to the extent of future temperature rise. With the 1995 Intergovernmental Panel on Climate Change (IPCC) report, decision making under uncertainty became a broadly discussed topic.[13] For insurance, this was not new as the business is based on dealing with uncertainty. As Franklin W. Nutter, President of the Reinsurance Association of America (RAA) put it, the "uncertainties associated with climate change are probably the insurance industry's best reason for taking this seriously."[14] Warren Buffett in his 2006 letter to shareholders expressed it as follows

> Was this onslaught of more frequent and more intense storms merely an anomaly? Or was it caused by changes in climate, water temperature or other variables we don't fully understand? [We] don't know the answer. . . . What we do know is that our ignorance means we must follow the course prescribed by Pascal in his famous wager about the existence of God. As you may recall, he concluded that since he didn't know the answer, his personal gain/loss ratio dictated an affirmative conclusion.[15]

Climate risks are not new. The industry acted on such risks long before modern climate change debates started. What changed with global warming was a rather gradual than abrupt shift from monitoring the weather to climate variability and then to acknowledging climate change. The uncertainties associated with climate change are inherent in weather uncertainty, which already early on provided sufficient reason for the industry to act. In the 1990s, climate change did, though, shift re/insurers' public agendas making them engage in public and even political debates. The decision to communicate already in the 1980s about climate change was mostly driven by concern over insurability but also by difficulties with adjusting prices.[16]

This chapter follows the discussions about weather and climate from the first climate change theories in insurance in 1955 to the 1995 UNEP Statement of Environmental Commitment by the Insurance Industry (UNEP Insurance Statement). A first part deals with early concerns in the United States about the insurability of natural catastrophes and the possible role of climate change as well as the start of risk modelling techniques. A second part traces the emerging interest of European insurers in climate change during the 1970s back to broader environmental liability concerns and looks at the role of the first climate research centre at the University of East Anglia in combining academic and private industry efforts. A third part investigates the background for industry-wide co-ordinated efforts and the emerging role of reinsurers and large insurers in climate debates and conferences.

11.1.1 Review of Literature on Insurance and Climate Change

Literature on insurance and climate change is abundant. The best overviews and discussions of literature so far are by Starominski-Uehara et al. and Bellman et al.[17] Dobes et al. provide an historical overview of the development of the literature on economics and climate change.[18] Nobanee et al. look at the development of literature from 1986 to 2020.[19] According to the latter, insurance-related climate change research can be traced back to the mid-1980s but publishing only picked up from circa 2005 onwards. Prior to the mid-1980s, some publications dedicated to natural hazards and environmental pollution occasionally mention climate change.

It was the industry itself, or more precisely, the reinsurance industry, that first publicly communicated on climate change in the 1970s, albeit as a sub-topic to natcats and broader environmental issues.[20] Munich Re's head of Geo Risk Research, Gerhard Berz, then published an account of how the insurance and reinsurance industry was affected by rising losses due to climate change.[21] His main findings were that traditional loss statistics lost value in the face of faster-changing weather patterns, yet the slow shift left time to adapt prices. The question, Berz further argues, was how much of a price increase the direct insurance industry and ultimately policyholders were prepared to pay. In that context, Berz also pointed out that direct insurers at the time did not see the need for preparing for climate change. The one exception was the German insurer Allianz, then still closely linked to Munich Re, on whose behalf Berz published the probably first instance of direct insurance communicating on climate change.[22]

Berz, along with the less than a handful of other available papers on insurance and climate change, provided some of the background for the first IPCC reports.[23] The IPCC identified impacts of climate change on various parts of the economy and stated that insurance was "[p]erhaps the most significant of these industries."[24] Despite this, and despite the UNEP Insurance Statement in 1995, the role of insurance in climate change remained virtually unexplored in academic publishing during the 1990s. The topic was, though, taken up by Greenpeace in conjunction with the 1992 Rio Summit.[25]

Together with Greenpeace, the industry was ahead of academia in researching insurance and climate change. Andrew Dlugolecki, Chief Manager at the Scottish General Accident Fire and Life Assurance, addressed insurance implications of climate change at the Chartered Insurance Institute's 1990 conference and the 1991 Geneva Association Conference.[26] The Geneva Association, an insurance industry think tank, then approached James Rasmussen, Director of the World Weather Watch Department at the World Meteorological Organization, also based in Geneva, for "An Analysis of the Climate Change Issue."[27] Dlugolecki provided the insurance analysis for the Association, pointing to clear signals of climate impacts such as increasing frequency of events and the evidence provided by meteorological institutions.[28] With this, he aimed at contradicting the industry argument popular back then that increasing losses were exclusively due to geographical concentration of values, new insurance products, and inflation. Insurers, according to Dlugolecki, had already resorted to adapting underwriting and limiting exposures, controlling the damage, transferring more risks to reinsurers, and reviewing their pricing, albeit without explicitly acknowledging climate change. Dlugolecki was to become a constant in climate and insurance literature with his contributions to the IPCC reports[29] and later publishing for the Chartered Insurance Institute.[30] The few papers that appeared in the wake of the Geneva Association's publications and the IPCC reports focused on the problem of cost and rising insurance premiums.[31] This was largely due to a series of superstorms of the late 1980s and the 1990s, most notably Hurricane Andrew in 1992 that created the by then largest insured losses and a subsequent hard market. By the 2000s, the industry had largely accepted global warming and its consequences as facts, yet with considerable regional differences.[32] Johnson, one of the few comprehensive works on climate change and (re)insurance, researched reinsurers' ways to measure and manage the impacts of climate change and the necessity of uncertainty for insurance.[33]

The literature reviews of Starominski-Uehara et al. and Bellman et al.[34] point to the following sub-disciplines of climate change having emerged in the twenty-first century as separate research fields: impact of increasing losses from natcats, difficulties in insuring climate change-related disasters in developing economies, increasing public-private partnerships, and the use of alternative risk transfer solutions such as insurance-linked securities that tap capital markets for risk capacity. These topics largely reflect advances in understanding climate change through publications such as IPCC reports.

A further field has developed with political economy studies that adopt a critical approach to insurers' climate agendas.[35] In this research area, the role and impact of insurance and reinsurance in dealing with climate change led to debates about the respective responsibilities of governments, international organisations, and private industries. Some authors see the prominence of insurance, and particularly reinsurance, in what they consider a matter of state sovereignty as problematic, as it may establish a market-oriented governance in climate issues. Others see exactly the lack of governmental action and the non-binding character of international agreements as an obligation for private industries, especially international entities such as reinsurers, to assume a global role in the climate debate.[36] Collier, Elliott, and Lehtonen dedicated an issue of the journal *Economy and Society* to climate change and insurance, focusing on what they term an "economisation" of climate change through insurance, brought about, among other factors, by an increased cooperation of the private and the public sector.[37] Meanwhile, in politically right-wing quarters, insurers' climate coalitions sparked criticism, for example with the UN Net Zero Insurance Alliance that has been accused of violating anti-trust law.[38]

In the mid-2010s, the failure of climate adaptation and mitigation strategies led countries most affected to call for alternative responses. The United Nations Framework Convention on Climate Change (UNFCCC) addressed this in 2013 by creating the Warsaw International Mechanism (WIM) to deal with what has been dubbed "loss and damage" in vulnerable countries.[39] This triggered ample research into the role and limits of insurance in dealing with climate change in low-income countries.[40]

Historical accounts of insurance and climate change are still scarce. Kohl (forthcoming) looks at how top management in the reinsurance industry has dealt with climate change since the 1970s. Two books look at the role of climate considerations in specific risk contexts and show how basic structures to deal with climate variability were put in place long before global warming became a topic: Rohland shows how climatic considerations in fire risks were common practice in reinsurance already in the nineteenth century.[41] Mauelshagen researched the impact of increasing hail damages in the 1920s and 1930s on reinsurers' underwriting strategies.[42]

11.2 Weather or Climatic Variability?

Weather and climate awareness is arguably highest in agriculture, where so-called "weather insurance" included cover against windstorm, hail, rain, or frost.[43] Hail damage, especially, was difficult to insure due to the covariate nature of the risk (referring to the uncertainty of time, place, and severity of the impact). To adapt to climate variability, statistics of hail damage were recorded in the early nineteenth century, long before the advent of meteorological institutions in the 1850s. Underwriting relied on such loss histories to attempt realistic future pricing. In the mid-nineteenth century, hail insurance associations in France started creating risk zones based on historical loss data to justify differing premiums and control exposure due to the accumulation of risks in certain areas.[44] This kind of risk zoning was to become essential when dealing with climate related risks. Life insurance proceeded in a similar way as insurers in the nineteenth century relied on ample medical research into the impact of climate on health and created risk zones according to favourable or detrimental climates. Life policies would often contain exclusions or charge extra premiums for "climate risks" or what was considered "unusual hazard to life arising from residence in foreign climates."[45]

Already in the 1920s, meteorological observations prompted climatological reflections. A continuous increase in hail damage led the Swiss Reinsurance Company (Swiss Re) to temporarily abandon this line of business as it pondered if "the underlying causes are to be sought in certain climatic changes."[46] The observation was remarkable both for providing the probably earliest quote on insurance and climate change and even more so as an example of decision making under conditions of climate uncertainty, a dilemma that has earmarked climate change debates up to the present.

The reflections were purely based on the company's own observations from loss statistics. Science, at the time, did not provide much help in understanding climatic variability, let alone prepare for its negative impact. Throughout the late nineteenth and early twentieth centuries, some research into climate change had been done, most prominently by Svante Arrhenius who "discovered" anthropogenic climate change. But many scientists deemed warming to have positive outcomes as it might increase agricultural yields, make colder regions inhabitable and open the Arctic to maritime transport.[47] Research into human action causing the climate to change was scarce until Guy Stewart Callendar, a British steam engineer and amateur researcher on climate, collected global temperature data and attempted to prove that carbon-dioxide emissions caused a rise of worldwide temperatures.[48] His theory was discussed but largely dismissed in meteorological circles.[49]

During World War II, meteorology became important for military air force and naval purposes.[50] This provided a boost to research, often supported by governmental aids. After the war, the Massachusetts Institute of Technology (MIT) endeavoured to capitalise on such research by making meteorology appealing to private industries.[51] In 1951, the *Compendium of Meteorology*, a landmark in climate and meteorological studies, was published out of MIT under the auspices of Thomas Francis Malone.[52] Offering a detailed overview of the state of meteorology and climate studies at the time, the over 1300-page strong publication reflected the pride of a profession that had finally gained respect. The volume also reveals the ongoing interest of the U.S. Army in weather research. It had been commissioned by the American Meteorological Society under the tenure of retired Navy Officer Howard T. Orville and was supported by the U.S. Air Force Cambridge Research Center, headed by Helmut Landsberg, an MIT colleague of Malone's.

Climate change was discussed in various chapters. Opinions as to its causes differed. Charles Ernest Pelham Brooks of the Meteorological Office in London who contributed a chapter on *Geological and Historical Aspects of Climate Change*, dismissed Callendar's carbon-dioxide theory on the basis that

> during the past 7000 years there have been greater fluctuations of temperature without the intervention of man, and there seems no reason to regard the recent rise as more than a coincidence. [Callendar's] theory is not considered further.[53]

Eugen Glueckauf of the Atomic Energy Research Establishment in Harwell, however, quoted Callendar's theory as justifiable and called for more research into the matter and "experimental verification with modern techniques."[54] Research intensified during the International Geophysical Year (IGY) in 1957, the first global scientific research programme in the cold war period.

This was largely thanks to Roger Revelle, director of the Scripps Institution of Oceanography at the University of California, who helped create the IGY. He published a paper together with Hans Suess, his colleague at Scripps, that finally corroborated Callendar's findings, although with reservations as to the extent of carbon-dioxide being responsible for climate warming. Yet the paper famously stated that, with carbon-dioxide emissions, "human beings are now carrying out a large scale geophysical experiment" (Revelle and Suess, 1957: 19). Revelle's testimonies to the U.S. Congress during the 1956 hearings on the IGY (U.S. Congress: 1957) established the topic on the U.S. political agenda. It also brought catastrophes related to climate change into public limelight as *Time Magazine* reported on their research: "When all their data have been studied, they may be able to predict whether man's factory chimneys and auto exhausts will eventually cause salt water to flow in the streets of New York and London."[55]

11.3 1950s and 1960s: Travelers Insurance and the War on Weather

In the 1950s, floods were making headlines in the United States. Increasing losses had already for some years questioned the insurability of natural phenomena. A year before Revelle's testimony, insurance turned meteorological concerns to climatological worries. The Great Flood of the Kansas River in 1951 had created economic losses of between USD 1 and 2.5 billion (circa USD 8.5 to 21.5 billion in 2024 values), the latter figure roughly the equivalent of the entire U.S. property insurance industry's free assets at the time.[56] After another devastating flood in the Northeastern States, the Senate in 1955 held hearings on national flood insurance.[57] There, the American Insurance Association (AIA) officially declared flood to be uninsurable.[58] Partly, this declaration was influenced by a report that the AIA had commissioned to estimate the impact of floods on the economy.[59] The AIA presented preliminary results to the U.S. Senate. It noted that, since 1938, U.S. storms had been considerably more frequent than in the previous hundred years. There were two possible explanations for this. Either it was an unusual simultaneous occurrence of factors or there "have been changes in general climatic or atmospheric conditions in recent years that tend to increase the frequency and intensity of storms of this type along the North Atlantic seaboard."[60] Although the AIA stressed that such "climatic changes" were not likely to be felt within the coming 50 or 100 years, it cautioned that, if the theory of climatic change was justified, "any estimates of flood probability based on previous records would have to be revised to include the effect of the assumed changes in meteorological conditions."[61]

At the time, probability calculations and loss estimates for weather events relied on extrapolating historical loss data. But recent experience, as the AIA pointed out, suggested that neither estimating impact nor frequency of floods was possible anymore. Adding climate change to the equation was actuarially not feasible. An alternative was seen in improving weather forecasting that would allow for better loss prevention with storms and floods. Yet meteorological instruments at the time were limited. Hurricane alerts could be sent out only 48 to 36 hours in advance and failed to reach remote areas.[62] Weather satellites that gradually allowed for large-scale weather information and insights into changing weather patterns were only launched in the 1960s.

Hartford-based Travelers Insurance decided to take matters into their own hands. In 1954, they founded the *Travelers Weather Research Center* (TWRC), the first privately owned research institute for weather studies.[63] As its first director, they recruited Thomas F. Malone, editor of the aforementioned compendium. What predestined him for the job was his experience with teaching the naval and air forces weather forecasting during the war as well as his

MIT courses on industrial meteorology, a term for what later came to be called private sector meteorology. Travelers Insurance's decision to investigate weather and climate first hand was very much a rational one, yet also an experiment in dealing with uncertainty. Backward-looking statistics provided incontestable, albeit often unreliable evidence. Switching to forecasting with assumed factors implied a radical shift in risk management. This led to developing loss scenarios and simulations, allowing science, even if uncertain, to be put to practical use in underwriting by adjusting premiums to probable and estimated losses. Malone was joined by his earlier student at MIT, Don. G. Friedman with whom he had conducted research into weather forecasting. Together, they produced the first risk modelling techniques. Friedman "adopted a modular approach that represented separately the hazard process [. . .] and the damage producing process."[64] Loss data were put into a different perspective in combination with the severity of the weather event. This allowed to create scenarios or loss simulation. Weather variability thus triggered the development of risk modelling, a technique now ubiquitous in the insurance industry and essential for dealing with climate risks. However, in their infancy the models were not yet reliable. As the many Senate Hearings show, the damage caused by the unusually high number of floods in the 1950s and 1960s painfully demonstrated the deficiencies and triggered more political debates.

Yet, U.S. politics had by then discovered a seemingly more promising route to deal with natural catastrophes and the climate. Under the heading "experimental meteorology," Malone's Compendium pointed to research that suggested that the weather could be modified.[65] The first experiments with cloud seeding were carried out in 1946.[66] Weather modification was of immense interest to the insurance industry. But it required vast resources to research. Travelers imagined that interesting government organisations such as the CIA might secure such funds. In 1960, a TWRC report was delivered to the Deputy Director of the CIA, General Charles P. Cabell, summarising the results of a meeting between Malone and the CIA's Research Chairman, Admiral Luis de Florez.[67] It suggested that the centre's research might be able to change the climate with a range of measures such as altering surface albedo by covering polar areas with soot. It also promoted weather engineering as a modern way of warfare. At the request of Admiral Florez, Malone sent the report directly to President Roosevelt (CIA, 1960b). Congress' interest surged.[68]

At the 1966 and 1967 Weather Modification Senate Hearings, Malone explained the connection with the insurance industry. He said that his company's interest in saving on losses got him involved, arguing that of the average USD 300 million in damages paid due to weather events per year, about USD 187 million might be saved.[69] This corresponded to roughly three times the annual costs of the suggested bill to co-ordinate weather research on a federal level. Avoiding drought was one of the ambitious goals. The Hearings quote French research into artificial rain sponsored by the insurance industry.[70] Quoting billion-dollar figures that could be saved in areas such as agriculture, water resources management, or transportation, provided the bills with further economic reason.[71]

Meteorologists and weather or cloud modifiers relied heavily on data collected by hail insurers and the Crop Hail Insurance Actuarial Association to analyse occurrence of hail according to topography or geographical spread and size of hailstorms. Such research in turn served to justify modification experiments in hail-exposed areas. Weather engineering proponents promised nothing less than "the elimination of such weather hazards as hail, tornadoes, hurricanes, and floods."[72] As insurers tended to shun away from extremely hail-prone areas or resorted to pricing risks out of affordability for farmers, the latter joined the combative mood of weather engineers. Some endeavours by farmers aimed at deliberate and permanent man-made climate change. Mr. Ostenberg, "farmer, rancher, and cattle feeder" from Scottsbluff Nebraska, declared that the cloud seeding airplanes of the Valley Hail Suppression Inc., a joint operation by farmers and businesspeople, had finally solved the problem of weather and climate aberrations.[73]

Yet they hadn't. Some doubts as to the safety of such endeavours had arisen within the TWRC (by then renamed Travelers Research Center, TRC). Malone's colleagues Robert Hendrick and Don Friedman investigated adverse impacts of weather modification on insurance. They remarked that, in theory, entirely eliminating the risk would also eliminate the need for weather insurance, although this was not very likely.[74] But an even bigger problem was identified by the TRC. Weather modification blurred the distinction between natural and man-made catastrophes. If, for example, a storm was diverted from an urban to a rural area, "the hazards become in part, man-made, or politically made, and new bases for rating, underwriting and claim adjustment might be required."[75] This had legal implications, both nationally and internationally. Questions arose as to who, if anyone, could own the weather and what kind of liability that implied. Maryland, for example, issued a law making all kinds of weather modification illegal.[76] All this alarmed the insurance industry even more than hail and drought losses, or the rather unlikely event of modification making weather insurance redundant. During the 1958 Hearings, liability issues were not yet addressed, even though an insurance broker had already in 1946 suggested that General Electric should take out liability insurance to be covered against accident liabilities when producing artificial snowfall.[77] In 1966 and 1967, one of the main questions was whether liability traceable to weather modification could at all be covered by insurance.

Hendrick's and Friedman's observations about natural phenomena turning man-made were of importance beyond insurance and legal questions. Some, including Malone, were concerned that weather modification could, inadvertently, turn into irreversible climate modification. The release of carbon dioxide, according to Malone, presented an "inadvertent intervention." He noted a rise in average temperature of 0.2° C since 1900, due to "particularly the ejection of submicroscopic particles of lead from automobile exhausts."[78] Cloud seeding had the potential to have equally inadvertent effects and exacerbate warming. Gordon J. F. MacDonald of the National Academy of Sciences, for example, cautioned that with "the burning of fossil fuels [. . .] an increase of temperature of similar magnitude might also result from the water vapor introduced into the high atmosphere by the planned supersonic transports" that were needed for cloud seeding.[79]

But the progress of weather modification was boosted by a more powerful argument. Russian activity during the IGY, especially the launch of Sputnik 1 in 1957, had increased fears that the Russians were technologically ahead of the United States. A lengthy report by the U.S. Department of Commerce suggested that they could already control precipitation.[80] At the 1958 Hearings, fears arose that the Russians could wilfully deprive the United States of rainfall. Strong views hence came from military quarters that saw weather engineering as a solution for the climate change problem. One of the most drastic warnings thus came from Malone's acquaintance, the former president of the American Geophysical Society and retired Navy Officer Howard T. Orville. He advocated geophysicist Joseph Kaplan's gloomy predictions that sea levels would rise over 40 feet within the course of 50 or 60 years due to "man's tremendous use of oil and gas."[81] Yet, rather than taking this as an incentive to reduce oil and gas consumption, the feasibility of using atomic bombs to defeat tornadoes and hurricanes was discussed, a measure that Orville doubted would be effective but nevertheless considered worthy of being pursued.

Military interest was also behind project *Stormfury*, started in 1962 as a joint venture of the U.S. Navy and the Weather Bureau, both of which were clients of the TRC. The aim was to develop a method for modifying hurricanes. Malone was a commission member of *Stormfury*, and in this function as well as a representative of Travelers Insurance he testified at the 1966 hearings. But by then, Malone's scepticism had grown and created a dilemma. He delivered a somewhat convoluted testimony that only partly acknowledged the need for and possible success of *Stormfury*. His later activism against global warming clearly transpired as he cautioned against the potential dangers of weather modification. The Federal programme, he argued, would have to be

> a balanced program which addresses itself to the problems of cloud physics, to hurricanes, tropical convection, hail, lightning, water budget, all conscious attempts at intervention, as well as the inadvertent modification caused by an increase of CO_2, air pollution, and rocket exhaust at high levels.[82]

His vision, he concluded, was a "world in which the benefits of weather and climate control are allocated equitably among nations." His nightmare was

> a world in which the "race to the moon" is replaced by a race to extend mastery over the atmosphere, the problem of nuclear proliferation replaced by the proliferation of environmental modification capability. A world in which scientists meet interminably in Geneva to put the genii back in the bottle. A world we do not want to see. A world that need not be,[83,84]

It took another ten years before Congress started broad-scale investigation into the changing climate. Congress' *Primer on Climatic Variation and Change* endeavoured to establish "a coordinated national program relating to climate."[85] The report dwelled on different climate theories but mostly adhered to Hubert Lamb's prediction of a coming ice age, noting that "today most experts agree that the Earth is in a cooling phase."[86] It listed agriculture, water availability, and energy consumption as the areas where the most severe impact would be felt. The concerns of insurers did not make it into the report, even though an attached *Fortune* article suggested that climate cooling was responsible for increasing droughts and floods.[87] Congress hearings in 1976 on the National Climate Act, the first U.S. climate hearings, were presided by George Edward Brown Jr., a Californian Democrat. In the opening speech, he cited opposing theories and concluded that "perhaps we should be satisfied if we can determine what additional research is needed to answer some of the most important and pressing questions.[88] Eventually, this led to the 1978 National Climate Act (Public Law 95-367) which was concerned with food production, energy use, and water resources. Insurance had made decisions to deal with climate change already when politics was still discussing its probability.

But in 1969, Travelers had started reducing its research arms, and in 1970, at the age of 53, Malone re-joined academia. With this, climate change discussions seem to have disappeared from the agenda of U.S. insurers. The AIA did not pursue climate research. The final flood report, published shortly after the 1955 Senate hearings, had dropped the references to climate change made in the preliminary version.[89] Apart from the AIA's short-lived concerns about climate change, Travelers Insurance appears to have been an unusually early and isolated pioneer.

We can only speculate as to why the topic did not cause further alarm within the industry. Scientific discord possibly played a role, although the controversy about warming versus cooling as well as about the causes of climate change most probably did not percolate down to underwriting practices. Overall, it seems that climate awareness was driven more by individuals or small groups within Travelers than by the company as such. Without Malone, there would most likely not have been any climate debate in the U.S. insurance industry. And Malone found better channels to become a climate advocate when he returned to academia. His concerns were instrumental in establishing the IPCC in 1988. In 1992, he was invited to the Rio Earth Summit as a special honorary guest.[90]

The insurance industry had no need yet to be outspoken on climate issues. The question of anthropogenic causes was to some degree irrelevant. Indeed, insurers had no pressing need to acknowledge climate change as a fact. It sufficed to realise that losses from weather events were rising, regardless of causalities. Insurance had by then achieved some means and methods to deal with climate change, even if the changes were made without directly referring to a warming climate. Risk zoning had found its way into several insurance lines allowing to avoid accumulating exposure, the nature of covariate risks was better understood, and risk modelling was in the making. Another calmative came from the government in the wake of Hurricane Betsy in 1965, the most violent U.S. storm in decades. It led to worries that future storms might go beyond combined insurance capacity, which revived political action.[91] This eventually resulted in the 1968 National Flood Insurance Program (NFIP) and took some pressure off the private industry to deal with the extreme events that climate change could cause. Perhaps most importantly, climate change was still a fringe topic and theories often inchoate. As the remainder of this chapter will argue, it took broader environmental concerns and associated liability issues to revive the discussions among insurers and reinsurers.

11.4 1980s: Reinsurance

At the start of the 1970s, some scientists saw the climate as being overall stable, others believed in long-term warming, and for some, a recent dip in temperatures signalled an overall cooling. Malone's warnings had failed to impress all scientists, some of whom by then had further developed methods for climate modification. They suggested solutions such as artificially melting polar ice caps, cutting down rainforests to create arable land, or detonating thermonuclear devices over Arctic seas to combat temperature decrease.[92] But warming theories gradually got the upper hand. Stanford's Stephen Schneider a younger, but increasingly important scientific and political voice in the debate, initially argued that the warming effect of burning fossil fuels was more than compensated by the cooling effect of aerosols, which warranted theories of a coming ice age.[93] In later writings, he admitted to miscalculations in his early theories. Overall, the scientific debate about climate change surged, not least due to new disciplines such as sociology joining in.[94] Public awareness was driven by a surge in general environmental concerns.[95] With public attention, the topic became more politicised. However, scientific controversy deprived politics of actionable arguments.

Another, less noticed, debate emerged. Should scientists go public with their concerns? Should they become politically engaged? For some, uncertainty suggested that they should not. Helmut Landsberg, one of the people who had commissioned Thomas Malone's Compendium, was sceptical about science being abused in other quarters. In 1972, he wrote that "the claims of omniscience in this field by some in other professions [. . .] such as biologists or even politicians are indeed astonishing and very misleading."[96] In 1976, he further argued that there were no signs of recently increased climate variability. Rather, "our economic (by poor planning) has become continually more vulnerable to the normal climatic fluctuations."[97] But Stephen Schneider held that

> Uncertainty does not simply mean that there are no problems, nor does it deserve a "wait and see" attitude [. . .] Nor does political action require knowledge of the exact location of each tree behind which a wolf may be hiding. Rather, knowledge of the probability that wolves do lurk in the forest should be sufficient information for deciding whether to take preventive action.[98]

Both views corresponded to insurers' respective positions. Landsberg's arguments later resonated with insurers who attributed the increase in losses to growing density of risks. Schneider's view very much corresponded to the reaction, rather than the view, of insurers—decision making under conditions of uncertainty. Schneider consequently started looking towards insurance for solutions. He and his colleague William Kellogg were not only early adopters of modelling techniques but also suggested creating a global "no-fault climate disaster insurance" that might mitigate the liabilities of climate modification or a "no-fault famine insurance" to provide food security in face of climate disasters.[99] In 1976, Schneider elaborated the problems of inactivity when confronted with uncertainty to the Senate at the National Climate Program Act Hearings and suggested that some form of disaster insurance should be adopted on an international level.[100]

From this point on, insurance gradually emerged as an important industry voice, for some decades the only industry to express concerns. But this happened at a slow pace as climate change was for some time merely a sub-topic to broader environmental issues and associated liability claims. And with some theories predicting an ice age and others a greenhouse, insurers were as much puzzled as politicians and the public. Yet, whether the earth was cooling or warming was, from an insurance technical perspective, irrelevant. For insurers it was the excessive deviation from an average norm that defined risk. In that sense, climate change in any direction was very much a textbook definition of risk. And although sea level rises of several yards did not appear imminent, storm frequency and severity had clearly been deviating from the norm as was felt with Hurricane Betsy in 1965. Natural disasters suggested that insurance should join the debate.

But insurance-related research moved from the United States to Europe. Shortly after Travelers had divested its research units, Hubert Horace Lamb, former Chair of the United Nations Meteorological Organisation, in 1972 founded Europe's first Climate Research Unit (CRU) at the University of East Anglia. Lamb had gained a reputation as *the ice man*, due to his predictions that a new ice age was on its way, although, according to East Anglia University's historiographer, the 1975/76 heat wave in Britain made him somewhat revise the cooling theory.[101] The CRU had been established with the funding of BP, Shell, and the Nuffield Foundation, a charitable trust initiated in 1943 by the founder of Morris Motors, Lord Nuffield. One project, started in 1975 and funded by Shell Exploration and the Leverhulme Trust, investigated storms in the North Sea, a subject that was also of interest to the insurance industry, as the preface noted.[102]

By the mid-seventies, the CRU was facing financial difficulties as government support was running out. Lamb then managed to get financial support from seven leading UK insurance companies that were also investigating the climate, seeking to control their weather-related losses.[103] Commercial Union, for example, in the 1970s funded a PhD at the UEA that investigated variations and changes in tropical cyclone behaviour over a century.[104] Lamb was then able to recruit Philip Jones, later director of the CRU. Jones launched the magazine *Climate Monitor* funded by an Insurance Companies' Consortium of the other six companies led by Sun

Alliance. The publication aimed at "keeping a finger on the pulse of world climate" by publishing summaries of climatic events and listings of "exceptional climatic events" along with climate articles.[105] The CRU, in turn, was provided insurance event dates, although back then, such data was somewhat inaccurate, missing the actual date by several days.[106]

Climate change, or the weather as such, did, though, not make it onto insurers' communications agendas. Nor did insurers work together as an industry. Due to heavy competition, research was mainly supported to gain competitive advantages.[107] The Commercial Union's support of a climate-related PhD was briefly reported in the news[108] but insurers did not broadly communicate about such matters. The first to go public with their concerns about global warming were the two largest reinsurers, Munich Re and Swiss Re. Reinsurers were the industry's most affected players as they more than proportionally absorbed peak risks caused by natural disasters. Munich Re, in a brochure on flood risks, hinted at climate change by cautioning that climate variabilities were most significant when investigating trends of claims experience.[109]

In the late 1970s, Swiss Re decided to inform the public about environmental issues with a publication on *Ecology and Environmental Protection*.[110] Interestingly, they did not let their natural disasters and later environmental specialist, Herbert Tiedemann, author the brochure. Instead, they commissioned Walter Schiesser, a renowned journalist writing on environmental topics for the liberal newspaper *Neue Zürcher Zeitung*. The booklet called for waste recycling, protecting rainforests, and an end to oil spills, while noting that "the increasing release of carbon dioxide in the combustion of oil, coal and natural gas [. . .] disturbed the climate."[111] Ten years later, a sequel with a similar sense of urgency was published.[112] It called for a radical change of societal behaviour and reduction of consumption. To curb carbon dioxide emissions, Swiss Re argued, fossil fuels must be reduced along with methane emissions from cattle farming and rice plantations. The alternative, the company went on, was to be found in green energies such as solar and nuclear, even if the latter came with caveats.[113]

Why would a reinsurer adopt a view that back then might count as anti-capitalist? Environmental concerns presented an unexpected but welcome opportunity for reinsurers to communicate to both clients and investors as well as society as such and explain the positive impact of their business. The grounds had been prepared by several initiatives such as UNESCO's Man and Biosphere Programme launched in 1971[114] and the 1972 Stockholm UN Conference on the Human Environment. With a preamble calling for "qualitative growth" Swiss Re's publication echoed the Club of Rome's first report, *The Limits to Growth*.[115] The book had made scepticism about unlimited growth acceptable for corporations. A sidebar that goes throughout Swiss Re's publication draws heavily on all three initial Club of Rome reports and states that "it is thanks to the Club of Rome, that our eyes have been opened to ecology."[116]

What made it necessary for the company to question growth, a phenomenon it profited from just like the overall economy? Insurance had difficulties keeping up with the fast-growing size and complexity of risks that the rapid post-war developments brought about. The industry feared that risks would grow out of insurability. Concerns about the financial impact of environmental damage had started with the Torrey Canyon oil spill disaster in 1967.[117] In the 1970s, it led to cross-industry cooperation when a group of European reinsurers published a study of *Environmental Pollution and Insurance* that questioned the insurability of environmental damage:

> To the extent that [pollution] is accepted as such and no counter-measures are taken, it represents a continuing, uncontrolled impairment of the environment, which is allowed to occur

> deliberately, or at least consciously. Therefore, risks of this sort are not fortuitous and thus do not fulfill the classic conditions of insurability.[118]

The extent and significance of Swiss Re's and the industry's worries explain why Swiss Re's Chairman Max Eisenring personally signed the introduction of the environmental report, a novelty at the time:

> It is no coincidence that this publication is issued by the Swiss Reinsurance Company: Any mention of loss or damage calls insurance to mind. But in the classical relationship of loss and insurance, a fundamental change must come about. [. . .] What is needed is the realization that, although many losses can indeed be financially mitigated by insurance, they can not as such be completely redressed. One only has to consider the chain of sea pollution—fauna—human nutrition. Every effort must therefore be made to avoid those losses which are irreversible.[119]

Yet, uncertainty about the magnitude of the problem remained. As the 1979 report remarked,

> possible repercussions on the climate of the effects mentioned here can not, in the opinion of many experts, be reliably predicted at the moment [. . .] Agreement exists more readily to the fact that the problem has to be taken very seriously.[120]

Translated into practice, this meant resorting to estimating maximum losses for certain events and limiting exposure accordingly.

Contradictions transpired in the work of Herbert Tiedemann, engineering expert at Swiss Re and consultant to UNESCO and UNDRO on earthquake risks.[121] In the 1980s, Tiedemann extended his research to environmental issues and turned to Hubert Lamb's writings at CRU. In an expert publication on European storms, Tiedemann contradicted Swiss Re's earlier position and argued that the then current meteorological conditions would remain constant in the near future. In the long run, he continued, "[T]he next Ice Age is on its way, if we do not roast as a result of the greenhouse effect."[122] In a publication on environmental changes and catastrophe hazards that was written for a much broader audience he held that, so far, "it is impossible to prove that climatic changes are caused by the increase of carbon dioxide in the atmosphere," yet he conceded that there "is little doubt that further pollution, in particular by CO_2, will lead to a warmer climate because of the greenhouse effect. The increased energy content of the atmosphere will generate more frequent, more vehement and more destructive natural events."[123] Munich Re had already earlier sent contradictory signals. Gerhard Berz, head of Munich Re's GEO Risk Research department, had studied meteorology and geophysics and graduated with a dissertation on earth's temperatures and atmospheric movements.[124] After joining Munich Re in 1974, he soon became the probably most respected climate advocate for the industry. Yet Munich Re's director Klaus Gerathewohl in his seminal handbooks on reinsurance never quotes Berz and demotes climate change to a footnote, observing that theories about anthropogenic influence on the climate "are still very controversial."[125] Obviously, by the 1980s, reinsurers had not yet defined an official standpoint on climate change. Publications to quite some degree reflected personal opinions of the authors. There were also no concerted efforts to reach beyond the industry itself to the scientific community or to politics.

11.5 1990s—Greenpeace and the Road to UNEP

The 1990s had been declared the International Decade for Natural Disaster Reduction by the United Nations.[126] This caused Swiss Re's risk management unit to research the development of catastrophes from 1970 to 1989. The dramatic increase, it concluded, was due to "value concentrations, risk accumulations and a higher degree of insurance," but "climatic changes could also have an unfavourable effect on the development of catastrophes."[127] Business logic obliged the company to react, no matter how uncertain the causes and the extent of climate change were. The annual report for 1989 stated that

> [a] connection is assumed between the increasing frequency and vehemence of windstorm, hailstorm and floods observed in recent years and climatic changes linked with growing atmospheric pollution from emissions and the subsequent greenhouse effect. This is prompting insurers and reinsurers not only to raise their premiums for meteorological risks but also to reconsider traditional insurance concepts for natural hazards with a view to the fair distribution of such losses between policyholder, insurer and reinsurer.[128]

The press conference held in conjunction with the annual general meeting focused almost exclusively on climate change. An article of Swiss Re's head of reinsurance business about storm insurance[129] and a five-page press release on climate change were distributed. The main message was in the title of the press release, *Climate Change: Not a Threat but a Fact.*[130] Climate change was not a scenario anymore, it was happening. Discussions about the causes and the extent of climate change should not divert from the fact that insurance was already affected. The press conference was a great success, albeit only in the German speaking world where it received broad coverage. Internationally, it had no impact. Swiss Re at the time was still a Swiss international company with largely independent local operations throughout the world. But it was not a global company.

Munich Re's Gerhard Berz, on the other hand, had chosen to communicate via scientific publishers. His paper *Climatic Change: Impact on International Reinsurance*[131] was published by the Australian Commonwealth Scientific and Industrial Research Organisation and resonated with the IPCC where it was discussed in the 1990 report.[132] United Nations organisations now also started quoting Munich Re's natural catastrophe data to underline the severity of climate change impacts, for example UNEP's Climate System Monitoring project that referred to Munich Re's hurricane statistics.[133]

But neither Munich Re, nor Swiss Re, nor any insurance company or association was prominent at the next big event, the Earth Summit in Rio de Janeiro in 1992. Unlike the main business representatives in Rio, delegates from the Global Climate Coalition (GCC) representing mainly oil, coal, and automobile companies (see Bergquist and Jones in this volume), the insurance industry had no international lobby. The fact that insurance as the probably most affected industry was virtually absent, caught the attention of Jeremy Leggett, an earth scientist who had joined Greenpeace in 1989. Soon after, he edited a volume on Greenpeace's position on global warming.[134] In his own contribution on the "Greenpeace View," Leggett suggested a precautionary principle akin to (but unaware of) what Swiss Re's 1990 press release had argued:

> The prospects of future environmental security being compromised by global warming are now so real as to make adoption of a precautionary response imperative—in other words,

> policy should focus on buying insurance and policymakers should be clearly aware that waiting in perpetuity for better scientific data entails the real risk of waiting until it is too late.[135]

Returning from the Rio conference, Leggett decided to pay reinsurers a visit to investigate the reasons for their absence. The climate change problem had been identified by the parties he spoke to: Munich Re, Lloyd's, and Swiss Re. The 1990 IPCC report was available, and 1992 was not only the year of the Rio Summit, it was also the year of Hurricane Andrew, the so far financially most devastating storm ever. Combined with a series of other unusually severe storms across the globe such as U.S. Hurricane Hugo in 1989, European Windstorm Daria in 1990, and Japanese Typhoon Mireille in 1991, claims wiped out a number of insurance companies.[136] Andrew finally brought the connection of global warming and natural catastrophes into the headlines. As Newsweek wrote, "look at Andrew; that may be what a greenhouse world would be like."[137] But it seemed that reinsurers could only analyse impacts in general terms. Data and modelling techniques had advanced since the Travelers Research Center, but they did not yet produce reliable risk scenarios for particular regions and estimate possible losses.[138]

Post-Andrew, insurers faced a dilemma. The massive losses justified increasing premiums which made it unattractive to exit the market. But finding risk capital was challenging as much of it had been wiped out by the catastrophe. Reserves for future events would have to be even more substantial. For each large event, at least double the amount of capital at risk had to be reserved in order to survive one catastrophe and have enough financial resources to insure the next one. Worst of all, the market did not accept substantial premium increases, even if justified. Critics accused insurers of exaggerating climate risks in order to cash in on higher rates (Leggett, 2001: 119).[139] Another problem was that reserves had to be kept for so-called long-latent claims from asbestos and environmental pollution legacies from the 1970s and 1980s. Such problems combined left exiting markets as a main option.

Lloyd's had been especially hard hit by liability issues. This made Richard Keeling, deputy chairman of Lloyd's, open up to do "a lot of work with Greenpeace on ways to avoid climate change,"[140] particularly collaborating with the NGO on their Climate Change and the Insurance Industry publication.[141] Lloyd's had been warned about the consequences of climate warming as one of the syndicates at the beginning of the 1990s had commissioned a "leading UK climate research centre" to provide a prognosis.[142] It subsequently restricted catastrophe underwriting in hurricane-exposed areas, just in time to avoid Andrew-related losses.[143] Swiss Re had already in the 1980s reduced its exposure to natcats. But, when insurers exited business, the brunt would be on governments and ultimately on taxpayers, as the representatives of Lloyd's told Leggett.[144] In some cases such as in Florida, governments therefore forced insurers to stay. An alternative to exiting the natural catastrophe market was to seek risk capital in other areas. Capital markets could offer capacity that by far surpassed insurers' means. This led to a surge in alternative risk transfer via so-called insurance-linked securities[145] and the foundation of new reinsurance companies and Bermudan entities specialising in natural catastrophes, the so-called "Bermuda class of 1992."[146] More risk capital brought some relief but no solution to deal with the causes of climate change. What was still lacking was an international body addressing the causes.

In 1994, Swiss Re presented its first publication dedicated exclusively to global warming. Again, the writing was commissioned to an external author, Christian Brauner, who had already written numerous risk publications for the company. Rudolf Kellenberger, responsible for reinsurance markets and special lines and member of Swiss Re's executive board, signed the preface. The brochure was "intended for insurance specialists and catastrophe protection experts, as well as decision-makers in the fields of politics."[147] The appeal to politicians was earmarked by Swiss Re's organisational change at the time. The company had just elected a new chairman, Ulrich Bremi who put a new CEO in charge, Lukas Mühlemann, a McKinsey consultant. Together, they set about rejuvenating the old-school company. Bremi, a prominent Swiss liberal politician, welcomed the idea of a private corporation joining the political debate.[148] Most importantly, Bremi's and Mühelmann's plan was to turn the internationally active company with local offices into a global company with one brand and one strategy, a prerequisite to deal with global phenomena such as climate. The new spirit was soon felt. The sequel of the Rio conference was held in Berlin in 1995.[149] Swiss Re sent two employees with the official Swiss delegation to Berlin.[150]

Now, Swiss Re's engagement had an impact. Shortly after the 1994 brochure, Bremi was approached by Elizabeth Dowdeswell, Executive Director of UNEP and Åge Korsvold, President of Norwegian insurer Uni Storebrand. Together with Scottish General Accident's Andrew Dlugolecki, Gerling Global, and later Japanese insurer Sumitomo Fire and Marine, Swiss Re was invited to be a founding member of a UNEP Insurance Sector Steering Committee that should draft a UNEP *Statement of Environmental Commitment by the Insurance Industry* along the lines of the commercial banks' 1991/1992 UNEP Initiative. Bremi delegated Kellenberger who recruited Brauner to participate in the UNEP meetings before hiring environmental engineer Ivo Knöpfel as Swiss Re's full-time Climate Adviser in October 1995.[151]

Greenpeace was delighted that the insurance industry finally started tackling the issue. The NGO had acted both as a facilitator to unify insurers in preparing the UNEP statement and as a critical voice on investment policies,[152] one of the controversial agenda points of the UNEP statement as most companies were still reluctant to give up investments in fossil fuels. Already in 1993, Greenpeace had started organising a series of climate events with Lloyd's of London, the US College of Insurance, and the British Bankers Association respectively.[153] On the eve of the Berlin UNFCCC Conference in March 1995, Greenpeace organised a gathering of the financial sector to discuss the threat posed by climate change. Also invited were representatives of the solar energy industry. The event was held under the auspice of Rolf Gerling, heir of the Gerling insurance company and environmental campaigner, once referred to as the "Mother Teresa of environmental protection" by German weekly *Der Spiegel*.[154] Greenpeace and Gerling Insurance was a combination that secured international media attention. According to Gerling, this was the first time that insurance, reinsurance, banking, and pension funds gathered to discuss climate change.[155] Speakers from insurance included Andrew Dlugolecki from the Scottish General Accident Assurance who by then had become a leading figure in climate campaigning. Uni Storebrand delegated Carlos Joly, Senior Vice President for Environmental Policy and Investments and later proponent of ESG investment. Franklin W. Nutter was the only U.S. representative. He had just started his second term as President of the Reinsurance Association of America (RAA).[156]

The UNEP Insurance Initiative was signed on November 23, 1995, in Geneva by over 70 insurance companies from 25 different countries, though without any from the U.S. Frank Nutter had fought a lonely battle. He had invited Greenpeace to speak to the annual industry gathering in 1994, but the participants remained unimpressed.[157] Together with several U.S. insurance associations, Nutter even managed to add insurance to the agenda of Al Gore.[158] Still, resonance in the industry, from both direct insurers and reinsurers, was limited. One American exception to sign the UNEP statement later on was Employers Re, a reinsurance company at the time still owned by General Electric.[159] Yet they withdrew shortly after signing. But for European and some Asian insurers, the UNEP statement was the starting point for making insurance concerns heard. Several initiatives followed such as the 1997 Insurance Industry Initiative for the Environment to create working groups on practical questions and prepare a position paper for the COP 3 in Kyoto.[160] This was when the hard work started.

11.6 Conclusions

Insurance and reinsurance were somewhat well prepared to deal with climate change thanks to techniques already adopted in the nineteenth century. From the 1970s on, the industry, however, concentrated on the consequences of global warming rather than on its causes. Methods to tackle climate variability such as risk zoning, risk modelling, and limiting maximum exposures with cautious underwriting were introduced before the current climate debate, already since the nineteenth century. The subsequent leap from climate variability to climate change was a small one. But the traditionally high awareness of weather- and climate-related issues led to a paradox. On the one hand, it sensitised insurers to global warming. But at the same time, it eliminated the need to acknowledge it. Climate change became another known unknown. The uncertainty associated with this paradox forced insurers to make decisions in their underwriting strategies without a final answer as to what caused climate change or whether it really happened. The industry improved risk modelling since the 1950s, yet some companies reduced underwriting or withdrew from natcat markets, others tried to increase premiums. The latter was difficult, both for direct insurers and for reinsurers, a fact which may have prompted reinsurers to raise both client and public awareness through publications since the 1970s. The grounds for such climate communication had been prepared by environmental liability issues in the 1960s and 1970s. The associated financial liabilities catapulted green topics to top management levels, with climate change as part and parcel.

Being outspoken about climate change eventually became a choice. The choice to do so was to a large degree driven by individuals such as Malone, Berz, Eisenring, Dlugolecki, and Nutter from the 1950s until recent times.

Reasons to not communicate are difficult to explain. Many companies may have avoided a politically charged topic. It appears that mainly large companies such as Munich Re, Swiss Re, and Storebrand engaged in forward-looking risk management research, usually the companies that also acknowledged climate change and communicated about it. Smaller companies did not have the resources to engage in research, they could choose to either remain silent, as most of them did, or adhere to the positions adopted by larger companies, continental reinsurers especially. The latter were often felt to speak on behalf of the direct industry. This explains why climate change topics in the 1970s and 1980s trickled down from reinsurers to direct insurers. But only the two leading continental reinsurers, Munich Re and Swiss Re, chose to communicate extensively on the subject. This may be explained by the business model of continental reinsurers who always offered services such as risk research and analyses along with reinsurance cover in order to obtain an actuarially fair price. The London Market and U.S. companies, on the other hand, sold reinsurance based mainly on a market-fair price. Further arguments to communicate included an interest in understanding the issue and adapting underwriting strategies, preparing the grounds for price increases, and seizing the opportunity to communicate about the positive impact of re/insurance. However, the industry's efforts lacked co-ordination until the advent of UNEP. Unlike other industries, insurance and reinsurance had no lobbying bodies. It was only with UNEP in the 1990s that insurance finally got to be heard.

The predominance of European companies leading the debate cannot be readily explained. The U.S. insurance industry addressed the issue long before it was recognised in Europe. Research by the TWRC in the 1950s and 1960s identified problems such as international climate liabilities and the blurring of natural and man-made disasters. It took decades for these to make it back onto U.S. insurance and political agendas. But it appears that with increasingly contradictory scientific research in the 1970s and rising public awareness and political connotations, climate change vanished from U.S. insurers' agendas. Frank Nutter's lonely U.S. mission did reach politically influential quarters when he became an advisor to Al Gore, but apparently did not yet convince many U.S. reinsurers to speak publicly about the topic. It is also difficult to determine how far the main U.S. industry organisations, the AAR and the AIA, represented the opinion of their members. Sympathising with climate campaigns could, with some clients, be seen as controversial, denying climate change could be bad for an insurer's public image. One advantage of allowing the industry associations to adopt a clear position against climate change was that it provided some anonymity for individual companies' respective and potentially opposing views. This to some degree alleviated the pressure to engage in what in many U.S. quarters is still seen as scientifically and politically controversial. Eventually, though, Frank Nutter managed to have AAR members adopt a climate change policy in 2008.[161]

Notes

1 The author would like to thank Thierry Corti, Swiss Re; Bridget Gillies, University of East Anglia Archives; Philip Jones, University of East Anglia; Ivo Knöpfel, onValues; Jeremy Leggett, Highlands Rewilding; Frank Nutter, RAA; Jean Palutikof, Griffith University; and Benjamin Rüber at Swiss Re for their help in writing this chapter.
2 Munich Re, NatCatSERVICE - The natural catastrophe loss database; Swiss Re, Natural catastrophes and inflation in 2022.
3 Swiss Re, *The Risk of a Lifetime,* 3.
4 Swiss Re, *Natural Catastrophes in Times of Economic Accumulation.*
5 Blazey and Govind, *Financial Adaptation*; Arruda, *The P&C Insurance Industry.*
6 Dlugolecki, *Climate Change and the Insurance Sector.*
7 Thistlethwaite, *The ClimateWise Principles.*
8 Haufler, *Insurance and Reinsurance.*
9 Haueter and Jones, *From City Fires.*
10 Mills, *Global Review.*
11 Phelan et al., *Insurance and Climate Change.*
12 Brogi et al., *Determinants of Insurance Companies*, 1357. See Chapter 15 on ESG and climate change.
13 For example Heal and Kriström, *Uncertainty.*
14 Leggett, *Carbon War,* 194.
15 quoted from Johnson, *Insuring Climate Change*, 37.
16 For example Swiss Re, *126th Annual Report 1989.*
17 Starominski-Uehara et al., *Natural Hazard Insurance Literature*; Bellman et al., *Climate Change, Insurance, and Households.*
18 Dobes et al, "The Economics of Global Climate Change."
19 Nobanee et al., "Insurance for Climate Change and Environmental Risk."
20 Munich Re, *Flood/Inundation*; Schiesser, *Ecology and Environmental Protection.*
21 Gerhard Berz, *Climatic Change.*
22 Berz, "Schäden durch Naturgefahren."
23 Peele, "Insurance and the Greenhouse Effect"; Friedman, "Implications of Climate Change."
24 IPCC, *IPCC Impact Assessment*, 5–25.
25 Leggett, *Climate Change and the Insurance Industry.*
26 *"Climatic Change and Their Effects on Risks" (Within the Fourth International Conference on Strategy Issues, London, October 28–30, 1991).*

27 Rasmussen, "Analysis of the Climate Change Issue."
28 Dlugolecki, "Insurance Implications of Climate Change."
29 Dlugolecki, "1995 Report on the IPCC"; IPCC, 1996, *Climate Change 1995*.
30 For example CII, *Coping with Climate Change*.
31 For example Tucker, "Climate Change and the Insurance Industry"; White and Etkin, "Climate Change, Extreme Events."
32 Dlugolecki and Keykhah, "Climate Change and the Insurance Sector"; Haufler, "Insurance and Reinsurance."
33 Johnson, *Insuring Climate Change?*
34 Starominski-Uehara et al., *How Does Natural Hazard Insurance Literature*; Bellman et al., *Climate Change, Insurance, and Households*.
35 For example Booth et al., *Climate, Society and Elemental Insurance*.
36 for a discussion, see Haueter, *Impact of Reinsurance*.
37 Collier, Elliott, and Lehtonen, "Climate Change and Insurance."
38 Ian Smith, "European Insurers Say US Backlash Has Damaged Climate Change Push," *Financial Times*, August 12, 2023.
39 James et al., "Characterising Loss and Damage."
40 see Mechler et al., *Loss and Damage*.
41 Rohland, *Sharing the Risk*.
42 Mauelshagen, "Sharing the Risk of Hail."
43 Hoffman, "Weather Forms of Insurance."
44 Walford, *Insurance Cyclopædia, Volume IV*, 285–88.
45 Walford, *Insurance Cyclopædia, Volume III*, 142–44.
46 quoted from Mauelshagen, " Sharing the Risk of Hail," 35.
47 Weart, "The Development of the Concept of Dangerous Anthropogenic Climate Change."
48 Callendar, "Artificial Production of Carbon Dioxide"; see Anderson et al., " CO2, the Greenhouse Effect and Global Warming" for a history of theories.
49 A discussion of the research is added to the first publication in the *Quarterly Journal Royal Meteorological Society*.
50 Jacobs, "Wartime Developments in Applied Climatology."
51 Good, "Thomas F. Malone."
52 Malone, *Compendium of Meteorology*.
53 Brooks, *Geological and Historical Aspects,* 1016.
54 Glueckauf, "Composition of Atmospheric Air," 9.
55 "Science: One Big Greenhouse," *Time Magazine*, May 28, 1956.
56 U.S. Congress, *Federal Disaster Insurance*.
57 U.S. Congress, *Federal Disaster Insurance*.
58 U.S. Congress, *Federal Disaster Insurance,* 1069.
59 Foster et al., *Report on Floods*.
60 U.S. Congress, *Federal Disaster Insurance,* 1091.
61 U.S. Congress, *Federal Disaster Insurance,* 1091.
62 U.S. Congress, *Federal Disaster Insurance,* 887.
63 "The Travelers Weather Research Center," *Weatherwise,* no. 7, 6, 159.
64 Pita, 2022, "Foundation and Development."
65 For a history of weather modification see Fleming, *Fixing the Sky*.
66 Schaefer, "Snow and Its Relationship," 221–34.
67 CIA, *Memorandum for: General Charles P. Cabell*.
68 U.S. Congress, *Bills Relating to Weather Modification* and *Comprehensive Program in the Field of Weather Modification*.
69 U.S. Congress, *Comprehensive Program in the Field of Weather Modification,* 91.
70 U.S. Congress, *Bills Relating to Weather Modification,* 269.
71 U.S. Congress, *Bills Relating to Weather Modification,* 425.
72 U.S. Congress, *Bills Relating to Weather Modification,* 423.
73 U.S. Congress, *Weather Modification Research*, 72.
74 Hendrick and Friedman, *Potential Impacts of Storm Modification*, 244.
75 Hendrick and Friedman, *Potential Impacts of Storm Modification*, 244.
76 U.S. Congress, *Comprehensive Program in the Field of Weather Modification,* 90.

77 Fleming, *Fixing the Sky,* 149.
78 U.S. Congress, *Comprehensive Program in the Field of Weather Modification,* 88.
79 U.S. Congress, *Bills Relating to Weather Modification,* 148–49.
80 U.S. Department of Commerce, *Soviet Bloc International Geophysical Year Information.*
81 U.S. Congress, *Weather Modification Research,* 50.
82 U.S. Congress, *Bills Relating to Weather Modification,* 136.
83 United States Senate, *Bills Relating to Weather Modification,* 142.
84 Sadly, Malone's predictions came true shortly after when the U.S. military launched Operation Sober Popeye in Vietnam in 1967 (for weather modification and warfare see Fleming, *Fixing the Sky,* 165–88).
85 U.S. Congress, *Primer on Climatic Variation.*
86 U.S. Congress, *Primer on Climatic Variation,* 28.
87 Alexander, "Ominous Changes."
88 U.S. Congress, National Climate Program Act, 2.
89 Foster et al., *Report on Floods.*
90 Good, "Thomas F. Malone."
91 Cf Werner and Rauscher chapter in this volume.
92 Kellogg and Schneider, "Climate Stabilization."
93 Schneider, *Genesis Strategy.*
94 Henderson, "Dilemma of Reticence."
95 Weart, "The Development of the Concept of Dangerous Anthropogenic Climate Change."
96 Quoted from Henderson, "Dilemma of Reticence," 73.
97 Quoted from Henderson, "Dilemma of Reticence," 74.
98 Quoted from Henderson, "Dilemma of Reticence," 69.
99 Kellogg and Schneider, "Climate Stabilization."
100 U.S. Congress, *Primer on Climatic Variation and Change,* 58.
101 Sanderson, *History of the University of East Anglia,* 285; Author's correspondence with Phil Jones, April 30, 2023.
102 Lamb, *Historic Storms,* ix.
103 Mörth, "Climatological Research"; Sanderson, *History of the University of East Anglia*, 285.
104 Raper, *Variations in the Incidence of Tropical Cyclones*; Author's correspondence with Phil Jones, April 27, 2023.
105 University of East Anglia, *Climatic Research Unit,* 12–13.
106 Author's correspondence with Phil Jones, April 27, 2023.
107 Author's correspondence with Phil Jones, April 24, 2023.
108 For example *Eastern Daily Express*, March 24, 1975.
109 Munich Re, *Flood/Inundation.*
110 Schiesser, *Ecology and Environmental Protection*.
111 Schiesser, *Ecology and Environmental Protection*, 18.
112 Swiss Re, *Naturgefahr.*
113 Swiss Re *Naturgefahr,* 42.
114 From the 1970s on, Swiss Re was an adviser to UNESCO on natural catastrophes.
115 Meadows et al., *Limits to Growth.*
116 Schiesser, *Ecology and Environmental Protection*, 22.
117 Niels Viggo Haueter, "Insuring Fossil Fuels," in *The Routledge Handbook of Natural Resources*, ed. Espen Storli et al. (forthcoming).
118 Swiss Re, *Environmental Pollution*, 3.
119 Schiesser, *Ecology and Environmental Protection,* 5.
120 Schiesser, *Ecology and Environmental Protection,* 23.
121 Swiss Re Historical Archives: SRCA 10.106 683.08.
122 Tiedemann, *Storm in Europe,* 4.
123 Tiedemann, *Environmental Changes,* 6.
124 Berz, *Untersuchungen.*
125 Gerathewohl, *Rückversicherung*, 230.
126 International Decade for Natural Disaster Reduction | UNDRR.
127 Swiss Re, *Natural Catastrophes and Major Losses*, 6.
128 Swiss Re, *126th Annual Report,* 9.

129 Kaufmann, *Sturmschadenversicherung*.
130 Klimaveränderung: Nicht Bedrohung, sondern Tatsache.
131 Berz, *Climatic Change*.
132 IPCC, *Climate Change*, 5–26.
133 UNEP, *Global Climate System*, 60.
134 Leggett, *Global Warming*.
135 Leggett, *Global Warming*, 460.
136 Lynne McChristian, "Hurricane Andrew and Insurance: The Enduring Impact of an Historic Storm," *Insurance Information Institute* 16 (2012).
137 Sharon Begley, "Was Andrew a Freak—or a Preview of Things to Come?" *Newsweek*, September 6, 1992.
138 Leggett, *Carbon War*, 102–8; Johnson, *Insuring Climate Change?* 54–55.
139 Johnson, *Insuring Climate Change?* 67, investigated this but dismissed accusations that "insurers and reinsurers intentionally exaggerate risks and adopt scare tactics about global warming out of profit-seeking. There are probably instances of this occurring, but as a general explanation, such a claim misses the point."
140 Luessenhop and Mayer, *Risky Business*, 104.
141 Leggett, *Climate Change and the Insurance Industry*; Author's correspondence with J. Leggett, April 27, 2023.
142 Leggett, *Carbon War*, 201.
143 Leggett, *Carbon War*, 104.
144 Leggett, *Carbon War*, 104.
145 Kunreuther and Michel-Kerjan, "Catastrophe Risk Markets."
146 Lechner et al., "Continuity and Change," 286.
147 Brauner, *Global Warming*, 7.
148 Oral History Interviewee 1.
149 Berlin Climate Change Conference, March 1995 | UNFCCC.
150 Swiss Re, *Progress Report*.
151 Swiss Re Company Archives 010632; Authors' Oral History interview with Ivo Knöpfel, July 12, 2022.
152 Melchett, *Climate Change and the Insurance Industry*.
153 Leggett, *Climate Change and the Financial Sector*, 15.
154 Christoph Pauly, "Der Geist des Hauses Gerling," *Der Spiegel*, June 22, 2003.
155 Leggett, *Climate Change and the Financial Sector*, 193.
156 Author's email correspondence with Frank Nutter, April 29, 2023.
157 Leggett, *Carbon War*, 155–61.
158 Leggett, *Carbon War*, 186.
159 GE Insurance Solutions: Rebranding GE's Employers Reinsurance | GE News.
160 Founding Meeting of the "Insurance Industry Initiative for the Environment in Association with the United Nations Environment Programme"—United Nations Environment—Finance Initiative (unepfi.org)
161 Author's correspondence with Frank Nutter, April 30, 2023.

Bibliography

Alexander, Tom. "Ominous Changes in the World's Weather." *Fortune Magazine* 89, no. 2 (1974): 90–95.

Anderson, Thomas R., Ed Hawkins, and Philip D. Jones. "CO2, the Greenhouse Effect and Global Warming: From the Pioneering Work of Arrhenius and Callendar to Today's Earth System Models." *Endeavour* 40, no. 3 (2016): 178–87.

Arruda, Philip. "The P&C Insurance Industry's Role in Sustainable Development and Commerce." MBA study, University of Guelph, 2013.

Bellman, Lina, Sara Ekholm, Katarina Giritli Nygren, Olov Hemmingsson, Karin Jarnkvist, Linda Kvarnlöf, Minna Lundgren, and Anna Olofsson. "Climate Change, Insurance, and Households: A Literature Review." RCR Working Paper Series, 2016:2. Risk and Crisis Research Centre, 2016.

Berz, Gerhard. "Untersuchungen zum Wärmehaushalt der Erdoberfläche und zum bodennahen atmosphärischen Transport." PhD diss., University of Munich, 1969.

Berz, Gerhard. "Schäden durch Naturgefahren—eine weltweit zunehmende Bedrohung." *Der Maschinenschaden* 63, no. 5 (1990): 173–204.

Berz, Gerhard. "Climatic Change: Impact on International Reinsurance." In *Greenhouse: Planning for Climate Change*, edited by Gil I. Pearman. East Melbourne: CSIRO Publications, 1988.

Blazey, Patricia, and Paul Govind. "Financial Adaptation Challenges for the Insurance Industry Due to Climate Change." *Macquarie Journal of Business Law* 4 (2007): 15–48.

Booth, Kate, Chloe Lucas, and Shaun French, eds. *Climate, Society and Elemental Insurance*. Milton Park: Routledge, 2022.

Brauner, Christian. *Global Warming: Element of Risk*. Zurich: Swiss Re, 1994.

Brogi, Marina, Antonella Cappiello, Valentina Lagasio, and Fabrizio Santoboni. "Determinants of Insurance Companies' Environmental, Social, and Governance Awareness." *Corporate Social Responsibility and Environmental Management* 29, no. 5 (2022): 1357–69.

Brooks, Charles Ernest Pelham. "Geological and Historical Aspects of Climatic Change." *Compendium of Meteorology: Prepared Under the Direction of the Committee on the Compendium of Meteorology* (1951): 1004–18.

Callendar, Guy Stewart. "The Artificial Production of Carbon Dioxide and Its Influence on Temperature." *Quarterly Journal of the Royal Meteorological Society* 64, no. 275 (1938): 223–40.

Chartered Insurance Institute (CII). *Coping with Climate Change*. London: CII, 2009.

Central Intelligence Agency (CIA). "Letter to Mr. Roosevelt." CIA-RDP78-03425A002100020009-8, 1960.

Central Intelligence Agency (CIA). "Memorandum for: General Charles P. Cabell. Subject: Climate Control." CIA-RDP78-03425A002100020014-2, 1960.

Collier, Stephen J., Rebecca Elliott, and Turo-Kimmo Lehtonen. "Climate Change and Insurance." *Economy and Society* 50, no. 2 (2021): 158–72.

Dobes, Leo, Frank Jotzo, and David I. Stern. "The Economics of Global Climate Change: A Historical Literature Review." *Review of Economics* 65, no. 3 (2014): 281–320.

Dlugolecki, Andrew F. "Insurance Implications of Climatic Change." *Geneva Papers on Risk and Insurance. Issues and Practice* (1992): 393–405.

Dlugolecki, Andrew F. "The 1995 Report on the IPCC Working Group II—Chapter 17—Financial Services." *Geneva Papers on Risk and Insurance* 17 (1995): 496–500.

Dlugolecki, Andrew F. "Climate Change and the Insurance Sector." *The Geneva Papers on Risk and Insurance-Issues and Practice* 33 (2008): 71–90.

Dlugolecki, Andrew F., and Mojdeh Keykhah. "Climate Change and the Insurance Sector: Its Role in Adaptation and Mitigation." *Greener Management International* 39 (2002): 83–98.

Duguid, Andrew. *On the Brink: How a Crisis Transformed Lloyd's of London*. London: Springer, 2014.

Eastern Daily Express. "Weather Wise Insurance Group." March 24, 1975.

Fleming, James Rodger. *Historical Perspectives on Climate Change*. Oxford: Oxford University Press, 1998.

Fleming, James Rodger. *Fixing the Sky: The Checkered History of Weather and Climate Control*. New York: Columbia University Press, 2010.

Foster, Alden, Curtis J. Cooper, and A.A. Koch. *Report on Floods and Flood Damage of 1955 in the Northeastern States, Prepared for American Insurance Association*. New York: Parsons, 1955.

Friedman, Don G. "Implications of Climate Change for the Insurance Industry." In *Coping with Climate Change*, edited by John C. Topping. Washington, DC: Climate Institute, 1989.

Gerathewohl, Klaus, Wolf Otto Bauer, Peter Glotzmann, Ernst Hosp, Julius Klein, Harold Kluge, and Werner Schimming. *Rückversicherung: Grundlagen und Praxis, Band II*. Karlsruhe: Verlag Versicherungswirtschaft, 1979.

Glueckauf, Eugen. "The Composition of Atmospheric Air." *Compendium of Meteorology: Prepared Under the Direction of the Committee on the Compendium of Meteorology* (1951): 3–10.

Good, Gregory, A. "Thomas F. Malone (1917–2013)." *Eos* 94 (2013): 486–87.

Haueter, Niels Viggo. "The Impact of Reinsurance." In *Oxford Research Encyclopedia of Business and Management*. Oxford: Oxford University Press, 2021.

Haueter, Niels Viggo, and Geoffrey Jones, eds. *Managing Risk in Reinsurance: From City Fires to Global Warming*. Oxford: Oxford University Press, 2016.

Haufler, Virginia. "Insurance and Reinsurance in a Changing Climate." *Changing Climates in North American Politics: Institutions, Policymaking, and Multilevel Governance* (2009): 241–62.

Heal, Geoffrey M., and Bengt Kriström. "Uncertainty and Climate Change." Columbia Business School Working Paper, 2002.

Henderson, Gabriel. "The Dilemma of Reticence: Helmut Landsberg, Stephen Schneider, and Public Communication of Climate Risk." *History of Meteorology* 6 (2014): 53–78.

Hendrick, Robert L., Don G. Friedman, and W. Sewell. "Potential Impacts of Storm Modification on the Insurance Industry." In *Human Dimensions of Weather Modifications*, edited by Derrick W.R. Sewell, 227–48. Chicago: University of Chicago, 1966.

Hill, Alice C., and Leonardo Martinez-Diaz. *Building a Resilient Tomorrow: How to Prepare for the Coming Climate Disruption*. Oxford: Oxford University Press, 2020.

Hoffman, G. Wright. "Weather Forms of Insurance." *The Annals of the American Academy of Political and Social Science* 130, no. 1 (1927): 121–30.

IPCC. *Climate Change: The IPCC Impact Assessment*. Geneva: World Meteorological Organization and United Nations Environment Program, 1990.

IPCC. *Climate Change: The IPCC Response Strategies*. Geneva: World Meteorological Organization and United Nations Environment Program, 1990.

IPCC. *Climate Change 1995, Impacts Adaptation and Mitigation of Climate Change: Scientific-Technical Analysis*: Cambridge: Cambridge University Press, 1996.

Jacobs, Woodrow C. "Wartime Developments in Applied Climatology." In *Wartime Developments in Applied Climatology*, 1–52. Boston, MA: American Meteorological Society, 1947.

James, Rachel, Friederike Otto, Hannah Parker, Emily Boyd, Rosalind Cornforth, Daniel Mitchell, and Myles Allen. "Characterizing Loss and Damage from Climate Change." *Nature Climate Change* 4, no. 11 (2014): 938–39.

Johnson, Leigh Taylor. "Insuring Climate Change? Science, Fear, and Value in Reinsurance Markets." PhD diss., University of California, Berkeley, 2011.

Kaufmann, Hans Rudolf. "Sturmschadenversicherung—quo vadis?" *Versicherungswirtschaft* 45 (1990).

Kellogg, William W., and Stephen H. Schneider. "Climate Stabilization: For Better or for Worse? Even if We Could Predict the Future of Our Climate, Climate Control Would Be a Hazardous Venture." *Science* 186, no. 4170 (1974): 1163–72.

Kunreuther, Howard C., and Erwann O. Michel-Kerjan. "The Development of New Catastrophe Risk Markets." *Annual Review of Resource Economics* 1, no. 1 (2009): 119–37.

Lamb, Hubert, and Knud Frydendahl. *Historic Storms of the North Sea, British Isles and Northwest Europe*. Cambridge: Cambridge University Press, 1991.

Lechner, Roman, N.V. Haueter, and Lawrence Kenny. "Continuity and Change in Reinsurance 1990–2016. In *Managing Risk in Reinsurance: From City Fire*, edited by Niels Viggo Haueter and Geoffrey Jones. Oxford: Oxford University Press, 2016.

Leggett, Jeremy K., ed. *Global Warming: The Greenpeace Report*. Oxford: Oxford University Press, 1990.

Leggett, Jeremy K. *Climate Change and the Insurance Industry: Solidarity Among the Risk Community?* London: Greenpeace, 1993.

Leggett, Jeremy K. *The Carbon War. Global Warming and the End of the Oil Era*. New York: Routledge, 2001.

Leggett, Jeremy K., ed. *Climate Change and the Financial Sector: The Emerging Threat—The Solar Solution*. Munich: Gerling Akademie Verlag, 1996.

Luessenhop, Elizabeth, and Martin Mayer. *Risky Business: An Insider Account of the Disaster at Lloyd's of London*. New York: Scribner, 1997.

Malone, Thomas F., ed. *Compendium of Meteorology*. Boston: American Meteorological Society, 1952.

Mauelshagen, Franz. "Sharing the Risk of Hail: Insurance, Reinsurance and the Variability of Hailstorms in Switzerland, 1880–1932." *Environment and History* 17, no. 1 (2011): 171–91.

Meadows, D.H., D.L. Meadows, J. Randers, and W.W. Behrens. *The Limits to Growth*. New York: Universe Books, 1972.

Mechler, Reinhard, Laurens M. Bouwer, Thomas Schinko, Swenja Surminski, and JoAnne Linnerooth-Bayer. *Loss and Damage from Climate Change: Concepts, Methods and Policy Options*. Cham: Springer Nature, 2019.

Melchett, Peter. *Climate Change and the Insurance Industry*. Open Letter to the UK Insurance Industry. London: Greenpeace, June 1995.

Mills, Evan. "A Global Review of Insurance Industry Responses to Climate Change." *The Geneva Papers on Risk and Insurance-Issues and Practice* 34 (2009): 323–59.

Morth, H.T. "Climatological Research: An Interdisciplinary Study." *New Scientist* 80 (1978): 691–93.

Munich Re. *Flood/Inundation*. Munich: Munich Re, 1973.

Nobanee, Haitham, Mehroz Nida Dilshad, Omar Abu Lamdi, Bashaier Ballool, Saeed Al Dhaheri, Noura AlMheiri, Abdalla Alyammahi, and Sultan Salah Alhemeiri. "Insurance for Climate Change and Environmental Risk: A Bibliometric Review." *International Journal of Climate Change Strategies and Management* 14, no. 5 (2022): 440–61.

Peele, B.D. "Insurance and the Greenhouse Effect." In *Greenhouse: Planning for Climate Change*, edited by Gil I. Pearman. East Melbourne: CSIRO Publications, 1988.

Phelan, Liam, Cameron Holley, Clifford Shearing, and Louise Du Toit. "Insurance and Climate Change." In *Routledge International Handbook of Green Criminology*, edited by Nigel South and Avi Brisman. London: Routledge, 2020.

Pita, Gonzalo L. "Foundation and Development of Natural Catastrophe Modeling." *Natural Hazards Review* 23, no. 4 (2022): 04022025.

Raper, Sarah Christian Broun. "Variations in the Incidence of Tropical Cyclones and Relationships with the Global Circulation." PhD diss., University of East Anglia, 1978.

Rasmussen, J.L. "An Analysis of the Climate Change Issue." *The Geneva Papers on Risk and Insurance-Issues and Practice* 17 (1992): 371–85.

Revelle, Roger, and Hans E. Suess. "Carbon Dioxide Exchange Between Atmosphere and Ocean and the Question of an Increase of Atmospheric CO2 During the Past Decades." *Tellus* 9, no. 1 (1957): 18–27.

Rohland, Eleonora. *Sharing the Risk: Fire, Climate and Disaster, Swiss Re 1864–1906*. Lancaster: Crucible Books, 2011.

Sanderson, Michael. *The History of the University of East Anglia, Norwich*. New York: Hambledon and London Ltd., 2002.

Schaefer, Vincent J. "Snow and Its Relationship to Experimental Meteorology." In *Compendium of Meteorology*, edited by Thomas F. Malone, 221–34. Boston: American Meteorological Society, 1952.

Schiesser, Walter. *Ecology and Environmental Protection*. Zurich: Swiss Re, 1979.

Schneider, Stephen Henry, and Lynne E. Mesirow. *The Genesis Strategy: Climate and Global Survival*. New York: Plenum Press, 1976.

Starominski-Uehara, Marvin, and E. Carina H. Keskitalo. "How Does Natural Hazard Insurance Literature Discuss the Risks of Climate Change?" *Journal of Insurance Regulation* 35, no. 6 (2016).

Swiss Re. "Environmental Pollution and Insurance." *sigma* 12 (1972).

Swiss Re. *Naturgefahr und Ereignisschaden*. Zurich: Swiss Re, 1989.

Swiss Re. *Umweltschutz—Lebensschutz*. Zurich: Swiss Re, 1989.

Swiss Re. *126th Annual Report 1989*. Zurich: Swiss Re, 1990.

Swiss Re. "Natural Catastrophes and Major Losses in 1989: Dramatic Increase in Insured Damage to Over US-$ 13 Billion." *sigma* 2 (1990).

Swiss Re. *Environment and Climate Progress Report*. Zurich: Swiss Re, 1996.

Swiss Re. "Natural Catastrophes in Times of Economic Accumulation and Climate Change." *sigma* 2 (2020).

Swiss Re. "Natural Catastrophes and Inflation in 2022: A Perfect Storm." *sigma* 1 (2023).

Swiss Re. *The Risk of a Lifetime: Mapping the Impact of Climate Change on Life and Health Risks*. Zurich: Swiss Re, 2023.

Thistlethwaite, Jason. "The ClimateWise Principles: Self-Regulating Climate Change Risks in the Insurance Sector." *Business & Society* 51, no. 1 (2012): 121–47.

Tiedemann, Herbert. *Environmental Changes and Catastrophe Hazards*. Zurich: Swiss Re, 1986.

Tiedemann, Herbert. *Storm in Europe: Losses and Scenarios*. Zurich: Swiss Re, 1986.

Tucker, Michael. "Climate Change and the Insurance Industry: The Cost of Increased Risk and the Impetus for Action." *Ecological Economics* 22, no. 2 (1997): 85–96.

UNEP. *The Global Climate System: Climate System Monitoring. December 1988—May 1991*. Geneva: United Nations Environment Programme and World Meteorological Organization, 1992.

United States Congress, Senate Committee on Banking and Currency. *Federal Disaster Insurance. Eighty-Fourth Congress. 1955. First Session on Bills to Provide Insurance Against Natural and Manmade Disasters, and for other Purposes*. Washington, DC: U.S. Government Printing Office, 1956.

United States Congress. House of Representatives, Committee on Appropriations. *March 8, 1956. National Science Foundation. Report on the International Geophysical Year*. Washington, DC: U.S. Government Printing Office, 1957.

United States Congress. *Weather Modification Research. Hearings Before a Subcommittee of the Committee on Interstate and Foreign Commerce House of Representatives. Eighty-Fifth Congress, Second Session*. Washington, DC: U.S. Government Printing Office, 1958.

United States Congress. *Hearings Before the Committee on Commerce. Eighty-Ninth Congress. First and Second Sessions on S. 23 and S. 2916 Bills Relating to Weather Modification*. Washington, DC: U.S. Government Printing Office, 1966.

United States Congress. *Hearings Before the Committee on Interstate and Foreign Commerce. House of Representatives, Ninetieth Congress, First Session. A Bill to Authorize the Secretary of Commerce to*

Carry Out a Comprehensive Program in the Field of Weather Modification. Washington, DC: U.S. Government Printing Office, 1968.
United States Congress. *A Primer on Climatic Variation and Change*. Washington, DC: U.S. Government Printing Office, 1976.
United States Congress. *The National Climate Program Act. Hearings Before the Subcommittee on the Environment and the Atmosphere*. Washington, DC: U.S. Government Printing Office, 1976.
United States Department of Commerce. *Soviet Bloc International Geophysical Year Information*. Washington, DC: Office of Technical Services, 1958.
University of East Anglia. *Climatic Research Unit Sixth Annual Report*. Norwich: UEA, 1977.
Walford, Cornelius. *Insurance Cyclopædia*. Vol. III. London: Charles and Edwin Layton, 1874.
Walford, Cornelius. *Insurance Cyclopædia*. Vol. IV. London: Charles and Edwin Layton, 1876.
Weart, Spencer R. "The Development of the Concept of Dangerous Anthropogenic Climate Change." In *The Oxford Handbook of Climate Change and Society*, edited by John S. Dryzek, Richard B. Norgaard, and David Schlosberg. Oxford: Oxford University Press, 2012.
White, Rodney, and David Etkin. *Climate Change, Extreme Events and the Canadian Insurance Industry*. Dordrecht: Springer, 1998.

12

BUSINESS FACING ACTIVISM

Organised Business and Civil Society Movements in Germany and Switzerland Since the 1970s[1]

Sabine Pitteloud

12.1 Introduction

In 2023, Holger Lösch, Deputy Director General of the *Bundesverband der deutschen Industrie* (BDI—the Federation of German Industries) and a former journalist and marketing consultant, discussed "how the climate movement and the industry could become best friends" with a panel of young industry leaders.[2] Lösch's statement is revealing in that it acknowledges important trends characterising the evolution of the relationship between business and civil society movements. First, it shows that business leaders and their organisations consider that social movements matter to corporate strategy. Second, it implies that while relationships between businesses and climate activists are often initially confrontational, they can evolve into win-win situations. Consequently, the topic seems particularly well suited to a business history approach to the climate movement as it provides insight into the various and evolving tactics that companies and their business associations have deployed to address social demands and criticism.

Climate activism, its impact on, and the response of, corporations has so far mainly attracted the attention of international business, management, political science, and sociology of organisations scholars, which have offered a variety of case studies and interesting conceptual frameworks. The lack of business history insight on the theme is not surprising since climate activism is a rather new phenomenon that started in the late 1980s and has intensified in the last decade. Nevertheless, thanks to existing historical studies that have focused on the relationship between corporations and various forms of activism since the late 1960s, it is possible to historicise recent trends.[3] Such long-term assessment is sorely needed since the literature usually acknowledges how difficult it is to fully grasp activism's impact as it dually confronts firms and governments with repercussions which might have both short- and long-term effects. Moreover, using recent companies' public communications might lead to an overemphasis on activists' influence, as it is extremely difficult to untangle firms' public relations strategies from meaningful changes in management.[4]

Relying on the archives of the main business associations in Switzerland and Germany, namely the *Schweizerischer Handels- und Industrieverein* (SHIV—The Swiss Federation of Commerce and Industry) and the *Bundesverband der deutschen Industrie* (BDI—the Federation of German

DOI: 10.4324/9781003478089-13

Industries), this chapter analyses the complex and evolving relationships between business and activists and offers insights on general patterns regarding corporate attitudes towards environmental and climate demands. Globally, Switzerland and Germany have often been singled out for their track records in fighting pollution and had been at the forefront since the late 1980s in the push for international agreements to reduce CO_2 emissions. Since the SHIV and the BDI archives amass minutes of meetings and correspondence with regional chambers of commerce, sectoral associations, and individual companies, they offer a good vantage point from which to study how the business community perceived social movements and the general political climate at various periods. These archival documents are complemented with public sources such as reports, communications, and the websites of individual companies, consulting firms, NGOs, activist groups, and business associations, as well as online newspaper archives. Mixing business associations and firm-level examples helps to distinguish general patterns from the exceptional cases which usually attract more media and scholarly attention. Moreover, triangulating the views and strategies of companies and activists since the 1970s and connecting them to the broader political context contributes to avoiding presentism and provides a critical perspective on recent trends emphasising communication and cooperation with NGOs, and the potential of such public/private partnerships to solve the current climate crisis.

The chapter is organised as follows. Section 12.2 provides a literature review of the scholarship, engaging with the theme of business and activism, and shows why a business history perspective is valuable. Section 12.3 analyses the reactions and strategies of the main business associations in Switzerland and Germany regarding the increased environmental activism since the 1970s. It also highlights emerging trends since the late 1980s such as the proliferation of voluntary guidelines, the rise of corporate environmentalism, and of NGO/company partnerships. Section 12.4 focuses on business response to climate change activism.

12.2 Business History and Social Movements

While not new, activism, which can be defined as "the use of direct and noticeable action to achieve a goal," is nowadays a distinctive feature of social and political life.[5] Literature overviews of the relationships between business and activists often underlines that both companies and social movements encompass a variety of realities.[6] Activism deploys in a continuum, ranging from small, informal and short-lived grassroots mobilisations to much more organised and enduring factions of civil society such as non-governmental organisations (NGOs). The business world is also very diverse and comprised companies of different sizes and from numerous sectors whose managers have various cultural and educational backgrounds and operate either within the jurisdiction of a single nation or internationally. Adding another dimension, both companies and activists often coalesce in networks, which is an important feature to consider.[7] Moreover, companies should not be considered as unitary units, but need to be studied as organisations gathering a variety of sub-units. Some studies have indeed highlighted that once sustainability divisions are created, their employees can be perceived as activists from within and whose goals might be at odds with the general views of companies' management which are oriented towards profit maximisation.[8]

Activists' actions can potentially have significant implications for companies as they might undermine a firm's legitimacy and therefore damage its brand, diminish its market value, destabilise its employees, limit its access to financial markets, or dissuade consumers from buying.[9] Legitimation spillovers are key, as the delegitimisation of an individual company can affect similar types of companies and sectors, and, possibly, the entire capitalist system.[10] Moreover, activists and NGOs can also affect corporations through their involvement in politics by influencing regulations and subsidies or in legal battles.[11] Consequently, activism increases both economic and political risks. However, the impact need not always be negative from a business perspective. Activists and NGOs can create business opportunities by developing competitive advantages for virtuous business alternatives. From a corporate standpoint, cooperation with NGOs can also help to meet stakeholder expectations and improve the brand.[12]

Management and social science scholarship has highlighted several factors that have made companies more prone to being targeted by activists such as working in socially sensitive sectors, producing morally questionable goods, being at the center of a global value chain, having a well-known brand, or having experienced corporate scandals in the past.[13] It has also identified the variety of techniques they have deployed such as compromise, ignoring, defending, or counter-attacking.[14] Given such complexity, it is consequently not surprising that the studies assessing activists' impact are careful when presenting generalisations.[15] Legitimation spillovers and the effect of activists on the general business environment should certainly be further investigated, as well as the complex triangular relationships between activists, companies, and regulators. As a result, authors like Bereni & Dubuisson-Quellier have rightly advocated for dynamic, constructivist, and contextualised approaches to analyse the ongoing and porous relationships between enterprises and social movements.[16]

Following such a call, there is an opportunity to meaningfully tackle this topic in historical perspective. Business history has, so far, paid limited attention to the relationships between business and social movements, with the exception of well-publicised cases such as the Nestlé boycott and activists targeting companies doing business in South Africa during the apartheid regime.[17] Several historical studies have nevertheless uncovered the wide-ranging business counteroffensives that took place in the 1970s' context of social unrest, highlighting the value-added of having access to internal documents from both companies and business associations.[18] Moreover, this body of literature has made the case that a variety of individual and collective corporate strategies co-existed, for example ideological counteroffensives, dialogues with moderate NGOs and the isolation of radical critics, as well as multiple channels of influence towards the public and policymakers.[19] Authors have also underlined how corporate social responsibility (CSR) and voluntary agreements has been a corporate device in the past, which did not primarily aimed at fulfilling corporate self-regulation goals, but was destined to decrease overall social pressure.[20]

Regarding environmental activism, Lubinski and Jones, in their seminal study of the German pharmaceutical industry, showed the usefulness of studying waves of contestations to assess the long-term greening of industries and to what extent public relations strategies also resulted in changes to corporate culture.[21] Similarly, Keetie Sluyterman who, as a business historian, was the first to focus on three waves of environmental contestation from Shell's perspective since the 1920s, highlighted the need for studying firms' responses and their complex interplay with international competition dynamics, consumers' attitudes, and governmental regulations.[22]

The business history scholarship has also provided important insights by showing how business leaders have sometimes presented themselves as activists with the hope that other companies will follow their lead.[23] Geoffrey Jones, through his systematic study of responsible business over a century, has identified structural difficulties preventing such corporate activists from being the norm such as financialisation and shareholder-value maximisation.[24] Adam Rome has also pointed to the proliferation of corporate environmentalism since the 1980s, with many companies emphasising the need to go beyond compliance, and the limited effects actually produced by such a stance.[25]

Other scholars have underlined the ability of the capitalist system to co-opt its own critics and have shown that partnership ideology became increasingly important.[26] Such trends are closely associated with the progressive corporatisation of NGOs which have multiplied since the 1970s and are therefore in competition with each other to attract funding and portray themselves as relevant organisations.[27] LeBaron and Dauvergne warn against such trends, stating that "achieving current gains is requiring activist organisations to conform with, rather than work to transform, global capitalism."[28] Moreover, Western-born NGOs have been criticised for not sufficiently taking into account past responsibilities regarding environmental damage and the need for some countries to alleviate poverty through economic development.[29]

As these pioneer historical studies show, activists' impacts and firms' responses gain to be understood as context-dependent and assessed both in the long run and beyond public discourse through the use of internal archival documents. All these reasons explain why a business history approach is particularly well suited and useful.

12.3 Organised Business Against "Emotional" Politicisation Since the 1970s

Environmental damage due to industrial activities has been under public and political scrutiny since at least the Industrial Revolution.[30] One can think, for example, of Engels' description of the environmental degradation of the River Irk in nineteenth-century Manchester.[31] Protests were mostly local however, where water and air pollution were visible as direct consequences of industrial activity. Scientist and activist Rachel Carson is remembered for raising public awareness of the global consequence of business impacts. In her 1962 book *Silent Spring*, she denounced the effects on insect and bird populations by the use of pesticides. Carson's actions were also remarkable, as they showed that activism can have a global and meaningful impact since, by raising public and political awareness, she contributed to the creation of the Environmental Protection Agency in the United States, and to a ban on DDT in several countries.[32] Her findings were also disseminated in the European press; in Germany, *Die Zeit* und *Der Spiegel* soon published articles on pesticides, emphasising that "the new dangers had no geographic, temporal and sensory boundaries."[33]

In the 1970s, activism rose to another level. The Council of Europe declared 1970 to be European Conservation Year, aimed at raising environmental awareness in the "Old World."[34] Conservationist US Senator Gaylord Nelson, along with peace and environmental activists John McConnell and Denis Hayes, organised a nationwide mobilisation in the United States called *Earth Day*, bringing 20 million people to the streets. The Club of Rome's *Limits to Growth* publication in 1972 and the United Nations Conference on the Human Environment organised in Stockholm the same year also precipitated a general acknowledgement of the planetary dimensions of the impact of economic activities and the potential irreversibility of the damage.[35] While several environmental organisations predated the 1970s, the emphasis was mostly on conservationism, that is, the protection of natural areas and landscape.[36] In Germany, the *Bundesverband Bürgerinitiativen Umweltschutz* (the National Association of Citizens' Environmental Protection Initiatives, BBU) was created in 1972 and the Bund für Umwelt und Naturschutz Deutschland (League for Environment and Nature Protection) in 1975, with increased focus on environmental pollution.[37] Similar trends were visible in Switzerland with the creation of the *Schweizerische Gesellschaft für Umweltschutz* (Swiss Society for Environmental Protection) in 1971.[38] The 1970s was also a period of increasing institutionalisation of the environmental question within the state apparatus. The Swiss *Office Fédéral de l'Environnement* was created in 1971, while Germany developed an environmental programme in the same year and established the *Umweltbundesamt* (Environment Agency) in 1974. The ancestors of the Green Party were created in Switzerland at the cantonal (state) levels in 1971, and some of their members were elected to cantonal parliaments in 1973. In Germany, electoral lists of environmental parties were presented in some *Länder* (state-level) in the mid-1970s.[39]

From historical records, it is obvious that, in general, the business community was concerned by these trends. In response, the two main business associations from Germany and Switzerland, the BDI and the SHIV, both created dedicated task forces. In October 1972, the BDI executive committee established the *BDI-Ausschuss für Umweltfragen* (BDI-Committee for Environmental Issues), which aimed at "dealing with the economic and socio-political aspects of environmental protection."[40] The SHIV studied the possibility of better co-ordination within business circles regarding environmental issues in 1972 and organised a Conference on the Environment (*Umweltkonferenz*) to promote exchanges of views between industrialists and government officials.[41] In the mid-1970s, it established the *Vorort-Gremiums für Fragen der Umweltschutzpolitik* (Vorort-Committee for Environmental Protection Policy Issues), a task force dedicated to following all relevant environmental regulation.[42] In addition, several sectoral organisations in the plastic, chemical, and machine tool industries established groups to co-ordinate business action regarding environmental technical challenges and regulation.[43] The International Chamber of Commerce (ICC) did the same internationally, with the creation of the International Center for Industry and the Environment (ICIE) whose goal was to collaborate with the United Nations Environment Program (UNEP).[44] The ICIE paralleled the political engagement of environmental NGOs at the international level.

Such business mobilisation was primarily motivated by the activities carried out in newly created governmental environmental bodies, whose civil servants, often having a scientific and technical background, did not always consider environmental regulations in their economic dimensions or, at least, not to the extent that the SHIV and BDI wished for. Social movements such as the BBU and the *Schweizerische Gesellschaft für Umweltschutz* nevertheless have played an important role in two respects. First, by raising awareness about several dimensions of environmental degradation, they have put pressure on the political process and prompted action by politicians since the early 1970s. The "emotional" dimension of the debate, as well as its "dramatisations," was, for instance, often emphasised by the BDI.[45] It noted that it no longer had any influence on the political agenda because "public opinion discusses [environmental issues] with great passion and, unfortunately, just as much polemical confusion."[46] The SHIV also underlined that due to political pressures, the cooperation between the federal civil servants in charge of environmental protection and the private sector seemed to be "too little consistent and too much dependent on random processes," leading to the "danger of creating a legal system that takes too little account of technical and economic realities."[47] When an issue gains high political saliency in the public sphere, business influence decreases because journalists and politicians have more incentive to invest time in gathering information from various parties.[48] Consequently, business cannot position itself as the unique source of expertise, and its views are more prone to be challenged. The BDI indeed noted the increased importance of scientific experts and activists in, for instance, disseminating the idea of limits to growth which, if translated into politics, could "transform our social market economy into the straitjacket of the planned economy."[49]

Overall, in the 1970s and early 1980s, companies' concerns related to the influence that social movements might have on politics, that is, regulation. During this period, direct confrontation between NGOs and corporations was just the tip of the iceberg and often not related to environmental issues. One of the first iconic cases of the power of transnational activists' networks against corporations was the Nestlé boycott launched in 1974 from the United States to denounce its commercialisation practices for infant nutrition in the Global South.[50] Divestment campaigns, including those in Switzerland and Germany, mostly targeted companies conducting business in the segregationist apartheid regime in South Africa. And when individual companies came under public opinion's scrutiny, it was often due to their own course of action, as was the case of the Seveso accident in Italy, which involved an Italian subsidiary of the chemical group Hoffmann-La Roche whose release of a pesticide cloud in 1976 led to the European Economic Community (EEC)'s 1982 adoption of the Seveso directive.[51] Environmental activists like Greenpeace used direct action and media campaigns that mainly focused on big infrastructure (nuclear plants, roads, and so on) and environmentally-adverse practices (littering, nuclear testing, whale hunting, and so on).[52] Overall, in the early 1970s, the main driving principle behind the actions of environmental NGOs was to raise awareness and to foster political change that would force corporations to comply with the new norms rather than have their CEOs address the environmental requests directly.

Consequently, what was at stake for business circles was to decrease the political saliency of the environmental questions and contain social movements' political outreach.[53] To do that, two main strategies were pursued: the limitation or prevention of opportunities for social movements to engage in the political process, and improvements in communication. Regarding consultation procedures, business associations were already well-established partners, which left limited space for environmental organisations, especially in the case of Switzerland. For example, the Swiss Water Commission, which had been established in 1972 to document and monitor water pollution and to make policy recommendations to the Swiss government, consisted overall of 30 members and was composed of seven business representatives and only a single representative from an environmental protection NGO.[54] In Germany, the formal political participation of social movements played a comparatively more important role, even if antagonism between the state and leftist NGOs was not always easily sidestepped. The Federal Minister of the Interior had even provided support for the establishment of the BBU, whose goal was to act as "the umbrella organization for mushrooming environmental initiatives."[55] The BDI regretted that the BBU was preventing the development of nuclear power plants and was making other necessary energy projects extremely uncertain.[56]

In both countries, the participation of some environmental NGOs was nevertheless limited to consultation regarding environmental norms but was not institutionalised with respect to other domains that might have important consequences for the environment such as trade treaties. Business associations were eager to maintain their prerogatives in the matter. In 1972 for instance, the Swiss chemical industry lamented that the *Schweizerische Gesellschaft für Umweltschutz* was requesting environmental sovereignty regarding treaties between Switzerland and the EEC, which could jeopardise trade.[57] In retrospect, it seems that in the 1970s, business circles were often successful in advocating for free trade over environmental principles and that environmental NGOs remained outsiders with respect to international economic policies. For example, the Swiss Foreign Economic Affairs Department often supported the SHIV when it argued against the implementation of national environmental standards as this might be interpreted by the EEC as protectionist.[58]

Despite such limited direct political participation, other institutional instruments helped the new movements make their voices heard. In Switzerland, popular initiatives enabled any group of citizens to put an issue to the vote as soon as it collected 50,000 signatures. This possibility was used on several occasions in the 1970s to encourage public debate on the externalities of road traffic or to change regulatory standards and pollution limits.[59] The result was that businesses had to engage in public debate and it was no longer sufficient for them to simply place experts on governmental commissions. The use of popular initiatives was therefore far from pleasing to the SHIV and Swiss right-wing political parties who advocated tripling the required number of signatures before Swiss citizen could vote on a particular topic, with the official justification being that due to population growth, it had become too easy to collect 50,000 signatures.[60] Moreover, when some initiatives did collect the required number of signatures, some business representatives and parliamentarians toyed with the idea of declaring them invalid because the foreseen pace of implementation was not feasible due to limitations in the current technology. Such ideas led for instance to contentious discussions during the examination of a popular initiative aimed at introducing U.S. vehicle emission norms in Switzerland but, in the end, even parliamentarians hostile to environmental movements considered it too risky politically to advocate for the initiative's dismissal as this would obviously be interpreted as a restriction of democratic rights.[61] Within the parliament, the Swiss Socialist Party was alone in fighting the augmentation of signatures requirement, regretting "a weakening of political rights."[62] In the end, the 100,000 signatures requirement was submitted to a vote and implemented in 1977. For grassroots movements which had limited financial and organisational resources, such an increase meant that it was more difficult to take advantage of direct democracy.

In Germany, allowing interest groups the possibility of filing complaints (*Verbandsklage*) against state authorities was under consideration in the early 1970s.[63] Business circles strongly opposed the ability of social movements to file collective complaints as this would allow environmental NGOs to put pressure on government officials and to further mobilise public opinion. The BDI fought to avoid such a provision being included in the Federal Law for Nature Protection.[64] Despite fierce opposition, *Verbandskalge* possibilities were ultimately introduced at the beginning of the 1980s in the nature conservation laws of the states of Bremen, Hamburg and Hesse.[65] Since the introduction in 1967 of the Federal Law for the Protection of Nature and Landscapes, environmental associations in Switzerland could use a right of appeal (*droit de recours*) to oppose decisions regarding state planning and large infrastructures. Further discussions around introducing such rights in the new Federal Law for Environment Protection, which would have empowered environmental associations to go to court if they considered that the law was not being properly applied, led to heated parliamentarian debates. The ability to appeal was finally introduced in the 1983 law but there were restrictions regarding which organisations could resort to it.[66] Overall, the extension of legal and democratic rights to civil society movements was opposed by right-wing political parties as well as by business associations.

During those years of high political saliency of pollution issues, the SHIV and the BDI also increasingly became involved in public debates. Both associations emphasised that, on the one hand, businesses should behave as environmentally-aware actors and show that they had already made significant efforts and, on the other hand, that businesses were not the problem but were likely the actors most able to solve the problem through technical innovation. To spread such rhetoric, both the SHIV and the BDI relied on specific specialised communication organisations respectively the *Wirtschaftförderung* (Society for the Promotion of the Economy, Wf) in the Swiss case and the *Institut der Deutschen Wirtschaft* (IW) in the German case to improve corporate image and market economy principles in the public's eyes *(Aufklährungsarbeit)* by publishing relevant facts and studies, and defending business views during political campaigns. In 1972, the SHIV therefore declared that it would "consider [it] particularly important that the great efforts of the private sector in the field of environmental protection are also reflected in the press, radio and television." The SHIV also encouraged its section "to report false or one-sided representations in the press, radio and television to the Wf without delay" and to provide "names of expert personalities from the private sectors" who could take a stand in the media.[67] The BDI acknowledged the need to strengthen the IW's environmental communication, "which had for years failed to fulfill its task of providing scientific input and effective, continuous public relations work in the area of environmental protection, which is politically so explosive and persistently topical in its many facets" and noted that the BDI therefore suffered from "an open flank in this area."[68]

Over the years, the BDI and the SHIV produced several brochures for widespread public distribution. While the environmental question did not disappear in the late 1970s, the feeling of urgency seemed to have diminished in comparison to the previous decade. As the BDI summarised in 1979:

> The environmental policy of the first ten years was essentially a crisis policy in which the protection aspect was in the foreground. This was characterized by a lack of intensive discussion of goals, strategies, instruments and organizational structures, as well as a hectic legislative phase. This approach was favored by the broad support that environmental protection met with among the population and in parliaments.[69]

The SHIV and the BDI welcomed the fact that the discussions were less emotional and were increasingly framed to take technical and economic constraints into account. The economic crisis following the two oil shocks certainly contributed to restoring the importance of industrial activities to a central position on the political agenda.

The respite was short-lived, however. In the early 1980s, the "forest dieback" (*Waldsterben*) became the new salient issue which "brought a stagnant policy apparatus back to full speed."[70] By 1983, German power plant operators had been compelled to install desulfurisation devices within five years, and three-way catalytic converters became mandatory to reduce cars' toxic emissions. Switzerland, along with Sweden, was also at the forefront in pushing for the introduction of unleaded gasoline and three-way catalytic converters in Europe.[71] Regarding such issues, the SHIV and the BDI had an ambivalent attitude. They were not opposed *per se* to the new rules, as German car manufacturers had mastered the catalytic converter technology while Switzerland had no important domestic carmakers, but both organisations fought the unilateral introduction of the standards (a so-called *Alleingang*) if all members of the EEC were not in agreement as that would have created market fragmentation.[72] Regarding the *Waldsterben* issue, there is little doubt that environmental mobilisations and public outcry played a significant role in speeding up regulation.

While the 1980s was a crucial decade in establishing Switzerland as an environmental frontrunner and Germany as the "greenest nation in the world," to quote the words of German environmental historian Frank Uekötter,[73] the global ideological climate progressively shifted with the accession of Reagan and Thatcher to power in the United States and the United Kingdom respectively, both of whom wished to free economic forces from regulation and were critical of NGOs.[74] Business backlash against environmental movements therefore found a more ideologically friendly terrain.[75] More moderate and wide-encompassing business associations such as the International Chamber of Commerce worked efficiently, thanks to its networks and the financial weight of its members, to re-establish business as a key partner in assessing environmental challenges and to disseminate the idea that growth should not be limited but made "sustainable" instead.[76] As a result of these trends, while pollution limits and technical standards had dominated environmental governance in the 1970s and early 1980s, voluntary guidelines, self-discipline statements about corporate social responsibility, and market-driven instruments started proliferating in the following decade.[77] In parallel to traditional business associations' political involvement and co-ordination efforts, individual companies and groups of CEOs increasingly developed corporate environmentalism beyond compliance, sometimes motivated by genuine aims, sometimes to argue against the necessity of regulation.[78] Multi-stakeholder discussions became popular, especially at the international level where reaching intergovernmental agreements proved difficult.[79] Moreover, in the context of economic globalisation being dominated by large multinationals, and the strengthening of European regulation over national policies, NGOs were incentivised to target corporations instead of states.[80]

12.4 From Politics to Multi-stakeholder Partnerships in Solving Climate Change

In 1979, for the first time ever, climate change was discussed at a global level on the occasion of the first World Climate Conference which was organised in Geneva under the auspices of the World Meteorological Organization.[81] Nevertheless, it was only in the late 1980s that climate change and the ozone hole became prominent environmental issues. At first, awareness was mostly raised by scientists such as the Intergovernmental Panel on Climate Change (IPCC) established in 1988 by the UNEP, which published scientific reports.[82] Consequently, the first round of business mobilisations regarding climate change mainly concerned the petroleum industry. Exxon infamously engaged in "producing doubts" about climate science by funding alternative studies and groups involved in climate change denial.[83] While European firms such as Total and Shell supported some of Exxon's efforts in the late 1980s, they soon acknowledged the growing scientific consensus.[84] Shell primarily advocated for energy-saving measures and warned against "overreaction" that could impede economic development.[85] Scientists were backed by NGOs which coalesced into the Climate Action Network in 1989 and the International Council for Local Environmental Initiatives in 1990. Since the adoption of the Single European Act in 1987, NGOs increasingly became involved in European Union (EU) politics as well. The largest such as Greenpeace and the World Wide Fund for Nature (WWF) became affiliated through the Green 8 (now Green 10) network and even received subsidies from the European Commission for specific projects and operational costs.[86] In 1990, the EEC and the EFTA agreed to stabilise their CO_2 emissions by 2000 to the levels of 1990, without much specification regarding the concrete measures which each state would be responsible for. In 1991, Germany fixed a reduction target of 25-30% less emissions by 2005 compared to 1990 levels. The BDI and the SHIV acknowledged the importance of climate in their annual reports of the early 1990s. The BDI organised a conference in May 1990 which gathered together scientific experts, policymakers, and industrialists. From the start, both business associations emphasised the need for global solutions and the avoidance of national regulations which would negatively impact the competitiveness of national industries. They also pointed to the positive effect that industries might have thanks to energy-saving measures and green technologies.[87]

During the 1992 Rio Conference, a first Framework Convention on Climate Change was adopted, with the aim of stabilising greenhouse gas concentrations at a level that would prevent dangerous anthropogenic interference with the climate system. Both business and NGOs were perceived as key by the United Nations to solving the most pressing problems. The International Chamber of Commerce (ICC), of which the BDI and the SHIV were members, had succeeded in becoming a privileged interlocutor. During the Rio summit, the ICC somewhat neglected the issue of climate change, while contributing to putting the idea of sustainable development at the center of the discussions.[88] The BDI adopted similar rhetoric, stating its support for "a policy that constructively combines environmental protection and economic development."[89]

The SHIV, in an internal meeting of its executive committee, warned against a populist takeover of the climate change issue and the need for a political dialogue in which business associations would have a say:

> In environmental policy in particular, there are cases in which populist propaganda has led to poorly prepared political initiatives and proposals for actions are launched without careful examination. However, the CO_2 problem is not an issue that can be handled on a whim, but an issue that needs to be explored in depth on all sides.[90]

By the mid-1990s, the idea of introducing an ecotax to impel the achievement of the emissions abatement targets, be it at the national or European level, gained traction. Both Swiss and German industry opposed the implementation of an ecotax and promoted voluntary agreements to avoid such outcomes.[91] In March 1995, the BDI announced voluntary commitments to reduce CO_2 emissions to at least a minimum of 20% by 2005 to align with 1987 levels.[92] Internationally, the next milestone occurred when the Kyoto Protocol was adopted by 191 countries in 1997, which received official support from the SHIV and the BDI. In Germany, following the 1998 accession of the Green Party to government, a national energy tax was introduced despite the opposition of industry, whereas NGOs and the Environment Ministry had been promoters of such a measure. In Switzerland, a compromise was found in the 1999 Climate Law: an ecotax would only be introduced should voluntary agreements remain unfulfilled.[93] The SHIV established a private organisation called *Agence de l'économie pour l'énergie* (AEnEC) which was sponsored by the Swiss government to be responsible for negotiating emissions reductions with companies.

Overall, those sectors of the German economy which had set voluntary targets had either met or superseded them by the early 2000s, while their lobbying had introduced many exemption regimes.[94] Moreover, the existing voluntary agreements were also used by the BDI to oppose the establishment of an EU Emissions Trading System, which was nevertheless introduced in 2005. In Switzerland, it became clear in the mid-2000s that voluntary agreements would not be enough to meet the 2010 targets that had been set in the 1999 Climate Law. Again, due to the strong opposition by industry and the Swiss Petroleum Union to any mandatory ecotax, a new voluntary solution was introduced—the so-called "climate cent"—which was a 1.5 cent levy on each litre of fuel sold at the pump, so that the cost would primarily fall upon consumers. The collected amount was invested in emission offsetting projects, 90% of which were being implemented abroad rather than in Switzerland, leading to criticism by NGOs.[95] Given the divergent views of parties and economic circles on the right wing and the Greens and NGOs on the left, Switzerland adopted a wait-and-see attitude in the 2000s, aligning itself with EU positions, and could no longer be depicted as a "pusher" with respect to climate change mitigation.[96]

The publicity of climate change issues increased during the mid-2000s, with, for instance, the distribution of Al Gore's 2006 movie *An Inconvenient Truth* as well as the political pressure exerted when, between 2008 and 2011, many industrialised countries, including Switzerland and Germany, committed to further greenhouse gas emissions reduction in line with the goals set out by the Kyoto protocol.[97] New NGO coalitions such as *Climate Justice Now!* and *Climate Justice Action* were founded in 2007 and 2009 respectively, acknowledging the climate debt that industrialised countries owe to the South. In 2009, there were about 100,000 demonstrators in Copenhagen for the UN Climate Change Conference (COP 15). Because scientific evidence regarding climate change had accumulated, the social movements' campaigns had been framed as a "now or never" opportunity which, since COP15 did not result in a strong political commitment, soon faded.[98] Consequently, climate activism partly refocused on smaller-scale targets, including spotlighting individual companies and increasingly mobilising e-communication techniques. Such strategies might, however, prove uncertain since, as Bartley and Child argued, "shaming a corporation mobilizes attention, but it also carries risks—of becoming more attentive to images than to on-the-ground practices, letting other forms of exploitation go unnamed, or obscuring the contributions of states."[99]

In 2010, Greenpeace targeted Swiss-based Nestlé Company for its use of palm oil which, according to the NGO, "endangered orangutans to the brink of extinction and was accelerating climate change."[100] It also targeted the German car manufacturer Volkswagen (VW), which was accused of greenwashing and lobbying the European Union against policies aimed at bettering the climate performance of vehicles.[101] Both campaigns bore similarities, focusing on high-profile companies and turning their advertising campaigns into ones of derision to create a buzz on the internet. Greenpeace turned the slogan "Have a break, have a KitKat" into "Give the Orangutan a break," showing orangutan fingers and blood incorporated into the Nestlé bars. It also parodied the VW commercial with a kid disguised as Darth Vader in a Star Wars pastiche, showing VW as being on the dark side of The Force. Nestlé first tried to file a web complaint to the YouTube platform for infringement of the brand identity, but the attempt only led to a momentary withdrawal of the video which then went even more viral. At the same time, Nestlé announced in the press that it had ended its contract with Smart, a controversial Indonesian palm oil producer, and its intension to use only Certified Sustainable Palm Oil by 2015.[102] This objective would finally be postponed for almost ten years.[103] VW, in response to attacks, mainly underlined its past efforts in reducing CO_2 emissions by 13% between 2006 and 2010.[104] In 2015, the International Council on Clean Transportation (ICCT), an NGO which had conducted tests in real-time driving, led the US Environmental Protection Agency to carry out its own investigation which revealed that VW was using a device to reduce its emissions during the testing phase, while pretending to be a world leader in emissions mitigation.[105] This scandal, nicknamed "dieselgate" impugned VW's reputation even further and led to various lawsuits.

The cases of Nestlé and VW are interesting as they illustrate some general features of NGO/big business relationships. Both companies had developed their own climate agenda by late 2020, partnering with business schools and consulting groups, promoting multi-stakeholder engagements and advocacy, developing carbon certificates and carbon offset projects.[106] Nestlé, for instance, joined the Lowering Emissions by Accelerating Forest Finance (LEAF) Coalition, which is a public/private initiative to raise "global climate ambition and contribute to tropical forest protection."[107] Neither Nestlé's efforts nor VW's were considered meaningful by several NGOs, which still criticised their role in producing greenhouse emissions.[108] In 2022, investor groups such as the Swedish public pension fund and the Church of England Pensions expressed their concern about VW's climate lobbying, while ACTARES, a Swiss-based shareholder organisation critically assessed Nestlé's efforts regarding climate goals.[109] In addition, social movements turned to legal options. Greenpeace sued VW in 2021 for "failing to decarbonize the company in line with the 1.5°C goal agreed in Paris" and tried to capitalise on "the landmark Karlsruhe constitutional court ruling of April 2021, when the judges ruled that future generations have a fundamental right to climate protection."[110] A large NGO coalition also launched a popular initiative in Switzerland which was aimed at forcing Swiss multinationals to exert due diligence in the domain of the environment, including deforestation and human rights. Nestlé fought the initiative in the public arena, while still promoting its own sustainability agenda.[111] Consequently, even if stakeholder dialogue was praised, it often appeared to be a conversation with the deaf when studied over several decades. Consultants were privileged over NGOs, and when partnership did occur, this did not prevent many other actors from continuing to be very critical although they were mostly still ignored. Moreover, while almost all major global companies have developed a climate agenda within the last decade, they usually have been simultaneously fighting binding commitments, which begs the question of the sincerity of their engagement. Given such trends, it is therefore not surprising that both the short- and long-term effects of climate activism directed at corporations are extremely difficult to untangle and that studies focusing on partnerships or an individual case over a short-term period appear much more optimistic than historical accounts.

Although seldom targeted directly by social movements and NGOs, business associations like the BDI and Economiesuisse (which resulted from the merging of the SHIV and the Wf in 2000) had to also take climate activism into consideration. Their attitude is to some extent a reminder of the one they had regarding social movements in the 1970s, when they resisted activists' contributions to the increased political saliency of environmental issues. For instance, Economiesuisse explained on its webpage that despite its acknowledgement of the importance of climate change, "excessive government activism or hasty regulations are unnecessary."[112] The BDI also underlined:

> Non-governmental organizations (NGOs) have become a challenge for trade associations. They are increasingly dealing with traditional topics of trade associations and, through increasing professionalization and internationalization, have become players to be taken seriously on the national as well as on the international political stage. Their influence on political decision-making processes is thus growing directly; but it is also growing indirectly through their impact on public opinion.[113]

NGOs sometimes appeared to be more effective, since they could specialise on single issues such as climate change, often had a shorter reaction time thanks to their organisational flexibility, and could combat business associations' political lobbying, especially at the international level.[114] While both the BDI and Economiesuisse per se preferred harmonised international solutions with respect to climate change instead of national solutions, they monitored closely the foreseen enforcement mechanisms and the potential influence of their political opponents. In response, the BDI promoted the creation of an expert pool from the business community which could be mobilised quickly to counterbalance NGOs' influence.[115]

In addition to improvements in communication and their traditional political mobilisation, Economiesuisse and the BDI also developed new strategies, especially concerning the promotion of solutions developed jointly by businesses, governments, and NGOs. Economiesuisse for instance underlined its "desire for more effective and constructive collaboration among stakeholders and companies," while highlighting that a commissioned study had shown that "cooperation with NGOs was seen by companies as partly "critical and constructive" and "sometimes as unnecessarily antagonistic."[116] As early as 2000, the BDI also emphasised that past experience had proved that "it is possible to hold roundtable discussions with NGOs on many topics without having to give up one's own positions, which enables a better assessment of NGO demands and helps to reduce fears of contact." It added that, "in the long term, thought could also be given to joint high-profile events or campaigns with NGOs on various topics."[117]

In addition to a declared willingness to cooperate more with NGOs, the BDI and Economiesuisse turned to strategies similar to those of individual companies such as showing businesses' present and future efforts at reducing their impact on the climate and turning to consulting groups such as McKinsey to do so. The very same consulting companies that had advocated methods for hard-nosed shareholder value maximisation therefore found themselves as consultants promoting sustainable initiatives and stakeholder engagement.[118] In this vein, McKinsey was commissioned by the BDI in 2007 to publish a report about the "costs and potentials of greenhouse gas abatement in Germany," which investigated several hundred individual technical solutions for the abatement of greenhouse gas through the participation of 70 companies and associations, while simultaneously "intentionally avoid[ing] any assessment of policies, political implementation programs, and other governmental interventions."[119] It is therefore striking to note how depoliticisation and technical solutions underlie such consulting approaches. Economiesuisse, partnering with the WWF, also mandated McKinsey to produce a 2022 report on decarbonising opportunities based on contact with 180 companies and sectoral business associations, highlighting the opportunities to learn from best practices and the role NGOs might have in sensitising companies to the existing climate-friendly solutions.[120]

Additionally, business associations increasingly played the role of corporate environmentalists in the domain of climate change. The BDI, which had a climate-dedicated task force (*Ausschuss Energie-und Klimapolitik)* aimed at identifying and establishing "the common energy and climate policy interests of German industry, in order to then effectively incorporate them into the political processes"[121] also, in 2000, partenered with Econsense, the *Forum for Sustainable Development of German Business.* Econsense assembled a network of about 30 big businesses with a strong German presence including VW, Lufthansa, Bayer, Thyssenkrupp, and HeidelbergCement. The goal of the organisation was "in exchange with business, politics, and civil society, [to] proactively address sustainability challenges and advocate frameworks and policies that enable innovation and competitiveness."[122] In a 2016 paper on the future contributions by business to prevent climate change, Econsense identified four types of risk: market, physical, regulatory and reputational. Regarding reputational risk, they explained that these are "the result of a perceived insufficient commitment of companies in the context of climate protection, at both company and product levels. Companies can improve their reputation through climate-friendly solutions and technologies."[123] The perception by civil society (consumers and NGOs) was therefore clearly considered as a risk related to climate change, alongside physical risks stemming from natural phenomena. Such an initiative emanating from the BDI is also interesting since it shows how business associations coached their members to become corporate environmentalists to prevent legitimation and regulatory issues.

12.5 Conclusion

Since 2018, activism and its scope seem to have reached another level in the wake of the school strike movement started by Swedish activist Greta Thunberg. Under the banners of Fridays for Future (FFF) and Extinction Rebellion (XR), the September 2019 strike took place around the world, gathering an estimated 1.4 million people in Germany and about 100,000 people in Switzerland. In addition, litigation has become an important course of action for social movements in forcing states to take a more active role in achieving effective climate mitigation policies for future generations, as is shown by the recent judgment of the European Court of Human Rights which condemned Switzerland for violating the right to climate protection. Such trends are caused by the "disillusionment with multilateral environmental governance" and represent once again a shift from insider tactics such as advocacy to renewed forms of outsider activist strategies.[124] Overall, with the climate crisis worsening, it is likely that climate activism in its various forms will continue to be of concern for corporations and their business associations.

This chapter makes the case for the relevance of a business history approach to complement the literature developed by management, international business, and scholars of the sociology of organisations. Indeed, studying activist/business relationships over the long-term and through corporate and business associations' archives helps to untangle greenwashing from the actual implementation of managerial practices aimed at diminishing corporations' carbon footprints. Moreover, it considers not only direct business-NGO relationships but also the latter's struggles to gain influence over policymakers.

The study of business associations from Germany and Switzerland in the 1970s indeed uncovers how business circles considered that strong environmental movements had an impact on the political climate. The movements' influence on public opinion through the "dramatisation" of the environmental question and their direct involvement in politics was particularly noteworthy. In response, the SHIV and the BDI created dedicated task forces and intensified their communication efforts. The political actions of NGOs still appear to be important for climate change mitigation, especially when directed at international organisations. The BDI indeed recognised that, to some extent, NGOs played a role similar to that of business associations, and their specialisation on a single issue and their flexible structure made them serious opponents.

The example of the SHIV and the BDI also indicates that there is a learning curve over time, with an increased emphasis on multi-stakeholder partnerships in solving societal problems. A global assessment of such cooperation nevertheless shows that the dialogue often seems limited to mainstream and well-established NGOs such as the WWF, while grassroots movements and NGOs using high impact tactics such as Greenpeace are ignored. Moreover, business associations tend to turn more often to consulting firms and business schools than to NGOs for advocacy. Finally, there is also a trend towards developing corporate environmentalism. Demonstrating concrete examples of business contributions to greenhouse reduction can be virtuously portrayed as addressing the climate crisis, but simultaneously can decrease the sense of urgency in policymaking. In the long run, despite some periodic victories, the consequence of the primacy of voluntary agreements, corporate environmentalism, and related stakeholder partnerships, is that positive global impacts on climate change have become dependent on the aggregated goodwill of individual corporations to respond to social demands. Moreover, this leaves very little room for addressing global climate justice issues and the historically ignored responsibilities of industrialised countries.

Notes

1 The author would like to thank the Schmidheiny Foundation for its financial support for the conducted archival research.
2 BDI, "Ohne technologischen Fortschritt wird das Klimaproblem nicht gelöst," https://bdi.eu/spezial/forward-to-the-new/emission-thinking#/artikelwide/news/klimadebatte-holger-loesch-und-ottmar-edenhofer.
3 For an overview, see Bereni and Dubuisson-Quellier, "Au-delà de la confrontation"; Soule, *Contention and Corporate Social Responsibility*; Rome, "Plan C for Greening Business."
4 See Duguid and Lopes, Chapter 14 of this book on the emergence and evolution of greenwashing.
5 See for instance: Cambridge Dictionary, "Activism," https://dictionary.cambridge.org/fr/dictionnaire/anglais/activism.
6 Bereni and Dubuisson-Quellier, "Au-delà de la confrontation," 506; Yaziji and Doh, *NGOs and Corporations,* 4.
7 Yaziji and Doh, *NGOs and Corporations,* 10.
8 Wright, Nyberg, and Grant, " 'Hippies on the Third Floor'," 1452.
9 Yaziji and Doh, *NGOs and Corporations,* xiii.
10 Kostova and Zaheer, "Organizational Legitimacy."
11 Soule, *Contention and Corporate Social Responsibility*, 3.
12 Pedersen and Pedersen, "The Rise of Business-NGO Partnerships," 8.
13 Yaziji and Doh, *NGOs and Corporations,* 62–65.
14 Balsiger, "Managing Protest."
15 Pedersen and Pedersen, "The Rise of Business-NGO Partnerships," 14.
16 Bereni and Dubuisson-Quellier, "Au-delà de la confrontation," 509.
17 Minefee and Bucheli, "MNC Responses"; Levy, "Black Power"; Pitteloud, "Have Faith."
18 Phillips-Fein, *Invisible Hands*; Pitteloud, "Let's Coordinate"; Rowell, *Green Backlash.*
19 Chamayou, *La société ingouvernable*; Pitteloud, "Unwanted Attention."
20 Kaplan, "Who Has Been Regulating Whom," 126; Stutz, "History in Corporate Social Responsibility," 177.
21 Jones and Lubinski, "Making 'Green Giants'."
22 Sluyterman, "Royal Dutch Shell," 225–26.
23 Jones, *Profits and Sustainability.*
24 Jones, *Deeply Responsible Business.*
25 Rome, "Beyond Compliance"; Rome, "DuPont"; Hoffman, *From Heresy to Dogma.*
26 Boltanski and Chiapello, *Le nouvel esprit*; Aloisi de Larderel et al., "Des pratiques autonomes," 149; Aggeri and Cartel, "Le changement climatique."
27 LeBaron and Dauvergne, *Protest Inc;* Pedersen and Pedersen, "The Rise of Business-NGO Partnerships," 10.
28 LeBaron and Dauvergne, *Protest Inc,* 156.
29 Herman, "Greenpeace Goes South."
30 Jarrige and Le Roux, *La contamination du monde.*
31 Engels, *The Condition of the Working-Class*, 59.
32 US Environmental Protection Agency, "EPA," https://www.epa.gov/archive/epa/aboutepa/rachel-carson.html.
33 Uekötter, *The Greenest Nation*? 78–79.
34 Meyer, "L'européanisation."
35 Aloisi de Larderel et al., "Des pratiques autonomes des entreprises," 146.
36 For overviews of the evolution of environmental awareness in Switzerland and Germany, see Walter, *Les Suisses et l'environnement*; Uekötter, *The Greenest Nation?*
37 Bund, "40 Jahre Bund, 40 Jahre ein starker Freund der Erde 1975–2015," https://www.bund.net/fileadmin/user_upload_bund/_migrated/publications/150721_bund_ueber_uns_40_jahre_bund_geschichte_des_bund.pdf.
38 Damir Skenderovic, "Ökologische Bewegung," in *Historisches Lexikon der Schweiz*, 2012, https://hls-dhs-dss.ch/de/articles/016515/2012-03-27/.
39 Jungjohann, "The German Green Party."
40 BDI, BDI-Ausschuss für Umweltfragen, December 1972, UP 7 K104, BDI Archiv, Berlin.

41 SHIV, Umweltschutz (Möglichkeiten der Koordination im Rahmen von Handel und Industrie) an die Mitglieder des Vororts, November 3, 1972, 118.3.1.1, AfZ IB-Vorort Archiv, Zürich.
42 SHIV, Aktennotiz über die 1. Sitzung des Vorort-Gremiums für Fragen der Umweltschutzpolitik, April 6, 1976, 118.3.1.2, AfZ IB-Vorort Archiv, Zürich.
43 Dr. A Camani, Arbeitsgruppe der Schweizerischen Kunstoffindustrie, Umweltschutz (Möglichkeiten der Koordination im Rahmen von Handel und Industrie), an den Vorort des SHIV, October 2, 1972, 118.3.1.1, AfZ IB-Vorort Archiv, Zürich.
44 Bergquist and David, "Beyond Planetary Limits!," 490.
45 BDI, Jahresbericht des Bundesverbandes der Deutschen Industrie, June 1, 1970–May 31, 1971, BDI-Archiv, Berlin, 102.
46 BDI, Anlage 2 Zu Punkt 1 der Tagesordnung: Konstituierung des Ausschusses. Ausführungen von Herrn Dr. Neef, Köln, November 2, 1972, UP 7 K104, BDI Archiv, Berlin.
47 Vorort des SHIV, Umweltschutz (Möglichkeiten der Koordination im Rahmen von Handel und Industrie) an die interessierten Fachverbände, November 3, 1972, 1–2, 118.3.1.1, AfZ IB-Vorort Archiv, Zürich.
48 Culpepper, *Quiet Politics*.
49 Dr. Hans Karl Vellguth, an Herrn Dr, Hellmuth Wagner, Geschäftsführendes Präsidentialmitglied des BDI, Düsseldorf, April 25, 1972, UP 7 K104, BDI Archiv, Berlin.
50 Sasson, "Milking the Third World?"
51 Centemeri, "The Seveso Disaster."
52 Zelko, "Scaling Greenpeace."
53 Of course, the BDI and the SHIV did not limit their strategies to depoliticisation. The analysis of the environmental lobbying efforts of the BDI and the SHIV, as well as their role in helping co-ordinate the views of the business community, goes beyond the scope of this chapter.
54 Vorort des SHIV, Eidg. Kommission für Gewässerschutz, an den Vorsteher des Eidg. Departements des Innern Herrn Bundesrat H.P.Tschudi, March 1, 1973, 118.3.1.1, AfZ IB-Vorort Archiv, Zürich.
55 Uekötter, *The Greenest Nation?* 91.
56 BDI, Geschäftführungskonferenz, Punkt 2 Probleme des Umweltschutzes, September 20, 1977, A219 (1), BDI-Archiv, Berlin.
57 Schweizerische Gesellschaft für Chemische Industrie, Ergebnisnotiz über die Sitzung der Arbeitsgruppe „Boden-Wasser_Luft" (BWL) im Gebäude 200, der Sandoz AG, December 13, 1973, 7–8, 118.3.1.1, AfZ IB-Vorort Archiv, Zürich.
58 Pitteloud, "Delay and Dilution."
59 Walter, *Les Suisses et l'environnement*, 250.
60 Gazette de Lausanne, Initiative et referendum, Nette tendance en faveur d'un doublement du nombre de signatures, February 5, 1973.
61 Pitteloud, "Delay and Dilution."
62 Gazette de Lausanne, Augmentation du nombre de signatures pour initiatives et referendum, le PSS est déçu, December 19, 1976.
63 Dr. Gässler, Ergebnisvermerk über die Besprechung mit dem Bundesinnenministerium am 27.05.1975 in Bonn Köln, June 2, 1975, 4, UP8/2, K104, BDI Archiv, Berlin.
64 Dr. Gässler, Abt. III Umweltfragen, In welchen Bereichen der Umweltpolitik hat der BDI in den letzten Jahren Erfolg gehabt? an Herrn Franz, Büro der HGF 11, Köln, May 9, 1980, UP009/2 K105, BDI Archiv, Berlin.
65 Hill, BDI, Sitzung des BDI-Ausschusses für Umweltfragen am April 20, 1982, Diskussion mit dem Bundesminister des Innern Herrn Gerhardt Rudolf Baum, Köln, February 10, 1982, UP 006 K 103, BDI-Archiv, Berlin.
66 Thurnherr, "Droit de recours," 187–98.
67 Vorort des SHIV, Umweltschutz (Möglichkeiten der Koordination im Rahmen von Handel und Industrie) an die interessierten Fachverbände, November 3, 1972, 5, 118.3.1.1, AfZ IB-Vorort Archiv, Zürich.
68 Dr. Gässler, Kabelitz, Abt. II Umweltfragen, Zusammenarbeit BDI III/3 und IW-Umweltfragen, Besprechung BDI/IW, an Herrn Dr. Walther, Köln, January 24, 1980, UP009/2 K105, BDI Archiv, Berlin.
69 BDI, Bedeutsame Entwicklungslinien der Umweltpolitik- im Rückblick und in der Vorausschau-, 1979, UP 006 K 103, BDI-Archiv, Berlin.
70 Uekötter, *The Greenest Nation?* 114.
71 Näsman and Pitteloud, "The Power and Limits of Expertise," 241–60.

72 Näsman and Pitteloud, "The Power and Limits of Expertise."
73 Uekötter, *The Greenest Nation?* 124–25; Mieg and Haefeli, *La politique environnementale.*
74 Pestre, "La mise en économie," 36.
75 Rowell, *Green Backlash*, 4.
76 Bergquist and David, "Beyond Planetary Limits!"
77 Pestre, "La mise en économie de l'environnement," 37.
78 Rome, "Beyond Compliance," 410.
79 Bergquist and David, "Beyond Planetary Limits!," 27.
80 Soule, *Contention and Corporate Social Responsibility*, 8; Zelko, "Scaling Greenpeace."
81 Gupta, "A History," 636.
82 Moser, "Communicating Climate Change," 32.
83 Oreskes and Conway, *Merchants of Doubt*; Supran and Oreskes, "Rhetoric and Frame Analysis."
84 Bonneuil, Choquet, and Franta, "Early Warnings."
85 Sluyterman, "Royal Dutch Shell," 222.
86 Parks, della Porta, and Portos, "Environmental and Climate Activism," 104–5.
87 SHIV, Jahresbericht, 1990, 101; BDI, Bericht 1988–1990, 159.
88 Bergquist and David, "Beyond Planetary Limits!," 16.
89 BDI, Bericht 1990–1992, 144.
90 SHIV, Protokoll der 5. Sitzung des Vorstandes, June 24, 1991, 30.
91 Michaelowa, "The Politics of Climate Change," 316–17.
92 Watanabe and Mez, "The Development of Climate Change Policy," 116.
93 Dupuis, "Le climat," 130–31.
94 Watanabe and Mez, "The Development of Climate Change," 118.
95 Dupuis, "Le climat," 131–32.
96 Kammerer, Ingold, and Dupuis, "Switzerland," 235–36.
97 Gupta, "A History," 636–53.
98 Moor de et al., "New Kids," 620.
99 Bartley and Child, "Shaming the Corporation," 674.
100 Greenpeace, "Nestlé KitKat Video," https://media.greenpeace.org/archive/Give-the-Orangutan-a-Break-Nestle-KitKat-Spoof-Video-27MZIF20N4P6.html.
101 Greenpeace, Le côté obscure de Volkswagen, 2010.
102 Letemps, "Sous pression de Greenpeace, Nestlé rompt avec un fournisseur," https://www.letemps.ch/economie/pression-greenpeace-nestle-rompt-un-fournisseur.
103 Nestlé, "Responsibly Sourced Palm Oil," https://www.nestle.com/sustainability/sustainable-sourcing/palm-oil.
104 20 minutes, "Greenpeace contre-attaque contre VW," https://www.20min.ch/fr/story/greenpeace-contre-attaque-contre-vw-431661960488.
105 Mujkic and Klingner, "Dieselgate"; Zhang et al., "Corporate Social Irresponsibility."
106 Volkswagen, "Climate Action with Added Social Value—Promoting Sustainable Technology," https://volkswagen-climatepartner.com/en/mission; Nestlé, "Climate Action," https://www.nestle.com/sustainability/climate-change.
107 Nestlé, "Nestlé Supports the LEAF Coalition to Protect Tropical Forests and Enhance Global Climate Action," https://www.nestle.com/media/news/leaf-coalition-protect-tropical-forests-global-climate-action.
108 Coalition Multinationales Responsables, "Un produit Nestlé contribue à la déforestation illégale de la forêt tropicale," https://responsabilite-multinationales.ch/etudes-de-cas/produit-nestle-contribue-deforestation-illegale/.
109 Actares, "Nestlé: Une feuille de route ambitieuse—quelques interrogations sur sa mise en œuvre," https://www.actares.ch/fr/news/210413-mm-gv-nestle.
110 Greenpeace, "Greenpeace Sues Volkswagen for Fuelling the Climate Crisis and Violating Future Freedom and Property Rights," https://www.greenpeace.org/international/press-release/50625/greenpeace-sues-volkswagen-for-fuelling-the-climate-crisis-and-violating-future-freedom-and-property-rights/.
111 RTS, "Si l'initiative passe, Nestlé ne pourra peut-être pas rester partout," https://www.rts.ch/info/suisse/11720669-si-linitiative-passe-nestle-ne-pourra-peutetre-pas-rester-partout.html.
112 Economiesuisse, "Développement durable," https://www.economiesuisse.ch/fr/focus/developpement-durable.

113 BDI, Aussenwirtschaftspolitik, Nichtregierungsorganisationen- Herausforderung für die Wirtschaftsverbände, September 5, 2000, BDI-Archiv, Berlin.
114 Newell, "Civil Society."
115 Newell, "Civil Society."
116 Economiesuisse in Zusammenarbeit mit WWF, Klimastandort Schweiz. Schweizer Unternehmen als globale Treiber für netto-Null, 2022, 23.
117 BDI, Aussenwirtschaftspolitik, Nichtregierungsorganisationen- Herausforderung für die Wirtschaftsverbände, September 5, 2000, BDI-Archiv, Berlin.
118 Mazzucato and Collington, *The Big Con*.
119 BDI, Business for Climate Initiative, Costs and Potentials of Greenhouse Gas Abatement in Germany, A report by McKinsey & Company Inc., 2007.
120 Economiesuisse in Zusammenarbeit mit WWF, Klimastandort Schweiz. Schweizer Unternehmen als globale Treiber für netto-Null, 2022.
121 BDI, BDI-Gremienstruktur, Ausschüsse und Arbeitskreise, 2023, 22.
122 Econsense, Accelerating the Race to Net Zero. How Companies Set, Implement, and Benefit from Science—based Targets, 2021, 2.
123 Edition Econsense, Zukunftsfähig wirtschaften, Der Beitrag von Unternehmen zur Bewältigung des Klimawandels, Ein Beitrag zur COP22, 2016, 8.
124 Parks, della Porta, and Portos, "Environmental and Climate Activism," 100.

Bibliography

Aggeri, Franck, and Mélodie Cartel. "Le changement climatique et les entreprises: Enjeux, espaces d'action, régulations internationales." *Entreprises et histoire* 86, no. 1 (2017): 6–20.

Aloisi de Larderel, Jacqueline, Richard Armand, Michel Capron, Bertrand Collomb, Geneviève Massard-Guilbaud, Éric Pezet, Franck Aggeri, and Aurélien Acquier. "Des pratiques autonomes des entreprises aux systèmes de régulation à multiples parties prenantes." *Entreprises et histoire* 45, no. 4 (2006): 142–56.

Balsiger, Philip. "Managing Protest: The Political Action Repertoires of Corporations." In *The Oxford Handbook of Social Movements*, edited by Donatella della Porta and Mario Diani, 653–66. Oxford University Press, 2015.

Bartley, Tim, and Curtis Child. "Shaming the Corporation: The Social Production of Targets and the Anti-Sweatshop Movement." *American Sociological Review* 79, no. 4 (2014): 653–79.

Bereni, Laure, and Sophie Dubuisson-Quellier. "Au-delà de la confrontation: saisir la diversité des interactions entre mondes militants et mondes économiques." *Revue française de sociologie* 61, no. 4 (2020): 505–29.

Bergquist, Ann-Kristin, and Thomas David. "Beyond Planetary Limits! The International Chamber of Commerce, the United Nations, and the Invention of Sustainable Development." *Business History Review* 97, no. 3 (2023): 481–511.

Boltanski, Luc, and Eve Chiapello. *Le nouvel esprit du capitalisme*. Paris: Gallimard, 1999.

Bonneuil, Christophe, Pierre-Louis Choquet, and Benjamin Franta. "Early Warnings and Emerging Accountability: Total's Responses to Global Warming, 1971–2021." *Global Environmental Change* 71 (2021): 102386.

Centemeri, Laura. "The Seveso Disaster Legacy." In *Nature and History in Modern Italy*, edited by Marco Armiero and Marcus Hall, 251–73. Athens: Ohio University Press & Swallow Press, 2010.

Chamayou, Gregoire. *La société ingouvernable: Une généalogie du libéralisme autoritaire*. Paris: La Fabrique, 2018.

Culpepper, Pepper D. *Quiet Politics and Business Power: Corporate Control in Europe and Japan. Cambridge Studies in Comparative Politics*. Cambridge: Cambridge University Press, 2011.

Dupuis, Johann. "Le climat: La Suisse, pionnière des politiques climatiques?" In *La politique environnementale en Suisse: De la police des forêts aux écobilans*, edited by Harald A. Mieg and Ueli Haefeli, 127–47. Berne: Neue Zürcher Zeitung NZZ Libro, 2020.

Engels, Frederick. *The Condition of the Working-Class in England*. London: Panther Book, 1969.

Gupta, Joyeeta. "A History of International Climate Change Policy." *WIREs Climate Change* 1, no. 5 (2010): 636–53.

Herman, Rebecca. "Greenpeace Goes South: The Promise and Pitfalls of Global Environmentalism in Argentina." *Environmental History* 29, no. 1 (2024).

Hoffman, Andrew J. *From Heresy to Dogma: An Institutional History of Corporate Environmentalism*. Expanded ed. Stanford: Stanford University Press, 2001.

Jarrige, François, and Thomas Le Roux. *La contamination du monde: Une histoire des pollutions de l'âge industriel*. Paris: Éditions du Seuil, 2017.

Jones, Geoffrey. *Profits and Sustainability: A History of Green Entrepreneurship*. Oxford: Oxford University Press, 2019.

Jones, Geoffrey. *Deeply Responsible Business: A Global History of Values-Driven Leadership*. Cambridge, MA: Harvard University Press, 2023.

Jones, Geoffrey, and Christina Lubinski. "Making 'Green Giants': Environment Sustainability in the German Chemical Industry, 1950s–1980s." *Business History* 56, no. 4 (2014): 623–49.

Jungjohann, Arne. "The German Green Party: From a Broad Social Movement to a Volkspartei." In *Environmental Sustainability in Transatlantic Perspective: A Multidisciplinary Approach*, edited by Manuela Achilles and Dana Elzey, 69–90. London: Palgrave Macmillan, 2013.

Kammerer, Marlene, Karin Ingold, and Johann Dupuis. "Switzerland: International Commitments and Domestic Drawbacks." In *Climate Governance Across the Globe*, edited by Rüdiger K.W. Wurzel, Mikael Skou Andersen, and Paul Tobin, 235–56. London: Routledge, 2020.

Kaplan, Rami. "Who Has Been Regulating Whom, Business or Society? The Mid-20th-Century Institutionalization of 'Corporate Responsibility' in the USA." *Socio-Economic Review* 13, no. 1 (2015): 125–55.

Kostova, Tatiana, and Srilata Zaheer. "Organizational Legitimacy Under Conditions of Complexity: The Case of the Multinational Enterprise." *Academy of Management Review* 24, no. 1 (1999): 64–81.

LeBaron, Genevieve, and Peter Dauvergne. *Protest Inc.: The Corporatization of Activism*. Hoboken: John Wiley & Sons, 2014.

Levy, Jessica Ann. "Black Power in the Boardroom: Corporate America, the Sullivan Principles, and the Anti-Apartheid Struggle." *Enterprise & Society* 21, no. 1 (2020): 170–209.

Mazzucato, Mariana, and Rosie Collington. *The Big Con: How the Consulting Industry Weakens Our Businesses, Infantilizes Our Governments and Warps Our Economies*. London: Allen Lane, 2023.

Meyer, Jan-Henrik. "L'européanisation de la politique environnementale dans les années 1970." *Vingtième Siècle: Revue d'histoire* 113, no. 1 (2012): 117–26.

Michaelowa, Axel. "The Politics of Climate Change in Germany: Ambition versus Lobby Power." *WIREs Climate Change* 4, no. 4 (2013): 315–20.

Mieg, Harald A., and Ueli Haefeli. *La politique environnementale en Suisse: De la police des forêts aux écobilans*. Berne: Neue Zürcher Zeitung NZZ Libro, 2020.

Minefee, Ishva, and Marcelo Bucheli. "MNC Responses to International NGO Activist Campaigns: Evidence from Royal Dutch/Shell in Apartheid South Africa." *Journal of International Business Studies* 52, no. 5 (2021): 971–98.

Moor, Joost de, Michiel De Vydt, Katrin Uba, and Mattias Wahlström. "New Kids on the Block: Taking Stock of the Recent Cycle of Climate Activism." *Social Movement Studies* 20, no. 5 (2021): 619–25.

Moser, Susanne C. "Communicating Climate Change: History, Challenges, Process and Future Directions." *WIREs Climate Change* 1, no. 1 (2010): 31–53.

Mujkic, Edin, and Donald Klingner. "Dieselgate: How Hubris and Bad Leadership Caused the Biggest Scandal in Automotive History." *Public Integrity* 21, no. 4 (2018): 365–77.

Näsman, Mattias, and Sabine Pitteloud. "The Power and Limits of Expertise: Swiss–Swedish Linking of Vehicle Emission Standards in the 1970s and 1980s." *Business and Politics* 24, no. 3 (2022): 241–60.

Newell, Peter. "Civil Society, Corporate Accountability and the Politics of Climate Change." *Global Environmental Politics* 8, no. 3 (2008): 122–53.

Oreskes, Naomi, and Erik M. Conway. *Merchants of Doubt: How a Handful of Scientists Obscured the Truth on Issues from Tobacco Smoke to Climate Change*. Reprint ed. New York, NY: Bloomsbury Publishing, 2011.

Parks, Louisa, Donatella della Porta, and Martín Portos. "Environmental and Climate Activism and Advocacy in the EU." In *Handbook on European Union Climate Change Policy and Politics*, edited by Tim Rayner Kacper Szulecki, Andrew J. Jordan, and Sebastian Oberthür, 98–112. Edward Elgar Publishing.

Pedersen, Esben Rahbek Gjerdrum, and Janni Thusgaard Pedersen. "Introduction: the rise of business—NGO partnerships." *The Journal of Corporate Citizenship* 50 (June 2013).

Pestre, Dominique. "La mise en économie de l'environnement comme règle." *Ecologie politique* 52, no. 1 (2016): 19–44.

Phillips-Fein, Kim. *Invisible Hands: The Making of the Conservative Movement from the New Deal to Reagan*. New York: Norton, 2009.

Pitteloud, Sabine. "Delay and Dilution in the Implementation of Environmental Norms: Business Groups and the Regulation of Car Emissions in Switzerland in the 1970s–1980s." *The Paul Bairoch Institute of Economic History Working Paper Series*, (2020): 1–14.

Pitteloud, Sabine. "Unwanted Attention: Swiss Multinationals and the Creation of International Corporate Guidelines in the 1970s." *Business and Politics* 22, no. 4 (December 2020): 587–611.

Pitteloud, Sabine. "Have Faith in Business: Nestlé, Religious Shareholders, and the Politicization of the Church in the Long 1970s." *Enterprise & Society* 25, no. 3 (2024): 699–727.

Pitteloud, Sabine. "Let's Coordinate! The Reinforcement of a 'Liberal Bastion' Within European Industrial Federations, 1978–1987." *Business History* 65, no. 2 (2023): 345–65.

Rome, Adam. "DuPont and the Limits of Corporate Environmentalism." *Business History Review* 93, no. 1 (2019): 75–99.

Rome, Adam. "Beyond Compliance: The Origins of Corporate Interest in Sustainability." *Enterprise & Society* 22, no. 2 (2021): 409–37.

Rome, Adam. "Plan C for Greening Business." *Stanford Social Innovation Review*, June 2024.

Rowell, Andrew. *Green Backlash: Global Subversion of the Environment Movement*. London: Routledge, 2017.

Sasson, Tehila. "Milking the Third World? Humanitarianism, Capitalism, and the Moral Economy of the Nestlé Boycott." *The American Historical Review* 121, no. 4 (2016): 1196–224.

Sluyterman, Keetie. "Royal Dutch Shell: Company Strategies for Dealing with Environmental Issues." *Business History Review* 84, no. 2 (2010): 203–26.

Soule, Sarah A. *Contention and Corporate Social Responsibility*. Illustrated ed. New York: Cambridge University Press, 2009.

Stutz, Christian. "History in Corporate Social Responsibility: Reviewing and Setting an Agenda." *Business History* 63, no. 2 (2021): 175–204.

Supran, Geoffrey, and Naomi Oreskes. "Rhetoric and Frame Analysis of ExxonMobil's Climate Change Communications." *One Earth* 4, no. 5 (2021): 696–719.

Thurnherr, Daniela. "Droit de recours des organisations environnementales." In *La politique environnementale en Suisse: de la police des forêts aux écobilans*, edited by Harald A. Mieg and Ueli Haefeli, 187–98. Berne: Neue Zürcher Zeitung NZZ Libro, 2020.

Uekötter, Frank. *The Greenest Nation? A New History of German Environmentalism*. Cambridge: MIT Press, 2014.

Zelko, Frank. "Scaling Greenpeace: From Local Activism to Global Governance." *Historical Social Research/Historische Sozialforschung* 42, no. 2 (160) (2017): 318–42.

Zhang, Michael, Glyn Atwal, and Maya Kaiser. "Corporate Social Irresponsibility and Stakeholder Ecosystems: The Case of Volkswagen Dieselgate Scandal." *Strategic Change* 30, no. 1 (2021): 79–85.

Walter, François. *Les Suisses et l'environnement: Une histoire du rapport à la nature, du XVIIIe siècle à nos jours*. Carouge-Genève: Ed. Zoé, 1990.

Watanabe, Rie, and Lutz Mez. "The Development of Climate Change Policy in Germany." *International Review for Environmental Strategies* 5, no. 1 (2004): 109–26.

Wright, Christopher, Daniel Nyberg, and David Grant. "'Hippies on the Third Floor': Climate Change, Narrative Identity and the Micro-Politics of Corporate Environmentalism." *Organization Studies* 33, no. 11 (2012): 1451–75.

Yaziji, Michael, and Jonathan Doh. *NGOs and Corporations: Conflict and Collaboration*. Cambridge: Cambridge University Press, 2010.

13

FROM WHITEWASHING TO GREENWASHING (AND BACK AGAIN)

Paul Duguid and Teresa da Silva Lopes

13.1 Introduction

In 2004, Adam Werbach, formerly the youngest head of the Sierra Club and a Greenpeace activist, declared environmentalism "dead" and that he was "done calling [himself] an environmentalist."[1] Seven years earlier, Werbach had been urging climate activists to "fight" for 'radical change.'"[2] His "autopsy" of 2004 argued, in contrast, that activists' radicalism was not merely ineffective but positively damaging, undermining environmental progress. Despite its earlier victories, he claimed, the environmental movement had begun "to lose the war of ideas." Thus, though Werbach had played a part in shaping the movement's radicalism, he now denounced it. He sought to develop a new approach, encapsulated in the title of a book written five years after his autopsy, *Strategy for Sustainability: A Business Manifesto* (2009). The book continues Werbach's denunciation of "greens," insisting that "a green strategy is not a strategy for sustainability."[3] Werbach's title echoed the subtitle of another influential book challenging environmentalists that came out in 2008 and took a similarly "corporatist" approach: E. Bruce Harrison's *Corporate Greening 2.0: Create and Communicate Your Climate Change and Sustainability Strategies.*

Echoing titles and promotion of corporate "sustainability" are not the only similarities between these two authors. Though Werbach does not acknowledge Harrison's work, in 1993, Harrison too had announced that environmental activism had "died." "[R]ag-tag activists" of the past, he maintained, were being replaced by people like himself, "polished" figures in a movement "increasingly managed like a business." These figures engaged in "sustainable communication" and formed "sustainable relations."[4] In promoting these new visions for corporate environmental projects, Werbach and Harrison shared one more commonality. In the 1960s, Harrison had helped the Manufacturing Chemists Association counter the environmental challenge to corporate chemistry in Rachel Carson's *Silent Spring*. Then in 1970, he founded his own public relations (PR) firm and developed "sustainable" PR strategies for corporations. An obituary for Harrison referred to him as the "Dean of green PR," a reflection in part of Harrison's influential *Going Green: How to Communicate Your Company's Environmental Commitment*, which came out in 1993.[5] Similarly, Werbach followed his denunciation of environmentalism

DOI: 10.4324/9781003478089-14

by turning to PR to shape his future work. He joined Saatchi & Saatchi and worked with clients like Walmart and Amazon to help them project themselves as environmentally sustainable businesses, becoming known as one of "the big dirty hands behind Wal-mart's Greenwashing."[6]

Corporate attempts to promote themselves as "green" companies have long been shaped by PR figures such as Harrison and Werbach. In 2022, a U.S. Congress committee questioned six major PR companies about the tactics they deployed to support polluting clients, despite claims to be engaging in "Green PR."[7] It is the goal of this chapter to fit the tactics of such companies and of figures like Harrison and Werbach into a larger, business-historical account of the shaping of greenwashing. To do this, it draws on a history of PR promoted by a figure known as "the father of public relations," Edward Bernays.[8] Bernays saw himself as helping to separate the fledgling field of PR from more conventional practices of advertising and publicity. From early on, he insisted that the PR counsel "is not an advertising man." While the latter primarily seeks to persuade potential consumers to buy a particular product, the PR counsel's role is the more complex one of making "the producer . . . understand what the public wants and to make the public understand the objectives of the producer."[9] From this perspective, while marketing tends to be associated with products and services, PR is aimed at enhancing the credibility of organisations or the individuals it represents.

While the field of business history has engaged with environmentalism, it has paid comparatively little attention to the contribution of PR.[10] Though Bernays had numerous exchanges with Norman Gras, a founding figure of business history who thought that business history was essential to PR, neither Bernays, though a major figure in the history of PR, nor Harrison, though significant in the development of greenwashing PR, received much attention from the field.[11] What attention business history journals did pay to Bernays declined from the 1980s, just as greenwashing was developing as a concept.[12] Hence as the field increases its engagement with climate change, it seems important to address the topic from the perspective of PR.

In what follows, we examine enduring PR strategies aimed at dismissing the critics of their clients while presenting those clients to the public in a favourable light. We follow Bernays' argument that these strategies grew out of earlier attempts at corporate "white-washing." In Bernays' view, corporations (or their PR strategists) deployed whitewashing, the practice of concealing faults and errors by painting a picture of an entity in a positive light, in response to the growth of what U.S. president Theodore Roosevelt had called "muckraking" journalism. One critical whitewashing strategy was to paint critics as unrestrained radicals engaged in spreading scandal.[13] Over time, however, from Bernays' perspective, PR strategies became more subtle and more reputable. The field achieved this reputability, he claimed, through a series of progressive "revolutions" in the first half of the twentieth century. We will outline these, then move beyond Bernays' account into the second half of the century and PR's engagement with environmentalism. Although Bernays' own PR work did not focus on environmental issues *per se*, viewed in retrospect, connections can be seen between his work and environmental topics. Promoting meat, tobacco, and banana companies, he managed to conceal problematic health issues for consumers and workers. He also worked to distract attention from environmental damage caused by firms he worked with.

More generally, Bernays' work both highlighted companies' genuine achievements and capabilities and disguised or changed the narrative relating to unfavourable aspects of their work. In so doing, he also misrepresented, dismissed, or just denounced critics. In these ways, it is possible to see Bernays as an important progenitor of greenwashing tactics. Putting his account of PR in the context of figures like Harrison and Werbach also raises questions of not only whether

Bernays' revolutions were far less progressive than he claimed but also whether, as the proponents of the "new" environmental PR like Harrison and Werbach denounce contemporary critics of greenwashing as being politically inept radicals, PR has retreated to the whitewashing tactics that Bernays claimed the field had abandoned.

We begin this argument by looking, in the following section at the development of notions critical to the argument such as "green," "environmental," "sustainable," and "greenwashing." We next introduce Bernays' and his influential role before turning to his argument about the "revolutionary" development of PR. We then turn to Harrison, who carried many Bernaysean ideas into the world of environmental PR, before returning, via figures like Werbach, to environmental PR at the beginning of the twenty-first century and ways in which it echoes the whitewashing of the early twentieth century.

13.2 Background and Literature Review

Greenwashing can seem to come from all directions. Wherever notions of "green" and "environmental" are deployed, greenwashing soon follows. Thus, it reaches out from every shelf in the local supermarket with brands such Body Shop maintaining that they did not "believe in profits without [environmental] principles."[14] It reaches onto your "desktop," as corporations like Amazon, running huge, energy-consuming server farms, talk instead in terms of the ethereal "cloud."[15] Meanwhile, research shows how Apple and Google rely on "green magic" and environmental ambiguity to distract from the carbon footprints of their Silicon Valley campuses.[16] But greenwashing also has a global reach, raised in central environmental events such as the COP 27 United Nations environmental conference in 2022. For this, as they had so often before, oil companies deployed PR agencies to frame for them an environmentally positive or "sustainable" identity in the context of the conference.[17] Such framing was also deployed at events less directly linked to environmental issues such as the FIFA World Cup held in Qatar—a country that is said to have the highest per-capita production of CO_2—in 2022.[18] FIFA nonetheless claimed the event to be "fully carbon neutral," pushing back against accounts that saw it as "a climate catastrophe."[19]

Following a Swiss court's condemnation of this carbon neutral claim, a representative of one organisation that had taken FIFA to court argued confidently that "[t]his ruling is further proof that people are seeing through greenwash and that flaky carbon-neutral claims will get quickly called out."[20] Such optimism is naive. Despite figures like Harrison endorsing green "transparency," greenwashing in general succeeds through information asymmetry, obscurity, ambiguity, and opacity.[21] Many consumers are keen to make environmentally responsible choices, but the evidence they rely on is regularly the product of manipulation by PR and marketing, which promote deceptive claims of climate sustainability in order to achieve financial sustainability for their clients.[22] A 2009 report on environmental labelling claimed that 98% of nearly 800 products surveyed in the UK committed "at least one Sin of Greenwashing."[23] Potential for confusion and obfuscation is evident in the US Securities and Exchange 2022 report concerning companies that deploy "Environmental, Social, and Governance" (ESG) strategies. This showed that many such deployments are little more than greenwashing, providing misleading data about environmental practices and the terms used to describe them.[24] And in 2024, a press release by Democrats in the U.S. Congress included a report on the extensive greenwashing of major oil companies.[25]

The notion of "environmentalism," central to this discussion, spread in the 1960s in the wake of Carson's *Silent Spring*. It can be marked in such things as the 1967 founding of the

Environmental Defense Fund (EDF) by a group of lawyers who, like Carson, had exposed environmental damage caused by DDT. Three years later U.S. President Richard Nixon created the Environmental Protection Agency (EPA). EPA regulation was seen as important by environmentalists, while, as we shall see, strongly opposed—often with the help of PR companies—by those whom the EPA sought to regulate.

As environmental pollution reached across the globe, such regulation, which tended to be limited by national borders, was seen by many as inadequate. In the 1970s, the Greenpeace project, developed initially as an anti-nuclear movement, increasingly took on polluters and portrayed environmentalism as a global issue. Thus, by the 1970s, *green* had become a term espousing international environmental concerns and activism. Consequently, it became a worrying concept for corporations, concerns reflected in George Fisk's *Marketing and the Ecological Crisis* (1974). This was an early attempt to proclaim corporate probity in this area through deft corporate marketing, though these two activities—probity and marketing—are often in significant tension with one another.[26]

Unsurprisingly, given the potential of such marketing to build a positive reputation for entities acting negatively, in the following decade, the term *greenwashing* emerged. It was coined by a young environmentalist Jay Westerveld in 1983 to dismiss hotel notices telling guests not to expect their towels to be washed every day as doing so was bad for the environment. Westerveld realised that daily washing was also bad for hotels' bottom lines, but that was not mentioned in the notices. By the end of the decade the term was widely used to designate organisations misleadingly projecting themselves as environmentally friendly.[27] One of the first published explorations of the notion came in 1992 from Greenpeace in *The Greenpeace Book of Greenwash.* In keeping with Greenpeace's international perspective, this proclaimed that "greenwash is going global" through such things as the participation of transnational polluting corporations in the United Nations Conference on the Environment in Rio that year. Among the companies denounced were predictable candidates such as Dupont, Shell, General Motors, and Westinghouse. But it also included the PR firm of Burson-Marsteller, accused of playing a significant role in disguising contributions of particular corporations to major climate disasters such as the nuclear explosion at Three Mile Island, the Union Carbide disaster in Bhopal, and the Exxon Valdez oil spill. For these and many other environmentally damaging events, Burson-Marsteller claimed to have helped "clients to manage such issues" and "counteract activist-generated . . . concerns." Denouncing PR agencies as "Greenwash Professionals," the Greenpeace book argued that such firms are hired to target "the hearts and minds of the global public" regardless of truth.[28]

In the 1990s, documents from tobacco industry archives show Philip Morris fearing that "anti-corporate organizations" such as Greenpeace would attack the company for "attempting to 'greenwash' its image." The company saw this as "serious" threat, fearing that such critics could "disseminate its propaganda worldwide."[29] Within the corporate world more generally, "greenwashing" was seen as an accusation to be evaded. In response, PR agencies and their clients appropriated positive notions of "environmentalism," "sustainability" and "green" for their clients. Hence, for example, while critics sought to label Harrison as "the father of greenwashing," the PR industry, as noted, labelled him "the Dean of Green PR."[30]

Another concept that came to play an important role in the context of environmentalism and greenwashing was the notion of "sustainable" (and, by extension, "sustainability"). Though often dated to the 1980s, *sustainable* as an environmental term emerged in the 1970s. It appears in the title of a 1976 report by the Georgia Conservancy, encouraging "sustainable energy development." This "Wolfcreek Statement" was circulated to, among others, the U.S. President-elect Jimmy Carter, who was also a member of the Conservancy.[31] In 1977, Carter commissioned the Council on Environmental Policy to investigate the efficacy of such policy. This resulted in the *Global 2000 Report*, which, among other things, explored whether such policy was likely to be "sustainable."[32] Carter wrote an introduction to the 1991 edition, which concluded with assertions about what "we can do collectively, as men and women committed to sustainable development." He also deplored the inefficiency of such things as the US Agency for International Development, which he noted was "supposed to be designed for sustainable development," but which he saw as "almost totally incompetent" and, by implication, playing a role in greenwashing military policies of the government.[33]

The term *sustainable* was more widely adopted following the report of the Brundtland Commission, sponsored by the UN World Commission on Environment and Development in 1987.[34] The report promoted the notion of "sustainable development" as that which "meets the need of the present without compromising the ability of future generations to meet their own needs." The notion was quickly and deftly appropriated by Harrison, who promoted "sustainable communication," something that pushed PR strategists to create an osmosis between environmental and corporate sustainability, allowing them to vindicate themselves and their clients despite questionable environmental activity.[35]

Today, the literature around these concepts is extensive. But, as noted earlier, analysis was relatively late coming into the field of business history. Thus, in trying to trace the historical development of greenwashing and strategies it embraced, what follows will look outside business history and into the history of PR, beginning, in the following section, by looking through the eyes of one of its early progenitors, Edward Bernays.

13.3 Edward L. Bernays

Bernays was the nephew of Sigmund Freud, whose influence can be detected in Bernays' approach to manipulating the human subconscious.[36] His early work engaged with medical research, bringing new medical findings to a larger audience, leading Bernays later to lean on doctors to endorse clients' products, from bacon and tobacco to bananas.[37] When the United States entered World War I, Bernays joined the Foreign Press Bureau of the U.S. Committee of Public Information (known after its chairman, George Creel, as the "Creel Commission"). The committee oversaw war-time propaganda and garnered public support for government and military policy by, as Bernays put it, "shaping the morale of our own people."[38] From this work, Bernays embraced the idea of "propaganda," which, as it had led to victory, was generally seen positively at the time.[39]

In an influential book written ten years after the war, Bernays acknowledged that, though the need to shape the public mind remained critical, the term *propaganda* carried "to many minds an unpleasant connotation" and was better dropped.[40] Nonetheless, he continued to embrace propaganda-like practices as, conveniently, essential not merely to marketing but also to democracy:

> If we understand the mechanisms and motives of the group mind, it is now possible to control and regiment the masses according to our will without their knowing it. . . . Those who manipulate this unseen mechanism of society constitute an invisible government which is the true ruling power of our country.[41]

Thus, after World War II, but still in step with the work of the Creel Commission, Bernays in 1945 came to describe the work of his field as the "engineering of consent" and his role as a "societal technician," intimating that consumers were machines that could be tuned with the help, in particular, of advances in the social sciences.[42]

Fighting against the inherent criticism of terms like *propaganda*, Bernays sought to elevate the field, implicitly aligning PR with the legal profession by arguing that "business organizations . . . often rely on an outside counsel on Public Relations, comparable to the practice that prevails in legal relations" while fending off likely criticism of its propagandist activities by insisting that it was driven by "public interest and not pecuniary motivation" and that it played a "constructive social role." In the same vein, he suggested that as a "profession," like law and medicine, PR "deserved validation by the state."[43] In all such writings, Bernays sought to lend credibility not only to his clients, the usual focus of PR activity, but also to himself and his field, presenting that as a profession with high ethical standards and socially beneficial outcomes.[44]

Bernays' clients included a number of U.S. presidents: Calvin Coolidge, Woodrow Wilson, Herbert Hoover, and Dwight Eisenhower.[45] Such political connections probably fuelled Bernays' concern when, in 1933, he was told that the German fascist Heinrich Goebbels acknowledged basing Nazi anti-Semitic propaganda strategies on Bernays' *Crystallizing Public Opinion*. In response, Bernays quickly noted "any human activity can be used for social purposes or misused for antisocial one."[46] In clearing his name (and his profession's reputation) from Nazi-like associations, Bernays drew on strategies developed around the same period as he worked to prevent corporations from being blamed for their role in bringing about the Great Depression. In so doing, and in an ongoing tradition for corporate PR, he helped clients resist government regulation. In such ways, and throughout his life, Bernays generally managed to tout what he saw as the positive effects of his work, without acknowledging negative ones. Thus, when he wrote for *Business History Review* about his work with the American Tobacco Company to encourage women to smoke, he made his major contribution seem to be persuading women to change the colour of their clothing to avoid clashing with cigarette packaging.[47]

Overall, Bernays work portrays PR as manufacturing and manipulating public opinion for the good of the individual and society. He acknowledged this has risks, but he claimed that he and the field managed over time to raise themselves above the unethical and antisocial and instead to pursue positive "social purposes." As this suggests, one of the challenges in understanding Bernays' work and his reflections upon it, is deciding whether his mandate for enlightened PR promotes ethical behaviour or the *appearance* of ethical behaviour. This distinction between practice and portrayal is, of course, central to the notion of greenwashing.

13.4 PR Revolutions

Bernays' vision of the development of PR can be explored through "The Revolution in Publicity," an article he wrote in 1941. The account lays out four periods: 1900–1914, 1914–1918, 1919–1929, then 1929 and beyond (bringing his readers up to 1941). As the article's title suggests, each new period is separated from its predecessor by transformative (and beneficial) "revolutions."[48]

Bernays describes the first period as the era of "white-washing," developed in response to muckraking journalism. Such journalism, in Bernays' view, turned the "white glare of publicity on the 'curse of bigness in business.'"[49] The journalists had revealed inequities of government and industry, uncovering disturbing but generally hidden features of slums, prisons, factories, and the like. A central "muckraking" text was Upton Sinclair's *The Jungle*, revealing horrifying practices in the meat industry in the U.S. Midwest. Bernays was later to become the champion of the bacon trade. He encouraged doctors to publicly endorse healthy breakfasts, then suggested that bacon and eggs fulfilled this goal.[50]

Another significant muckraking work was Ida Tarbell's *History of the Standard Oil Company* (1904). Compiled from earlier, widely circulated articles, it is probably the first major attack on the oil industry, looking in particular at Standard Oil's growing monopolistic power and the processes by which a particular industry passes from "the control of the many to that of the few."[51] Standard Oil was also a pioneer in PR. When the company's resistance to union activity led to a mass shooting of strikers at the "Ludlow Massacre" in Colorado in 1914, the company's founder, John D. Rockefeller, hired the early PR strategist Ivy Lee to salvage the company's image and his own. Lee deftly distracted from the Massacre by promoting Rockefeller's reputation as a philanthropist.[52]

One consequence of muckraking revelations was antitrust legislation and other forms of regulation. *The Jungle* was swiftly followed by the Pure Food and Drug Act and the Meat Inspection Act, both in 1906, while Tarbell's work on monopolies helped encourage the strengthening of antitrust regulations and the establishment of the Federal Trade Commission in 1914. Such regulation, controlling labour practices in food production and restricting corporate monopolies, was inimical to many corporations. Abetted by PR pioneers, the companies fought vehemently against government intervention.[53] Such work included "counter-attacks" by proto-PR agencies, denouncing critics of their clients as dishonest. In so doing, in Bernays' view, corporations became aware of the "power of words" and learnt to say "what they wanted to and what they thought would fool the public."[54] This account of relatively crude strategies in the first period of PR allowed Bernays to enhance those like himself that overthrew such crudity in the ensuing PR "revolutions." Of course, Bernays' account is itself a significant act of whitewashing of the PR profession. Throughout the succeeding periods laid out by Bernays and well after it, up to the greenwashers of our time, whitewashing those criticised while dismissing and denouncing the critics has continued to play a major strategic role.

Bernays brief comments on the second period, 1914–18, move from corporations to nations, following governmental recognition of "the importance of managing public opinion" and the ensuing efforts towards the selling of "arms and ideals."[55] The evident success of the "Words that Won the War" led, in Bernays' view, to the post-war, third period, 1919–1929, when the "lessons of war" were applied to peace-time pursuits, so that strategists could "regiment the public mind every bit as much as an army regiments its soldiers."[56] In all, Bernays argues that from World War I to World War II, propaganda was as justifiable in peace time as in war, because the "conscious and intelligent manipulation of the organized habits and opinions of the masses is an important element in democratic society."[57]

With his final revolution, Bernays is eager to resist the condemnation of PR that might follow such analysis. In Bernays' view, the Great Depression and its causes led, in 1929, to a "revolution towards public responsibility" and an "enlightened new view towards publicity and public relations."[58] In an earlier article, he had already noted that "[t]he public relations profession enlarged its activities through the depression, because business realized that in addition to selling its products under unfavourable conditions it also needed to sell itself to the public."[59] In this vein of accentuating the positive (and in anticipation of the development of greenwashing), he told the American Petroleum Institute in 1933, that "to counteract ill will from the past; the oil industry must devise extraordinary means of informing the public about its actual constructive policies and actions."[60] Thus in writing about the "revolutions" in 1941, Bernays was eager to put this "selling" in a positive light. PR is "no long whitewashing; it no longer pulls the wool over anybody's eyes," he insists. Rather, it "keep[s] us . . . safe."[61]

Bernays portrays the new age of PR as transcending the whitewashing of the past to emerge as a new, "enlightened" and essential aspect of society. This shift from the third to the fourth age of PR is reflected in Bernays' own work. In 1928, the year that he wrote *Propaganda* for a general audience, he wrote a more academic article for the *American Journal of Sociology*. This exonerates PR by stressing its educational role: "every man who teaches the public how to ask for what it wants is at the same time teaching the public how to safeguard itself against his own possible tyrannous aggressiveness."[62] And he limits on the possibility of bad behaviour by the profession by concluding that "[h]e who seeks to manipulate public opinion must also heed it."[63] The article is called "Manipulating Public Opinion." Nearly twenty years later, now writing in PR's post-revolutionary fourth age, he abandons the notion of "manipulating." Just as he left "propaganda" behind, so now, after the manipulative work of Goebbels and the like had been exposed, he moves away from "manipulation." Instead, as he had aligned PR agents with doctors and lawyers, he brings in the respectable roles of the "engineer" and "technician." And rather than playing with "opinion," the PR agent now gives more agency to the consumer by working for "consent." Thus, his profession now seeks "to accomplish purposefully and scientifically what we have termed 'the engineering of consent,' driven by 'scientific principles.'"[64] This approach he portrays as the "very essence of the democratic process," now not part of, but rather acting to "supplement the educational process." As such PR becomes one of "our most valuable contributions to modern society."[65] Though at the time he was working on promoting corporate and national hegemony over "middle America", he here suggests that "proper use" will promote "racial and religious tolerance." And he concludes by putting his profession at the heart of modern society: "in a democracy [such ideas] can be accomplished principally by the engineering."[66]

Such an argument fits within the tradition of PR deftly trying to whitewash itself and needs to be read as perhaps deceptive as much as descriptive. In later interviews, Bernays sought to distinguish good PR agents and "proper" action from bad, but, as greenwashing reveals, PR continues to manipulate opinion in the interests of those who fund PR, creating favourable images of businesses in order to resist criticism that might damage reputations and invoke regulations.

As Bernays' revolutions suggest, the period after World War I was particularly significant because PR played a major role in familiarising the public with political and scientific advances of that war. And though he did not mark a further revolution, something similar can be seen in the period after World War II. Such scientific and political challenges and appropriate responses are encapsulated in Bernays' post-war work. In the 1940s he began to work for the United Fruit Corporation (UFC), forming a close alliance with its president Sam Zemurray. Zemurray sought, in particular, to resist indigenous opposition to the banana industry and its expropriation and despoilation of Central American agriculture. In response, Bernays created the "Middle America Information Bureau" (MAIB). The MAIB flooded the media with what looked like credible scholarly research vindicating the activities and benefits of United Fruit and helped bureaucrats write White Papers in support of U.S. government and corporate activity in the region while ignoring the profoundly detrimental effects of the industry on people, politics, and the environment. (The chemicals United Fruit used were, among other things, known to make male workers sterile).[67] Indigenous political resistance was dismissed as "communism," a powerful ploy in the context of the developing cold war.[68] In Guatemala, this led to the coup of 1954, when the CIA, supported by the MAIB, helped overthrow Jacob Árbnenz, the democratically elected president, and returned power over the environment to United Fruit.[69]

UFC had to wrestle with science as well as politics. Based on research from the 1920s, UFC had portrayed bananas as healthy, suggesting they could help children resist celiac disease.[70] Twenty years later, Bernays helped UFC promote this claim, arranging large gatherings to celebrate the research and mailing medical endorsements to representatives of the press, dietitians, and other influential groups. Families were encouraged to feed their children bananas every day, and it was said that some children consumed up to 200 bananas each week. At the time, however, new research was emerging indicating that it was not bananas but a gluten-free diet that affected celiac disease. Commercial interests now came into confrontation with scientific evidence as well as political resistance. Zemurray and Bernays worked to distract from the unfavourable science.[71] Thus, though Bernays, in extolling his final revolution in PR, had talked about abandoning whitewashing and moving towards "public responsibility" and an "enlightened new view towards publicity and public relations," in the period after World War II, Bernays' work (though not his writings) suggests that it was often as important for PR to conceal relevant information as it was to reveal it.[72]

We hope this brief survey of Bernays and his revolutions can help clarify the development of PR across the first half of the twentieth century. As some of his strategies are at the heart of greenwashing today, such a survey should help not only to look back but also to look forward. The next section explores the second half of the twentieth century and the emergence of environmentalism as a critical challenge for PR. We turn to E. Bruce Harrison, exploring how he promoted notions of the environment and sustainability to distract from other, more damaging corporate activities.

13.5 In Bernays' Footsteps

Bernays' own account of the profession's four revolutions ended in 1941, when his article was published. It is, however, possible to trace another revolution in the wake of World War II. As noted earlier, when societies attempted to return to peace, the surge of scientific innovation that war-time research had produced led to significant social and political tensions. Confrontations arose between industries drawing on that science and others voicing concerns about the effects of those industries on society. Particularly significant were the confrontations that arose around Rachel Carson and her work on military and industrial pollution. In 1951, Carson's *The Sea Around Us* drew a widely praised picture of ocean life and its importance. The first edition revealed anxieties about the warming ocean. Then, a decade later in a "revised edition," Carson added a preface that sought to correct her "previously held supposition" that the ocean was "too big . . . to be much affected by human activity." She had since become aware of the "ominous problem" of atomic pollution, through the "deep-sea dumping of nuclear waste."[73] The following year, *Silent Spring* raised further concerns over long-term damage to the environment. Now, Carson investigated the effects of agricultural chemicals, in particular DDT, previously championed as an unproblematic way to keep mosquitoes and thus malaria under control.

Silent Spring quickly drew a wide and deeply concerned audience. Some of that concern came from the chemical manufacturers that the book indicted. One saw the book and its wide circulation as no less than a "Pearl Harbor" for their business.[74] Action was called for. Yet, though Carson had worked since 1936 as a scientist for the U.S. Bureau of Fisheries (later Fish and Wildlife) and intermittently at the Woods Hole Marine Biological Laboratory, the industry representatives saw their task "not as a scientific challenge, but as a public relations problem." Thus, they launched a suitably aggressive PR "counterattack."[75] The National Agricultural Chemicals Association "expanded its public relations program and published a number of new brochures which reaffirmed the unadulterated blessings of chemical pesticides." Monsanto, sensing "an opportunity to wield [its] public relations power" circulated a parody, "The Desolate Year."[76] Any notion of "a scientific challenge'" was resisted by figures such as Robert White Stevens, a spokesman for the Chemical industry. On television he spoke from a laboratory wearing a white coat (whereas Carson was interviewed from a desk in her house). Drawing on the scientific credibility these gave him, White Stevens dismissed Carson's claims as "gross distortions of the actual facts, completely unsupported by scientific, experimental evidence."[77] Elsewhere Carson was portrayed disparagingly as, among other things, a woman, a spinster, and a communist, thereby implicitly dismissing her work as little better than muckraking journalism.[78]

Shortly before this display of PR power, Harrison had joined the Manufacturing Chemists Association (MCA), becoming their "director of environmental information" in 1962. Thus, as Harrison put it in 1993, "In the summer of 1962, when *Silent Spring* exploded on the chemical industry, I was tapped . . . to help . . . the industries' management put together a rational, responsive program."[79] As corporate attacks grew more vicious, Harrison came to realise that the industry's counterattack was counterproductive, increasing attention to Carson and sympathy for her work. It is not unreasonable to trace links from Carson's celebrity to the Clean Air Act (1963), the Wilderness Act (1964), National Environmental Policy Act (1969), the Occupational Safety and Health Act (1970), and major reforms to the Federal Water Pollution Control Act (1972).[80] As the industry he worked for faced such regulation, Harrison sought a more subtle strategy than denial and dismissal. Thus, in the PR company he formed in 1973, Harrison devised more evasive strategies rather than direct confrontation.[81] Through such strategies, companies increasingly sought to embrace environmentalism rather than resist it, developing a "climate friendly" image that changed firms' environmental pronouncements if not necessarily their environmentally damaging practices.

As Carson's and subsequent environmental work prompted international gatherings to address environmental issues, Harrison developed cross-border alliances to respond. In advance of UN Conference on the Environment in Stockholm in 1972, Harrison helped form an alliance of chemical, petroleum, agricultural, and mining companies that he called the National Environmental Development Association (NEDA), acting as its PR agent. Then, in 1989, ahead of the 1992 Rio Earth Summit, he helped assemble the Global Climate Coalition (GCC) and went to the Summit as its representative.[82] Between Stockholm and Rio, Harrison worked on defensive strategies for himself, his firm, and its clients, whom he portrayed as threatened by attacks from "environmental activists . . . by the likes of Rachel Carson" who would put them "high on the list of presumed villains."[83] But he sought to avoid simple denial or aggressive counterattack. His work, instead, was, in the words of the environmentalist John Stauber, "filled with environmentally correct sounding jargon."[84]

Harrison's developing strategies can also be seen in his work at the same time on behalf of tobacco companies such as R.J. Reynolds and Philip Morris.[85] For this sector, he similarly encouraged more subtle approaches than simple confrontation. As regulation moved to control indoor smoking, Harrison encouraged the companies to acquire "the flexibility to track, monitor and respond to emerging IAQ [Indoor Air Quality] issues," to "outreach to Science groups," to build "inroads into the mainstream press," and thereby "become an effective player in the legislative and regulatory process."[86] For his NEDA and GCC clients, Harrison adopted similar strategies. And he was quite successful, triumphantly announcing in 1994 that "more than 1,200 companies throughout the world have now formally embraced [the notion of] 'sustainable development.'" He saw such adoption as essential for "every company that wants to be viewed favorably by an environmentally conscious public."[87]

His route towards this championing of "sustainability" can in part be traced through the titles of papers he wrote in the previous decade. Early papers, with names like "In the Grip of the 'Green Giant'" (1982) and "Cancer from Cake Mix?" (1984) indicate a continuing tendency to disparage his environmental opponents. But later titles such as "Environment/Energy: Looking for Co-Communication" (1987), "Ploughing New Ground in Environmental Affairs" (1991), "A Quality Approach to Environmental Communication"(1992), and "Achieving Sustainable Communication" (1992) reflect a steady shift to less confrontational approaches, responding to his fear that "negative images of industrial accidents and environmental disasters coupled with the media's bias towards green advocacy are shaping public perception."[88] He insisted that for PR, "the bottom line is 'sustainable communication', a term I have coined to indicate a new level of environmental communication."[89] "Sustainable," as noted earlier, spread after the report of the Brundtland Commission in 1987.[90] Harrison realised that if his clients deployed the notion themselves, rather than leaving it in the hands of opponents, they could cultivate a "Green Reputation," while maintaining a financially sustainable business model. Fusing the environmental and economic meanings of "sustainable," it became possible to suggest that corporations and the environment could unproblematically grow together. For this, Harrison's primary goal was to build "business that will last because its financial performance is not handicapped by social and political factors."[91]

Despite his desire to present a positive image, Harrison did not simply bow before environmental science. In his plan for communications by the GCC, he also encouraged drawing attention to "continued uncertainty of the science."[92] Moreover, going green, from his perspective was more a communications challenge than an environmental one. His account of the catastrophic Exxon Valdez oil spill reads more as a communication than an environmental disaster.[93]

To maintain "sustainable communication," Harrison encouraged alliances with environmentally "friendly" groups like the NEDA, endorsing "best practices of partnering" and "pursuit of 'green awards'."[94] This approach is in part based on Harrison's premise that the conventional "style of environmentalism died." Its death, he concluded in *Going Green*, was a result of the institutionalization of "sustainable development."[95] While corporations adopted his sustainable approach, making public proclamations about their sustainability, in his view, radical opponents sought to "criminalise" the sorts of organisations that Harrison and his ilk were defending. By contrast, Harrison insists, "WE ARE THE GOOD GUYS."[96] Those accusing such "guys" of greenwashing can only be extremists. Just as Bernays had sought to separate PR from whitewashing, so Harrison seemed to establish a similar separation from such things as greenwashing, in part by condemning those who made such accusations. This approach was highly influential. As we argue in the following section, it led to a series of influential figures separating themselves from established, but now-condemned, environmental movements.

13.6 Back to Whitewashing?

In 1979, Harrison engaged the lawyer John R. Quarles to work with the NEDA. Previously, Quarles had worked as deputy administrator of the US Environmental Protection Agency. His EPA credentials gave his views, and by extension those of the NEDA, an environmental gravitas. They also helped to distract from the implicit anti-government and anti-regulation positions of the NEDA. In this move, Harrison set a precedent for what, over the next 30 years, would become a convention for corporations: establishing a sustainable identity by hiring from more recognisably environmental organisation.

Harrison becomes clearer about such moves in his book *Going Green*. There he talks about "Best Practices of Partnering," setting up as his prime example the alliance between McDonald's and the Environmental Defense Fund.[97] This he sees as an example of "an increasing tendency for some of the more moderate environmental activist organization to move towards cooperation with business."[98] Such cooperation, of course, tends to imply that activists that don't form such partnerships are extremists unable to engage with their opponents. This implication can be seen in a memo for Philip Morris in 1995. The memo raises concerns of "anti-corporatist activists [who] charge that Philip Morris is engaging in 'greenwashing.'" The memo then assesses, through quite detailed analysis of each, various organisations (Greenpeace, Public Citizen, Earth First! and others) that seem likely to "become active . . . against Philip Morris on an anti-corporate or anti-greenwashing campaign."[99] The survey then highlights EDF for its change "from one that was confrontational and anti-corporate to one that is more tactful and which emphasizes cooperation and mutual understanding," promoting a "realistic, economics-based approach to environmentalism."[100]

In these pieces, Harrison and the memo (which was possibly produced by Harrison's company) display a growing interest in environmentalists willing to accommodate corporations. Twenty years later in 2014, Harrison calmed anxious corporations by telling them that a "number of green former antagonists, may already be looking your way."[101] Partnerships with moderate environmentalists were growing, leaving the radicals isolated. One figure who turned in this direction between these two pieces by Harrison was Adam Werbach, mentioned at the beginning of this chapter. As noted, he was a well-known environmental champion. In the 1990s, when Harrison wrote his earlier piece, Werbach projected a quite different figure. Writing for *Stanford Environmental Law Journal*, he promoted "radical change" and encouraged environmentalists not to compromise but to "fight for a sweeping plan."[102] In the same year, Werbach published the book *Act Now, Apologize Later*, attacking corporations for their environmental blindness and for their greenwashing (which implied that they were not blind but duplicitous). Among these companies was the U.S. retailer Walmart, which Werbach labelled as a "new breed of toxin" for its environmental footprint.[103]

Over the next seven years Werbach's viewpoint moved in Harrison's direction. In 2004, he gave the speech noted at the beginning of this chapter. The environmental movement, he then complained, was locked in the past, and he called for "a new language, a new set of strategic initiatives, a new set of institutions, and a new metric for evaluating our success." When this was done, he declared "[W]e cease to be 'environmentalists' in any meaningful sense of the term and open ourselves up to the possibility of becoming progressive Americans." Despite this radical change of perspective, he maintained some of his old enemies. Indicting the environmental movement for ignoring working-class voters, he also indicted companies "like Wal-Mart" that paid half as much [as industry jobs] and offered no health care or retirement benefits." Nonetheless, in 2006, having raised the question of "Where the Environmental Movement Can and Should Go from Here," Werbach went from Greenpeace to become a consultant at Walmart, working on its "Environmental Health and Wellness Program." From that position he wrote the book *Strategy for Sustainability*, while working for the advertising and PR organisation Saatchi & Saatchi. Many of the conventional environmentalists of the sort that Werbach denounced, still saw Walmart as fundamentally anti-environmental.[104] His denunciation, however, made it easier to portray those environmentalists as extreme and out of touch.[105]

Another figure who can be seen as following Werbach's path is Patrick Moore. Abandoning a long-term membership of Greenpeace (of which he was a co-founder) in 2010, Moore too denounced the conventional environmental movements in his book *Confessions of a Greenpeace Dropout* (2010). This, he declared, was an "effort, after 50 years as a scientist and environmental activist, to expose the misinformation and outright lies used to scare us and our children about the future of the Earth." He formed the Clean and Safe Energy Coalition, and in a later book he condemned *Fake Invisible Catastrophes and Threats of Doom* (2021). In 2023 he contributed to *Climate the Movie: The Cold Truth* made by Martin Durkin, who had previously made *The Great Global Warming Swindle* (2007). Both endorse climate-change denial.

Moore's strategy included accusing environmentalists of obfuscation and obscurity, while people like himself were open-eyed and transparent. Thus, in an article in 2018 for *Energy Processing California*, he stressed the importance of visibility in assessing claims of damaging climate change. Because coral reefs and salmon are below the water and most people don't go to the Arctic to look at polar bears, he argued, they should not succumb to arguments others make about these proclaimed victims of climate change. And because you never see nuclear radiation, there is no reason to fear it. In general, his arguments suggest that people are not victims of climate change so much as of fake narratives and disinformation. The biography accompanying the article says he criticises the environmental movement "for what he sees as scare tactics and disinformation" and that he believes the environmental movement has "abandoned science and logic in favor of emotion and sensationalism."[106]

In this way, members of Moore's Coalition, like Werbach's followers, could claim to be developing a new and enlightened form of environmentalism, one that was leaving old and corrupt science and old politics behind. This position allows them to denounce opponents as being behind the times. Such a stance also allows them and collaborators such as the pipeline industry and oil sands technologists who run advertisements alongside Moore's article mentioned earlier, to imply that any opponents are merely muckrakers and that, three (or more) revolutions later, we still need PR to engineer a less sensational view of society and its sustainability. As Michael Mann put it in his book the *New Climate War* (2021), in this "new" war, a critical divide-and-conquer strategy is to get "climate advocates fighting with each other so that [they] don't present a unified voice."[107] This strategy is not so new. In 1995, Stauber and Rampton noticed the trend of "sectors of the environmental movement to enter into 'partnerships' with major polluters."[108] Similarly, in 2002 the Australian environmentalist Sharon Beder noted figures moving from "Green Warriors to 'Greenwashers'" through a "revolving door" (though in the context of PR we might call this a spinning door).[109] Similarly, Cary Gillam's book on Monsanto notes the company "recruiting and paying experts who would lend credibility to Monsanto's claims of product safety" despite increasing concerns about the effects of GMO on agriculture. Among these "experts" were former EPA officials.[110] These new alliances allow corporations to dismiss activists as, inherently, contemporary "greenrakers," who deserve not attention but contempt. The corporations can rely on the reputations of these former green figures who have gone through the spinning door to whitewash, as Bernays put it—or to greenwash, as we might put it—the not-so-green businesses with whom they have formed alliances.

As this chapter shows, Bernays and the PR industry in general have attempted to present the evolution of the profession as progressive. While admitting to the whitewashing practices of early practitioners, Bernays claimed that the succeeding revolutions led to a profession that was inherently honest and beneficial to society. But a historical analysis of the actual practices of Bernays and successors suggests that this was in itself whitewashing. Thus, looking at the emergence of greenwashing in the hands of PR agents like Harrison and Werbach reveals enduring attempts to portray agents and their clients as deeply engaged in environmentally sustainable practices. To validate such portrayal, they have solicited figures from environmental groups. The environmental credentials these figures have brought with them help the agents and clients whitewash concerns about the potentially harmful effects they can have on society. Despite Bernays' denials of its death, PR whitewashing appears to have returned.

Notes

1 Werbach, "Where the Environmental Movement Can and Should Go from Here."
2 Werbach, "Save Headwaters Forrest Now," xiv.
3 Werbach, *Strategy for Sustainability*.
4 Harrison, *Going Green*, 8–9.
5 McCauley, "E. Bruce Harrison Dies at 88."
6 Burton, "The Big Dirty Hands."
7 Budryk, "Two House Democrats Question PR-Awards Firms on Work with Fossil Fuel Companies."
8 This phrase can be traced back at least to a 1957 newspaper article noting that Bernays "claims to be the father of public relations" (Davie, "Trouble-Shooter," 6). For continued recognition see, for example, Bernays' obituary in the *New York Times*, March 10, 1995; or his entry in the *Dictionary of Marketing*, 4th ed., 2016.
9 Bernays, *Propaganda*, 44.
10 Looking across more than 30 years of academic research, Carey (1976) and Miller and Dinan (2008) suggest that academia more generally has been reluctant to engage with PR.
11 Some of Bernays' exchanges with Gras, the head of Business History at Harvard Business School, can be found in the Norman S.B. Gras Papers in the Special Collections of the Baker Library. For Gras' account of business history and PR, see Gras, "Shifts in Public Relations."
12 A search of *Business History Review's* website shows 20 distinct entries for Bernays, of these only 4 occur after 1983.
13 Bernays, *Revolution in Publicity*, 1941.
14 Jones, *Deeply Responsible Business*, 243ff.
15 Hu, *A Prehistory of the Cloud.*
16 Sims, "Green Magic"; Sims and Sivakumar, "Making Sustainability Concrete."
17 Hsu, "A PR Giant Is Caught."
18 Ritchie, "Where in the World do People Emit the Most CO_2?"
19 Boykoff, "The World Cup in Qatar Is a Climate Catastrophe."
20 *Badvertising*, "FIFA Shown Red Card."
21 Harrison, *Corporate Greening 2.0*, chapter 4.
22 Hameed et al., "Greenwash and Green Purchase Behavior."
23 Terrachoice, "Greenwashing Affects 98% of Products."
24 ESG, "ESG Disclosures for Investment Advisors"; see also Liu et al, "Why Greenwashing," and Chapter 5 in this volume.
25 Joint Staff Report, *Denial, Disinformation, and Doublespeak.*
26 For continuing tensions in this area, see Polonsky, "Transformative Green Marketing."
27 See "Greenwashing" in Idowu et al., *Encyclopedia*.
28 Bruno, *The Greenpeace Book of Greenwash*, 6–7.
29 Philip Morris Records, 1–2.
30 Stauber, "Going . . . Going . . . Green," 3; McCauley, "E. Bruce Harrison Dies."
31 Claxton, *The History of the Georgia Conservancy*, 135, 141. The report was made public in 1977 through an article in *The Ecologist* journal. Orr and Phillips, "Toward a Sustainable Energy Society."
32 Barney, *The Global 2000 Report.*
33 Barney, *The Global 2000 Report*, xi–x.
34 Borowy, *Defining Sustainable Development*; Jones, *Profits and Sustainability*.

35 Harrison, "Managing for a Better Green Reputation," 26.
36 See Bernays, *Propaganda*, 52; Saranillo, *Unsustainable Empire,* 132.
37 Clients included Beech Nut (bacon), American Tobacco (cigarettes), and United Fruit (bananas). For a fuller list of Bernays' clients, see Bernays, *Biography of an Idea*, 817ff.
38 Bernays, "The Engineering of Consent," 115.
39 Recognition of Bernays as the father of PR overlooks other significant figures, among them Ivy Ledbetter Lee, who promoted coal, oil, car, and railroad companies. In contrast to Bernays, however, Lee did not put similar energy into promoting himself. He died in 1934 with relatively few publications. Bernays lived for another 60 years and, with a copious bibliography, became a primary historical source for accounts of the field and its development. By 1938 *Time Magazine* described Bernays as "the nation's No. 1 publicist" (January 24, p. 24). It is perhaps further evidence of Bernays' craft that while that phrase is widely quoted, the following phrase—"dark, Machiavellian Edward L. Bernays"—is usually overlooked.
40 Bernays, *Propaganda*, 20.
41 Bernays, *Propaganda*, 9.
42 Lippmann, *Public Opinion*, 248; Bernays, "Public Relations Counsel," 9.
43 Bernays, "The Emergence of Public Relations," 299.
44 Bernays, "Public Relations Counsel," 6–7.
45 See Bernays, *Biography of an Idea.*
46 Bernays, *Biography of an Idea*, 652.
47 Bernays, "Emergence," 305.
48 Bernays, "The Revolution in Publicity." Bernays' potential to influence business history might be seen in Gras' account of PR across the twentieth century, written five years later after multiple exchanges with Bernays. Gras' "transformations" map tightly onto Bernays' "revolutions." Gras, "Shifts in Public Relations," 127–28.
49 Gras, "Shifts in Public Relations," 3.
50 Bernays, *Biography of an Idea*, 370.
51 Tarbell, *The Standard Oil Company*, vii.
52 Stauber and Rampton, *Toxic Sludge*, 18–21.
53 Bernays, "The Emergence of the Public Relations Counsel," 300.
54 Bernays, "The Revolution in Publicity," 2.
55 Bernays, "The Revolution in Publicity," 6.
56 Bernays, "The Revolution in Publicity," 4.
57 Bernays, *Propaganda*, 196.
58 Bernays, "The Revolution," 4.
59 Bernays, "Recent Trends" quoted in Larson, *Public Relations*, 27.
60 Larson, *Public Relations*, 44.
61 Bernays, "The Revolution," 18.
62 Bernays, "Manipulating Public Opinion," 960.
63 Bernays, "Manipulating Public Opinion," 971.
64 Bernays, "The Engineering of Consent," 113.
65 Bernays, "The Engineering of Consent," 115.
66 Bernays, "The Engineering of Consent," 120.
67 Schlesinger and Kinzer, *Bitter Fruit.*
68 Immerman, "Guatemala as Cold War History"; Saranillo, *Unsustainable Empire.*
69 See, for example, Middle American Information Bureau, 1945.
70 Haas, "The Value of the Banana."
71 Gottschall, "Whatever Happened"; Anderson et al., "Celiac Disease"; Westfall, "Persuasion for Profit."
72 Bernays, "Revolution," 18.
73 Carson, *The Sea Around Us*, Second Edition, xi–xiii. Atomic scientists, after defending their earlier atomic research and its implications against challenges such as Carson's, later lent their scientific credentials to help polluting companies fight against environmental and health regulation. Oreskes and Conway, *Merchants of Doubt*, 25–31.
74 Aronczyk, "Public Relations," 844.
75 Graham, *Since Silent Spring*, Chapter 4.
76 Graham, *Since Silent Spring*, 65.
77 "The Silent Spring of Rachel Carson," CBS Reports, April 3, 1963.

78 Markowitz and Rosner, *Deceit and Denial*, 146.
79 Harrison, *Going Green*, xiv.
80 Lepore, "The Right Way."
81 When Harrison sold the company in 1997, long-time clients included many companies fighting to sustain a green reputation, among them General Motors, BP America, Monsanto, Ford Motor, American Petroleum Institute, R.J. Reynolds, Union Carbide, Phillips Petroleum, Philip Morris, Unocal, and Uniroyal Chemical. McCauley, "E. Bruce Harrison Dies."
82 E.B. Harrison, Inc, *Global Climate Coalition*.
83 Harrison, *Going Green*, 257.
84 Stauber, "Going . . . Going . . . Green!" 2.
85 This is easier work to trace as lawsuits against tobacco companies led to the release of data that would otherwise be carefully guarded. See the University of San Francisco Tobacco Archive, https://www.industrydocuments.ucsf.edu/tobacco.
86 Harrison, "A Proposal to Serve RJR."
87 Harrison, "Achieving Sustainable Communication," 244–45.
88 Harrison, "Ploughing New Ground," 32.
89 Harrison, "Achieving Sustainable Communication," 244.
90 The phrases "sustainable business" and "corporate sustainability" do not appear on JSTOR's archive of academic journals before 1990. Between 1990 and the writing of this chapter in April 2024, the former has 2,931 hits and the later 2,175.
91 Harrison and Mühlberg, *Leadership Communication*, 243.
92 Brulle, "Advocating Inaction," 199.
93 Harrison, *Going Green*, Chapter 18.
94 Harrison, *Going Green*, Chapter 21.
95 Harrison, *Going Green*, 8–11.
96 Harrison, *Going Green*, 21.
97 Harrison, *Going Green*, Chapter 23.
98 Harrison, *Going Green*, 189.
99 See "Truth Tobacco Industry Documents," UCSF Archive, 1, https://www.industrydocuments.ucsf.edu/tobacco/docs/#id=nhck0177.
100 "Truth Tobacco Industry Documents," 8.
101 Harrison and Mühlbert, *Leadership Communication*, 244.
102 Werbach, "Save Headwaters Forest Now," xiv, xvii.
103 Werbach, *Act Now*.
104 See Michell, "Walmart's Greenwash"; Sierra Club, "What Is Wal-Mart's True Environmental Footprint." It is not only conventional environmentalists that have seen Walmart in this light. The Federal Trade Commission has successfully prosecuted Walmart for selling products made of rayon as bamboo. *U.S. (FTC) v Walmart*, 2022.
105 Humes, *Force of Nature*.
106 Moore, "Twelve Invisible Eco-Catastrophes."
107 Mann, 2021, 15.
108 Stauber and Rampton, *Toxic Sludge*, 67.
109 Beder, *Global Spin*, 267.
110 Gillam, *Whitewash*, 12.

Bibliography

Anderson, Charlotte M., A.C. Frazer, J.M. French, J.W. Gerrard, H.G. Sammons, and J.M. Smellie. "Celiac Disease: Gastro-Intestinal Studies and the Effect of Dietary Wheat Flour." *The Lancet* 259, no. 6713 (1952): 836–42.

Aronczyk, Melissa. "Public Relations, Issue Management, and the Transformation of American Environmentalism, 1948–1992." *Enterprise & Society* 19, no. 4 (2018): 836–63.

Badvertising. "FIFA Shown Red Card for Potentially 'Criminal' Misleading Claims Over World Cup." *Badvertising*, June 7, 2023. https://www.badverts.org/latest/fifa-shown-red-card-for-potentially-criminal-misleading-green-claims-over-world-cup?rq=FIFAshownredcard.

Barney, Gerald O., ed. *The Global 2000 Report to the President: Entering the Twenty-First Century*. Washington, DC: U.S. Government Printing Office, 1991 (first edition 1980).

Beder, Sharon. *Global Spin: The Corporate Assault on Environmentalism*. Totnes, Devon: Green Books, 2002.

Bernays, Edward L. *Propaganda*. New York: Liveright Publishing Corporation, 1928.

Bernays, Edward L. "Manipulating Public Opinion: The Why and the How." *American Journal of Sociology* 33, no. 6 (1928): 958–71.

Bernays, Edward L. "Recent Trends in Public Relations Activities." *Public Opinion Quarterly* 1 (1937): 147–51.

Bernays, Edward L. "The Revolution in Publicity." *The Saturday Review of Literature* 28 (1941): 3–4, 18.

Bernays, Edward L. "Public Relations Counsel: Evolution of a Profession." In *Public Relations Directory and Yearbook*. Vol. I, edited by Karl E. Ettinger, 6–11. New York: Public Directions Directory and Yearbook Inc., 1945.

Bernays, Edward L. "The Engineering of Consent." *The Annals of the American Academy of Political and Social Sciences* 250 (March 1947): 113–20.

Bernays, Edward L. *Biography of an Idea: Memoirs of a Public Relations Counsel*. New York: Simon and Schuster, 1965.

Bernays, Edward L. "The Outlook for Public Relations." *Public Relations Quarterly* X, no. 3–4 (1966): 34–39.

Bernays, Edward L. "The Emergence of the Public Relations Counsel: Principles and Recollections." *Business History Review* 45, no. 3 (1971): 296–316.

Borowy, Iris. *Defining Sustainable Development for our Common Future: A History of the World Commission on Environment and Development (Brundtland Commission)*. London: Routledge, 2013.

Boykoff, Jules. "The World Cup in Qatar Is a Climate Catastrophe." *Scientific American*, November 23, 2022. https://www.scientificamerican.com/article/the-world-cup-in-qatar-is-a-climate-catastrophe/.

Bruno, Kenny. *The Greenpeace Book of Greenwash*. Washington, DC: Greenpeace, 1992.

Budryk, Zack. "Two House Democrats Question PR-award Firms on Work with Fossil Fuel Companies." *The Hill*, February 2022. https://thehill.com/policy/energy-environment/593552-two-house-democrats-question-pr-firms-on-work-with-fossil-fuel/.

Carson, Rachel. *The Sea Around Us*. New York: Golden Press, 1951 (second edition, 1961).

Carson, Rachel. *Silent Spring*. New York: Houghton Mifflin, 1962.

Claxton, Robert H. *The History of the Georgia Conservancy, 1967–1981*. Atlanta: The Georgia Conservancy, 1985.

Davie, Michael. "Trouble-Shooter in Anglo-U.S. Relations." *Goble and Mail*, August 7, 1957, 6.

Delmas, Magali A., and Vanessa Cuerel Burbano. "The Drivers of Greenwashing." *California Management Review* 54, no. 1 (2011): 64–87.

E.B. Harrison Inc. *Global Climate Coalition Communications Program, October 1994–December 1995*. Washington, DC: E.B. Harrison Inc, 1995.

ESG. 2002. "ESG Disclosures for Investment Advisors and Investment Companies." Securities and Exchange Commission Fact Sheet. https://www.sec.gov/files/ia-6034-fact-sheet.pdf.

Ferguson, Niall. *The Pity of War: Explaining World War I*. New York: Basic Books, 2008.

Fisk, George. *Marketing and the Ecological Crisis*. New York: Harper & Row, 1974.

Gillam, Carey. *Whitewash: The Story of Weed Killer, Cancer, and the Corruption of Science*. Washington, DC: Island Press, 2017.

Gottschall, Elaine. "Whatever Happened to the Cure for Coeliac Disease." *Nutritional Therapy Today* 7, no. 1 (1997): 8–11.

Gras, N.S.B. "Shifts in Public Relations." *Bulletin of the Business Historical Society* 19, no. 4 (1945): 97–148.

Haas, Sidney V. "The Value of the Banana in the Treatment of Celiac Disease." *American Journal of Diseases of Children* 28, no. 4 (1924): 421–37.

Hameed, Irfan, Seeshan Hyder, Muhammad Imran, and Kashif Shafiq. "Greenwash and Green Purchase Behavior." *Environment Development and Sustainability* 23, no. 13 (2021): 113–34.

Harrison, E. Bruce. "In the Grip of the 'Green Giant'." *PR Journal* 38, no. 12 (1982): 18.

Harrison, E. Bruce. "Cancer from Cake Mix? The Public Demands Personal Answers." *PR Journal* 41, no. 1 (1984): 32.

Harrison, E. Bruce. "Environment/Energy: Looking for Co-Communication." *PR Journal* 43, no. 6 (1987): 5.

Harrison, E. Bruce. "A Quality Approach to Environmental Communication." *Environmental Quality Management* 1, no. 3 (1992): 225–31.

Harrison, E. Bruce. "Ploughing New Ground in Environmental Affairs." *PR Journal* 47, no. 4 (1991): 32.

Harrison, E. Bruce. *Environmental Communication and Public Relations Handbook*. Rockville, MD: Government Institutes, Inc., 1992.

Harrison, E. Bruce. "Achieving Sustainable Communication." *The Columbia Journal of World Business* 27, no. 3–4 (1992): 242–45.

Harrison, E. Bruce. *Going Green: How to Communicate Your Company's Environmental Commitment*. Homewood, IL: Business One Irwin, 1993.

Harrison, E. Bruce. "A Proposal to Serve RJR: ETS Strategies." Presented by E. Bruce Harrison Company. University of California, San Francisco, Industry Documents Library, September 7, 1994.

Harrison, E. Bruce. "Managing for a Better Green Reputation." *International Public Relations Review* 17, no. 3 (1994): 25–30.

Harrison, E. Bruce. "Corporate Greening 2.0: Factors in Play as Executives Zero in on Climate Change." *Corporate Communications* 14, no. 3 (2008): 280–85.

Harrison, E. Bruce. *Corporate Greening 2.0: Create and Communicate Your Company's Climate Change and Sustainability Strategies*. Exeter, NH: Publishing Works Inc., 2009.

Harrison, E. Bruce, and Judith Mühlberg. *Leadership Communication: How Leaders Communicate and How Communicators Lead in Today's Global Enterprise*. New York: Business Expert Press, 2014.

Hsu, Tiffany. "A PR Giant Is Caught Between Climate Pledges and Fossil Fuel Clients." *New York Times*, December 10, 2021, B4.

Hu, Tung-Hui. *A Prehistory of the Cloud*. Cambridge, MA: MIT Press, 2015.

Humes, Edward. *Force of Nature: The Unlikely Story of Wal-Mart's Green Revolution*. New York: Harper Business, 2011.

Idowu, Samuel O., Nicholas Capaldi, Liangrong Zu, and Anada Das Gupta. *Encyclopedia of Corporate Responsibility*. New York: Springer, 2013.

Immerman, Richard H. "Guatemala as Cold War History." *Political Science Quarterly* 95, no. 4 (1980–81): 629–53.

Joint Staff Report. "Denial, Disinformation, and Doublespeak: Big Oil's Evolving Efforts to Avoid Accountability for Climate Change." https://oversightdemocrats.house.gov/sites/evo-subsites/democrats-oversight.house.gov/files/evo-media-document/2024-04-30.COA%20Democrats%20-%20Fossil%20Fuel%20Report.pdf.

Jones, Geoffrey. *Profits and Sustainability: A History of Green Entrepreneurship*. Oxford: Oxford University Press, 2017.

Larson, Keith A. *Public Relations, Edward L. Bernays and the American Scene: Annotated Bibliography of and Reference Guide to Writings by and About Edward L. Bernays from 1917 to 1951*. Concord, NH: Rumford Press, 1951.

Lepore, Jill. "The Right Way to Remember Rachel Carson." *New Yorker*, March 19, 2018. https://www.newyorker.com/magazine/2018/03/26/the-right-way-to-remember-rachel-carson.

Liu, Yupei, Weian Li, Lixiang Want, and Qiankun Meng. "Why Greenwashing Occurs and What Happens Afterwards? A Systematic Literature Review and Future Research Agenda." *Environmental Science and Pollution Research* 30 (2023): 118102–16.

Mann, Michael E. *The New Climate War: The Fight to Take Back the Planet*. New York: PublicAffairs, 2021.

Markowitz, Gerald E., and David Rosner. *Deceit and Denial: The Deadly Politics of Industrial Pollution*. Berkeley: University of California Press, 2002.

McCauley, Kevin. "E. Bruce Harrison Dies at 88." *O'Dwyers*, January 25, 2021. https://www.odwyerpr.com/story/public/15557/2021-01-25/e-bruce-harrison-dies-at-88.html.

Middle America Information Bureau. *Every American Has a Personal Stake in Our Relations with Middle America*. New York: Middle America Information Bureau, 1945.

Mitchell, Stacey. "Walmart's Greenwash: How the Company's Much-Publicized Sustainability Campaign Falls Short, While Its Relentless Growth Devastates the Environment." Institute for Local Self-Reliance Report, May 2012. https://cdn.ilsr.org/wp-content/uploads/2012/03/walmart-greenwash-report.pdf.

Moore, Patrick. *Confessions of a Greenpeace Dropout; The Making of a Sensible Environmentalist*. New York: Beatty Street, 2010.

Moore, Patrick. "Twelve Invisible Eco-Catastrophes & Threats of Doom that are Actually Fake." *Energy Processing Canada* 50, no. 6 (2018): 4–9.

Moore, Patrick. *Fake Invisible Catastrophes*. New York: Ecosense Environmental Inc., 2021.

Oreskes, Naomi, and Erik M. Conway. *Merchants of Doubt: How a Handful of Scientists Obscured the Truth on Issues from Tobacco Smoke to Global Warming*. New York: Bloomsbury, 2011.

Orr, David W., and Cecil R. Phillips. "Toward a Sustainable Energy Society." *The Ecologist* (August–September 1977): 294–97.

Philip Morris Records. "Truth Tobacco Industry Documents." University of California San Francisco Archives. https://www.industrydocuments.ucsf.edu/docs/nhck0177.

Polonsky, Michael Jay. "Transformative Green Marketing: Impediments and Opportunities." *Journal of Business Research* 64 (2011): 1311–19.

Ritchie, Hannah. "Where in the World Do People Emit the Most CO_2?" 2019. https://ourworldindata.org/per-capita-co2.

Saranillo, Dean Itsuji. *Unsustainable Empire: Alternative Histories of Hawai'i Statehood*. Durham, NC: Duke University Press, 2011.

Schlesinger, Stephen C., and Stephen Kinzer. *Bitter Fruit: The Story of the American Coup in Guatemala*. Boston, MA: Harvard University Press, 1999.

Sierra Club. "What Is Wal-Mart's True Environmental Footprint?" https://fdocuments.net/document/what-is-wal-marts-sierra-is-wal-marts-true-environmental-2-whereas.html.

Sims, Christo. "'Green Magic': On Technologies of Enchantment at Apple's Corporate Headquarters." *Public Culture* 34, no. 2 (2022): 291–317.

Sims, Christo, and Akshita Sivakumar. "Making Sustainability Concrete: Designs for Green Architecture in Silicon Valley." *Public Culture* 35, no. 3 (2023): 367–78.

Sinclair, Upton. *The Jungle*. New York: Doubleday, Page & Company, 1906.

Stauber, John C. "Going . . . Going . . . Green." *PR Watch* 1, no. 3 (1994): 1–3.

Stauber, John C. "Smoke and Mirrors: How Tobacco and PR Grew Up Together." *PR Watch* 1, no. 3 (1994): 6–7.

Stauber, John, and Sheldon Rampton. *Toxic Sludge Is Good for You: Lies, Damn Lies, and the Public Relations Industry*. Monroe, Main: Common Courage Press, 1995.

Tarbell, Ida M. *The History of Standard Oil Company*. New York: McClure Phillips & Co., 1904.

Terrachoice. "Greenwashing Affects 98% of Products Including Toys, Baby Products, and Cosmetics." May 2009. https://p2infohouse.org/ref/44/43576.pdf.

Werbach, Adam. *Act Now, Apologize Later*. New York: Cliff Street Books, 1997.

Werbach, Adam. "Save Headwaters Forest Now." *Stanford Environmental Law Journal* 16 (1997): xi–xvii.

Werbach, Adam. "Where the Environmental Movement Can and Should Go from Here." Speech presented at the Commonwealth Club of San Francisco, December 8, 2004. https://grist.org/politics/werbach-reprint/.

Werbach, Adam. *Strategy for Sustainability: A Business Manifesto*. Boston, MA: Harvard Business Press, 2009.

Westfall, Irene. "'Persuasion for Profit': Public Relations and Corporate Power at Midcentury." Theis submitted to Wesleyan University, April 2022.

14

THE CONSTRUCTION OF CLIMATE AS A BUSINESS RISK THROUGH ESG AND SUSTAINABLE FINANCE

Judith Stroehle

14.1 Introduction

Climate change is one of the biggest challenges our society is facing today. In view of this, the United Nation's (UN) Sustainable Development Goal (SDG) 13 calls upon "urgent action to combat climate change and its impacts"[1] and the Intergovernmental Panel on Climate Change (IPCC) outlines that "risks and projected adverse impacts and related losses and damages from climate change escalate with every increment of global warming."[2] Yet, not just the UN recognises climate change as a significant risk to society. Today, climate change is also recognised as a business risk which investors, companies, and regulators seek to mitigate to ensure long-term market viability and stability.[3] Smart firms, sounds the mantra, manage climate risks to survive and thrive in the long-term.

Many international developments supported this development. In 2015, for example, the world witnessed the emergence of the first international treaty on climate change, the Paris Agreement. In the same year, the SDGs were adopted as a universal call to increase action to ensure planetary health and thriving human societies. Their adoption was driven by mounting and undeniable evidence of climate change, as well as a loudening narrative around the need for increased social responsibility of business and financial markets after the financial crisis in 2008. The sustainable investment movement noticeably benefitted from these developments. Organizations like the UN-backed Principles of Responsible Investments (PRI), founded in 2006, saw exponential growths in investor members which pledged to the integration of environmental, social, and governance (ESG) issues into their investment decisions.[4] The PRI's "blueprint for sustainable investing" thus outlines how sustainable markets are those which "address unsustainable aspects of the markets that investors operate in, to achieve the economically efficient, sustainable global financial system that responsible investors and beneficiaries need"[5] and identifying climate change as their members' "highest priority issue"—counting over 4000 institutions with a combined 120 trillion USD in assets under management.[6]

The enthusiasm for climate change as a business risk, however, is not without challengers. For example, while the IPCC was already set up in 1988 to provide governments with

DOI: 10.4324/9781003478089-15

regular scientific assessments of changes in the climate, only the most recent derivations of these reports (2023 was the sixth assessment cycle) have gained unequivocal attention from the media. In addition, flagship news sources such as the BBC still thought it appropriate to interview a climate-sceptic geologist to show "both sides" of the story as late as 2013, in the fifth assessment cycle of the IPCC.[7] And while in Europe many would find this unimaginable ten years later, globally, critical voices about climate change and the associated sustainability agenda remain. According to the U.S. political right, for example, integration of ESG into investment decisions, including considerations of climate change, is nothing more than the fruits of a leftist, "woke" political agenda, harming liberal market ideas.[8] In South America, far-right political forces also continuously challenge climate protection. Interestingly, however, the political backlash against ESG and climate seems to be countered by a market convinced that climate risk is important. Various news sources thus report that these campaigns have so far failed to fully deter companies and investors from considering climate risk in their decisions.[9]

What has changed? A principal component of the continued acceptance and protection of climate change as an acknowledged business risk has been the broad buy-in from financial markets. Indeed, by declaring climate risk as "financially material"—meaning the recognition of potential effects of climate change on companies' business models, cash flows, financial position, and financial performance,[10]—capital markets today seem to have fully embraced the climate risk narrative. Particularly through investor-backed frameworks such as the Financial Stability Board's Taskforce on Climate-Related Financial Disclosures (TCFD) and standard-setting accounting bodies such as the International Financial Reporting Standard (IFRS) Foundation's International Sustainability Standards Board (ISSB), this financial materiality has been institutionalised, stressing the important of its integration into financial statements.[11] The chairman of the ISSB, Emmanuel Faber, even went so far as to call the establishment of climate-related financial risk the organisations' ultimate "obsession."[12] On the investor side, these developments have led to a widespread use of so-called climate stewardship, where investors actively engage with companies on climate risk awareness and action at firms.[13] Also active management of investment portfolios in relation to climate[14] and even climate activism of large investors in public markets[15] have become increasingly normalised. Ilhan, et al.[16] corroborate these insights with systematic evidence from a survey with more than 400 institutional investors, discussing how institutional investors now not only value but *demand* corporate climate risk disclosures, with impacts on ownership structure and holding periods for firms that don't disclose.

This chapter discusses one element of how this "awakening" came about. Through an analysis of the changing measurements and narratives around climate change, it discusses why and how especially the introduction and growth of ESG criteria for sustainable investing have had an important impact on the recognition of climate change as business risk, and how it influenced its measurement and management at the firm level accordingly. Looking at how climate measures have been captured and evolved through time at one of the globally largest ESG rating agencies—MSCI ESG Inc.—the analysis highlights that the tools for managing climate change such as carbon accounts are a product of historical and social construction, influenced by ratings, the rating agencies, and the investors that use ratings alike. The chapter concludes with a discussion of the relevance of this insight for ESG for climate change management, and the challenges that it poses in a continuously contested space.

14.2 The Origins of Business-Related Climate Change Measurement

In the absence of regulated standards for corporate climate measurement and reporting, much of what is today common knowledge and practice in business-related climate change measurement has been developed by NGOs (such as the GHG Protocol, the Carbon Disclosure Project (CDP), the Global Reporting Initiative (GRI), and the Sustainability Accounting Standards Board (SASB) based on recommendations from international organisations (such as the UN IPCC), not by professional accounting bodies or regulators as one might expect.[17] Particularly the release of the Greenhouse Gas (GHG) Protocol in 2001 marked an important milestone for the harmonisation of emissions reporting. While not officially mandated, it serves as a quasi-standard until today, marking also the measurement base for the climate-part of international, regional, and national reporting standards—such as the ISSB, the U.S. Securities and Exchange Commission (SEC) or the European Union Corporate Sustainability Reporting Directive (CSRD).[18]

While climate reporting has increased significantly in the last two decades,[19] structured climate *risk* reporting and assessments only developed after the release of the TCFD recommendations from 2015 onwards. TCFD recommendations are structured around four thematic areas—governance, strategy, risk management, and metrics and targets—and "designed to solicit decision-useful, forward-looking information that can be included in mainstream financial filings."[20] Since 2023, the TCFD forms part of the international accounting body IFRS. Before the TCFD, however, corporate climate risk reports were often absent, and investor climate risk assessments were, as a consequence, often much less structured. Indeed, investors often relied on the measures proposed in ESG ratings and ranking products. Through the ESG assessment, climate change was framed and ultimately accepted as a key factor for business risk assessment.[21]

14.2.1 The Measurement of "ESG" and Its Divergence

Historically, the concept of ESG can be traced back to two influential reports from UN-backed institutions in the early 2000s. Concretely, in 2004, the United Nations Global Compact released the report "Who Cares Wins—Connecting Financial Markets to a Changing World: Recommendations to better integrate environmental, social and governance issues in financial analysis, asset management and securities brokerage."[22] This report is important, as it is the first public document to mention E, S and G as a conjunct concept for investment consideration. It is followed by the equally important "Freshfield Report," published by the United Nations Environmental Programme's Finance Initiative (UNEP-FI) one year later in 2005. Here, the asset management working group of UNEP-FI set out a "legal framework for the integration of environmental, social and governance issues into institutional investment," in order to answer the question about whether "the integration of environmental, social and governance issues into investment policy (including asset allocation, portfolio construction and stock-picking or bond-picking) [is] voluntarily permitted, legally required or hampered by law and regulation."[23]

At the point of publication of these reports, financial markets had been playing with the idea of integrating sustainability concerns into financial analysis for some years.[24] Some, to create a more holistic fundamental analysis, some to integrate clients' normative wishes into investment decisions. The idea to create a holistic measurement construct made up of social, environmental, and governance topics therefore fell on fertile ground. Perceptive of this interest on the market, it was mainly the rating and ranking industry that subsequently popularised the acronym ESG by creating new offerings of data products, security ratings, indices, and rankings to now reflect a more holistic view of sustainability for investor's portfolio construction. The goal was to provide an "easy" solution to investors who sought to integrate sustainability concerns into their investment decisions.

Today, the ecosystem of organisations that provide ESG data, often called ESG data vendors or rating agents, is vast. According to the Global Initiative for Sustainability Ratings, over 100 organisations were collecting data, analysing, and rating or ranking company ESG performance in 2018. Most well-known are assessments from global financial rating players such as MSCI ESG Inc., Morningstar's Sustainalytics, Bloomberg, S&P, ISS-ESG, Moody's ESG (formerly Vigeo-Eiris), or LSEG (formerly Refinitiv) who have, since 2010, systematically bought into the ESG market.[25] While in the early 2000s, the organisations collecting ESG data were small research agencies with relatively low scalability, their consolidation into the financial rating market enabled scale, scope and ultimately usefulness of ratings, increasing their attractiveness to investors.[26] Additionally, technologies like the internet and machine learning impacted the rating game, particularly targeting to increase the frequency of ratings.[27] Due to this, tech-powered ratings, like Clarity AI, RepRisk or Truvalue Labs (now within Factset), have rapidly captured parts of the market in recent years.

Concerned with the rapid rise in importance of ESG, various groups of scholars have shown that the construct of ESG—often used as a synonym for sustainability—is, in fact, not a uniform measurement concept at all.[28] The main criticism from this hinted at the fact that ESG as a construct was not a good substitute or proxy for sustainability concerns, and that it was unclear how the different black-box methodologies of rating agents led to the different "shades of green" in ratings. Addressing the debate, Berg, et al.[29] unravelled this "aggregate confusion" by showing that the divergence of ESG ratings from six of the biggest rating agents was due to differences in measurement scope, aggregation, and weights. Measurement divergence, contributing to 56% of difference in ratings, results from two raters using different indicators to measure the same attribute. Scope divergence, contributing 38%, results from two raters considering a different set of attributes. Whereas weight divergence (6%) results from two raters aggregating the same indicators using different weights. Their paper clearly positioned ESG as an analytical product, not a single source of truth for sustainability.

Building on this, Eccles and Stroehle[30] argued that the reason for these technical divergences lies in the social and historical origins of rating agents. They outline how differences in measurements can be traced back to differences in the early (1980–1990s) conceptualisation of sustainability and materiality of the rating agents, distinguishing between those with a more normative, values-oriented lens, and those with a more financial value-seeking purpose. Normative, values-oriented data vendors sought to create data to change the world by enabling investors to push companies towards more sustainable behaviour. Financial, value-oriented ESG data vendors sought to identify more competitive and successful companies. With time and consolidation, most data vendors however adopted a financial, value-seeking lens, considering more quantitative and financial indicators to be able to serve what they believed to be the biggest interest of investors at the time: risk mitigation.[31]

14.2.2 The Measurement of Climate Within ESG

The emergence of ESG and its divergence in measurement is relevant for us, as it had an important influence on the development of business-related climate change assessment in the financial markets. In particular, there is often the mistaken belief that climate-related measurement doesn't suffer the same divergence problems as other topics within the aggregate of ESG. This has to do with two common misconceptions. The first is that climate change measurement starts and ends with GHG emissions, and that these can tell us everything we need to know about a company's climate change management. The second is that, even if this were true, all GHG emissions are standardised in disclosure and thus almost always comparable. Believing this, one would expect that the GHG emissions incorporated in ESG measures from different rating firms are at base always the same.

Regarding misconception number one, the online appendix of the paper from Berg et al.[32] does us a service in showing the different issue categories and numbers of indicators for each that raters use to capture their notion of ESG. Table 14.1 displays issue categories and number of measures by different data vendors. The six chosen ESG data vendors represent some of the largest and most well-known raters in the market. KLD, the only data vendor not still active, is included as it is the predecessor of MSCI. Refinitiv has since been renamed to LSEG. The issue categories selected are those that could hold relevant information of a company's climate change management. The diversity in indicators within each issue category highlights just how diverse the assessment of these issues can be. Each data vendor thus makes prerogative choices in relation to the measurement of topical categories—both in number and focus of indicators. This can lead to widely different measurement results. Climate change–relevant information could therefore be included in a variety of topics such as energy management, ecological impacts, air quality, physical impacts of climate change, and GHG emissions. Other environmental and governance topics, like waste management and materials sourcing, can have a big relevance, too; both for the way a company manages

TABLE 14.1 ESG issue categories with potential importance for climate change

Selected ESG issue categories	*KLD***	*Sustainalytics**	*Moody's ESG*	*S&P Global*	*Refinitiv****	*MSCI*
Access and affordability	3	9		3	2	3
Air quality		2			3	
Critical incident risk mgmt.	1	1			2	
Ecological impacts	3	11	3	6	9	1
Energy mgmt.	1	3	1	6	5	2
GHG emissions	1	8	1	2	9	
Materials sourcing and efficiency		1		3	6	
Physical impacts of climate change	2			2	1	1
Product design and lifecycle mgmt.	3	16	1	6	23	4
Supply chain mgmt.	6	21	4	3	4	3
Systemic risk mgmt.	1			1		1
Waste and hazardous Materials mgmt.	3	4	1	3	5	1
Water and wastewater mgmt.	2	2	1	2	3	1
	26	**78**	**12**	**37**	**72**	**17**

Source: Table adapted from Berg et al. (2022).

Note: Including only those issues from which was assumed that relevant information about climate change management could be included. Issue categories are modelled by the authors after the SASB framework.

* The data vendor KLD has been bought by MSCI in 2011 and is not used for commercial purposes anymore;
** Sustainalytics is part of Morningstar;
*** Refinitiv is now called LSEG, forming part of the London Stock Exchange analytical offer.

its impact on the climate and for the risk which climate poses to its business. Finally, when managing climate risks to a business model, there can be social spillover risks to topics such as access and affordability. This means that corporate climate change management can also become important in understanding how companies address potentially important social aspects of the "just transition."[33] Particularly in the social aspects, however, measurement can vary widely. For example, whereas KLD would consider topics such as "product benefits to economically disadvantaged" in their categorisation of "access and affordability," MSCI would rather focus on the broader assessment of how a business' actions affects peoples' "access to healthcare."

It is important to remark that the actual information about how many of the indicators in Table 14.1 really do hold relevant information about climate change is not available for this chapter. The main point is that all of these criteria may include relevant information of climate change, and that GHG emissions alone are certainly not enough to capture a complete picture of a firm's relationship with climate change. In addition, and even without knowing the exact content of indicators, Table 14.1 gives us a flavour of just how differently the rating agents must conceptualise climate change within ESG. For example, while some datasets don't include indicators in relation to air quality, others don't include specific indicators for physical impacts of climate change. And while some measure GHG emissions with nine indicators, others don't have a GHG indicator at all.

This leads us to misconception number two: the measurement of GHG emissions is always the same. While most (albeit not all) data vendors abide by the logic of "Scopes" from the GHG Protocol,[34] it is important to note that (1) not all companies report GHG emissions in a standardised way and (2) not all data vendors capture GHG emissions in the same way. The biggest problem for measurement poses certainly the concept of "Scope 3," which comprises "all indirect emissions (not included in scope 2) that occur in the value chain of the reporting company, including both upstream and downstream emissions."[35]

While the GHG Protocol is widely understood as a quasi-standard for carbon accounting, much remains to be desired from current corporate reporting on emissions. Created as a collaborative effort between businesses (represented by the World Business Council on Sustainable Development, WBCDS) and the non-profit sector (specifically, the NGO World Resource Institute WRI), since 2001, the GHG Protocol gives guidance on how companies can capture, report on, and manage their GHG emissions. Standards for project emissions (2006), public sector emissions (2010), and supply chain emissions (2011) followed in subsequent years. Today, the protocol's website states that nine out of ten Fortune 500 firms report according to its guidelines. Despite this concerted effort and seeming success, implementation challenges and different levels of transparency remain.

While this is the result of a variety of reasons, the biggest problem is that GHG reporting is still largely voluntary around the world. While the WRI reported that over 40 countries worldwide have GHG reporting included in some legal format,[36] many of these provisions only apply to companies listed on the stock exchange of these countries, specific industries (such as the extractives or energy sector), and companies of very large size. In addition, they mostly only require Scope 1 and 2 reporting and are often quasi-voluntary on a "comply or explain" basis. While recent years have seen an encouraging increase in mandatory regimes for carbon reporting, historically, we are looking at a patchwork picture of global GHG reporting. The result of this: What is reported in relation to GHG can differ greatly between sectors, countries, sizes, and ownership types of firm. It is therefore no surprise that studies like Berg, et al.[37] found extremely low correlations, even between the categories of ESG rating specifically focused on emissions (see Table 14.2).

TABLE 14.2 Correlations between emission categories of different ESG ratings, 2014

Data vendors in comparison	*KL-SA*	*KL-MO*	*KL-RE*	*SA-MO*	*SA-SP*	*SA-RE*	*MO-RE*	*SP-RE*	*Average*
Categories scores compared									
SASB GHG Emissions	0	− 0.03	−0.06	0.32	0.63	0.59	0.56	0.36	0.3
GHG emissions	0	− 0.03	−0.06	0.28		0.31	0.5		0.17
GHG policies					0.48	0.62		0.41	0.5
Non-GHG air emissions						0.28			0.28
Ozone-depleting gas						0.44			0.44

Source: Data and calculations from Berg et al. (2022).

Note: SA, SP, MO, RE, KL, and MS are abbreviations for Sustainalytics, S&P Global, Moody's ESG, Refinitiv, KLD, and MSCI respectively. Correlations of category scores for pairs of rating agencies. Category scores are an average of the indicators assigned to the same category. SASB GHG Emissions describes the category of emissions classified as SASB conform. GHG Policies describes the number of policies counted at the company. Category scores are an average of the indicators for firm "f" and rater "k" assigned to the same category. Empty cells reflect differences in scope or missing data. The negative and positive correlations indicate the magnitude and direction of the overlap in rating agent's assessment. On average, all raters predicted in the same direction (no disagreement between positive and negative assessments), yet at a relatively low level.

14.3 The Construction of Climate Change as a Business Risk

The diversity of categories and indicators in the measurement of climate change within ESG shows us that climate measurement—like any other social evaluation[38]—is ultimately a social construct. In accepting this, one can shift the focus from asking about the "right" measurements to focus on the question about how measurement was constructed in the first place, and who influenced the narrative around it. Today, markets consider climate change as an important business risk.[39] In the following the chapter focuses on the evolution of the climate change narrative in ESG and in sustainable finance to try and understand how this acceptance has evolved. First, it looks at the case of one of the globally largest ESG rating agents—US-based MSCI, Inc.– and how they have evolved their measurement and framing of climate change over time. Second, it looks at the use of the information they produce, asking how this measurement of ESG has influenced the use of climate change information in investment decisions at the firm level accordingly.

14.3.1 Constructing Climate as Financial Risk: The Case of MSCI

In their article "The social origins of ESG: An Analysis of KLD and Innovest," Eccles, Lee, and Stroehle[40] tell the story of how MSCI Inc.—one of the largest index and data providers for both financial and nonfinancial analysis today—came to the decision to create their own ESG products. It outlines how both KLD and Innovest, two of the oldest social and environmental rating agents, created in 1988 and 1992 respectively and therefore existing way before the ESG acronym, were acquired by MSCI through a series of acquisitions in 2009 and 2010. Both ratings seized to exist under their own name at that point. In 2011, MSCI ESG launched their own ESG rating, based on the old Innovest methodology.

The story of this acquisition outlines how each the two historical data vendors had a very different philosophy for the creation of ESG data. Whereas KLD's mission[41] was to "influence corporate behaviour toward a more just and sustainable world," Innovest sought to make the case for eco-efficiency as a core component to competitiveness and profitability of firms.[42] They are thus an illustrative example of how rating agents could be more value-oriented, or financial value-seeking respectively when creating ESG information.[43] This also shows how such different perspectives of rating agents were not merely symbolic but led to tangible differences in the methodological and measurement decisions that went into the construction of ESG scores.[44] Figure 14.1 gives an overview of the different environmental categories used by KLD and Innovest. Tables 14.3 and 14.4 in the appendix give details to the indicators in them.

TABLE 14.3 Environmental indicators used by KLD and Innovest

	KLD		*Innovest*
	Beneficial products and services		Market risk—including environmental sensitivities of customers
	Clean energy		Regulatory/legal risk
	Climate change—climate change vulnerability		Other emissions risk
	Climate change—energy efficiency		Other sustainability risk—operations
	Climate change—financing environmental impact		Performance improvement vector
	Climate change—product carbon footprint		Strategic competence
	Environment other strength	**Risk assessment**	Energy efficiency
	Environmental opportunities—opportunities in renewable energy		Environmental business Development strategy/planning
Environmental strengths	Environmental opportunities—opportunities in green building		Organisational structure
	Management systems strength		Environmental sensitivity of geographic regions served
	Natural capital—biodiversity and land use		Environmental sensitivity of demographic groups served
	Natural capital—raw material sourcing		Phase-out risk of products and services
	Natural capital—water stress		Other operating risk
	Pollution prevention		Resource use efficiency/recycling
	Pollution and waste—electronic waste		Environmental opportunity
	Property, plant, equipment	**Opportunities**	Environmental improvement potential
	Recycling		Environmental positioning within sector

(*Continued*)

TABLE 14.3 (Continued)

	KLD		*Innovest*
	Agriculture chemicals		Current environmental businesses
	Climate change		Environmental businesses under development
	Environment other Concerns		Contaminated site liabilities
	Hazardous waste		Other historic liabilities
	Land use and biodiversity		Spills and releases
Environmental concerns	Negative impact of products and services		Regulatory compliance* scores include NYU data
	Non-carbon releases	**Outcomes**	Toxic emissions
	Ozone depleting chemicals		Hazardous waste
	Regulatory problems		
	Substantial emissions		
	Supply chain management		
	Water management		

Source: Wharton Research Database.

Note: Indicators are a snapshot from the full dataset of 2011. Information on Innovest data was taken from the Innovest IVA methodology brochure (2005); this is the latest available list of Innovest indicators before its acquisition by Risk-Metric in 2009 and subsequent acquisition by MSCI in 2010.

TABLE 14.4 Governance indicators used by KLD and Innovest

	KLD		*Innovest*
	High compensation		Board structure
	Ownership concern		Senior environmental officer level
	Accounting concern		Environmental factor in compensation
	Transparency concern		Number and qualifications of environmental staff
	Political accountability concern		ISO 14000 or other certified EMS
	Public policy concern		Environmental performance indicators
	Governance structures controversies		Audit existence
Governance strengths and concerns	Controversial investments		Audit adequacy and frequency
	Business ethics		Audit impartiality
	Corp. gov.—number of concerns		Environmental reporting
	Corp. gov. other concerns	**Environmental governance**	Environmental accounting
	Limited compensation		Environmental training and development
	Ownership strength		Use of CERES/GRI guidelines
	Transparency strength		Other outside code
	Political accountability strength		Voluntary EPA programmes
	Public policy strength		Life cycle analysis
	Corruption and political instability		Suppliers—environmental screen
	Financial system instability		Eco-labels

(*Continued*)

TABLE 14.4 (Continued)

KLD	*Innovest*
	Policies Integration with core business Profitability linkages Consistency—all operations/ international

Source: Wharton Research Database.

Note: Indicators are a snapshot from the full dataset of 2011. Information on Innovest data was taken from the Innovest IVA methodology brochure (2005); this is the latest available list of Innovest indicators before its acquisition by RiskMetric in 2009 and subsequent acquisition by MSCI in 2010.

Table 14.5 Summary of indicators used by KLD and Innovest

KLD		
Env. strengths	17	
Env. concerns	12	
Governance strengths and concerns	18	**47**
Innovest		
Env. risk assessment	14	
Env. opportunities	5	
Env. outcomes	6	
Env. governance	22	**47**

Source: Constructed from data from Wharton Research Database; Innovest IVA Methodology brochure (2005).

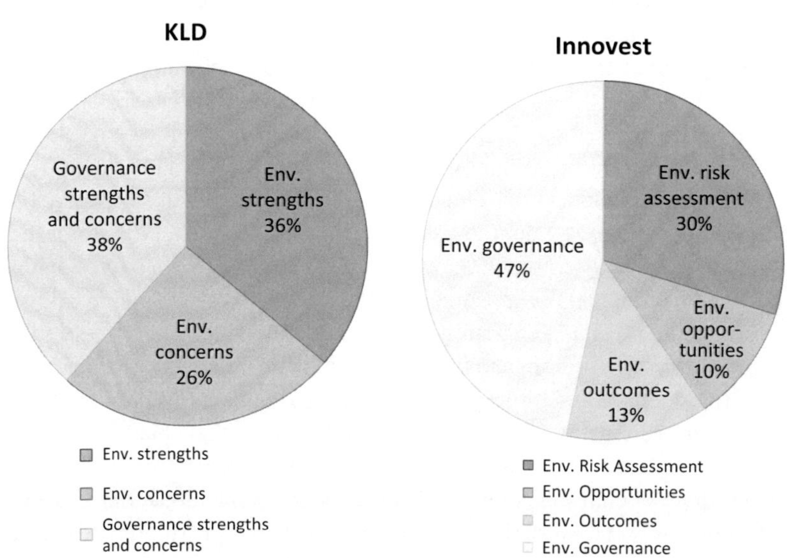

FIGURE 14.1 Innovest and KLD environmental assessment criteria

One quickly sees that the logic between the ratings differs tremendously. On the one hand, KLD focused on rating the strengths and weaknesses of firms in relation to environmental and climate change topics. Both operational measures (e.g. environmental management) and outcome measures (e.g. water stress, emissions, clean energy use) were included in the assessment of strengths and weaknesses, yet with a clear bias in number and interest towards the latter. KLD explicitly sought *not* to create aggregate ratings for firms, but rather to enable a more holistic understanding of firms' exposure to systemic changes like climate change.[45] Investors using this information would then have to engage in a construction exercise themselves if they were interested in a more aggregate picture of a firm's environmental performance.[46] Most governance and social topics were seen as separate from the environmental rating. The different namings of categories give an indication of just how different the assessment of environmental topics were framed at each respective data vendor.

Innovest, on the other hand, created a readily available Intangible Value Assessment (IVA) which rated firms from AAA to CCC.[47] The environmental dimension played an overwhelming role for the overall rating. Yet it shows that the rating focused less on actual environmental outcomes (13% of environmental indicators) and more on risk, opportunity, and environmental governance and efficiency. Governance was therefore not separate but core to the environmental assessment of Innovest. This describes why the concept of "eco-efficiency" was at the heart of its rating strategy.

Understanding the difference between KLD's and Innovest's ratings is important, as it informs us about how the market developed subsequently. Eccles, Lee, and Stroehle[48] describe how MSCI, after acquiring RiskMetrics in 2010, which had acquired KLD and Innovest the year before, had to decide how to build its own rating based on the acquired methodologies. Given that MSCI—as one of the oldest and largest financial index providers—had always been a very financial-market-oriented firm, priorities were outlined as twofold. First, the new product needed to meet the clients' (investors') needs for financially focused assessments. Second, the methodology needed to be scalable, to provide clients with ever larger assessment universes. It comes as no surprise, then, that today's MSCI ESG Ratings are built upon the more financial value-oriented Innovest methodology, whereas the more values-oriented KLD methodology was discontinued. Abiding by the logic of financial markets likely has been an important factor in driving the acceptance and success of ESG factors, and with its climate factors, in mainstream investment practice. In other words, by going the business risk/value-route, firms like MSCI—and later Moody's, Morningstar, and S&P—made these "new" investment criteria relevant for investors.[49]

With the decision to continue with Innovest's rating logic, MSCI made an explicit decision to framing climate change as business risk and contributed to installing this narrative in the market. This narrative still holds today. Under the header "Climate Solutions" MSCI[50] describes that its data "is designed to support investors seeking to achieve a range of objectives, including measuring and reporting on climate risk exposure, implementing low carbon and fossil fuel-free strategies, and factoring climate change research into their risk management processes." Until today, every investor who uses MSCI ratings to inform their climate-related investments will therefore get an explicitly risk-focused assessment and analysis. The impact of this is not to be underestimated. ESG ETFs linked to MSCI ESG indexes had 73.3% market share in August 2021, with $185bn assets under management globally. This means that three out of four passive investors use MSCI ESG ratings as their benchmark. This shows the tremendous impact which the framing and narrative of ESG ratings can have on the market. The topic of climate change is hereby particularly top of mind. According to one executive director at MSCI, climate change indexes are where MSCI receives the most enquiries regarding its ETFs.[51]

14.3.2 *Climate-Risk in the Financial Markets*

After seeing how ESG rating agencies framed climate change as risk, what evidence is there that this indeed had an influence on the perception of climate as risk factor in the market? While it is difficult to claim direct causality in the matter, there are various pieces of anecdotal evidence which hint at just how these ratings may have influenced investor perception on climate change and, in turn, firm-level climate reporting and management. It is particularly worth discussing (1) the climate risk reporting that ESG induced through its data collection and (2) the climate risk-related stewardship activities that grew to become an important part of the ESG investing practice.

In relation to point (1), one needs to understand that while ESG rating agents generally work for the investor clientele (although some also offer consultancy services to corporations), they also have a not-to-be-underestimated impact on the companies they rate. Rating agencies like MSCI, who have a large market power, may send signals that can be flattering or detrimental to a company. Companies therefore have a high incentive to manage their business in a way that will win them favourable scores with the MSCIs of the world. Indeed, many companies openly communicate that their internal sustainability targets include the attainment or sustainment (in the case of positive past ratings) of a high score in one of the top three to four ESG ratings (generally perceived to be MSCI, Morningstar's Sustainalytics, and ISS ESG or S&P). This kind of prioritisation shows how ratings can have an important impact on firm-level climate reporting and management practice.

In the case of MSCI, as discussed before, the focus on risk is pertinent. It therefore pushes companies to assess climate through a very coherent risk lens and to disclose information accordingly. Let's take the example of Volkswagen in 2016. The company had just undergone the Diesel Scandal and was facing tremendous pressure from governments as well as investors. In September 2015, the EPA revealed that Volkswagen had breached the Clean Air Act by incorporating illegal software into their diesel vehicles, prompting investigations by regulators in several countries and causing Volkswagen's stock price to plummet by one-third in the following days. As a result, VW's MSCI ESG rating in 2016 was CCC—the lowest attainable. Due to the scandal, MSCI defined business ethics and corporate governance as the most significant risks that year. However, it also did a rigorous assessment of climate-related topics in relation to categories such as product quality & safety, product carbon footprint, carbon emissions, and opportunities in clean tech (MSCI VW Rating Report 2016). In each of these categories, the assessment of risk is front and centre with a main matrix in each category plotting risk management (ranging from "limited management" to "strong management" on the y-axis) against risk exposure (ranging from "low risk" to "high risk" on the x-axis).[52]

Several things were striking about this rating. First, Volkswagen ranked fairly high in topics such as "product carbon footprint performance" and "opportunities in clean teach," despite these topics having a clear relevance for the scandal that had just occurred. In both areas, MSCI considered Volkswagen's management to be moderate to strong, ranking in the second quartile compared to its competitors. Similarly striking is the evaluation of risk exposure in these areas. MSCI considers the risk to be moderate for each of the topics, not high, as one might have assumed after the Diesel scandal. Also here, various other competitors are seen as higher-risk candidates.

The assessment of risk in this report tells us various things about the framing of risks in popular ESG ratings and their limits. First of all, it shows that the aggregate MSCI rating of CCC, which the company received in 2016, was apparently not due to flaws in risk management, nor due to increased risk exposure in key areas of product emission management. It suggests that even though risk was managed well, it is not a guarantee for the avoidance of a significant scandal. This bears the question of what *kind* of risk rating agencies are capturing in the first place. Second, it highlights that aggregate ratings can only tell us so much about what is really going on at a firm. If the risk management of this topics was mediocre but not terrible, what, then, drives the overall bad score? And how could the scandal still happen? It suggests that aggregate ratings themselves are not a complete, or even trustworthy signal. Third, the analysis opens questions about what kind of risks MSCI assesses in relation to these topics, and why they are not considered higher or worse managed at a firm that had just undergone such a public scandal. Since MSCI collects the information for these ratings only once a year, these risk assessments can also shift quickly, particularly for a firm in the context of a scandal.

Despite these questions, it becomes transparent that "risk" sits at the very core of MSCI's assessment. It seem, however, to be a very specific risk that is linked to financial detriment—despite it not having been able to gauge a scandal of this size. This begs the question of what other types of risk, or perhaps non-risk frames, investors would have to understand in order to assess business and climate change more holistically. Increasingly, MSCI therefore encourages and endorses specific climate-related risk reporting frameworks. Since 2022, for example, it offers a tailored TCFD-aligned risk reporting tool to help investors align with the recommendations and help them assess whether portfolio companies are aligning with their expectations around climate change risk management at the firm level.[53] While the foundation and wide acceptance of TCFD in and of itself was a strong signal that climate-risk management had become a dominant factor for both companies and investors to consider, its acceptance into the ESG and sustainable finance practice further fastened its recognition.

This leads us to point (2) mentioned earlier: the increase of climate-risk related investor stewardship activities since the early 2000s. Apart from influencing firm-level climate reporting and management directly, ESG rating agents also have an indirect influence through the investors that use their information. On the one hand, ESG information of course directly factored into investment decisions. It is, however, not always apparent how exactly investors include this information into their appraisals. A much more tangible source of information for firms to understand investor expectations is therefore through the increasing use of ESG stewardship activities. Investors use stewardship, often also referred to as "active ownership" or "ESG engagement," to proactively communicate with their portfolio companies about what they expect them to report and/or manage. Indeed, investor climate stewardship—both at an individual investment firm level and at a collective investor level—has seen a tremendous increase in popularity in recent years. A powerful example of this is the collaborative engagement effort of Climate Action 100+. Launched in 2017, the group now contains 700 investors, responsible for over $68 trillion in assets under management, who have teamed up to engage the world's largest emitters collectively on improving climate change governance, cutting emissions, and strengthening climate-related financial disclosures[54] (part of which is following the TCFD recommendation). Their focus on pushing companies to disclose more on climate as business risk has been a powerful show of force from the investor community. MSCI, as provider of climate information to most of these investors, actively encourages clients to use its information to inform stewardship activities.[55]

Some investors take their commitment to climate risk even further and engage in investor climate activism, arguing that companies that do not manage climate risk rigorously should change board or management composition. This narrative reached a high point in 2021 when a small U.S. activist hedge fund, Engine No. 1, engaged in a proxy fight with the mighty ExxonMobil with the goal of improving the company's climate governance and risk management practices.[56] In a campaign titled "reenergize Exxon,"[57] the hedge fund challenged the company to replace four of its board members citing that "ExxonMobil has not enunciated any kind of holistic strategy for navigating the carbon transition, which we believe is a fundamental threat to long-term shareholder value." The firm did *not* use ESG metrics for its argument but merely relied on financial fundamentals in the context of developing evidence around climate change as well as resulting market and regulatory developments. Engine No. 1 only held 0.02% of Exxon's shares it was able to win the support of some of the company's largest shareholders, among them BlackRock, Vanguard, CalPERS, and CalSTRS. In one of Wall Street's biggest climate upheavals in history, in May 2021, shareholders voted against Exxon's leadership and placed three of the Engine No. 1 proposed professionals on Exxon's board. Precisely because the activist's case was *not* based on ESG ratings but rather financial fundamentals, this case is an illustrative example of how climate change has become pushed into the realm of financial risk assessment.

14.4 Conclusion

This chapter reviews the history of climate-related measurement and discusses how the framing of climate through the construction of ESG and sustainable investment practices has shaped climate change as an accepted business risk factor. It traces the creation of influential NGO frameworks in the space—such as the GHG Protocol[58] that outlines Scope 1, Scope 2, and Scope 3[59] carbon emission reporting—and how their use was elevated by various influential UN-backed treaties (Paris Agreement) and investor organisations (UN PRI, UNEP-FI). It also discusses how in the early days, carbon risk was often framed by rating agents who sought to create meaningful products for their clients, the institutional investors. And how the emergence of the TCFD recommendations helped to institutionalise the consideration of climate as a business risk. Finally, the chapter highlights how today, national (such as the U.S. SEC rules on climate-related disclosure), regional (e.g. the European Union CSRD) and global (such as the IFRS/ISSB) regulation and standardisation have taken over to structure the space significantly. It also shows, however, how these standards still rely on established notions of business risk, particularly in relation to financial materiality, when it comes to climate change.

Through a closer examination of the history of one specific climate rating—the MSCI ESG Rating—the chapter gives an example of how climate assessments are structured at large ESG rating agencies. It highlights that, as much as any other social evaluation, climate change assessments have to be understood as social construct that are influenced by both dominant market logics and proprietary institutional interests. As such, relying on Berg, et al.'s[60] assessment of measurement divergence, the chapter reviews how different data vendors conceptualise not only climate risk differently but also how distinct the dimensions can be with which climate risk is assessed. It discusses the rating case of Volkswagen, which was rated markedly positively on key risk dimensions shortly after the Diesel scandal. This highlights different risk perspectives, and how climate "risk" can be assessed through risk management and risk exposure in relation to financial materiality, potentially lacking perspectives that would give it an indication of broader governance risks associated with climate failures. Due to these distinct risk frames, the paper concludes by highlighting how investors now use engagement and activism tools precisely to make sure that their risk understanding aligns with that of their portfolio firm.

In summary, the need to integrate environmental and particularly climate change concerns into corporate measurement and reporting is still on the rise, yet it is not without its challenges. While investors seem to have broadly accepted that climate change is a business risk, fundamental challenges in assessing and framing this risk appear to remain. These challenges mostly not only relate to data gaps and regulatory inconsistencies but also exist in relation to methodological uncertainties as well as inconsistent constructs and frames that exist to capture the notion of climate as a business risk. While the risk focus has generally aided the integration of climate change into the playbook of investors, more awareness is needed about the multidimensionality of these risks. Frameworks such as the TCFD have been particularly helpful in aligning these kinds of expectations and have supported the creation of a common frame for assessing climate change as a business risk.

Notes

1 United Nations, *Transforming Our World*, 14.
2 Intergovernmental Panel on Climate Change, *Synthesis Report of the IPCC Sixth Assessment Report,* 14.
3 Fiedler et al., "Business Risk and the Emergence of Climate Analytics," 87–94.
4 Gond and Piani, "Enabling Institutional Investors' Collective Action," 64–104.
5 PRI, *A Blueprint for Responsible Investment*, 2.
6 PRI, *A Blueprint for Responsible Investment*, 28.
7 Clark, "Years of Climate Scepticism Have Done Untold Damage."
8 Crowley and Eccles, "Rescuing ESG from the Culture Wars."
9 Jones, *ESG Backlash*; Cerulli, *U.S. Environmental, Social, and Governance Investing 2023 Regulation and Legislation.*
10 IFRS, "Effects of Climate-Related Matters on Financial Statements."
11 TCFD, *Recommendations of the Taskforce on Climate-related Financial Disclosures*; IFRS, "Effects of Climate-Related Matters on Financial Statements."
12 Mundy and Temple-West, "Emmanuel Faber Is on a Mission."
13 Barko et al., "Shareholder Engagement on Environmental, Social, and Governance Performance," 777–812; Hoepner et al., "ESG Shareholder Engagement and Downside Risk," 483–510.
14 Görgen et al., "Get Green or Die Trying?" 77–93.
15 Chambers et al., "To Divest or to Engage?" 10–20.
16 Ilhan et al., "Climate Risk Disclosure and Institutional Investors," 2617–50.
17 Bebbington and Larrinaga-González, "Carbon Trading: Accounting and Reporting Issues," 697–717.
18 Jia et al., "A Rapid Review of GHG Accounting Standards."
19 KPMG, *Big Shifts, Small Steps—Survey of Sustainability Reporting 2022.*
20 TCFD, *Recommendations of the Taskforce on Climate-related Financial Disclosures.*
21 PRI, *A Blueprint for Responsible Investment.*
22 Global Compact, *Who Cares Wins.*
23 UNEP Finance Initiative, *Freshfields Report—A Legal Framework for the Integration of Environmental, Social and Governance Issues into Institutional Investment*, 6.
24 Eccles, Lee, and Stroehle, "The Social Origins of ESG," 575–96.
25 GISR, "Global Initiative for Sustainability Ratings."
26 Eccles and Stroehle, "Exploring Social Origins in the Construction of ESG Measures."
27 TVL, "TruValue Labs™ Launches ESG Momentum™ Score in Suite of New Timely ESG Metrics."
28 Delmas et al., "Triangulating Environmental Performance," 255–67; Chatterji et al., "Do Ratings of Firms Converge?" 1597–614; Berg et al., "Aggregate Confusion," 1315–44.
29 Berg et al., "Aggregate Confusion," 1315–44.
30 Eccles and Stroehle, "Exploring Social Origins in the Construction of ESG Measures."
31 Eccles, Lee, and Stroehle, "The Social Origins of ESG," 575–96.

32 Berg et al., "Aggregate Confusion," 42 app.
33 International Labour Organization (ILO), *Frequently Asked Questions on Just Transition*, describes Just Transition as "greening the economy in a way that is as fair and inclusive as possible to everyone concerned, creating decent work opportunities and leaving no one behind."
34 The World Resource Institute, *GHG Protocol for Corporate Standard*, 25, defines Scope 1 Emissions as direct emissions that are produced by a firm's own operations (those from stationary and mobile combustion sources, as well as fugitive emissions). Scope 2 Emissions are indirect emissions which are produced through purchased electricity (and which occur through the operations of the energy provider), and Scope 3 Emissions are all other indirect emissions which are induced by the firm's upstream and downstream business activities, the customer's use of product, and the product end-life.
35 GHG Protocol, *The Corporate Value Chain Scope 3.*
36 Singh and Longendyke, *A Global Look at Mandatory Greenhouse Gas Reporting Programs.*
37 Berg et al., "Aggregate Confusion," 1315–44.
38 Rao et al., "Institutional Change in Toque Ville," 795–843; Sauder and Espeland, "The Discipline of Rankings," 63–82; Chatterji et al., "Do Ratings of Firms Converge?" 1597–614.
39 Sheldrick, "New Data Reveals Climate Change Risks to Corporate Assets in the United States."
40 Eccles, Lee, and Stroehle, "The Social Origins of ESG," 575–96.
41 KLD's archived website (date of capture: 2005; date of recovery: 2023) is online available under, web.archive.org/web/20050403203540/http://www.kld.com:80/about/mission.html.
42 Innovest's archived website (date of capture: 2000; date of recovery: 2023) is online available under, https://web.archive.org/web/20000620224819/http://www.innovestgroup.com:80/about_1.html.
43 Eccles, Lee, and Stroehle, "The Social Origins of ESG," 575–96.
44 Eccles and Stroehle, "Exploring Social Origins in the Construction of ESG Measures."
45 Sharfman, "The Construct Validity of the Kinder, Lydenberg & Domini Social Performance Ratings Data," 287–96.
46 Eccles, Lee, and Stroehle, "The Social Origins of ESG," 575–96.
47 Innovest, *Research and Rating Methodology.*
48 Eccles, Lee and Stroehle, "The Social Origins of ESG," 575–96.
49 While MSCI still provides KLD data for research purposes today, all data is in fact collected through the Innovest methodology and subsequently mapped to the original KLD model of ESG strengths and weaknesses. This data is not anymore commercialised in the financial market.
50 MSCI, *ESG Climate Change Solutions*, 1.
51 Chandler, "MSCI dominates ESG indices and enjoys long history in the emerging markets."
52 EPA, "Learn About Volkswagen Violations."
53 Cote and Hulme, *TCFD-Aligned Climate Risk Reporting.*
54 The full list of members, company targets as well as the group's approach to engagement and their asks can be found online under, https://www.climateaction100.org/about/. Due to the evolving strategy of the group, in 2023 three of the large asset managers—Vanguard, JP Morgan and State Street—have however backed out of this group, while BlackRock, the world's largest investor—announced to take a more passive role in the future.
55 MSCI, *ESG Climate Change Solutions.*
56 Eccles and Mayer, "Can a Tiny Hedge Fund Push ExxonMobil Towards Sustainability?"
57 Full details about the campaign goals, its outcomes and the firm Engine No. 1 can be found on the campaign's website under https://reenergizexom.com/who-is-engine-no-1/.
58 GHG Protocol, *The Greenhouse Gas Protocol.*
59 GHG Protocol, *The Corporate Value Chain Scope 3.*
60 Berg et al., "Aggregate Confusion," 1315–44.

Bibliography

Barko, Tamas, Martijn Cremers, and Luc Renneboog. "Shareholder Engagement on Environmental, Social, and Governance Performance." *Journal of Business Ethics* 180 (2022): 777–812. https://doi.org/10.1007/s10551-021-04850-z.

Bebbington, Jan, and Carlos Larrinaga-González. "Carbon Trading: Accounting and Reporting Issues." *European Accounting Review* 17, no. 4 (2008): 697–717. https://doi.org/10.1080/09638180802489162.

Berg, Florian, Julian F. Kölbel, and Roberto Rigobon. "Aggregate Confusion: The Divergence of ESG Ratings." *Review of Finance* 26, no. 6 (2022): 1315–44. https://doi.org/10.1093/rof/rfac033.

Cerulli Associates. "U.S. Environmental, Social, and Governance Investing 2023 Regulation and Legislation." Cerulli Research Report, Cerulli Associates, 2023.

Chambers, David, Elroy Dimson, and Ellen Quigley. "To Divest or to Engage? A Case Study of Investor Responses to Climate Activism." *Journal of Investing* 29, no. 2 (2020): 10–20. https://doi.org/10.3905/joi.2020.1.114.

Chandler, Beverley. "MSCI Dominates ESG Indices and Enjoys Long History in the Emerging Markets." *ETF Express*, October 22, 2021. https://etfexpress.com/2021/10/22/msci-dominates-esg-indices-and-enjoys-long-history-emerging-markets/.

Chatterji, Aaron K., Rudolphe Durand, David I. Levine, and Samuel Touboul. "Do Ratings of Firms Converge? Implications for Managers, Investors and Strategy Researchers." *Strategic Management Journal* 37 (2015): 1597–614. https://doi.org/10.1002/smj.2407.

Clark, Pilita. "Years of Climate Scepticism Have Done Untold Damage—Erroneous Claims, Scientific Caution and Poor Media Coverage Held Back Policymaking on Global Warming." *Financial Times*, March 21, 2023. https://www.ft.com/content/bcc3e9db-96a8-4ae1-a536-5496d88720f4.

Cote, Chris, and Mike Hulme. *TCFD-Aligned Climate Risk Reporting—A Practical Guide for Institutional Investors Using MSCI Data and Metrics*. New York: MSCI Screening Guide, 2022.

Crowley, Daniel F.C., and Robert G. Eccles. "Rescuing ESG from the Culture Wars." *Harvard Business Review*, February 2023, online ed. https://hbr.org/2023/02/rescuing-esg-from-the-culture-wars.

Delmas, Magali A., Dror Etzion, and Nicholas Nairn-Birch. "Triangulating Environmental Performance: What Do Corporate Social Responsibility Ratings Really Capture?" *Academy of Management Perspectives* 27, no. 3 (2013): 255–67. http://www.jstor.org/stable/43822025.

Eccles, Robert G., Linda-Eling Lee, and Judith C. Stroehle. "The Social Origins of ESG: An Analysis of Innovest and KLD." *Organizations & Environment* 33, no. 4 (2019): 575–96. https://doi.org/10.1177/1086026619888994.

Eccles, Robert G., and Colin Mayer. "Can a Tiny Hedge Fund Push ExxonMobil Towards Sustainability?" *Harvard Business Review*, January 20, 2021. https://hbr.org/2021/01/can-a-tiny-hedge-fund-push-exxonmobile-towards-sustainability.

Eccles, Robert G., and Judith C. Stroehle. "Exploring Social Origins in the Construction of ESG Measures." Working Paper, 2019. https://ssrn.com/abstract=3212685.

EPA. "Learn About Volkswagen Violations." United States Environmental Protection Agency. https://www.epa.gov/vw/learn-about-volkswagen-violations.

Fiedler, Tanya, Andy J. Pitman, Kate Mackenzie, Nick Wood, Christian Jakob, and Sarah E. Perkins-Kirkpatrick. "Business Risk and the Emergence of Climate Analytics." *Nature Climate Change* 11 (2021): 87–94. https://doi.org/10.1038/s41558-020-00984-6.

GHG Protocol. *The Greenhouse Gas Protocol—A Corporate Accounting and Reporting Standard*. Revised ed. Washington, DC and Geneva: World Resource Institute & World Business Council for Sustainable Development, April 2001. https://ghgprotocol.org/standards-guidance.

GHG Protocol. *The Corporate Value Chain Scope 3 Accounting and Reporting Standard*. Revised ed. Washington, DC and Geneva: World Resource Institute & World Business Council for Sustainable Development, 2011. https://ghgprotocol.org/corporate-value-chain-scope-3-standard.

GISR. "Global Initiative for Sustainability Ratings." https://shift.tools/resources/1570.

Gond, Jean-Pascal, and Valeria Piani. "Enabling Institutional Investors' Collective Action: The Role of the Principles for Responsible Investment Initiative." *Business & Society* 52, no. 1 (2012): 64–104. https://doi.org/10.1177/0007650312460012.

Görgen, Maximilian, Andrea Jacob, and Martin Nerlinger. "Get Green or Die Trying? Carbon Risk Integration into Portfolio Management." *Journal of Portfolio Management* 47, no. 3 (2021): 77–93. https://doi.org/10.3905/jpm.2020.1.200.

Hoepner, A.G.F., Ioannis Oikonomou, Zacharias Sautner, Laura T. Starks, and Xiao Y. Zhou. "ESG Shareholder Engagement and Downside Risk." *Review of Finance* 28, no. 2 (2024): 483–510. https://doi.org/10.1093/rof/rfad034.

International Financial Reporting Standards Foundation (IFRS). "Effects of Climate-Related Matters on Financial Statements: IFRS Accounting—Educational Material." https://www.ifrs.org/content/dam/ifrs/supporting-implementation/documents/effects-of-climate-related-matters-on-financial-statements.pdf.

Ilhan, Emirhan, Philipp Krueger, Zacharias Sautner, and Laura T. Starks. "Climate Risk Disclosure and Institutional Investors." *The Review of Financial Studies* 36, no. 7 (2023): 2617–50. https://doi.org/10.1093/rfs/hhad002.

Innovest Strategic Value Advisors. *Research and Rating Methodology—Intangible Value Assessment IVA*. New York: Innovest, 2005.

Intergovernmental Panel on Climate Change. "Synthesis Report of the IPCC Sixth Assessment Report (AR6)." Summary for Policy Makers, Geneva, 2023.
International Labour Organization. "Frequently Asked Questions on Just Transition." https://www.ilo.org/global/topics/green-jobs/WCMS_824102.
Jia, Jimmy, Kaya Axelsson, Abrar Chaudhury, and Evan Taylor. "A Rapid Review of GHG Accounting Standards." Working Paper, 2023. http://dx.doi.org/10.2139/ssrn.4523132.
Jones, Andrew. "ESG Backlash: How Companies Can Turn Adversity into Advantage." The Conference Board Research Report, New York, 2023.
KPMG. "Big Shifts, Small Steps—Survey of Sustainability Reporting 2022." KPMG International Report, 2022. https://assets.kpmg.com/content/dam/kpmg/se/pdf/komm/2022/Global-Survey-of-Sustainability-Reporting-2022.pdf.
Mundy, Simon, and Patrick Temple-West. "Emmanuel Faber Is on a Mission." *Financial Times—Moral Money Newsletter*, May 24, 2024. https://www.ft.com/content/2b214790-21dc-40fa-9efe-9c43ed3a758d.
MSCI Inc. *Rating Report of Volkswagen AG*. New York: MSCI ESG Ratings, 2016.
MSCI, Inc. *ESG Climate Change Solutions—A Holistic Toolkit to Enable Clients to Build More Resilient Portfolios*. New York: MSCI ESG Ratings, 2019.
Principles of Responsible Investment. "A Blueprint for Responsible Investment." PRI Practice Report, New York, 2017.
Rao, Hayagreeva, Philippe Monin, and Rodolphe Durand. "Institutional Change in Toque Ville: Nouvelle Cuisine as an Identity Movement in French Gastronomy." *American Journal of Sociology* 108, no. 4 (2003): 795–843. https://doi.org/10.1086/367917.
Sauder, Michael, and Wendy Nelson Espeland. "The Discipline of Rankings: Tight Coupling and Organizational Change." *American Sociological Review* 74, no. 1 (2009): 63–82. https://doi.org/10.1177/000312240907400104.
Sharfman, Mark. "The Construct Validity of the Kinder, Lydenberg & Domini Social Performance Ratings Data." *Journal of Business Ethics* 15, no. 3 (1996): 287–96. http://www.jstor.org/stable/25072754.
Sheldrick Michael. "New Data Reveals Climate Change Risks to Corporate Assets in the United States." *Forbes*, July 16, 2023. https://www.forbes.com/sites/globalcitizen/2023/07/16/new-data-reveals-climate-change-risks-to-corporate-assets-in-the-united-states/?sh=44b97331463e.
Singh, Neelam, and Lindsey Longendyke. *A Global Look at Mandatory Greenhouse Gas Reporting Programs.* Washington, DC: World Resource Institute, 2015. https://www.wri.org/insights/global-look-mandatory-greenhouse-gas-reporting-programs.
TCFD. "Recommendations of the Taskforce on Climate-Related Financial Disclosures." Final Report, TCFD, 2017.
TVL. "TruValue Labs™ launches ESG Momentum™ Score in Suite of New Timely ESG Metrics." https://www.sustainability-reports.com/truvalue-labs-launches-esg-momentum-score-in-suite-of-new-timely-esg-metrics/.
UNEP Finance Initiative. "Freshfields Report—A Legal Framework for the Integration of Environmental, Social and Governance Issues into Institutional Investment." Produced for the Asset Management Group of the UNEP FI. Freshfields, Bruckhaus, Deringer, 2005.
United Nations. "Transforming Our World: The 2030 Agenda for Sustainable Development." Resolution Adopted by the General Assembly, Geneva, 2015.
United Nations Global Compact. *Who Cares Wins—Connecting Financial Markets to a Changing World.* Geneva: United Nations, 2004.

15

BIG BUSINESS AND CLIMATE CHANGE LITIGATION IN THE DAWN OF THE TWENTY-FIRST CENTURY

Bart Elmore

15.1 Introduction

As this book went to press in 2024, large multinational firms, including big oil companies such as ExxonMobil and Shell and major utility and energy firms such as American Electric Power (AEP) and Duke Energy faced an unprecedented array of legal threats in the form of climate change litigation. As I write, courts in Delaware, Hawaii, Massachusetts, Minnesota, Oregon, Washington, and several other states have allowed plaintiffs to proceed with charges that fossil fuel firms knowingly deceived and harmed the public by denying the realities of climate change. Complaints filed in these cases were bold and ambitious in their scope.[1] In the case of *The State of Minnesota v. American Petroleum Institute, et. al.* (filed in 2020), plaintiffs sought to hold an array of companies, including the American Petroleum Institute, ExxonMobil, and Koch Industries,

> accountable for deliberately undermining the science of climate change, purposefully downplaying the role that the purchase and consumption of their products played in causing climate change . . . and for failing to fully inform the consumers and the public of their understanding that without swift action, it would be too late to ward off devastation.

In the end, plaintiffs sought

> to ensure that the parties who have profited from avoiding the consequences and costs of dealing with global warming and its physical, environmental, social, and economic consequences, bear the costs of those impacts, rather than Minnesota taxpayers, residents, or broader segments of the public.[2]

At the federal level, lawyers representing left-leaning states and environmental non-profits filed cases in the 2010s and 2020s arguing that government agencies such as the US EPA and the National Highway Traffic Safety Administration (NHTSA) neglected their responsibilities to protect citizens from greenhouse gases emitted by automobiles and coal-fired power plants.

DOI: 10.4324/9781003478089-16

The legal approaches in these cases varied widely, some focusing on violations of statutory laws such as the National Environmental Policy Act (NEPA) and the Clean Air Act (CAA), while others made constitutional arguments, appealing to the due process clause and the Fifth Amendment in seeking relief from federal agencies that failed to protect citizens from the consequences of climate change. All these cases against federal agencies have implications for big businesses, as the regulations these lawsuits hope to spur would ultimately have major effects on corporate practices.

While most climate litigation focused on oil majors, energy firms, and various government regulatory agencies, fossil fuel companies and utilities were not the only private industry targets of climate litigants. JBS Foods, one of the biggest meat processing companies in the world, faced a summons in 2024 in New York Superior Court, with New York's attorney general alleging the firm had made misleading claims about its efforts to reduce its greenhouse gas footprint. Delta Air Lines faced a similar complaint charging greenwashing in the Central District Court of California in 2023. And even consulting firm McKinsey became a named defendant in an Oregon climate change lawsuit, with plaintiffs arguing that the firm played a critical role in assisting "fossil fuel and mining company clients in promoting themes to deny the existence and/or gravity" of anthropogenic climate change.[3]

Prior to 2007, there were very few climate change cases, but that changed as activists and lawyers—who were concerned about the worsening climate crisis and frustrated with the gridlock in Washington, D.C. and the glacial pace of international climate negotiations—adopted novel legal strategies to try and hold firms and government agencies accountable for failing to adequately address greenhouse gas pollution. In 2017, over 654 climate change cases had been filed in U.S. state and federal courts, up from just a handful a decade earlier. By 2020, that number had risen to 1,200, and two years later the total was 1,522.[4]

Though most climate change cases originated in United States courts, climate change lawsuits were also spreading internationally in the 2010s and 2020s. By 2017, the United Nations reported that there had been 230 climate change cases filed in some 24 legal jurisdictions outside the United States. By 2022, there were approximately 658 lawsuits that had originated out of 65 different non-U.S. jurisdictions. Some of these cases focused on government actors.[5] In 2024, for example, the European Court of Human Rights ruled in favour of several Swedish women who brought suit against their government for not doing enough to curb greenhouse gas emissions, thereby endangering their lives.[6] But other cases targeted automobile companies, airlines, utility firms, and fossil fuel companies.[7]

The 2010s and 2020s, in other words, represented a new era in climate change litigation, one that has not garnered much attention from business historians. As we have seen throughout this volume, scholars in the fields of business and environmental history have produced illuminating scholarship that helps us understand how major multinational firms have reshaped our climate and planetary ecosystems more broadly. Yet there is more work to be done to understand how businesses' failures to adequately address such pressing issues as climate change have changed American and even international legal norms.

This chapter begins by exploring the origins of climate change litigation in the United States in the 1990s and 2000s before turning to the novel legal strategies plaintiffs' attorneys adopted to attack big polluting firms and the government agencies that regulated them. Failures to make big business take accountability for their greenhouse gas emissions such as when the Supreme Court decided in the case of *AEP v. Connecticut* (2011) that federal common law could not provide relief to plaintiffs seeking to hold major energy firms accountable for the pollution they generated, led to new legal approaches that involved focusing on state law and even constitutional claims.[8] The chapter ends by considering the recent successes some litigants have had in the courts and what this might mean for multinational firms' future business practices.

15.2 The Origins of Climate Litigation in the United States

Businesses simply did not do enough to address climate change in the 1990s and the early years of the twenty-first century. To be sure, corporate social responsibility had become a hallmark of most business strategising by the 2000s. As historians Adam Rome and many others have shown, chemical giants such as DuPont and Monsanto began making commitments to go green as early as the 1980s.[9] The world's largest corporation, Walmart, also made a major promise in 2004 to address climate change and focus on environmental sustainability in the years ahead.[10] In the early 2000s, big banks such as Citi, Bank of America, and Wells Fargo, pressured by the activism of the Rainforest Action Network, joined the chorus, promising new plans to green their financial operations.[11] As business historians Ann-Kristin Bergquist and Geoffrey Jones have shown earlier in this volume, a whole host of new organisations emerged in the 1990s and onward, from the UNEP-FI to the Carbon Disclosure Project (CDP) to Ceres, which sought to marshal the enormous resources of multinational firms to address the pressing issue of climate change.[12] UN Secretary-General Kofi Annan gave birth to Environmental, Social, and Governance (ESG) strategies when he reached out in the early 2000s to more than a dozen financial institutions, including Goldman Sachs, BNP Paribas, Deutsche Bank, and Morgan Stanley, and tasked these investors with finding a way to embed environmental, social, and governance standards in the financial system. The report ultimately issued in 2004 and endorsed by 18 financial institutions made clear that the "formulation of specific standards should . . . rely on market-driven voluntary initiatives." The world's financial leaders were asserting that corporations were to be the ones making the important decisions about how to solve big social justice and environmental problems, not government regulators.[13]

Yet despite the boldness of the promises and the shininess of new corporate sustainability reports, businesses' voluntary efforts to address environmental problems produced less than stellar results. In 2013, a study conducted by *MIT Sloan Management Review* and the Boston Consulting Group determined that there was a "disconnect between thought and action" when it came to firm's ESG commitments.[14] That same year the Carbon Disclosure Project's Global Report made clear that the top 50 largest greenhouse gas emitters were witnessing increases, not decreases, in their greenhouse gas emissions compared to 2009 benchmarks.[15] And in 2020, Bain & Company announced the findings of a massive survey of big businesses that looked at corporations' sustainability track record over the past several years. Bain found that among the hundreds of companies they surveyed there was a "paltry 4%" success rate when it came to meeting sustainability promises. Such statistics made clear that apparently good-faith corporate volunteerism in the era of ESG was not working to radically reduce corporations' greenhouse gas emissions fast enough.[16]

A closer look at ESG metrics helps to explain why companies failed to meet their targets to reduce their carbon emissions. Part of the issue had to do with the strange formulas used to create ESG indices by organisations such as the S&P Global, Carbon Disclosure Project, Sustainalytics, and MSCI.[17] For example, a company's ESG score was often calculated by combining environmental, social, and governance criteria into one number, meaning a firm could release harmful pollutants into the environment but achieve a high ESG rating because it had a diverse governing board.[18] ESG scores also often reflected how good a business was in comparison to other firms in a particular industry, which meant some firms contributing a great deal to climate change ranked very high in some ESG charts.[19] For example, Exxon ended up at the top of the S&P 500 ESG index in the early 2020s despite being one of the largest fossil fuel firms contributing to climate change in the world.[20]

Businesses developed their own lexicon to measure progress in solving pressing environmental problems in the 2010s and 2020s. One phrase that became popular in ESG circles was "carbon emissions intensity." Carbon emissions intensity is a ratio of a firm's total greenhouse gas emissions divided by its total annual revenues. If revenues go up, greenhouse gas emissions can go up as well, and firms still appear as if they are radically addressing their carbon footprint. FedEx, for example, reported in its 2022 ESG report that it had seen a 45% reduction in its "carbon emissions intensity" between 2019 and 2021. Progress. But buried in the final pages of the report were the firm's absolute emissions, which increased by nearly 10% over the same period.[21]

People were paying attention to the fact that corporate volunteerism was not meeting the moment of climate change, and this led young activists, lawyers, and policymakers to seek new ways to hold companies accountable for the environmental damage they wrought. Many of these actors turned to the courts, perhaps frustrated by the fact that both domestic and international treaties and laws were doing little to force businesses to make more aggressive commitments to curbing climate change.

This was particularly true in the United States, which became by wide margin the world's leader in climate change litigation after 2006, when a flurry of legal cases emerged in American courts that tried to hold both government and corporate entities accountable for actions that were warming the planet. At that time, activists and policymakers, anxious to see the federal government take a bold stand on climate change, were demoralised. Since approving the United Nations Framework Convention on Climate Change (UNFCCC) at the Rio Earth Summit in 1992, the United States government had done little to address climate change. In 1997, at the UN climate summit in Kyoto, Japan, the Clinton Administration agreed to establish national commitments to substantially reduce carbon emissions, but Congress failed to support the Kyoto Protocol, largely due to concerns that the agreement would place the United States at a disadvantage vis-à-vis countries such as China, which would not have faced the same demands for greenhouse gas emissions reductions as the United States under the Protocol.[22] In the late 1990s, Republican lawmakers even went so far as to propose legislation to prevent the EPA from including greenhouse gases in its air pollution regulations.[23] And when George W. Bush won the controversial election of 2000 with widespread backing from the fossil fuel industry, and as the Republican Party became increasingly opposed to any expansion of environmental legislation, the courts increasingly appeared the best way to try and effect change in the United States.[24]

Though there were very few major court cases specifically about climate change in domestic U.S. courts prior to 2000, climate change litigants in the early aughts nevertheless drew on case law from the 1980s and1990s to build out their briefs. One landmark Supreme Court case that proved useful to plaintiffs was *Chevron, U.S.A. Inc. v. Natural Resources Defense Council* (1984). This case established the legal precedent that federal agencies such as the Environmental Protection Agency (EPA) had broad authority to make "reasonable" judgments on how best to implement legislation that fell directly under their purview. The justices were essentially deferring to agency experts on issues that the court did not have the expertise to adjudicate.[25] In an opinion issued in 2024 in the case of *Loper Bright Enterprises v. Raimondo*, the Supreme

Court overturned the decision in *Chevron*, but for nearly 40 years prior to this monumental shift in judicial precedent, *Chevron* boosted EPA power.[26]

In 1990, the city of Los Angeles came together with a series of environmental organisations to sue the National Highway Traffic Safety Administration (NHTSA), arguing that the transportation agency's Corporate Average Fuel Economy (CAFE) regulations did not adequately consider the ways in which new fuel standards would affect greenhouse gas emissions.[27] Specifically, attorneys argued that the NHTSA had violated the National Environmental Policy Act (NEPA) of 1969, which tasked federal agencies with completing extensive environmental impact statements (EIS) ascertaining all the various ways a government project could alter the environment. The city of Los Angeles and its co-litigants would lose this case, but it was not an entire failure for those hoping to use NEPA to address climate change. As legal scholar Michael Gerrard explained, "The court found that plaintiffs had standing to bring the lawsuit (itself a significant holding), but that the one-mile per gallon change in the CAFE standard at issue was not so significant as to require an EIS." In the end, Gerrard concluded, "This court, like all subsequent federal courts to address the question, did not doubt that global warming was a proper subject for analysis under NEPA; it merely found a particular action's impacts to fall below the threshold of significance."[28] Thus, the case revealed how lawyers, with refined legal arguments, might utilise the courts and NEPA to push federal agencies to take bolder regulatory positions on climate change.[29]

Environmentalists and public officials concerned about climate change increasingly turned to the courts in the early 2000s as President George W. Bush showed hostility to the Kyoto Protocol and his EPA failed to answer petitions from environmental organisations to use the Clean Air Act to regulate greenhouse gas emissions.[30] The majority of the early court battles involved statutory law, especially NEPA and the Clean Air Act (CAA). One of the biggest cases was *Massachusetts v. EPA* (2007), litigation brought by 12 states that argued the EPA had a responsibility to regulate greenhouse gases under the Clean Air Act.[31] In a close decision, 5–4, the Supreme Court found in favour of the plaintiffs, issuing a blockbuster opinion that held the EPA did indeed have statutory power to regulate greenhouse gas emissions so long as it issued an "endangerment finding" explicitly stating that increased carbon pollution represented a threat to public health. In the immediate aftermath of the case, Bush's EPA issued no such endangerment finding, but in 2009, President Obama's EPA administrator, Lisa Jackson, moved the EPA forward by regulating greenhouse gases under the CAA.[32]

Massachusetts v. EPA was a pivotal moment in American jurisprudence after which corporate and state actors had to take seriously the federal government's expanded regulatory powers to address air pollution contributing to climate change. In the years ahead, corporations would try and undo this precedent, filing their own cases in federal courts, but as of the publication of this book, the precedent set in *Massachusetts v. EPA* remained in place.[33]

And there were other wins in the early 2000s in cases involving statutory law, especially related to NEPA and the Endangered Species Act of 1973. In 2008, the Center for Biological Diversity won a key case in the Ninth Circuit Court of Appeals against the NHTSA, the court finding that the transportation agency had not conducted a complete environmental impact statement (EIS) before approving its CAFE rules for automobile manufacturers. In 1990, as noted earlier, judges had ruled against plaintiffs seeking relief in a similar case brought by Los Angeles, but this time the Ninth Circuit Court found that a federal agency had indeed violated NEPA by not considering greenhouse gases a critical pollutant worth assessing in an EIS.[34] Legal scholars Jacqueline Peele and Hari M. Osofsky noted the significance of this ruling, pointing out that these NEPA cases forced federal agencies to be more thoughtful about climate change when carrying out their environmental impact studies in the years ahead. This in turn promised to have widespread implications not only for the automobile industry but also for other large greenhouse-gas emitting industries regulated by the federal government.[35]

The Endangered Species Act also became a key statute around which climate litigation evolved in the early 2000s. By 2020, Endangered Species Act litigation was one of the most common form of climate change litigation, totalling 75 filings, right behind cases involving NEPA (163), the CAA (151), and state-level NEPA initiatives (139).[36] Environmental organisations, especially the National Resources Defense Council and the Center for Biological Diversity, filed suits in different jurisdictions after 2006, targeting government agencies that did not take effective measures to ensure habitat critical to the preservation of endangered species was protected from changing atmospheric conditions caused by greenhouse gas emissions.[37]

But while some environmental organisations and state governments decided to focus on statutory law in their court battles, others took a novel approach by turning to common law and nuisance suits to try and affect change. Common law is essentially law shaped by state and federal court decisions rather than by legislative statutes or the U.S. Constitution.[38] Public nuisance cases involve situations where an action has been taken that infringes upon the right of citizens living in a specific place, and its exact definition can vary from state to state.[39] In many of these early environmental common law cases, plaintiffs focused their sights on corporate polluters themselves rather than government agencies regulating those businesses. Prior to 2010, few litigants tried to use nuisance law or federal or state common law to sue big polluters that were altering the climate, but that began to change as the effects of climate change came to be felt by more and more Americans.[40]

15.3 Early Federal Common Law Climate Cases and Shifting Legal Strategies Thereafter

One of the first big common law cases grew out of an insurance lawsuit filed by Mississippians in the U.S. District Court for the Southern District of Mississippi in the devastating wake of Hurricane Katrina. In 2005, attorneys for Ned and Brenda Comer and several other residents of coastal Mississippi delivered a complaint to the court charging that Nationwide Mutual Insurance, State Farm, and a select group of other insurance companies and mortgage providers were at fault for not paying insurance money to homeowners harmed by the recent hurricane. The complaint also charged oil companies with burning fossil fuels that caused damage to plaintiffs' property. That case later became *Comer v. Murphy Oil*, which featured an explosive revised complaint filed by the same plaintiffs in 2011 that sought "punitive damages" against numerous firms, including Exxon, Shell, BP, American Electric Power (AEP), Georgia Power, and other oil and energy businesses. The complaint held that these firms had engaged in "gross negligence and an illegal conspiracy to prevent dissemination of scientific information regarding the specific hazards created by Global Warming." Plaintiffs also brought trespass claims, arguing that the defendants' "behavior and greenhouse gas emissions caused saltwater, debris, sediments, hazardous substances, and other materials to enter and remain" on plaintiffs' property.[41]

This was a new day in American courts. Plaintiffs were taking direct aim at fossil fuel and utility companies and trying to use common law precedents regarding nuisance, negligence, and trespass to hold big polluters accountable for their contributions to climate change. And it wasn't just oil and energy firms taking the heat. In 2006, the state of California also filed a common law case against General Motors, Ford, Toyota, and other auto manufacturers, "seeking monetary damages for defendants' past and ongoing contributions to global warming, a public nuisance."[42] California ultimately decided to voluntarily issue a motion to dismiss this case after EPA Administrator Lisa Jackson made her "endangerment finding" in 2009 and signalled a commitment to use EPA authority to regulate greenhouse gas emissions stemming from the automobile industry.[43] But the California General Motors case foreshadowed a new era in which nuisance suits against big corporate contributors to climate change would become more and more common in U.S. courts.

Plaintiffs in these first few common law cases experienced setbacks early on. In *Native Village of Kivalina v. ExxonMobil Corp* (2008), Inupiat community members in the northern reaches of coastal Alaska sought to recover damages from the Houston oil giant and other energy companies to cover costs of relocation of their village due to rising sea levels caused by climate change.[44] The U.S. District Court for the Northern District of California ultimately issued an order asserting that the Indigenous plaintiffs could not find relief in the courts because the issue at hand was one that could not be adjudicated by the courts because it was of a complex political nature and therefore best suited for the executive or legislative branches. There was also a concern over "traceability," the inability to directly tie the actions of Exxon and other defendants to the impairment plaintiffs experienced in Alaska.[45] The Ninth Circuit Court of Appeals would uphold this decision and the Supreme Court denied a writ of certiorari, a request by plaintiffs to have the Supreme Court review the lower court's decision.[46]

The path forward for federal common law cases was markedly impaired after the Supreme Court heard argument in *AEP v. Connecticut* (2011). This case emerged from a nuisance suit filed by eight states and the City of New York against major utility companies in 2004. In that suit, state attorneys asked for a court order "requiring defendants to reduce their emissions of carbon dioxide, thereby abating their contribution to global warming, a public nuisance."[47] When an appellate court ultimately vacated the ruling of a lower court that dismissed the case, AEP filed a petition to the Supreme Court, which issued an opinion in the case in 2011. In an 8-0 decision, the Supreme Court used the theory of "displacement" to find in favour of AEP. Justice Ruth Bader Ginsburg, writing the opinion for the court, determined that federal common law did not apply in this case because it was displaced by the Clean Air Act that clearly gave authority to the EPA to regulate greenhouse gas emissions, per the 2007 *Massachusetts v. EPA* case.[48]

It seemed that for the foreseeable future, litigants were going to have a hard time using federal common law to make corporations responsive to climate change, but as legal scholars pointed out then and now, the *AEP v. Connecticut* decision did not block litigants from filing climate change complaints against corporations using *state* common law. In the years ahead, state courts would become critical battlegrounds for those seeking to use litigation to bring about change. But as late as 2015, when the total number of climate change cases that had been filed or litigated in the United States totalled about 500 (up from very few in 2006), not a single corporation had been forced to pay damages due to a nuisance suit.[49] The truth was many of these legal arguments were quite novel. It would take time for plaintiffs' attorneys to refine their arguments and perfect their litigation strategies.

Still, these were hopeful times for environmentalists seeking to do more to address climate change. In 2015, the Obama Administration, three years off a second-term election win and freed from the big battle of getting the Affordable Care Act through Congress, put its weight behind the Paris Climate Accord, the most ambitious UN climate agreement to date in which both the United States and China, the largest greenhouse-gas emitters in the world, pledged to make nationally determined contributions (NDCs) to reduce emissions and promised to ratchet up commitments in the years ahead to keep global temperatures from rising 1.5 degrees Centigrade above pre-industrial levels. The mechanisms through which this bold ambition would be met had not been worked out and some environmental activists in the United States feared whether this agreement would lead to action, especially if the election of 2016 ushered in a new administration hostile to environmental regulation. But there was optimism.

In this pivotal moment, a new group of litigants took to the courts, opening another front in legal battles designed to make the promises of the Paris Accord a reality.

15.4 A Youth Movement Rises in the Courts

Alec Loorz was just 12 years old when film producer Laurie David and former Vice President Al Gore's climate change film, *An Inconvenient Truth* (2006) premiered. Loorz was moved by the film and became a committed advocate over the next several years for climate justice. When he was 15 years old, he founded the organisation Kids vs. Global Warming, an organisation designed to inform and mobilise young people around the issue of climate change.[50] Partnering with other young environmentalists, Loorz filed the first "public trust" lawsuit against the EPA in 2011, the Department of Interior, and other federal agencies arguing that these bodies had failed to take adequate action to prevent climate change from degrading natural resources held in the public trust for young people like Loorz. The case, *Alec L. v. Jackson* (2011), was really a test case to see whether public trust arguments could be used to push federal agencies to do more to address climate change. This case was ultimately unsuccessful because judges decided that the public trust doctrine in this case was a state, not a federal, matter and even if it was a federal matter, the Clean Air Act would have displaced the public trust doctrine. Nevertheless, it marked a new chapter in climate litigation driven by youth activism and legal organisations who wanted to put young people before judges and juries to make their case about inaction on climate change.[51]

These were critical years in the youth climate movement as college students and high school students sought to ensure their voices were heard. In 2009, youth activists petitioned the secretariat of the United Nations Framework Convention on Climate Change to approve YOUNGO as the official representative of youth organisers and activists at the annual Conference of the Parties (COP).[52] In the same way research institutions and non-governmental agencies were represented by RINGO at the UN climate summit, YOUNGO became a powerful institution representing young environmentalists at these meetings as well. In 2010 and 2011, students at Hampshire College and Unity College completed the first successful fossil fuel divestment campaigns on college campuses. Similar campaigns emerged at Swarthmore, Harvard, Stanford, and other major universities across the country.[53] Dozens of new youth-led or youth-inspired organisations, including Our Children's Trust (founded in 2010) and the Sunrise Movement (2017), gained widespread popularity in these years.[54]

And some of these organisations became committed to continuing climate change litigation in domestic courts. Our Children's Trust, founded by environmental lawyer Julia Olson, became one of the spearheads of the new youth-led litigation. In 2015, Olson represented some 21 young people, including Sophie Kivlehan, the granddaughter of James Hansen, the prominent former director of the National Aeronautics and Space Administration's (NASA) Goddard Space Institute who famously raised the alarm about climate change before the U.S. Congress in 1988.[55] Olson filed a complaint with the Oregon U.S. District Court in the case of *Juliana v.*

United States, claiming that President Barack Obama, the Department of Interior, the EPA, and other major federal agencies had "infringed on Plaintiffs' fundamental constitutional rights to life, liberty, and property." Olson stated that "Defendants' acts also discriminate against these young citizens, who will disproportionately experience the destabilized climate system in our country." Olson urged the court to

> order Defendants to swiftly phase-down CO_2 emissions aimed at atmospheric CO_2 concentrations that are no more than 350 ppm by 2100, develop a national plan to restore Earth's energy balance, and implement that national plan so as to stabilize the climate system. Plaintiffs come before this Court to secure their fundamental rights under the Constitution, before it is too late.[56]

Olson and her young plaintiffs saw some early wins. Judge Ann Aiken in the Oregon case asserted in her 2016 opinion and order denying the federal government's motions to dismiss that she had "no doubt that the right to a climate system capable of sustaining life is fundamental to a free and ordered society." She went on to say that when "governmental action is affirmatively and substantially damaging the climate system in a way that will cause human deaths, shorten human lifespans, result in widespread damage to property, threaten human food sources, and dramatically alter the planet's ecosystem" there is a viable claim for "a due process violation" of the U.S. Constitution. "To hold otherwise," she concluded, "would be to say that the Constitution affords no protection against a government's knowing decision to poison the air its citizens breathe or the water its citizens drink. Plaintiffs have adequately alleged infringement of a fundamental right." This was a historic moment for Olson and the young people she represented.[57]

Soon, American Petroleum Institute (API) and other fossil fuel interest requested to join in the case as co-defendants. This is a unique situation, where interested parties not initially named as defendants in a case can join a case if they feel they may well be adversely affected by a verdict. That request was granted by the court, but when President Donald Trump became president, they backed out. The American Fuel and Petrochemical Manufacturers, one of the industry groups that joined API as a defendant-intervenor in this case explained the decision to the *Washington Post*, saying, "AFPM has decided to withdraw from this case, as we are confident that the U.S. Department of Justice will rigorously defend its position and that the court will conclude that setting national environmental policy is the role of Congress and the president."[58]

President Trump's Justice Department asked the Supreme Court to intervene in the case, but in 2018 the highest court in the country refused to prevent the case from proceeding in the lower courts.[59] The Justice Department then requested to have the case moved to the Court of Appeals for the Ninth Circuit, and Judge Aiken agreed to what is known as interlocutory appeal.[60] In 2020, a three-judge panel for the Ninth Circuit admitted that "the plaintiffs have made a compelling case," noting that "climate change is occurring, that the government has had a role in causing it, and that elected officials have a moral responsibility to seek solutions." But in the end, the judges "reluctantly" concluded that "the plaintiffs' case must be made to the political branches or to the electorate at large, the latter of which can change the composition of the political branches."[61] The only path forward was to take this case to the Supreme Court, but given the recent appointment of conservative justices, it was not clear they would make much headway.[62]

But in the meantime, a lot was happening both globally and elsewhere in American courts. In 2015, attorneys for the Dutch environmental organisation Urgenda, partnering with hundreds of Dutch citizens, successfully argued a case before the Hague District Court in the Netherlands, the court finding that the Dutch government had not done enough to curb greenhouse gas emissions. In 2018, litigants in Pakistan won a critical case, *Leghari v. Federation of Pakistan*, the court finding that the Pakistani government's inaction on climate change had violated the basic human rights of its citizens. In other words, there were now new precedents in courts in multiple countries for litigants seeking to make human rights claims against governments that failed to meet their climate change mitigation targets.[63]

Back in the United States, plaintiffs were trying new tactics to hold corporations and the government agencies that regulate them accountable for their failures to address climate change. One increasingly popular approach was to focus on securities fraud. In the 2016 case of *Ramirez v. Exxon* in the Northern District of Texas, for example, Pedro Ramirez Jr., an Exxon stockholder, filed suit against the oil company he invested in, claiming the firm had made "material misstatements and omissions" in its financial statements and had failed to consider financial risks related to climate change, thereby overinflating the value of its assets.[64] This case was still ongoing as this book went to press. In 2018, Barbara Underwood, the Attorney General of New York, filed a similar securities fraud suit against Exxon on behalf of the citizens of New York. Underwood used the Martin Act, a powerful state anti-fraud law passed in the 1920s. After nearly two weeks of trial, the New York Superior Court ruled in favour of Exxon and asserted that plaintiffs had failed to persuade the court that the oil company had made financial misstatements that would have deceived investors.[65] But in Massachusetts, Attorney General Maura Heeley pursued similar claims against Exxon in Massachusetts Superior Court, and as this book went to press, state judges had refused to dismiss Exxon's motions to throw out the case.[66]

After 2017, cities, counties, and states were also filing a series of new state common law cases against big fossil fuel companies, especially BP and Exxon, taking advantage of the fact that the Supreme Court decision in *AEP v. Connecticut* (2011) had held that *federal* common law cases were not viable but did not address *state* common lawsuits.[67] Some of these cases involved governments seeking money to cover costs associated with adapting to environmental changes wrought by climate change.[68] Many suits proved unsuccessful, but they nevertheless represented an ongoing threat to energy and fossil fuel firms as this book went to press.[69]

And there was also the growing threat of litigation led by young people. Perhaps the biggest win for plaintiffs on this front was the case of *Held v. State of Montana* which started in 2020. Our Children's Trust was involved in this suit, which focused on the state of Montana's 1972 Constitution, which included a clause requiring the state to "maintain and improve a clean environment in Montana for present and future generations."[70] The plaintiffs' complaint, filed on behalf of some 16 children and teenagers in 2020, argued that Montana's State Energy Policy did not adequately address climate change risks to Montana citizens nor did the "Climate Change Exception" to Montana's Environmental Policy Act, which courts interpreted as preventing state agencies from considering "the impacts of climate change in their environmental reviews."[71] The Montana district court ruled in favour of the plaintiffs, finding that the actions of the state had violated Montana's Constitution. The case was headed to the Montana's Supreme Court as this book went to press.[72] Focusing on state constitutions, rather than state common law, seemed to be a promising strategy, at least for the moment.

While celebrating the win in Montana, Our Children's Trust sought to keep the momentum going with a lawsuit in federal court that sought to bring similar constitutional claims at the federal level. The 2023 complaint in the case of *Genesis B. v. United States Environmental Protection Agency* featured 18 young people who claimed that they were "actively being harmed and discriminated against by their government's affirmative allowance of dangerous levels of climate pollution." Attorneys for the youth were blunt: "In a plea for their safety and equal protection of the law, Plaintiffs bring their constitutional claims against the United States Environmental Protection Agency ("EPA"), its Administrator, Michael Reagan, and the United States federal government, each of whom is charged with controlling air pollution that harms human health and welfare."[73] The case was still pending when this book went to press.

15.5 Conclusion

By the spring of 2024, there were over 1,760 climate change–related cases in the United States—ranging across constitutional law, public trust, securities fraud, statutory violations, common law, and so on—that had been filed or were ongoing in federal and state courts in the United States.[74] No other nation came near to this amount of climate change litigation. The second closest country, Australia, had only 127 climate change lawsuits as of the beginning of 2023.[75] In other words, the United States was a unique case study in which corporations were facing increasing pressure through the courts to do something about climate change.

Just how big business would respond to this changing legal landscape remains unclear at the time of writing this chapter. Scholars and journalists who have been closely following the trend in climate litigation point out that at the very least, insurance companies are already beginning to take climate lawsuits into account when considering issuance of new policies. Some insurance companies may decide not to cover the potentially high costs of covering settlements or verdicts that come out of climate litigation, which in turn may reshape corporate risk assessments when considering future business operations.[76] Other industry analysts have noted that even if certain climate cases against big firms are unsuccessful, the discovery process—in which lawyers are able to extract sensitive documents from corporations in preparation for trial—may have broad effects on corporate behaviour in the future. As L. Delta Merner, writing for the Union for Concerned Scientists, explained,

> Much like its historic role in the Tobacco Master Settlement Agreement of 1998, discovery in climate-related litigation can disclose internal documentation—scientific studies, corporate communications, and strategy memoranda—of immense value in establishing causation and assigning responsibility for climate-related damages. Robust discovery processes not only shape public discourse but also act as a catalyst for transformative legal actions.[77]

It was increasingly clear in 2024 that a determined group of young people were motivated to fight in the courts against firms and state agencies that had been negligent in dealing with a pollution issue that dramatically affected their future well-being. Whether constitutional claims against state or corporate actors contributing to climate change would continue to hold up in court had yet to be seen. But early success in Montana suggested the potential for further progress for plaintiffs. Several other states in the United States besides Montana had constitutions with clauses that might lend themselves to climate change claims, including Pennsylvania, Illinois, Massachusetts, Hawaii, and Rhode Island.[78] And what was to prevent other states such as Ohio from proposing ballot initiatives that would codify into state constitutions similar environmental rights as those offered Montanans in their 1972 state constitution? Would litigation then spread to these states?

To be clear, American jurisprudence was hardly settled on these big constitutional questions, but the legal environment in which businesses were operating had clearly undergone considerable change in the early twenty-first century. Big cases such as *Massachusetts v. EPA* had secured broad authority for federal agents to regulate greenhouse gas emissions. And *Center of Biological Diversity v. National Highway Transportation Safety Administration* had also showed that government regulators would have the power to put pressure on automotive companies and other industries to address climate change through the National Environmental Policy Act. Corporate actors shaping financial statements also had to think about the potential for misstatements about climate change to emerge in securities fraud litigation like the one, noted earlier, pending in Massachusetts. And by the 2020s, the successes of youth litigants in the court made clear that American firms needed to at least be watchful about the potential for constitutional claims to threaten their ability to release greenhouse gases that warmed the planet. In short, American firms had clearly entered a new era of corporate liability shaped by two decades of new case law. Planetary ecological changes had reshaped the American legal landscape. Whether that new landscape would help in cooling the planet down was a history yet to be written.

Notes

1 For a database that contains information about all past and ongoing climate change litigation in the United States, see the Sabin Center for Climate Change Law at Columbia University, which is overseen by Columba law professor Michael B. Gerrard, among others, https://climatecasechart.com/. See also L. Delta Merner, "Climate Change Litigation: Reflection and Anticipation for 2024," The Equation blog hosted by the Union for Concerned Scientists, January 25, 2024, https://blog.ucsusa.org/delta-merner/climate-litigation-reflection-and-anticipation-for-2024/.

2 *State of Minnesota v. American Petroleum Institute*, Docket Number 21-8005 (8th Circuit, 2023). See Complaint filed in this case on June 24, 2020, 3–4, https://climatecasechart.com/wp-content/uploads/case-documents/2020/20200624_docket-62-CV-20-3837_complaint.pdf.

3 For the JBS Foods case, see the Summons filed on February 28, 2024, in *The People of the State of New York v. JBS USA Food Company*, Docket Number 450682/2024 (N.Y. Sup. Ct.), https://climatecasechart.com/wp-content/uploads/case-documents/2024/20240228_docket-4506822024_complaint.pdf. For Delta, see Complaint filed on May 30, 2023 in *Berrin v. Delta Air Lines Inc.*, Docket Number 2:23-cv-04150 (C. D. Cal.), https://climatecasechart.com/case/berrin-v-delta-air-lines-inc/. For the McKinsey case, see Complaint filed on June 22, 2023, 53, in *County of Multnomah v. Exxon Mobil Corporation*, Docket Number 23CV25164 (Or. Circ. Ct.), https://climatecasechart.com/wp-content/uploads/case-documents/2024/20241007_docket-23CV25164-_complaint.pdf. See also Noor, "Why 2024 Will Be a Crucial Year for Climate Litigation."

4 *Global Climate Change Litigation Report*, 12, https://wedocs.unep.org/bitstream/handle/20.500.11822/43008/global_climate_litigation_report_2023.pdf?sequence=3. The UN's numbers are based on cases catalogued by the Sabin Center for Climate Change Law in their litigation database, which includes both cases in which plaintiffs have sued corporations and lawsuits in which businesses have sued state actors and non-government organisations (NGOs) such as Greenpeace to insulate themselves from liabilities associated with generating climate changing pollution.

5 *Global Climate Change Litigation Report*, 12.

6 "Swiss Parliamentary Committee Rejects European Climate Ruling."

7 See the "Suits Against Corporations, Individuals" section of the "Global Climate Change Litigation" database maintained by the Sabin Center for Climate Change Law, https://climatecasechart.com/non-us-case-category/suits-against-corporations-individuals/.

8 *American Electric Power Company v. Connecticut*, 564 U.S. 410.

9 Rome, "Beyond Compliance"; Carroll, Lipartito, Post, and Werhane, *Corporate Responsibility*; Berghoff and Adam, *Green Capitalism?*

10 On Walmart and sustainability, see Humes, *Force of Nature*; Elmore, *Country Capitalism*.

11 Elmore, "The Environmental History of an American Bank." For more on the greening of American finance see Jones, "Can Finance Change the World?"; Jones, Grandjean, and Spadafora, "Financing Sustainability; Schmidheiny and Zorraquín, *Financing Change*; Weber and Feltmate, *Sustainable Banking*; Jeucken, *Sustainable Finance and Banking*.

12 See Chapter 2, "Business, Institutions, and Climate Change," in this volume.
13 United Nations Environmental Program Finance Initiative (UNEP-FI), *Who Cares Wins*, iv.
14 Kiron et al., "Sustainability's Next Frontier"; Peel and Osofsky, *Climate Change*.
15 Peel and Osofsky, *Climate Change Litigation*, 177–78.
16 Davis-Peccoud et al., "Sustainability is the Next Digital."
17 For more on various ESG rating entities, see *Rate the Raters 2020: Investor Survey and Interview Results*, a report prepared by SustainAbility (March 2020), https://www.sustainability.com/globalassets/sustainability.com/thinking/pdfs/sustainability-ratetheraters2020-report.pdf.
18 Lyon, "How a Sustainability Index Can Keep Exxon but Drop Tesla."
19 Pucker and King, "ESG Investing Isn't Designed to Save the Planet."
20 S&P 500 ESG Index, https://www.spglobal.com/spdji/en/indices/esg/sp-500-esg-index/#data. See also, "ExxonMobil ESG Score"; Eckett, "'Dirty Little Secret'"; Cao, "Tesla Is Kicked Off an Index for Socially Responsible Companies While Exxon Remains."
21 2022 FedEx ESG Report, 5, 15, 20, 36–37.
22 Kosolapova, "Liability for Climate Change-Related Damage in Domestic Courts," 189; Peel and Osofsky, *Climate Change Litigation*, 2, 17, 29, 60, 62.
23 Meltz, "Climate Change Litigation," 4.
24 On the shift in the Republican leadership's stance on issues like climate change and other environmental legislation, see Turner and Isenberg, *The Republican Reversal*. See also Coodley and Sarasohn, "The Years That Followed."
25 Farber, "Climate Change Litigation in the United States," 244.
26 See *Loper Bright Enterprises v. Raimondo* 603 US ___ (2024).
27 *City of Los Angeles v. National Highway Traffic Safety Administration* 912 F.2d 478.
28 Gerrard, "Climate Change and the Environmental Impact Review Process," 20.
29 Meltz, "Climate Change Litigation," 17; Peel and Osofsky, *Climate Change Litigation*, 19.
30 Meltz, "Climate Change Litigation," 7.
31 *Massachusetts v. EPA* 549 US 497 (2007).
32 Gerrard and Wannier, "United States of America," 565; Gerrard, "Climate Change Litigation in the United States," 35.
33 For lawsuits that sought to undue *Massachusetts v. EPA*, see, for example, *Coalition for Responsible Regulation, Inc. V. EPA*, 684 F.3d 102 (D.C. Cir. 2012). Farber, "Climate Change Litigation," 245. The Supreme Court's decision in *West Virginia v. EPA* 597 US 697 (2022) restricted EPA power significantly by limiting the agency's ability to push power plants towards renewable power generation, but it did not wholly overturn *Massachusetts v. EPA*.
34 *Center for Biological Diversity v. National Transportation Highway Traffic Safety Administration* 538 F. 3d.1172 (2008). See opinion of the Ninth Circuit that replaced the 2007 opinion, issued on August 18, 2008, https://climatecasechart.com/wp-content/uploads/case-documents/2008/20080818_docket-06-71891_opinion-and-order.pdf.
35 Peel and Osofsky, *Climate Change Litigation*, 74; Kosolapova, "Liability for Climate Change-Related Damage," 191–92.
36 Gerrard, "Climate Change Litigation," 36.
37 See, for example, *National Resources Defense Council v. Kempthorne*, 506 F. Supp. 2d 322 (E.D. Cal. 2007); In Re Polar Bear Endangered Species Act Listing and Section 4(D) Rule Litigation -MBL N0. 1993, 709 F.3d 1 (D.C. Circuit 2013); Meltz, "Climate Change Litigation,"14.
38 See "Common Law," Cornell Law School's Legal Information Institute, https://www.law.cornell.edu/wex/common_law.
39 See "Public Nuisance," Cornell Law School's Legal Information Institute, https://www.law.cornell.edu/wex/public_nuisance#:~:text=A%20public%20nuisance%20generally%20refers,in%20civil%20and%20criminal%20statutes.
40 Peel and Osofsky, *Climate Change Litigation*, 153.
41 *Comer v. Murphy Oil USA, Inc.*, 839 F. Supp 2d 849. See Class Action Complaint filed in this case on May 27, 2011, 12, 17, 19, https://climatecasechart.com/wp-content/uploads/case-documents/2011/20110527_docket-111-cv-00220_complaint-1.pdf.
42 *People of the State of California v. General Motors, et al.* (N. D. Cal 2007). See Complaint for Damages and Declaratory Judgment filed in this case on September 20, 2006, 2, https://climatecasechart.com/wp-content/uploads/case-documents/2006/20060920_docket-06-cv-05755_complaint.pdf; Kosolapova, "Liability for climate change-related damage," 190; Meltz, "Climate Change Litigation," 26.

43 On the California Attorney General's decision to withdraw this case, see https://climatecasechart.com/case/california-v-gm-corp/.

44 See *Kivalina v. ExxonMobil Corp, 663 F. Supp. 2d 863* (N.D. Cal. 2009). For court documents from this case, including the original opinion, see https://climatecasechart.com/case/native-village-of-kivalina-v-exxonmobil-corp/.

45 Kosolapova, "Liability for Climate Change-Related Damage," 192–93.

46 For *Kivalina v. ExxonMobil Corp*, see https://climatecasechart.com/case/native-village-of-kivalina-v-exxonmobil-corp/.

47 *Connecticut v. American Electric Power Co.*, 406 F.Supp.2d 265, 268 (S.D.N.Y. 2005). Complaint filed by plaintiffs in the case on July 21, 2004, 1, https://climatecasechart.com/wp-content/uploads/case-documents/2004/20040721_docket-04-Civ.-5669-04-Civ.-5670_complaint.pdf.

48 *American Electric Power Company v. Connecticut*, 564 U.S. 410; See https://climatecasechart.com/case/american-electric-power-co-v-connecticut/; Farber, "Climate Change Litigation," 241–42; Gerrard, "Climate Change Litigation," 39; Gerrard and Wannier, "United States of America," 601.

49 Peel and Osofsky, *Climate Change Litigation*, 19, 77; Kosolapova, "Liability for Climate Change-Related Damage," 189.

51 Peel and Osofsky, *Climate Change Litigation*, 82. See Memorandum Opinion in *Alec L. v. Lisa P. Jackson* dated May 31, 2012, https://climatecasechart.com/case/alec-l-v-mccarthy/.

52 Introduction to YOUNGO, https://unfccc.int/sites/default/files/resource/Introduction%20to%20YOUNGO.pdf.

53 Raji, "Timeline."

54 Sunrise Movement Website, https://www.sunrisemovement.org/about/; Our Children's Trust Website, https://www.ourchildrenstrust.org/.

55 Profile of Sophie Kivelhan on Our Children's Trust Website, https://www.ourchildrenstrust.org/sophie.

56 *Julianna v. United States* 217 F. Supp. 3d 1224 (D. Or. 2016). See Complaint for Declaratory and Public Relief filed in this case on August 12, 2015, 5, 7, https://climatecasechart.com/wp-content/uploads/case-documents/2015/20150812_docket-615-cv-1517_complaint-2.pdf; Peel and Osofsky, "A Rights Turn in Climate Change Litigation?" 57.

57 *Juliana v. United States* 217 F Supp 3d 1224 (D. Or. 2016). Opinion and Order in the case filed on November 10, 2016, 32–33, https://static1.squarespace.com/static/571d109b04426270152febe0/t/5824e85e6a49638292ddd1c9/1478813795912/Order+MTD.Aiken.pdf; Farber, "Climate Change Litigation," 19; Gerrard, "Climate Change Litigation," 40–42.

58 Harvey, "These Fossil-Fuel Groups Joined a Historic Climate Lawsuit."

59 Liptack, "The Supreme Court Lets Youth's Case Demanding Climate Action Proceed."

60 For a full timeline of the *Juliana* case, see Our Children's Trust Website, https://www.ourchildrenstrust.org/juliana-v-us.

61 *Juliana v. United States* 947 F 3d 1159 (9th Cir. 2020). Opinion in the case filed on January 17, 2020, 31–32, https://climatecasechart.com/wp-content/uploads/case-documents/2020/20200117_docket-18-36082_opinion.pdf; Gerrard, "Climate Change Litigation," 41–42.

62 Gerrard, "Climate Change Litigation," 42.

63 *Urgenda Foundation v. State of Netherlands, ECLI:NL:HR:2019:*2007 (Sup. Ct. Neth. Dec. 20, 2019). See https://climatecasechart.com/non-us-case/urgenda-foundation-v-kingdom-of-the-netherlands/; *Ashgar Leghari v. Federation of Pakistan* (W. P. No. 25501/2015), argued in the Lahore High Court Green Bench. See https://climatecasechart.com/non-us-case/ashgar-leghari-v-federation-of-pakistan/; Peel and Osofsky, "A Rights Turn in Climate Change Litigation?" 38, 52, 55–56; Farber, "Climate Change Litigation," 13. For a comprehensive catalog of climate litigation outside the United States, see the "Global Climate Change Litigation" database maintained by the Sabin Center for Climate Change Law https://climatecasechart.com/non-us-climate-change-litigation/.

64 *Ramirez v. Exxon Mobil Corporation* No. 3:16-cv 3111 (ND Texas). Complaint for Violations of the Federal Securities Law filed November 7, 2016, 1–2, 4, https://climatecasechart.com/wp-content/uploads/case-documents/2016/20161107_docket-316-cv-3111_complaint-1.pdf; Gerrard, "Climate Change Litigation," 44.

65 *People of the State of New York v. Exxon Mobil Corporation* 119 NYS 3d 829 (NY Sup Ct 2019). For filing in this case, see https://climatecasechart.com/case/people-v-exxon-mobil-corporation/. Gerrard, "Climate Change Litigation," 43.

66 *Commonwealth v. Exxon Mobil Corporation* No 19-3333 (Mass Super Ct). For filings in this case, see https://climatecasechart.com/case/commonwealth-v-exxon-mobil-corp/. Gerrard, "Climate Change Litigation in the United States," 43–44.
67 See, for example, *City of New York v. BP* 325 F. Supp. 3d 466 (Dist. Court. SD New York, 2018), https://climatecasechart.com/case/city-new-york-v-bp-plc/; *City of Oakland v. BP* 969 F.3d 895 (9th Cir. 2020), https://climatecasechart.com/case/people-state-california-v-bp-plc-oakland/. For a complete list of federal and state common law cases filed and/or ongoing as of publication of this book, see https://climatecasechart.com/case-category/common-law-claims/; Gerrard, "Climate Change Litigation," 39; Farber, "Climate Change Litigation," 243–44.
68 Farber, "Climate Change Litigation," 243.
69 Gerrard, "Climate Change Litigation," 39.
70 See Article IX, Section 1, of the Constitution of the State of Montana, https://leg.mt.gov/bills/mca/title_0000/article_0090/part_0010/section_0010/0000-0090-0010-0010.html; Gelles and Baker, "Montana Judge Rules for Youth in Climate Case Climate Case."
71 *Held v. Montana*, No. CDV-2020-307 (Mont. 1st Dist. Ct. 2020). See Complaint for Declaratory and Injunctive Relief filed March 12, 2020, https://climatecasechart.com/wp-content/uploads/case-documents/2020/20200313_docket-CDV-2020-307_complaint.pdf.
72 For case files in *Held v. Montana*, see https://climatecasechart.com/case/11091/.
73 *Genesis B. v. U.S. Environmental Protection Agency*, 2:23-cv-10345 (C. D. Cal.). Complaint for Declaratory Relief filed on December 10, 2023, 1, https://climatecasechart.com/wp-content/uploads/case-documents/2023/20231210_docket-223-cv-10345_complaint.pdf.
74 See the Sabin Center for Climate Change Law for up to date numbers, https://climatecasechart.com/about/.
75 Peel, "Climate Litigation Is on the Rise Around the World and Australia Is at the Head of the Pack." This article notes that looking at per-capita rates of climate litigation, Australia was actually ahead of the United States.
76 "Climate Litigation Cases Have Quadrupled Over Past Decade"; Bastit and Englerth, "Guest Post."
77 Merner, "Climate Change Litigation."
78 Gerrard, "Climate Change Litigation," 42.

Bibliography

Databases

Sabin Center for Climate Change Law at Columbia University. https://climatecasechart.com/.

Court Cases and Legal Documents

American Electric Power Company v. Connecticut, 564 U.S. 410.
Ashgar Leghari v. Federation of Pakistan (W. P. No. 25501/2015).
Berrin v. Delta Air Lines Inc., Docket Number 2:23-cv-04150 (C. D. Cal.).
Center for Biological Diversity v. National Traffic Highway Traffic Safety Administration, 538 F. 3d.1172 (2008).
City of Los Angeles v. National Highway Traffic Safety Administration, 912 F.2d 478.
City of New York v. BP, 325 F. Supp. 3d 466 (Dist. Court. SD New York, 2018).
City of Oakland v. BP, 969 F.3d 895 (9th Cir. 2020).
Coalition for Responsible Regulation, Inc. v. EPA, 684 F.3d 102 (D.C. Cir. 2012).
Comer v. Murphy Oil USA, Inc., 839 F. Supp 2d 849.
Commonwealth v. Exxon Mobil Corporation, No. 19-3333 (Mass Super Ct.).
Connecticut v. American Electric Power Co., 406 F.Supp.2d 265, 268 (S.D.N.Y. 2005). Constitution of the State of Montana.
County of Multnomah v. Exxon Mobil Corporation, Docket Number 23CV25164 (Or. Circ. Ct.).
Genesis B. v. U.S. Environmental Protection Agency, 2:23-cv-10345 (C. D. Cal.).
Held v. Montana, No. CDV-2020-307 (Mont. 1st Dist. Ct. 2020).
In Re Polar Bear Endangered Species Act Listing and Section 4(D) Rule Litigation-MBL, No. 1993, 709 F.3d 1 (D.C. Circuit 2013).

Juliana v. United States, 947 F 3d 1159 (9th Cir. 2020).
Julianna v. United States, 217 F. Supp. 3d 1224 (D. Or. 2016).
Kivalina v. ExxonMobil Corp, 663 F. Supp. 2d 863 (N.D. Cal. 2009).
Loper Bright Enterprises v. Raimondo, 603 US ___ (2024).
Massachusetts v. EPA, 549 US 497 (2007).
National Resources Defense Council v. Kempthorne, 506 F. Supp. 2d 322 (E.D. Cal. 2007).
People of the State of California v. General Motors, et. al. (N. D. Cal 2007).
People of the State of New York v. Exxon Mobil Corporation, 119 NYS 3d 829 (N.Y. Sup Ct 2019).
The People of the State of New York v. JBS USA Food Company, Docket Number 450682/2024 (N.Y. Sup. Ct.).
Ramirez v. Exxon Mobil Corporation, No. 3:16-cv 3111 (N. D. Texas).
State of Minnesota v. American Petroleum Institute, Docket Number 21-8005 (8th Circuit, 2023).
Urgenda Foundation v. State of Netherlands, ECLI:NL:HR:2019:2007 (Sup. Ct. Neth. December 20, 2019).
West Virginia v. EPA, 597 US 697 (2022).

Books, Articles, Corporate Reports, and Secondary Sources

"2022 FedEx ESG Report." https://fedex.com/content/dam/fedex/us-united-states/sustainability/gcrs/FedEx_2022_ESG_Report.pdf.
Bastit, Bruno, and Tommy Englerth. "Guest Post: Assessing the Impact of Climate Litigation Remains Complex." *ESG Today*, June 5, 2024. https://www.esgtoday.com/guest-post-assessing-the-impact-of-climate-litigation-remains-complex/.
Berghoff, Hartmut, and Adam Rome, eds. *Green Capitalism? Business and the Environment in the Twentieth Century*. Philadelphia: University of Pennsylvania Press, 2017.
Cao, Sissi. "Tesla Is Kicked Off an Index for Socially Responsible Companies While Exxon Remains." *The Observer*, May 20, 2022. https://observer.com/2022/05/tesla-removed-from-sp500-esg-index/.
Carroll, Archie B., Kenneth J. Lipartito, James E. Post, and Patricia H. Werhane. *Corporate Responsibility: The American Experience*. New York: Cambridge University Press, 2012.
Children's Environmental Literacy Website. "Young Voices for the Planet." https://celfeducation.org/yvfp/kids-v-global-warming/#:~:text=Kids%20vs%20Global%20Warming%20is,empowers%20them%20to%20take%20action.
"Climate Litigation Cases Have Quadrupled Over Past Decade." *Risk &Insurance*, May 14, 2024. https://riskandinsurance.com/climate-litigation-cases-quadruple-since-2013/.
Coodley, Gregg, and David Sarasohn, eds. *The Green Years, 1964–1976: When Democrats and Republicans United to Repair the Earth*. Lawrence: University Press of Kansas, 2021.
Davis-Peccoud, Jenny, Jean-Charles van den Branden, Chris Brahm, Gerry Mattios, and Joëlle de Montgolfier. "Sustainability Is the Next Digital." A Brief Published by Bain & Company on Their Website, September 8, 2020. https://www.bain.com/insights/sustainability-is-the-next-digital/.
Eckett, Tom. "'Dirty Little Secret': Exxon Inclusion in ESG ETFs Highlights Carbon Emissions Issue." ETF Stream Website, October 8, 2020. https://www.etfstream.com/articles/dirty-little-secret-exxon-inclusion-in-esg-etfs-highlights-carbon-emissions-issue.
Elmore, Bart. "The Environmental History of an American Bank." *Environmental History* 27, no. 1 (January 2022): 113–39.
Elmore, Bart. *Country Capitalism: How Corporations from the American South Remade Our Economy and the Planet.* Chapel Hill: Ferris & Ferris, 2023.
"ExxonMobil ESG Score." S&P Global Website. https://www.spglobal.com/esg/scores/results?cid=3007562.
Farber, Daniel A. "Climate Change Litigation in the United States." In *Climate Change Litigation: A Handbook*, edited by Wolfgang Kahl and Marc-Philippe Weller. Munich, Oxford and Baden-Baden: C. H. Beck, Hart, Nomos, 2021.
Gelles, David, and Mike Baker. "Montana Judge Rules for Youth in Climate Case Climate Case." *New York Times*, August 15, 2023.
Gerrad, Michael B., and Gregory E. Wannier. "Climate Change and the Environmental Impact Review Process." *Natural Resources and the Environment* 22, no. 3 (Winter 2008): 20–24.
Gerrad, Michael B., and Gregory E. Wannier. "Climate Change Litigation in the United States: High Volume of Cases, Mostly About Statutes." In *Climate Change Litigation: Global Perspectives*, edited

by Ivano Alogna, Christine Bakker, and Jean-Pierre Gauci. Leiden, Netherlands: British Institute of International Law and Comparative Law, 2021.

Gerrad, Michael B., and Gregory E. Wannier. "United States of America." In *Climate Change Liability: Transitional Law and Practice*, edited by Richard Lord, Silke Goldberg, Lavanya Rajamani, and Jutta Brunnée. Cambridge: Cambridge University Press, 2021.

Harvey, Chelsea. "These Fossil-Fuel Groups Joined a Historic Climate Lawsuit: Now, They Want to Get Out." *Washington Post*, May 26, 2017. https://www.washingtonpost.com/news/energy-environment/wp/2017/05/26/three-fossil-fuel-groups-joined-a-historic-climate-lawsuit-now-they-want-to-get-out-of-it/.

Humes, Edward. *Force of Nature: The Unlikely Story of Wal-Mart's Green Revolution*. New York: Harper Business, 2011.

Jeucken, Marcel. *Sustainable Finance and Banking: The Finance Sector and the Future of the Planet*. London: Earthscan, 2010.

Jones, Geoffrey. *Profits and Sustainability: A History of Green Entrepreneurship*. New York: Oxford University Press, 2017.

Jones, Geoffrey, Emily Grandjean, and Andrew Spadafora. "Financing Sustainability." In *Varieties of Green Business: Industries, Nations, and Time*, edited by Geoffrey Jones. Northampton, MA: Edward Elgar, 2018.

Kiron, David, Nina Kruschwitz, Holger Rubel, Martin Reeves, and Sonja-Katrin Fuisz Kehrbach. "Sustainability's Next Frontier: Walking the Talk on Sustainability Issues That Matter Most." *MIT Sloan Management Review*, December 16, 2013. https://sloanreview.mit.edu/projects/sustainabilitys-next-frontier/.

Kosolapova, Elena. "Liability for Climate Change-Related Damage in Domestic Courts: Claims for Compensation in the USA." In *Climate Change Liability*, edited by Michael Faure and Marjan Peeters. Cheltenham: Edward Elgar, 2011.

Liptack, Adam. "The Supreme Court Lets Youth's Case Demanding Climate Action Proceed." *New York Times*, November 2, 2018. https://www.nytimes.com/2018/11/02/us/politics/supreme-court-youth-climate-case.html.

Lyon, Tom. "How a Sustainability Index Can Keep Exxon but Drop Tesla—And 3 Ways to Fix ESG Ratings to Meet Investors' Expectations." University of Michigan Ross School of Business News and Research Page, May 26, 2022. https://michiganross.umich.edu/news/how-sustainability-index-can-keep-exxon-drop-tesla-and-3-ways-fix-esg-ratings-meet-investors.

Meltz, Robert. "Climate Change Litigation: A Growing Phenomenon." In *Climate Change Litigation and Law*, edited by Jean-Francois Masson. New York: Nova Science, 2010.

Merner, L. Delta. "Climate Change Litigation: Reflection and Anticipation for 2024." The Equation blog hosted by the Union for Concerned Scientists, January 25, 2024. https://blog.ucsusa.org/delta-merner/climate-litigation-reflection-and-anticipation-for-2024/.

Noor, Dharna. "Why 2024 Will Be a Crucial Year for Climate Litigation." *The Guardian*, January 22, 2024. https://www.theguardian.com/us-news/2024/jan/22/climate-change-cases-2024-lawsuits.

Peel, Jacqueline, and Hari M. Osofsky. *Climate Change Litigation*. Cambridge: Cambridge University Press, 2015.

Peel, Jacqueline, and Hari M. Osofsky. "A Rights Turn in Climate Change Litigation?" *Transnational Environmental Law* 7, no. 1 (2017): 37–67.

Peel, Jacqueline, and Hari M. Osofsky. "Climate Litigation Is on the Rise Around the World and Australia Is at the Head of the Pack." *The Conversation*, July 27, 2023. https://theconversation.com/climate-litigation-is-on-the-rise-around-the-world-and-australia-is-at-the-head-of-the-pack-210375.

Pucker, Kenneth P., and Andrew King. "ESG Investing Isn't Designed to Save the Planet." *Harvard Business Review*, August 1, 2022. https://hbr.org/2022/08/esg-investing-isnt-designed-to-save-the-planet.

Raji, Michelle Y. "Timeline: Fossil Fuel Divestment." *The Harvard Crimson*, October 2, 2014. https://www.thecrimson.com/article/2014/10/2/timeline-fossil-fuels-divestment/.

"Rate the Raters 2020: Investor Survey and Interview Results." A report prepared by SustainAbility, March 2020. https://www.sustainability.com/globalassets/sustainability.com/thinking/pdfs/sustainability-ratetheraters2020-report.pdf.

Rome, Adam. "Beyond Compliance: The Origins of Corporate Interest in Sustainability." *Enterprise and Society* 22 (June 2021): 409–37.

Rothberg, Peter. "Kids vs. Global Warming." *The Nation*, December 8, 2010. https://www.thenation.com/article/archive/kids-vs-global-warming/.

Schmidheiny, Stephan, and Federico Zorraquín. *Financing Change: The Financial Community, Eco-Efficiency, and Sustainable Development*. Cambridge, MA: MIT Press, 1996.
"Swiss Parliamentary Committee Rejects European Climate Ruling." *Reuters*, May 21, 2024. https://www.reuters.com/world/europe/swiss-parliamentary-committee-rejects-european-climate-ruling-2024-05-21/.
Turner, James Morton, and Drew Isenberg. *The Republican Reversal: Conservatives and the Environment: From Nixon to Trump*. Cambridge, MA: Harvard University Press, 2018.
United Nations Environmental Program Finance Initiative (UNEP-FI). "Who Cares Wins: Connecting Financial Markets to a Changing World." A report issued by Global Compact of the UNEP-FI, 2004. https://www.unepfi.org/fileadmin/events/2004/stocks/who_cares_wins_global_compact_2004.pdf.
United Nations Environment Programme. "Global Climate Change Litigation Report: 2023 Status Review." A report published by the United Nations Environment Programme, 2023. https://wedocs.unep.org/bitstream/handle/20.500.11822/43008/global_climate_litigation_report_2023.pdf?sequence=3.
Weber, Olaf, and Blair Feltmate, eds. *Sustainable Banking: Managing the Social and Environmental Impact of Financial Institutions*. Toronto: University of Toronto Press, 2016.

INDEX

Note: Please note that page numbers in *italics* indicate a figure and page numbers in **bold** indicate a table on the corresponding page.

MIX
Papier | Fördert
gute Waldnutzung
FSC® C083411

Kolibri 360 GmbH
Ferdinand-Jühlke-Straße 7
99095 Erfurt, Deutschland
produktsicherheit@kolibri360.de